Lecture Notes in Networks and Systems

Volume 5

Series editor

Janusz Kacprzyk, Polish Academy of Sciences, Systems Research Institute,
Warsaw, Poland
e-mail: kacprzyk@ibspan.waw.pl

The series "Lecture Notes in Networks and Systems" publishes the latest developments in Networks and Systems—quickly, informally and with high quality. Original research reported in proceedings and post-proceedings represents the core of LNNS.

Volumes published in LNNS embrace all aspects and subfields of, as well as new challenges in, Networks and Systems.

The series contains proceedings and edited volumes in systems and networks, spanning the areas of Cyber-Physical Systems, Autonomous Systems, Sensor Networks, Control Systems, Energy Systems, Automotive Systems, Biological Systems, Vehicular Networking and Connected Vehicles, Aerospace Systems, Automation, Manufacturing, Smart Grids, Nonlinear Systems, Power Systems, Robotics, Social Systems, Economic Systems and other. Of particular value to both the contributors and the readership are the short publication timeframe and the world-wide distribution and exposure which enable both a wide and rapid dissemination of research output.

The series covers the theory, applications, and perspectives on the state of the art and future developments relevant to systems and networks, decision making, control, complex processes and related areas, as embedded in the fields of interdisciplinary and applied sciences, engineering, computer science, physics, economics, social, and life sciences, as well as the paradigms and methodologies behind them.

More information about this series at http://www.springer.com/series/15179

Suresh Chandra Satapathy · Vikrant Bhateja
K. Srujan Raju · B. Janakiramaiah
Editors

Computer Communication, Networking and Internet Security

Proceedings of IC3T 2016

Editors
Suresh Chandra Satapathy
Department of Computer Science and Engineering
PVP Siddhartha Institute of Technology
Vijayawada, Andhra Pradesh
India

K. Srujan Raju
CMR Technical Campus
Department of Computer Science and Engineering
Hyderabad
India

Vikrant Bhateja
Shri Ramswaroop Memorial Group of Professional Colleges (SRMGPC)
Lucknow, Uttar Pradesh
India

B. Janakiramaiah
DVR & Dr. HS MIC College of Technology
Kanchikacherla, Andhra Pradesh
India

ISSN 2367-3370 ISSN 2367-3389 (electronic)
Lecture Notes in Networks and Systems
ISBN 978-981-10-9820-8 ISBN 978-981-10-3226-4 (eBook)
DOI 10.1007/978-981-10-3226-4

Printed on acid-free paper

This Springer imprint is published by Springer Nature
The registered company is Springer Nature Singapore Pte Ltd.
The registered company address is: 152 Beach Road, #21-01/04 Gateway East, Singapore 189721, Singapore

Preface

This volume contains the papers that were presented at the 3rd International conference on Computer and Communication Technologies (IC3T 2016) held in Department of Computer Science & Engineering, Devineni Venkata Ramana & Dr. Hima Sekhar MIC College of Technology, Vijayawada, during November 5–6, 2016. IC3T 2016 aims to provide a forum for exchanges of research results and ideas, and experience of applications among researchers and practitioners involved with all aspects of computing and communication technologies. Previous editions of the conference were held in CMR Group of Colleges, Hyderabad, under the guidance of their management and Dr. Srujan Raju. The IC3T 2016 received total 398 submissions. Each of which was peer-reviewed by at least two members of the Program Committee. Finally, a total of 63 papers were accepted for publication in this proceeding. The IC3T 2016 was technically supported by Div-V of Computer Society of India. Several special sessions were offered by eminent professors in many cutting-edge technologies. A pre-conference workshop on multi-disciplinary research challenges and trends in computer science was the highlight of this meet. Topics on sensor networks, wireless communications, big data, and swarm optimization were discussed at length for the benefits of the participants, and guidance to research direction was provided. An excellent author workshop on "How to write for and get published in Scientific Journal" was conducted by Mr. Aninda Bose, Senior Publishing Editor, Springer India Pvt. Ltd, Delhi, India.

We would like to express our appreciation to the members of the Program Committee for their support and cooperation in this publication. We are also thankful to the team from Springer for providing a meticulous service for the timely production of this volume. Our heartfelt thanks to secretary and other management members of Devineni Venkata Ramana & Dr. Hima Sekhar MIC College of Technology for extending wholehearted support to host this in their campus. Special thanks to all guests who have honored us in their presence in the inaugural day of the conference. Our thanks are due to all special session chairs, track managers, and reviewers for their excellent support. Last but certainly not the least, our special thanks go to all the authors who submitted papers and all the attendees for their contributions and fruitful discussions that made this conference a great success.

Vijayawada, India
Lucknow, India
Hyderabad, India
Kanchikacherla, India
November 2016

Suresh Chandra Satapathy
Vikrant Bhateja
K. Srujan Raju
B. Janakiramaiah

Organising Committee

Patrons

Chief Patron

Dr. M.V. Ramana Rao M.E., Ph.D., Chairman, MIC College of Technology, Vijayawada, India.

Patrons

Sri N. Srinivasa Rao, Vice Chairman, MIC College of Technology, Vijayawada, India.
Sri D. Panduranga Rao, CEO, MIC College of Technology, Vijayawada, India.
Sri M. Srinivasa Rao, Director (P&D), MIC College of Technology, Vijayawada, India.
Sri K. Janardhan, Director, MIC College of Technology, Vijayawada, India.
Prof. N. Krishna, Director (Academics), MIC College of Technology, Vijayawada, India.
Dr. K.B.K. Rao, MIC College of Technology, Vijayawada, India.
Dr. Y. Sudheer Babu, Principal, MIC College of Technology, Vijayawada, India.

Honorary Chairs

Dr. Swagatam Das, Indian Statistical Institute Kolkota, India.
Dr. B.K. Panigrahi, Indian Institute of Technology, Delhi, India.

Advisory Committee

Sri C. Gopal Reddy, Chairman, CMRTC, & Secretary CMR Group Hyderabad, India.
Smt C. Vasanth Latha, Secretary, CMR Technical Campus, Hyderabad, India.
Dr. A. Raji Reddy, Director, CMR Technical Campus, Hyderabad, India.

Organising Committee

Organizing Chair
Dr. B. Janakiramaiah, Professor of CSE, MIC College of Technology, Vijayawada, India.

Program Chairs
Prof. Vikrant Bhateja, SRMGPC, Lucknow, India.
Dr. Nilanjan Dey, TICT, Kolkata, India.

Special Session Chairs
Dr. R.T. Goswami, BITS Mesra, Kolkata campus, India.
Dr. Manas Sanayal, University of Kalyani, West Bengal, India.

Track Chairs
Dr. Seerisha Rodda, GITAM, Visakahapatnam, India.
Prof. Pritee Parwekar, ANITS, Visakahapatnam, India.
Dr. M. Ramakrishna Murty, Vignan Institute of Information Tech, Visakahapatnam.

Technical Chairs
Dr. A. Jayalakshmi, Professor & HOD of CSE, MIC College of Technology, Vijayawada, India.
Dr. P. Srinivasulu, Professor of CSE, MIC College of Technology, Vijayawada, India.

Technical Co-Chairs
Mr. A. Rama Satish, Assoc. Professor of CSE, MIC College of Technology, Vijayawada, India.
Mr. D. Varun Prasad, Assoc. Professor of CSE, MIC College of Technology, Vijayawada, India.
Ms. G. Kalyani, Assoc. Professor of CSE, MIC College of Technology, Vijayawada, India.
Mr. D. Prasad, Associate Professor of CSE, MIC College of Technology, Vijayawada, India.

Steering Committee
Dr. Suresh Chandra Satapathy, Professor of CSE, PVP Siddhartha Institute of Technology, Vijayawada, India.
Dr. K. Srujan Raju, Professor of CSE, CMR Technical Campus, Hyderabad, India.

Publicity Chairs
Dr. J.K. Mandal, Professor of CSE, University of Kalyani, Kolkota, India.
Mr. D. Prasad, Assoc. Professor of CSE, MIC College of Technology, Vijayawada, India.
Mr. C.S. Pavan Kumar, Asst. Professor of CSE, MIC College of Technology, Vijayawada, India.

Web Committee

Mr. K. Mahanthi, Asst. Professor of CSE, MIC College of Technology, Vijayawada, India.

Advisory Committee

Dr. T.S. Nageswara Rao, Professor & HOD of CIVIL, MIC College of Technology, Vijayawada, India.
Dr. A. Guruva Reddy, Professor & HOD of ECE, MIC College of Technology, Vijayawada, India.
Dr. T. Vamsee Kiran, Professor & HOD of EEE, MIC College of Technology, Vijayawada, India.
Dr. M. Srinivasa Rao, Professor & HOD of Mechanical, MIC College of Technology, Vijayawada, India.
Mr. C.V.V.D. Srikanth, Assoc. Professor & HOD of MBA, MIC College of Technology, Vijayawada, India.
Dr. A.V. Naresh Babu, Professor of EEE, MIC College of Technology, Vijayawada, India.
Dr. Sarath Babu, Professor of Chemistry, MIC College of Technology, Vijayawada, India.
Dr. B. Seshu, Professor of Physics, MIC College of Technology, Vijayawada, India.
Dr. K.V. Rao, Professor of MBA, MIC College of Technology, Vijayawada, India.
Dr. M. Kaladhar, Professor of Mechanical, MIC College of Technology, Vijayawada, India.
Dr. T. Sunil Kumar, Professor of Mechanical, MIC College of Technology, Vijayawada, India.
Dr. G. Chenchamma, Professor of ECE, MIC College of Technology, Vijayawada, India.
Dr. P.V. Srinivasa Rao, Assoc. Professor of Mathematics, MIC College of Technology, Vijayawada, India.
Dr. K. Trimula Prasad, Assoc. Professor of Chemistry, MIC College of Technology, Vijayawada, India.
Dr. R. Durga Prasad, Assoc. Professor of Mathematics, MIC College of Technology, Vijayawada, India.
Dr. K. Praveen, Assoc. Professor of Chemistry, MIC College of Technology, Vijayawada, India.
Dr. Prasanna Kumar, Assoc. Professor of MBA, MIC College of Technology, Vijayawada, India.
Mr. Ch. Vijaya Kumar, Assoc. Professor & HOD of BED, MIC College of Technology, Vijayawada, India.
Mr. R.J. Lakshmi Narayana, Chief Librarian, MIC College of Technology, Vijayawada, India.

International Advisory Committee

Maurice Clerc, France
Roderich Gross, England
L. Perkin, USA
Sumanth Yenduri, USA
Carlos A. Coello Coello, Mexico
Dipankar Dasgupta, USA
Peng Shi, UK
Saman Halgamuge, Australia
Jeng-Shyang Pan, Taiwan
X.Z. Gao, Finland
Dr. Kun-lin Hsieh, NTU, Taiwan
Dr. Ahamad J. Rusumdar, KIT, Germany
Dr. V.R. Chirumamilla, EUT, The Netherlands
Dr. Halis Altun, MU, Turkey
Juan Luis Fernández Martínez, California
Oscar Castillo, Mexico
Leandro Dos Santos Coelho, Brazil
Heitor Silvério Lopes, Brazil
Rafael Stubs Parpinelli, Brazil
Prof. A. Govardhan, SIT, JNTUH, India.
Prof. A. Rama Mohan Reddy, SVU, India.
Prof. A. Ananda Rao, JNTUA, India.
Prof. B. Eswara Reddy, JNTUA, India.
Prof. P.V.G.D. Prasad Reddy, AU, India.
Prof. J.V.R. Murthy, JNTUK, India.
Prof. K. Usha Rani, SPMVV, Tirupati, India.
Prof. S. Satyanarayana, JNTUK, India.
Prof. S. ViswanadhaRaju, JNTUHCEJ, India.
Prof. E. Sreenivasa Reddy, ANU, India.
Prof. P. Bala Krishna Prasad, India
Dr. Ch. Satyananda Reddy, AU, India.
Gerardo Beni, USA
Namrata Khemka, USA
G.K. Venayagamoorthy, USA
K. Parsopoulos, Greece
Zong Woo Geem, USA
Lingfeng Wang, China
Athanasios V. Vasilakos, Athens
S.G. Ponnambalam, Malaysia
Pei-Chann Chang, Taiwan
Ying Tan, China

Chilukuri K. Mohan, USA
M.A. Abido, Saudi Arabia
Saeid Nahavandi, Australia
Almoataz Youssef Abdelaziz, Egypt
Hai Bin Duan, China
Delin Luo, China
Oscar Castillo, Mexcico
John MacIntyre, England
Frank Neumann
Rafael Stubs Parpinelli, Brazil
Jeng-Shyang Pan, Taiwan
P.K. Singh, India
M.K. Tiwari, India
Sangram Samal, India
K.K. Mohapatra, India
Sachidananda Dehuri, India
P.S. Avadhani, India
G. Pradhan, India
Anupam Shukla, India
Dilip Pratihari, India
P.K. Patra, India
T.R. Dash, Kambodia
Kesab Chandra Satapathy, India
Amit Kumar, India
Srinivas Sethi, India
Lalitha Bhaskari, India
V. Suma, India, and many more

National Advisory Committee
Dr. G.V.S.N.R.V. Prasad, Professor of CSE, GEC, Gudlavalleru.
Dr. M.V.P. Chandra Sekhara Rao, Professor of CSE, RVR&JC, Guntur.
Dr. K. Subba Ramaiah, Professor, YITS, Tirupati.
Dr. M. Babu Rao, Professor & HOD of CSE, GEC, Gudlavalleru.
Dr. M. Suneetha, Professor & HOD of IT, VRSEC, Vijayawada.
Dr. B. Narendra, Professor & HOD of CSSE, Sree Vidyanikethan, Tirupati.
Dr. M. Sunil Kumar, Professor of CSE, Sree Vidyanikethan, Tirupati.
Dr. B. Thirumala Rao, Professor of CSE, KLU, Vijayawada.
Dr. N. Ravi Shankar, Professor & HOD of CSE, LBRC, Mylavaram.
Dr. D. Naga Raju, Professor & HOD of IT, LBRC, Mylavaram.
Dr. K. Hima Bindu, Professor of CSE, Vishnu, Bhimavaram.
Dr. V.V.R. Maheswara Rao, Professor of CSE, SVECW, Bhimavaram.
Dr. B. Srinivasa Rao, Professor & HOD of CSE, Dhanekula, Vijayawada.
Dr. P. Harini, Professor & HOD of CSE, SACET, Chirala.
Dr. A. Yesu babu, Professor & HOD of CSE, CRR, Eluru.
Dr. S. Krishna Rao, Professor & HOD of IT, CRR, Eluru.

Dr. D. Haritha, Professor & HOD of CSE, SRK, Vijayawada.
Dr. S.N. Tirumala Rao, Professor & HOD of CSE, NEC, Narasaraopeta.
Dr. G.V. Padma Raju, Professor & HOD of CSE, SRKR, Bhimavaram.
Dr. G. Satyanarayana Murty, Professor & HOD of CSE, AITAM, Tekkali.
Dr. G. Vijay Kumar, Professor of CSE, VKR,VNB & AGK, Gudivada.
Dr. P. Kireen Sree, Professor of CSE, SVECW, Bhimavaram.
Dr. C. Nagaraju, Professor of CSE, Y.V.U, Proddatur.

Local Organising Committee

Chair
Dr. A. Jayalakshmi, Professor & HOD of CSE, MIC College of Technology, Vijayawada, India.

Registration Chair
Ms. L. Kanya Kumari, Assoc. Professor of CSE, MIC College of Technology, Vijayawada, India.

Reception, Registration Committee
Ms. A. Anuradha, Asst. Professor of CSE, MIC College of Technology, Vijayawada, India.
Ms. K. Vijaya Sri, Asst. Professor of CSE, MIC College of Technology, Vijayawada, India.
Ms. G. Rama Devi, Asst. Professor of CSE, MIC College of Technology, Vijayawada, India.
Ms. B. Lalitha Rajeswari, Asst. Professor of CSE, MIC College of Technology, Vijayawada, India.
Ms. R. Srilakshmi, Asst. Professor of CSE, MIC College of Technology, Vijayawada, India.
Ms. P. Reshma, Asst. Professor of CSE, MIC College of Technology, Vijayawada, India.

Stage Management Chair
Dr. B. Seshu, Professor of Physics, MIC College of Technology, Vijayawada, India.

Decoration, Stage Management Committee
Ms. Sujata Agarwal, Asst. Professor of English, MIC College of Technology, Vijayawada, India.
Ms. T.P. Ann Thabitha, Asst. Professor of CSE, MIC College of Technology, Vijayawada, India.
Ms. K. Vinaya Sree, Asst. Professor of CSE, MIC College of Technology, Vijayawada, India.

IT Services Chair
D. Varun Prasad, Assoc. Professor of CSE, MIC College of Technology, Vijayawada, India.
Mr. J.V. Srinivas, Asst. Professor of CSE, MIC College of Technology, Vijayawada, India.
Mr. P. Srikanth, Asst. Professor of CSE, MIC College of Technology, Vijayawada, India.
Mr. Y. Narayana, Asst. Professor of CSE, MIC College of Technology, Vijayawada, India.

Chair
Dr. P. Srinivasulu, Professor of CSE, MIC College of Technology, Vijayawada, India.

Technical Sessions Chair
Ms. G. Kalyani, Assoc. Professor of CSE, MIC College of Technology, Vijayawada, India.

Technical Sessions Committee
Mr. J. Ranga Rajesh, Asst. Professor of CSE, MIC College of Technology, Vijayawada, India.
Ms. V. Sri Lakshmi, Asst. Professor of CSE, MIC College of Technology, Vijayawada, India.
Ms. N.V. Maha Lakshmi, Asst. Professor of CSE, MIC College of Technology, Vijayawada, India.
Mr. R. Venkat, Asst. Professor of CSE, MIC College of Technology, Vijayawada, India.
Mr. A. Prashant, Asst. Professor of CSE, MIC College of Technology, Vijayawada, India.
Ms. M. Madhavi, Asst. Professor of CSE, MIC College of Technology, Vijayawada, India.
Mr. G.D.K. Kishore, Asst. Professor of CSE, MIC College of Technology, Vijayawada, India.
Ms. Lakshmi Chetana, Asst. Professor of CSE, MIC College of Technology, Vijayawada, India.
Mr. N. Srihari, Asst. Professor of CSE, MIC College of Technology, Vijayawada, India.
Ms. G. Prathyusha, Asst. Professor of CSE, MIC College of Technology, Vijayawada, India.

Refreshment Chair
Mr. D. Varun Prasad, Assoc. Professor of CSE, MIC College of Technology, Vijayawada, India.

Refreshment Committee

Mr. T. Krishnamachari, Asst. Professor of CSE, MIC College of Technology, Vijayawada, India.
Mr. B. Murali Krishna, Asst. Professor of CSE, MIC College of Technology, Vijayawada, India.
Mr. Kamal Rajesh, Asst. Professor of CSE, MIC College of Technology, Vijayawada, India.

Transport Chair

D. Prasad, Assoc. Professor of CSE, MIC College of Technology, Vijayawada, India.

Transport and Accommodation Committee

Mr. D. Durga Prasad, Assoc. Professor of CSE, MIC College of Technology, Vijayawada, India.
Mr. C.S. Pavan Kumar, Asst. Professor of CSE, MIC College of Technology, Vijayawada, India.

Finance Chair

A. Rama Satish, Assoc. Professor of CSE, MIC College of Technology, Vijayawada, India.

Sponsoring Chair

D. Prasad, Assoc. Professor of CSE, MIC College of Technology, Vijayawada, India.

Press and Media Chair

G. Rajesh, Assoc. Professor of ME, MIC College of Technology, Vijayawada, India.

Contents

Editors and Contributors

About the Editors

Dr. Suresh Chandra Satapathy is currently working as a professor and head of the Department of Computer Science and Engineering, PVP Siddhartha Institute of Technology, Andhra Pradesh, India. He received his Ph.D. in computer science and engineering from Jawaharlal Nehru Technological University (JNTU), Hyderabad, and his Master's degree in computer science and engineering from the National Institute of Technology (NIT), Rourkela, Odisha. He has more than 27 years of teaching and research experience. His research interests include machine learning, data mining, swarm intelligence studies, and their applications to engineering. He has more than 98 publications to his credit in various reputed international journals and conference proceedings. He has edited many volumes from Springer AISC and LNCS in the past. In addition to serving on the editorial board of several international journals, he is a senior member of the IEEE and a life member of the Computer Society of India, where he is the national chairman of Division-V (Education and Research).

Prof. Vikrant Bhateja is an associate professor at the Department of Electronics and Communication Engineering, Shri Ramswaroop Memorial Group of Professional Colleges (SRMGPC), Lucknow, and also the head of Academics & Quality Control at the same college. His research interests include digital image and video processing, computer vision, medical imaging, machine learning, pattern analysis and recognition, neural networks, soft computing, and bio-inspired computing techniques. He has more than 90 quality publications in various international journals and conference proceedings to his credit. Professor Vikrant has been on TPC and chaired various sessions from the above domain in international conferences of IEEE and Springer. He has been the track chair and served on the core-technical/editorial teams for numerous international conferences: FICTA 2014, CSI 2014, and INDIA 2015 under the Springer-ASIC Series and INDIACom-2015, ICACCI-2015 under the IEEE. He is an associate editor for the International Journal of Convergence Computing (IJConvC) and also serves on the editorial board of the International Journal of Image Mining (IJIM) under Inderscience Publishers. At present, he is the guest editor for two special issues of the International Journal of Rough Sets and Data Analysis (IJRSDA) and the International Journal of System Dynamics Applications (IJSDA) under IGI Global publications.

Dr. K. Srujan Raju is a professor and head of the Department of Computer Science and Engineering (CSE), CMR Technical Campus. Professor Srujan earned his Ph.D. in the field of network security and his current research interests include computer networks, information security, data mining, image processing, intrusion detection, and cognitive radio networks. He has published several papers in referred international conferences and peer-reviewed journals. He was also on the editorial board of CSI 2014 Springer AISC series 337 and 338 volumes. In addition, he served as a reviewer for many indexed journals. Professor Raju has been honored with the Significant Contributor Award and Active Member Award by the Computer Society of India (CSI) and is currently the Hon. Secretary of the CSI's Hyderabad Chapter.

Dr. B. Janakiramaiah is currently working as a professor in the Department of Computer Science and Engineering, Devineni Venkata Ramana & Dr. Hima Sekhar MIC College of Technology (DVR & Dr. HS MIC College of Technology), Kanchikacherla, Vijayawada, Andhra Pradesh, India. He received his Ph.D. in computer science and engineering from the Jawaharlal Nehru Technological University (JNTU), Hyderabad, and Master's degree in computer science and engineering from the Jawaharlal Nehru Technological University (JNTU), Kakinada. He has more than 15 years of teaching experience. His research interests include data mining, machine learning, studies, and their applications to engineering. He has more than 25 publications to his credit in various reputed international journals and conference proceedings. He is a life member of the Computer Society of India.

Contributors

S. Abirami Department of Information Science and Technology, Anna University, Chennai, India

Jagannath Aghav Pune, India

Sudhakar Alapati Department of ECE, RVR & JC College of Engineering, Guntur, AP, India

Vijayasankar Anumala Department of ECE, V R Siddhartha Engineering College, Vijayawada, India

Fayeza Arif Muffakham Jah College of Engineering and Technology, Hyderabad, Telangana, India

Juluru Aruna Computer Science and Technology, Department of CSE, ANIL Neerukonda Institute of Technology and Sciences (ANITS), Visakhapatnam, India

D. Ashwani Computer Science and Technology, Department of CSE, ANIL Neerukonda Institute of Technology and Sciences (ANITS), Visakhapatnam, India

Lalit Kumar Awasthi National Institute of Technology, Hamirpur, India

Vijayalakshmi Baba ECE Department, B.S. Abdur Rahman University, Chennai, Tamilnadu, India

I. Bhargavi Department of Computer Science and Engineering, VFSTR University, Guntur, India

S.A. Bhavani Department of CSE, ANIL Neerukonda Institute of Technology and Sciences (ANITS), Visakhapatnam, India

D.L. Chaitanya GRIET, Hyderabad, India

Krishnaveer Abhishek Challa Department of Linguistics, Andhra University, Visakhapatnam, Andhra Pradesh, India

M.M. Chandane Department of Computer Engineering and I.T, Veermata Jijabai Technological Institute, Mumbai, Maharashtra, India

Aditya Chaudhuri Dr. Sudhir Chandra Sur Degree Engineering College, Kolkata, India

Naveen Chauhan National Institute of Technology, Hamirpur, India

Tanupriya Choudhury Amity University, Noida, Uttar Pradesh, India

Rajath. B. Das Department of Electronics and Communication, RV College of Engineering, Bengaluru, Karnataka, India

C. Dastagiraiah Department of CSE, K L University, Vijayawada, India

Deepali Computer Science Department, Guru Nanak College, Budhlada, India

K.S. Deepthi Computer Science and Technology, Department of CSE, ANIL Neerukonda Institute of Technology and Sciences (ANITS), Visakhapatnam, India

Shyam Deshmukh Pune, India

S. Prabhu Deva Department of Information Science and Engineering, Jawaharalal Nehru College of Engineering, Shivamogga, India

K. DeviPriya Department of Computer Science & Engineering, Aditya Engineering College, Surampalem, AP, India

P. Dhanunjaya Rao Department of ECE, Raghu Engineering College, Visakhapatnam, AP, India

N.V. Dharani Kumari Department of Computer Applications, Dr. Ambedkar Institute of Technology, Bengaluru, India

Parvathy Dharmarajan Department of Computer Science and Applicaion, Amrita School of Engineering Amritapuri, Amrita Vishwa Vidyapeetham, Amrita University, Amirtapuri, India

S.G. Divakara Department of Chemistry, RV College of Engineering, Bengaluru, Karnataka, India

Uma N. Dulhare Muffakham Jah College of Engineering and Technology, Hyderabad, Telangana, India

Divya Gavini Department of Computer Science and Engineering, Vignan's Lara Institute of Technology and Science, Guntur, India

M. Geetha Priya Centre for Incubation, Innovation Research and Consultancy, Jyothy Institute of Technology, Bangalore, Karnataka, India

K.S. Geetha Department of Electronics and Communication, RV College of Engineering, Bengaluru, Karnataka, India

K. Geethanjali Department of CSE, ANIL Neerukonda Institute of Technology and Sciences (ANITS), Visakhapatnam, India

Sayantani Ghosh Jadavpur University, Kolkata, India

Sutanu Ghosh Indian Institute of Engineering Science and Technology, Shibpur, Howrah, India

R.H. Gopalkrishna Department of ECE, RAGHU Engineering College, Visakhapatnam, AP, India

M. Govinda Raju Department of Electronics and Communication, RV College of Engineering, Bengaluru, Karnataka, India

Anjuman Gul Department of Computer Science and Engineering, Lovely Professional University, Phagwara, India

Hardik Gupta Department of Computer Engineering and I.T, Veermata Jijabai Technological Institute, Mumbai, Maharashtra, India

Kunal Gupta Amity University, Noida, India

Jagadish Gurrala Computer Science and Technology, Department of Computer Science and Engineering, ANIL Neerukonda Institute of Technology and Sciences (ANITS), Visakhapatnam, India

Prashanti Guttikonda Vignan's Lara Institute of Technology & Science, Guntur, Andhra Pradesh, India

Madusu Hanmandlu Department of Computer Science & Engineering, MVSR Engineering College, Hyderabad, Telangana, India

K. Hari Kishore Department of ECE, KLEF, K.L University, Guntur, AP, India

Anjana Jain Shri Govindram Seksaria Institute of Technology and Science, Indore, M.P, India

Nidhi Jain Department of Computer Science & Engineering, Amity University, Noida, Uttar Pradesh, India

Rashmi Jain Amity University, Noida, India

Satya V. Jampana Data and Information Management Group, Indian National Centre for Ocean Information Services (INCOIS), Hyderabad, India

Sindhubala Kadirvelu ECE Department, B.S. Abdur Rahman University, Chennai, Tamilnadu, India

R. Kanaka Raju GVP college for degree & Pg courses, Visakhapatnam, India

Anjaneyulu Katuru Department of ECE, ANU College of Engineering & Technology, Guntur, AP, India

Arvinder Kaur USICT, Guru Gobind Singh Indraprastha University, New Delhi, India

Dupinder Kaur Department of Computer Science and Applications, Chaudhary Devi Lal University, Sirsa, HR, India

Harleen Kaur Baba Farid College of Engineering and Technology, Bathinda, India

Harveen Kaur Baba Farid College of Engineering and Technology, Bathinda, India

Sandhya Rani Kaviti Vignan's Lara Institute of Technology & Science, Guntur, Andhra Pradesh, India

D. Khalandar Basha Institute of Aeronautical Engineering, Hyderabad, India

M. Kranthi Kiran Computer Science and Technology, Department of CSE, ANIL Neerukonda Institute of Technology and Sciences (ANITS), Visakhapatnam, India

Dileep Kumar Koda Anil Neerukonda Institute of Technology and Sciences (ANITS), Visakhapatnam, India

Morarjee Kolla CMR Institute of Technology, Hyderabad, Telangana, India

Rajesh Kolli Department of Electronics and Communication Engineering, SIR C. R. Reddy College of Engineering, Eluru, India

B. Koushik Department of Electronics and Communication, RV College of Engineering, Bengaluru, Karnataka, India

V. Krishna Reddy Department of CSE, K L University, Vijayawada, India

B.T. Krishna JNTU, Vizianagaram, AP, India

D. Krishnaveni Department of Telecommunication Engineering, A P S College of Engineering, Bangalore, Karnataka, India

Sukanya Kulkarni Sardar Patel Institute of Technology, Mumbai University, Mumbai, India

Abhinav Kumar Amity University, Noida, India

Ashish Kumar Department of Computer Science and Engineering, ITS, Greater Noida, Uttar Pradesh, India

Hitesh Kumar Amity University, Noida, Uttar Pradesh, India

Praveen Kumar Amity University, Noida, Uttar Pradesh, India

Shreeya Laad Department of Computer Engineering and I.T, Veermata Jijabai Technological Institute, Mumbai, Maharashtra, India

S. Lakshmi Prasad Department of Electronics and Communication, RV College of Engineering, Bengaluru, Karnataka, India

A.M. Lakshmikanth Department of Electronics and Communication, RV College of Engineering, Bengaluru, Karnataka, India

R.V.S.S. Lalitha Computer Science and Technology, Department of CSE, ANIL Neerukonda Institute of Technology and Sciences (ANITS), Visakhapatnam, India

K. Madhu Sudhana Rao Department of ECE, ANU College of Engineering & Technology, Guntur, AP, India; Department of ECE, KKR & KSR Institute of Technology & Sciences, Guntur, AP, India

Arun Malik Department of Computer Science and Engineering, Lovely Professional University, Phagwara, India

P. Mallikarjuna Rao Department of ECE, AUCE (A), Andhra University, Visakhapatnam, India

G. Manikandan Department of Information Science and Technology, Anna University, Chennai, India

K. Manjunatha Chari GITAM University, Visakhapatnam, India

H. Manohar Department of Electronics and Communication, RV College of Engineering, Bengaluru, Karnataka, India

T. Maruthi Padmaja Department of Computer Science and Engineering, VFSTR University, Guntur, India

Siddhesh Mhatre Department of Computer Engineering and I.T, Veermata Jijabai Technological Institute, Mumbai, Maharashtra, India

B.K. Mishra Thakur College of Engineering and Technology, Mumbai University, Mumbai, India

Tusar Kanti Mishra Computer Science and Technology, Department of CSE, ANIL Neerukonda Institute of Technology and Sciences (ANITS), Visakhapatnam, India

AyodhyaRam Mohanthy Computer Science and Technology, Department of Computer Science and Engineering, ANIL Neerukonda Institute of Technology and Sciences (ANITS), Visakhapatnam, India

Ramakanta Mohanty Computer Science and Engineering, Keshav Memorial Institute of Technology, Narayanaguda, Hyderabad, India

A. Murali Department of ECE, KLEF, K.L University, Guntur, AP, India; Department of ECE, RAGHU Engineering College, Visakhapatnam, AP, India

D. Nagajyothi ECE Department, Vardhaman College of Engineering, Shamshabad, Telangana, India

P.V. Naganjaneyulu Department of ECE, MVR College of Engineering & Technology, Vijayawada, India

Satuluri Naganjaneyulu Department of Information Technology, LBRCE, Mylavaram, AP, India

D. Nagaraju Matrusri Engineering College, Hyderabad, India

M. Nagaratna Computer Science and Engineering, JNTUH College of Engineering, Kukatpally, Hyderabad, India

Surendra Kumar Nanda C.V.Raman Computer Academy, Bhubaneshwar, Odisha, India

Vikram Narayandas Department of Computer Science & Engineering, MVSR Engineering College, Hyderabad, Telangana, India

B. Naresh Institute of Aeronautical Engineering, Hyderabad, India

J.S. Pahariya Computer Science and Engineering, Rustamji Institute of Technology, Tekanpur, Gwalior, India

Sambhu Prasad Panda C.V.Raman Computer Academy, Bhubaneshwar, Odisha, India

Babita Pandey Department of Computer Applications, Lovely Professional University, Phagwara, India

Hari Mohan Pandey Department of Computer Science & Engineering, Amity University, Noida, Uttar Pradesh, India

J. Pavan Kumar Data and Information Management Group, Indian National Centre for Ocean Information Services (INCOIS), Hyderabad, India

M. Pavan Department of ECE, Raghu Institute of Technology, Visakhapatnam, India

Ashok Kumar Popuri VFSTR University, Guntur, Andhra Pradesh, India; S.V.U. College of Engineering, S.V. University, Tirupati, Andhra Pradesh, India

D. Prabhakar Department of ECE, DVR & Dr. HS MIC College of Technology, Kanchikacherla, India

M.V.S. Prasad Department of ECE, RVR & JC College of Engineering, Guntur, AP, India

Niharika Pujari C.V.Raman Computer Academy, Bhubaneshwar, Odisha, India

Rajesh Kumar Pullakura Department of ECE, Andhra University College of Engineering, Visakhapatnam, India

M. Rajani Devi JNTU Kakinada, Kakinada, AP, India

B. Rajathilagam Department of Computer Science and Engineering, Amrita School of Engineering Coimbatore, Amrita Vishwa Vidyapeetham, Amrita University, Coimbatore, India

D. Rajendra Prasad Department of ECE, St. Ann's College of Engineering & Technology, Chirala, India

V. Ramachandra Raju RGUKT (IIIT), AP, India

K. Ramanjaneyulu PVP Siddhartha Institute of Technology, Vijayawada, AP, India

Ningampalli Ramanjaneyulu RGMCET, Nandyal, A.P, India

S. Rambabu Institute of Aeronautical Engineering, Hyderabad, India

B.S.S.V. Ramesh Babu Department of ECE, Raghu Institute of Technology, Visakhapatnam, India

R. Ramesh Department of Mechanical Engineering, MVGRCE, Vizianagaram, AP, India

Sangeeta Rani Department of Computer Science & Engineering, Amity University, Noida, Uttar Pradesh, India

B. Thirumala Rao AP, India

K. Thirupathi Rao AP, India

M. Srinivas Rao Department of Computer Science and Engineering, SVIET, Machilipatnam, AP, India

P.S.V. Srinivasa Rao Department of Computer Science and Engineering, SVECW, Bhimavaram, AP, India

Mamata Rath Department of IT, C.V.Raman College of Engineering, Bhubaneswar, Odisha, India

B. Ravikiran Computer Science and Technology, Department of CSE, ANIL Neerukonda Institute of Technology and Sciences (ANITS), Visakhapatnam, India

L. Srinivasa Reddy Computer Science and Technology, Department of CSE, ANIL Neerukonda Institute of Technology and Sciences (ANITS), Visakhapatnam, India

J. Roopa Department of Electronics and Communication, RV College of Engineering, Bengaluru, Karnataka, India

Umesh Prasad Rout Department of IST, Ravenshaw University, Cuttack, Odisha, India

Biswajith Roy Department of Electronics and Communication, RV College of Engineering, Bengaluru, Karnataka, India

Ajay Kumar Sahu Department of Computer Science and Engineering, Raj Kumar Goel Institute of Technology and Management, Ghaziabad, Uttar Pradesh, India

Sanjib Kumar Sahu Department of Computer Science and Application, Utkal University, Bhubaneshwar, Odisha, India

A. Sai Ramya Department of ECE, Raghu Institute of Technology, Visakhapatnam, India

D.J. Santosh Kumar Computer Science and Technology, Department of CSE, ANIL Neerukonda Institute of Technology and Sciences (ANITS), Visakhapatnam, India

M. Santoshkumar Department of Computer Science and Engineering, GM Institute of Technology, Davanagere, India

Kodati Satya Prasad Department of ECE, Jawaharlal Nehru Technological University, Kakinada, A.P, India

T. Satyanarayana Raju Computer Science and Technology, Department of CSE, ANIL Neerukonda Institute of Technology and Sciences (ANITS), Visakhapatnam, India

B.S. Satyanarayana Munjal University, Gurugaon, Haryana, India

Donti Satyanarayana RGMCET, Nandyal, A.P, India

M. Satyanarayana Department of ECE, M V G R College of Engineering (A), Vizianagaram, India

Akshita Shah Department of Computer Engineering and I.T, Veermata Jijabai Technological Institute, Mumbai, Maharashtra, India

Rajesh Sharma National Institute of Technology, Hamirpur, India

Shilpi Sharma Amity University, Noida, Uttar Pradesh, India

R. Venkat Shesu Data and Information Management Group, Indian National Centre for Ocean Information Services (INCOIS), Hyderabad, India

K. Shivanna Department of Computer Science and Engineering, GM Institute of Technology, Davanagere, India

Shweta Department of Computer Science & Engineering, Amity University, Noida, Uttar Pradesh, India

B.S. Shylaja Department of Information Science and Engineering, Dr. Ambedkar Institute of Technology, Bengaluru, India

P. Siddaiah Department of ECE, University College of Engineering and Technology, Acharya Nagarjuna University, Guntur, India

Amit Prakash Singh University School of Information and Communication Technology, Guru Gobind Singh Indraprastha University, Dwarka, Delhi, India

Dilbag Singh Department of Computer Science and Applications, Chaudhary Devi Lal University, Sirsa, HR, India

Ankita Singla Computer Science Department, Guru Nanak College, Budhlada, India

Ankit Soni Institute of Engineering & Technology, DAVV, Indore, M.P, India

E. Srikala Department of ECE, Raghu Institute of Technology, Visakhapatnam, India

L. Srikanth Department of ECE, RAGHU Engineering College, Visakhapatnam, AP, India

P. Naga Srinivasu Computer Science and Technology, Department of CSE, ANIL Neerukonda Institute of Technology and Sciences (ANITS), Visakhapatnam, India

Shalini Srivastava Integrated Institute of Technology, Dwarka, Delhi, India

L. Sumalatha Department of Computer Science & Engineering, UCEK, JNTUK, Kakinada, AP, India

E. Suresh Babu Department of Computer Science and Engineering, K L University, Guntur, AP, India

K. Suresh Department of CSE, ANIL Neerukonda Institute of Technology and Sciences (ANITS), Visakhapatnam, India

D. Surya Kumari Department of CSE, ANIL Neerukonda Institute of Technology and Sciences (ANITS), Visakhapatnam, India

A.V.S. Swathi Department of ECE, Raghu Institute of Technology, Visakhapatnam, India

Lakshmi Thimmareddy Department of Computer Science & Engineering, MVSR Engineering College, Hyderabad, Telangana, India

Anilkumar Tirunagari Department of Electronics and Communication Engineering, SIR C. R. Reddy College of Engineering, Eluru, India

Sujanavan Tiruvayipati Department of Computer Science & Engineering, MVSR Engineering College, Hyderabad, Telangana, India

L. Trinadh Computer Science and Technology, Department of CSE, ANIL Neerukonda Institute of Technology and Sciences (ANITS), Visakhapatnam, India

A. Trinadha Department of ECE, Chinthalapudi Engineering College, Guntur, AP, India

Saswat Tripathy Department of ECE, Raghu Engineering College, Visakhapatnam, AP, India

Venkateswara Rao Tumati Department of Electronics and Communication Engineering, SIR C. R. Reddy College of Engineering, Eluru, India

T.V.S. Udaya Bhaskar Data and Information Management Group, Indian National Centre for Ocean Information Services (INCOIS), Hyderabad, India

R.S. Umamaheswara Raju Department of Mechanical Engineering, MVGRCE, Vizianagaram, AP, India

P. Unita Department of ECE, Raghu Institute of Technology, Visakhapatnam, India

Raksha Upadhyay Institute of Engineering & Technology, DAVV, Indore, M.P, India

D. Veeraiah Department of Computer Science and Engineering, VFSTR University, Guntur, India

V. Venkataiah Computer Science and Engineering, CMR College of Engineering and Technology, Medchal, Hyderabad, India

D. Venkatesulu Department of Computer Science and Engineering, Vignan University, Guntur, India

T. Venu Gopal JNTUH College of Engineering Sultanpur, Sultanpur, Telangana, India

Vidhi Vig USICT, Guru Gobind Singh Indraprastha University, New Delhi, India

S.P. Vighneshwar Computational Facilities and Web Based Services Group (CWG), Indian National Centre for Ocean Information Services (INCOIS), Hyderabad, India

G. Vijaya Padma Department of ECE, KLEF, K.L University, Guntur, AP, India

P. Vinod Babu Anil Neerukonda Institute of Technology and Sciences (ANITS), Visakhapatnam, India

Shashidhar Virupaksha IT Deptartmen, VRSEC, Vijayawada, India; CSE Deptartmen, Vignan University, Guntur, India

Mounika Vurity Department of Journalism and Mass Communications, Andhra University, Visakhapatnam, Andhra Pradesh, India

Sk. Yakoob Department of CSE, K L University, Vijayawada, India

Approach Towards Increasing Efficiency of Communication Protocol in Wireless Sensor Network Using Modified Routing Protocol

Rashmi Jain, Abhinav Kumar and Kunal Gupta

Abstract Wireless Sensor Networks (WSNs), have served humankind to screen nature of the spots which are impossible. The sensor hubs have constrained vitality to sense and send the information. The utilization of vitality ought to be effective so that system lifetime and also the throughput is moved forward. A portion of the directing conventions have been conceived to course the information detected by the sensors in the WSN are pointed to be vitality proficient. Grouping based Energy effective Steering conventions principally harp on expanding the lifetime as well as execution of the system. Drain based conventions can be adjusted to give a system more lifetime and enhanced execution. In this paper, we have investigated and thought about two WSN conventions, Neural M-GEAR, MGear and Neural LEACH, Leach on the grounds of system lifetime and execution of the system.

Keywords LEACH · MGREARNEURAL · NEURALLEACH

1 Introduction

Upgrades in calculations open numerous ways to new advancements in current innovation. Remote Sensor Systems (WSN) are giving administrations to humankind to accumulate the data in basic ecological conditions. WSNs can be utilized to gather data from war field to screen family unit gadgets or robots. The proficient utilization of assets dependably enhances the execution of the system. The execution of the WSN is enhanced by expanding the lifetime of the WSN [3]. Low

R. Jain (✉) · A. Kumar · K. Gupta
Amity University, Noida, India
e-mail: msrashmijain@gmail.com

A. Kumar
e-mail: abhinav.kumar234@gmail.com

K. Gupta
e-mail: kgupta@amity.edu

S.C. Satapathy et al. (eds.), *Computer Communication, Networking and Internet Security*, Lecture Notes in Networks and Systems 5,
DOI 10.1007/978-981-10-3226-4_1

power devouring directing convention keeps up the vitality level of the system higher to enhance lifetime. The sensor hubs utilized as a part of WSNs have a restricted force source. The objective of a sensor hub is to sense the environment and send the detected information for further preparing. For each WSN there exists a Base Station (BS) which gets and gathers the information sent by the sensor hubs. The vitality which is spent on sending a k bit information is a great deal more than getting k bit information [3]. Vitality to send one piece of information relies on the separation between the sender hub and the collector hub. On the off chance that sensor hubs were to send information straightforwardly to the BS the hubs which are more remote need to spend more vitality to send k bit information to BS [3]. So the sensor hubs which are situated a long way from the BS would kick the bucket sooner than the sensor hubs which are found closer to BS (Fig. 1).

$$\mathrm{Etx}(\mathrm{d}) = \mathrm{Eelec} * \mathrm{k}\,\mathrm{Eamp} * \mathrm{k} * \mathrm{d2} = \mathrm{d}\,\mathrm{ERxEelec} * \mathrm{k}$$

To keep this imparity and enhance the lifetime of WSN LEACH was proposed by Heinzelman et al. [3]. Drain is a bunching based steering convention for WSNs. The sensor hubs are made to structure a few bunches in the detecting locale. Filter is a self arranging, versatile grouping convention that employments randomization to disseminate the vitality stack equally among the sensors in the system [3]. A bunch head (CH) is picked at each round in a bunch. All sensor hubs in a bunch transmit information specifically to the neighborhood CH. At that point CH assembles all information from sensor hubs in the group and sends the information to BS. After each cycle another bunch head is chosen. In this paper two directing conventions M-gear and MODLEACH have been broke down and thought about. M-gear [1] was proposed by Nadeem et al. is LEACH based directing convention for WSNs. MODLEACH [2] was proposed by Mahmood et al. is additionally a LEACH based convention for WSNs.

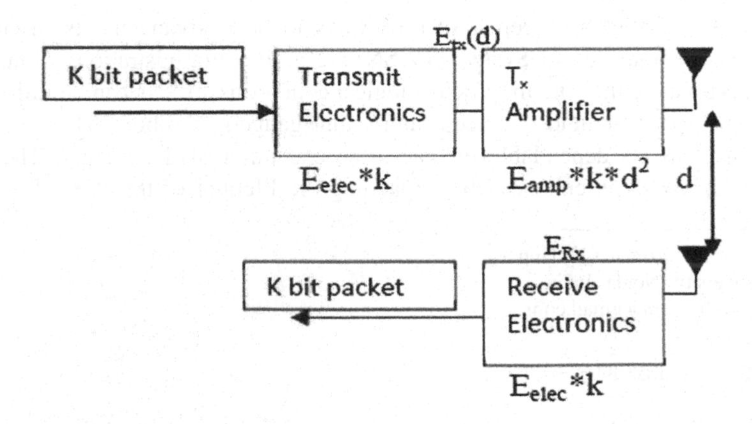

Fig. 1 Energy spend in radio transmission [3]

2 Related Work

There are a few directing conventions proposed that can be broke down in the context of remote sensor systems. We study and look at two such protocols, specifically M-gear furthermore, Mod-Leach utilizing our sensor network and radio models. The sensor hubs which are utilized as a part of WSNs have constrained figuring capacity and transmitting/accepting power. This force must be utilized as a part of such a way in this way, to the point that the sensor hubs live more. At first Direct Transmission to BS was talked about [3]. Hub specifically sends the detected information to BS. This methodology is secure yet prompts higher force utilization. Hubs which are found more amazing than the hubs which are found closer to the BS. To beat this issue least transmission vitality (MTE) was produced. In MTE the information is transmitted utilizing different jumps. This again climbed a comparative issue, the main distinction is that presently the hubs which are closer to BS started to kick the bucket prior. Estrin et al. [4] took a shot at a progressive grouping instrument managing a symmetric correspondence for power sparing in sensor hubs. As indicated by this instrument, every partaking hub of system are conveyed in 2-bounce bunch. Despite the fact that this convention is definitely not much vitality productive for remote sensor hubs be that as it may, it offers approach to various leveled grouping calculations. Bunching for vitality preservation is demonstrated as proficient system for remote sensor systems [5, 6]. At the point when a sensor system is conveyed, hubs build up groups and assign one hub from every bunch as a group head. These group head hubs are in charge of accepting information from different hubs of bunch, do information collection/combination of got information and transmit it to base station. The information transmission from sensor hubs in a bunch to group leader of that group is done utilizing TDMA. Every Node sits tight for now is the ideal time space to send the information to bunch head. Along these lines, data transfer capacity utilization and life time of system is upgraded [7]. Considering bunch based calculations, today a few conventions are produced, each having a few traits and upgrades predominantly in bunch head determination calculations. Despite the fact that one thing is basic, all conventions concentrate on vitality protection and information collection. Principle methodology of choosing a bunch head was given by LEACH and that is further upgraded by SEP and DEEC. This article harps on similar investigation of two such Drain based steering conventions for productive vitality utilization.

3 Neural LEACH and M-Gear

Drain opens scope for some steering conventions for WSNs. The strategies in LEACH manage homogeneous system. As indicated by LEACH new bunch head is chosen for each round. This prompts new bunch development for each round. New bunch arrangement for each round squanders a huge measure of restricted vitality.

On the off chance that present group head has more vitality than a portion of the groups in the bunch, then in the following cycle a hub with less vitality can be chosen as new bunch head. So lingering vitality of current bunch head must be mulled over prior to the race of new bunch head. Henceforth new bunch head substitution calculation was presented by D. Mahmood et al. The vitality required to transmit information from a hub to group head is straightforwardly corresponding to the square of the separation in the middle of hub and group head. Thus hubs dwelling close to the bunch head must utilize low enhancement of the sign than that of the hubs which are situated far from group head. So D. Mahmood et al. proposed a double transmitting power levels moreover.

A. Productive Cluster Head Replacement Algorithm

It is an edge in a group head arrangement for exceptionally next round. On the off chance that current group has not spent much vitality amid its residency and has more vitality than required edge, it will remain bunch head for the following round also. This is how, vitality squandered in steering bundles for new group head and bunch development can be spared. In the event that bunch head has less vitality than required limit, it will be supplanted as per LEACH calculation.

B. Double Amplification Levels For Data Transmission

There are three methods of transmission in a group based WSN.

1. Intra Cluster Transmission.
2. Entomb Cluster Transmission.
3. Group Head To Base Station Transmission.

Hub detects the information and sends it to group head specifically this transmission is intra group transmission. Group head totals the information got from hubs and transmits the information to base station this is Cluster head to Base station transmission. The transmission among Cluster Heads is termed as Inter Cluster Head Transmission. The least vitality required for each of the three sorts of transmissions can't be the same. So when a hub is chosen as the group head it employments high energy to open up the sign. What's more, in the following round at the point when other hub is chosen as the group head it utilizes little energy to open up the sign.

4 M-Gear

In most bunch based conventions Cluster Head is chosen on the base of likelihood. By and large Cluster Heads are dispersed consistently all through the field ofsensor. So it is conceivable that the Cluster Heads chose can be packed in one district of the system. Henceforth, a portion of the hubs would not get any Cluster Heads in their surroundings. Also a few conventions utilized unequal bunching and attempt to use recourses capably [1]. Sensor hubs have detected information for BS to process. Therefore, a programmed technique for joining or amassing the information into a little arrangement of pivotal data is required [8]. Information combination is the procedure of information accumulation. To enhance system lifetime and throughput, we send an extraordinary sort of hub, called portal hub, at the focal point of the sensor system field. The portal hub gathers and totals information from Group Heads and hubs close entryway, and sends to BS. The outcomes demonstrate that system lifetime and vitality utilization progressed. We include rechargeable entryway hub since it is on ground truth that the reviving of door hub is much less expensive than the cost of sensor hub [1]. The usage of M-GEAR convention is done in taking after stages [1]:

A. Introductory Phase

The sensor hubs are disseminated in the sensor field arbitrarily. The BS sends a solicitation bundle to all hubs to enquire their area. The sensor hubs react with their area to the BS. BS keeps this data about all hubs.

B. Setup Phase

The district is isolated into four intelligent areas. To begin with one is the one where the sensor hubs are closer to the BS. The hubs in first district direct transmit the detected information to the BS. Second district is the one where hubs are close to the portal hub. The hubs in second locale straightforwardly send information to the portal hub. Presently whatever remains of the area is separated into two bunches, where Cluster Heads are chosen for each round. Group heads in the bunched area get the information from the hubs in the group and total and send the information to the portal hub.

C. CH Selection

At first BS separates the system into locales. CHs are chosen in every district independently. Give r a chance to speak to the number of rounds to be a CH for the hub Si. Every hub choose itself as a CH once every r = 1/p rounds. Toward the begin of first round all hubs in both locales has rise to vitality level and has rise to opportunity to wind up CH. After that CH is chosen on the premise of the remaining vitality of sensor hub and with a likelihood p alike LEACH. In each round, it is required to have n x p CHs. A hub can ended up CH just once in an age and the hubs not chosen as CH in the current round fill right to the set C. The likelihood of a hub to (has a place with set C) choose as CH increments in each round. It is required to maintain adjusted number of CHs. Toward the begin of each cycle, a hub Si has a place with set C self-sufficiently pick an irregular number between 0 and 1. On the off chance that the produced irregular number for hub Si is not exactly a predefined edge T(s) esteem then the hub gets to be CH in the current round.

The limit worth can be found as:

$$T(s) = \begin{cases} \frac{p}{1-px\left(r\,mod\left(\frac{1}{p}\right)\right)} \text{ if s e C} \\ 0 \end{cases}$$

where P = the craved rate of CHs and r = the current round, C = set of hubs not chose as CH in current round. In the wake of choosing CHs in every area, CHs educate their role to neighbor hubs. CHs telecast a control parcel utilizing a CSMA MAC convention. Upon got control bundle from CH, every hub transmits recognize parcel. Hub who finds closest CH, gets to be individual from that CH.

D. Planning

After Cluster Head is chosen for the current round in the bunch, the Cluster Head makes TDMA openings for the sensor hubs in the bunch. Every sensor hub in the bunch sits tight for its TDMA time opening to send the information to the Bunch Head.

E. Relentless State Phase

In this form the Cluster Heads are chosen for each round and sensor hubs send their information to relating Bunch Heads in TDMA time space apportioned to them. Group Heads total and send the information to the door hub. The door hub gathers the information from the Cluster Heads furthermore, the hubs close to the portal hub then sends the collected information to the Base Station.

5 Test Results and Comparison

A. Reproduction

To evaluate the parameters to make examinations between the conventions MGEAR and MODLEACH MATLAB reproduction is utilized. A 100 × 100 sensor field is utilized to scatter the hubs. For homogeneity Base Station is put far from detecting field at (50, 120) for both conventions. The sensor hubs are haphazardly scattered in 100 × 100 detecting field. Same appropriation of sensor hub is supplied for both conventions (Fig. 2).

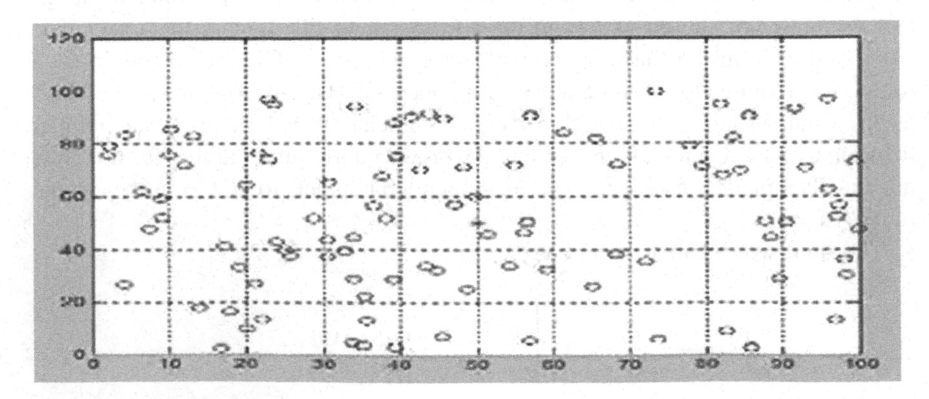

Fig. 2 Random nodes deployment

Table 1 System parameters

Parameters	Parameter's value
1. System size	100–100
2. Starting energy of node	0.5 J
3. Bundle size	4000 bits
4. Eelec	5 nJ/bit
5. Enhancement energy in MGear	Efs1 = 10 pJ/bit/m^2
6. Enhancement energy in MODLEACH (cluster to BS) for d > d0	Efs1 = 10 pJ/bit/m^2
7. Enhancement energy in MODLEACH (cluster to BS) for d < d0	Emp1 = 0.0013 pJ/bit/m^2
8. Enhancement energy in MODLEACH (intra group comm.) for d > d1	Efs2 = Efs1/10
9. Enhancement energy in MODLEACH (intra group comm.) for d < d1	Emp2 = Emp1/10

For M-GEAR passage hub is put at (50, 50) organizes. To begin with area, where hubs specifically transmit information to Base Station, constitute every one of the hubs which have Y coordinate more prominent than 80. Second district, where hubs send information straightforwardly to portal hub, are situated at the middle. What's more, the remaining district is partitioned into two groups where Cluster Head are chosen for each round (Table 1).

B. Execution Parameters

1) **Network lifetime**: It is the time interim from the begin of the system operation till the last incredible.
2) **Throughput**: To assess the execution of through put, the quantities of bundles got by BS are analyzed with the number of parcels sent by the hubs in each round.

C. Recreation Result And Analysis

1) **Network Lifetime:** After running the recreation a few times we generally discover that M-GEAR beats LEACH and MODLEACH both. By performing recreation in MATLAB taking after information was delivered (Table 2).

By breaking down the Fig. 3 we can see that hubs in the instance of M-GEAR convention begin to pass on prior yet add up to system keeps going longer than Neural LEACH and Leach.

2) **Throughput:** By breaking down Fig. 4 we see that the throughput of Neural M-GEAR and M-Gear is more noteworthy than that of Neural LEACH with Leach.

This is on the grounds that M-GEAR keeps up remaining vitality of sensor hubs to keep going long utilizing the passage hub.

Table 2 System lifetime measured by the maximum number of rounds till the last node was alive

No of run	NEURAL MGEAR MGEAR LEACH
1	2498 2312 1577
2	2492 2197 1687
3	2500 2071 1510
4	2496 2316 1660
5	2501 2205 1464
6	2468 2172 1669

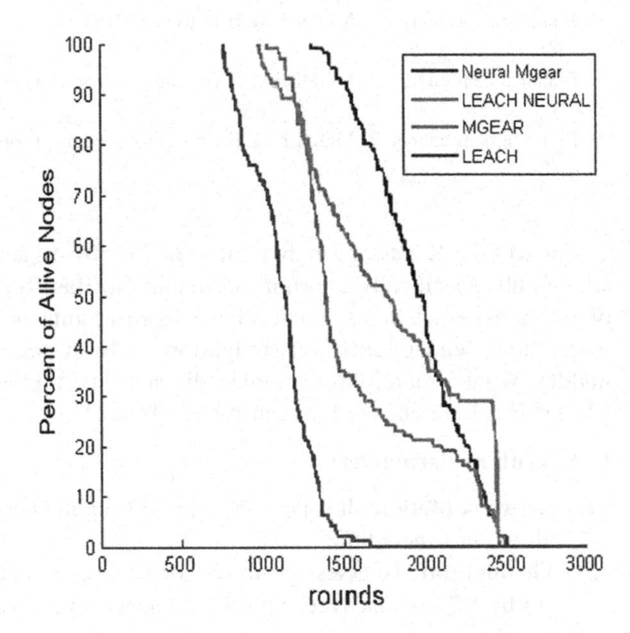

Fig. 3 Percentage of alive nodes

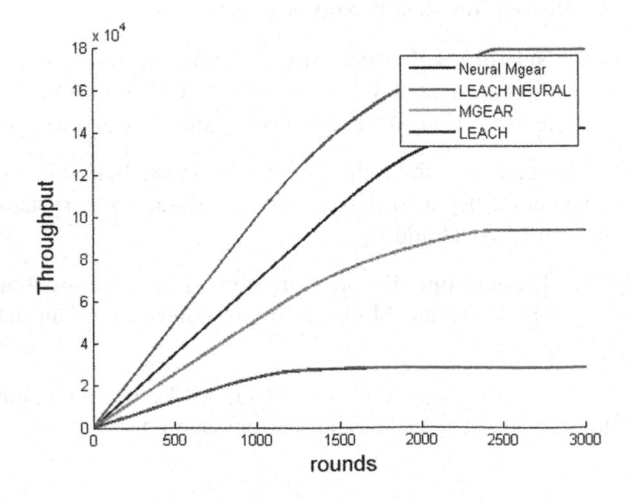

Fig. 4 Throughput of Mgear and Neural Mgear

6 Conclusion

We have dissected and thought about the exhibitions of two steering conventions MGEAR and Neural M-Gear the premise of system lifetime and throughput. In spite of the fact that, the execution of Neural LEACH is enhanced when contrasted with LEACH however the presentation of the entryway hub has enhanced the execution of the system. As indicated by the investigation in view of MATLAB reenactment we unmistakably see that passage hub which is conveyed on account of M-GEAR enhances the system lifetime and in addition the throughput of the system. Consequently we presume that to the detriment of the door hub one can without much of a stretch accomplish higher execution of the system.

References

1. Q. Nadeem, M. B. Rasheed, N. Javaid, Z. A. Khan, Y. Maqsood and A. Din, M-GEAR: Gateway-Based Energy-Aware Multi-Hop Routing Protocol for WSNs, Eighth International Conference on Broadband and Wireless Communication and Application, pp. 164–169, 2013.
2. D. Mahmood, N. Javaid, S. Mahmood, S. Qureshi, A. M. Memon and T. Zaman, MODLEACH: A Variant of LEACH for WSNs, Eighth International Conference on Broadband and Wireless Communication and Applications, pp. 158– 163, 2013.
3. W. Heinzelman, A. Chandrakasan, and H. Balakrishnan. Energy- Efficient Communication Protocols for Wireless Microsensor Networks, Hawaiian International Conference on Systems Science, January 2000.
4. D. Estrin, R. Govindan, J. Heidemann, and S. Kumar. Next Century Challenges: Scalable Coordination in Wireless Networks, 5th Annual ACM/IEEE International Conference on Mobile Computing and Networking (MOBICOM), pp. 263–270, 1999.
5. C. Y. Chong and S. P. Kumar, Sensor Networks: Evolution, Opportunities and Challenges, IEEE, 91, No. 8, pp. 1247–1256, Aug 2003.
6. M. Younis, P. Munshi, G. Gupta and S. M. Elsharkawy, On Efficient Clustering of Wireless Sensor Networks, Second IEEE Workshop on Dependability and Security in Sensor Networks and Systems, pp. 78–91, 2006.
7. L. M. C. Arboleda and N. Nasser, Comparison of Clustering Algorithms and Protocols for Wireless Sensor Networks, Canadian Conference on Electrical and Computer engineering, pp. 1787–1792, May 2006.
8. Klein, A. Lawrence, Sensor and data fusion concepts and applications, Society of Photo-Optical Instrumentation Engineers (SPIE), 1993.

Anomaly Detection System in a Cluster Based MANET

Vikram Narayandas, Sujanavan Tiruvayipati, Madusu Hanmandlu and Lakshmi Thimmareddy

Abstract This chapter presents the development of anomaly detection system (ADS) for locating a malicious node in a cluster based Manet. ADS makes use of AODV protocol that performs route discovery and data forwarding. Each node responds to root request (RREQ) messages and sends root reply (RREP) messages back to the source node. In a cluster based topology a threshold is applied to see if this root reply number is more than the threshold value. If so the node is malicious. Then each node sends an alert to the cluster head (CH) and its neighboring nodes. The proposed ADS avoids the routing to a malicious node thereby preventing high energy consumption of the associated nodes and safeguarding the data transfer in the Manet.

Keywords Manet · ADS · Cluster head · AODV · Energy discharge

1 Introduction

It is challenging to design routing protocols for Mobile Adhoc Networks (MANETs) because of their dynamic topologies and limited resources. Owing to open medium, dynamic topology and lack of central monitoring, Manets are more vulnerable to attacks like denial of services, black hole, gray hole and eavesdropping attacks. The transmission range of each node is limited. Thus, each node needs

V. Narayandas (✉) · S. Tiruvayipati · M. Hanmandlu · L. Thimmareddy
Department of Computer Science & Engineering, MVSR Engineering College, Nadergul, Hyderabad, Telangana, India
e-mail: vikramn_cse@mvsrec.edu.in

S. Tiruvayipati
e-mail: sujanavan_cse@mvsrec.edu.in

M. Hanmandlu
e-mail: hanmandlu_cse@mvsrec.edu.in

L. Thimmareddy
e-mail: tlakshmi_cse@mvsrec.edu.in

S.C. Satapathy et al. (eds.), *Computer Communication, Networking and Internet Security*, Lecture Notes in Networks and Systems 5,
DOI 10.1007/978-981-10-3226-4_2

to perform routing and transmit data packets from one node to others. The misbehaving nodes called attackers in an adhoc network are controlled by adversaries. They try to intrude the network with an intention to cause harm but are also capable of altering the data transfer between different nodes and make the packets difficult to reach their destination [1, 2].

The remainder of this chapter is organized as follows. The related works on Intrusion Detection System (IDS) in Manet are presented in Sect. 2. An overview of AODV and attacks in AODV are discussed in Sect. 3. Section 4 presents the proposed scheme and the derivation of the essential parameters for node description in Manet, and performance metric in routing. The simulation results of the proposed scheme are discussed in Sect. 5. Section 6 gives the conclusions followed by the future work.

2 Related Work

Wireless adhoc networks are configured as flat or multi layered network infrastructure. In a flat network, all nodes are equal but multi layered networks nodes can be considered as clusters with one CH for each cluster [3]. We have already several IDS [4] for Manets like standalone IDS architecture, distributed and cooperative IDS [5]. According to the distributed and cooperative wireless adhoc networks proposed by Zhonge and Lee [6], each node runs as IDS agent and makes local detection decision independently within radio range, here all nodes co-operate in decision making for global detection. Hierarchical IDS [7] is designed for multi-layered adhoc networks. This network is divided into clusters with a CH for each cluster that acts as a manage point similar to switches, routers or gateways. Zone based IDS Adhoc network [1] is partitioned into non-overlapping zones geographically.

3 AODV (Adhoc On-Demand Distance Vector)

AODV [2] used in wireless adhoc networks is capable of both unicast and multicast routing. AODV broadcasts HELLO messages to its neighbors in a network and it has two functions, viz. route discovery and route maintenance from Dynamic Source Routing (DSR) and uses hop-by-hop routing with a sequence number and the periodic beacons from Destination-Sequenced Distance-Vector (DSDV). AODV minimizes the number of required broadcasts by creating routes only on symmetric lines with different phases: (1) Path discovery, route maintenance and (2) Data forwarding. When a source node desires to send a message and initiates a path discovery process to locate its corresponding Mobile host; it broadcasts the route request (RREQ) packet to its neighbors and this request is forwarded to destination via successive neighbors. AODV utilizes a destination sequence number to ensure that all routes are loop free and contain the most recent data. Each node maintains its own sequence numbers as well as broadcast Id, which is incremented for every RREQ. Once RREQ reaches a destination with fresh route then destination node responds by uni casting route reply (RREP) packet back to the neighbor from which it first received RREQ. On receiving another request the RREQ message is discarded. When a source receives RREP message, link is established between the source node and the destination node. When the receiving node detects the disconnection of route between source node and destination node it generates a route error message (RERR) and sends it to source node. Now source node checks table whether it is in a route map or not. The result is sent to the neighboring nodes in a cluster network.

Types of attacks. The malicious node can misuse AODV [3] by forging source IP address, destination IP address, RREQ, sequence number and hop count. RREQ carries a fresh route in the adhoc network and based on this each node decides whether to forward an RREQ message to the next receiving node until it reaches the destination. The RREQ message is broadcast to select a new route in Manet. If a malicious node sends an excess number of RREQ messages, [8] then the network traffic will become congested with huge amount of RREQ traffic. This congestion leads to delays and packet drops [9]. To avoid this and to identify an excess number of RREQ messages sent by malicious node we go in for ADS. If there is a congestion in RREQ traffic, the neighbor node sends true positive alert to each neighbor and sink node (cluster head) in a cluster network. Each node encounters the traffic RREQ messages sent by source node and records the number of packets in an interval of time [10].

Objective. We focus on detecting the misbehaving nodes that participate in the route discovery and maintenance. To safeguard the data transmission from the onslaught of the misbehaving nodes in the adhoc networks the Anomaly detection method is applied on each cluster having fixed radius.

4 The Proposed Method

Different types of attacks are considered by the authors and an effort is made to quantify denial-of-service (DOS) attacks according to node density, mobility, and system size [11]. The behavior of malicious nodes committing in black hole attack in different routing algorithms like AODV and DSR is studied in [12]. AODV is an efficient protocol but is vulnerable to many security attacks [13]. When traffic is intrusion-prone false alarms are treated as one of the problems that IDS is facing in Manets [6]. Different secure schemes are in vogue to enhance the security in Manets [14, 15] that use Adhoc routing protocols. In the proposed scheme, we detect a malicious node clashing with other activities and frequently sending false RREQ. As a result, malicious node can drop packets for disturbing the working of a routing protocol by IDS. Malicious node may perform different types of attacks like, routing disruption attacks, resource consumption attacks, energy consumption attacks, passive attacks and active attacks [16]. For each cluster, a cluster head (sink node) is elected based on some quality of service (QOS) Criterion like highest battery power. All nodes send information periodically to the CH. Each node runs its own ADS and sends report to the cluster head and communication takes place subject to the node density and mobility. In the proposed system, we have used ADS for each cluster having a fixed radius. Anomaly Detection system is a process of detecting abnormal activities in a Manet. Major requirements for anomaly detection are depend on high True positive and True positive Rate.

Anomaly Detection system(ADS): An ADS defines the baseline profile of a normal system activity, where any deviation from the baseline is treated as a possible system anomaly. In anomaly IDS even when the traffic signature is unknown, the abnormality can be recognized. Major requirements for anomaly detection are based on high true positive. The true positive (TP) occurs when IDS raises true alerts in response to the detected malicious traffic comprising the total number of detected malicious activities.

Node description: Nodes consume energy while transmitting beacon signals to neighbor's nodes and listen to broadcast messages from neighbor nodes. When forwarding a RREQ message, each node keeps the destination IP address and the sequence number in its routing table. When an RREP message is received, the node checks the Routing table list for the presence of the same destination IP address; if so the sequence number is calculated, and this approach is implemented for every received RREP message to the destination. These nodes are expected to wait for predefined time t, between successive transmissions. Let node 1 start its transmission at time t_1 and reach to another node 2 at time t_2, then the time interval is $\Delta t = t_2 - t_1$. The average of all differences is calculated for each time slot Δt. Each node observes its own traffic and uses a time slot to record the number of packets for each time interval only once when RREQ is received by each receiving node. A Minimum value [17] for the time slot is preferred, and therefore, $\Delta\tau$ is set to a constant which is decided over several time intervals [18]. If RREQ is modified by any malicious node then it tries to send more RREQ messages continuously to the next hop node or other nodes in different time intervals. This is a sign of abnormality in the network. Next the average value is calculated for each time interval Δt [19]. Each node now snaps the connection to the malicious node and avoids the forwarding of the packets to that node. Transmission range can be calculated between two nodes as $\Delta d = d_{max} - d_{min}$. Receiving range can be calculated as $\Delta R = R_{max} - R_{min}$. Power can be calculated as the energy required to forward a packet to the next node as $E_c = E_i - E_{l,}$

Detection of malicious node: A node that sends RREP greater than a threshold of 2 and if energy discharge falls down rapidly then it is said to be malicious. If different networks are considered then threshold should be adjusted as per the network energy discharge and packet delivery parameters.

Pseudo code of the proposed algorithm

```
  while(route_request(node[i]) && node[i](energy[i])>0) {
   cluster_head=highest_energy(node[i]); set_threshold(value);
    if(node[i](rrep[i])>threshold(value)) {
       send_alert_to_cluster_head(node[malicious]);
       send_alert_to_neighboring_nodes(node[malicious]); }
       if(node[malicious]) {avoid_routing(node[malicious]);}
       }
send_packet(); calculate_energy_discharge(); }
```

Different routing parameters in Manet are: packet delivery ratio (PDR), Routing overhead, throughput, end to end delay. Packet delivery ratio (PDR). The number of packets received by the destination to the number of packets sent by a source is PDR = (TR − TS). where, TR is receive time, TS is sent time. Routing overhead = Route discovery + Route maintenance, where Route discovery = RREQ + RREP.

5 Experimental Results

5.1 Simulation Scenario

In our simulation using NS-2, the nodes of a network are grouped into a cluster comprising 1000 × 1000 in a random topology. A cluster head (access point) is elected for the cluster based on the energy parameter. In the cluster topology every node monitors its neighboring nodes and selects a threshold value based on RREQ for sending information to the neighbors. When RREQ exceeds this threshold, then the neighbor node is considered to be a malicious node and the neighboring node of ADS sends a True positive (TP) alert to its neighbors and CH. Each node then avoids the route path to the malicious node. CH is installed for each cluster and the primary objective of CH is to monitor communication between the trusted nodes within a cluster or beyond another CH of the neighbor cluster. ADS is used to detect the abnormal activities in a cluster network and between the clusters in a Manet.

During the cluster anomaly detection, the CH rotates while processing the workload among the neighboring nodes. It detects the routing attacks and monitors a large part of the network activity decision. CH equipped with the knowledge of all clusters has a bidirectional connectivity via a pair of unrelated unidirectional links. When a source has no route to destination, it forwards route request (RREQ) packet

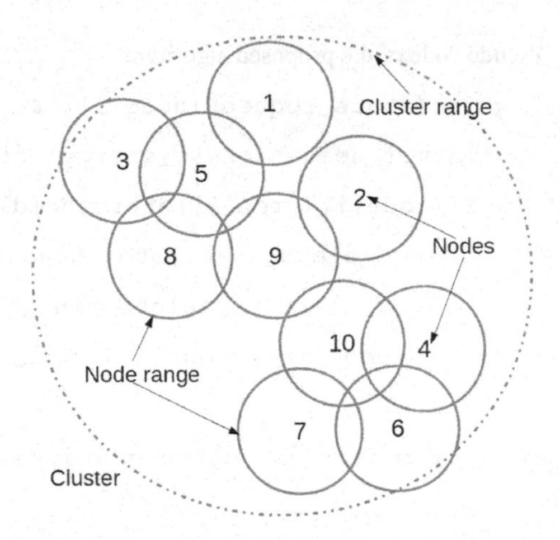

Fig. 1 Cluster setup for simulation

to the CH, When a CH receives request, it appends a request packet to its ID as well as a list of adjacent clusters and rebroadcasts it. A neighbor node also acts as a CH which is a gateway to one or more adjacent clusters and it makes a unicast request to the appropriate CH. Our simulation involves 10 nodes configured into a single cluster having a CH (sink node) dynamically varying based on the energy parameters as shown in Fig. 1. We have used AODV routing protocol for route discovery and route maintenance by hop-by-hop routing with a sequence number and grouped 10 nodes into a single cluster. The energy required for route request (RREQ) is ERREQ and route reply (RREP) is ERREP. The energy discharge of each node is calculated using the energy coefficients of ERREQ and ERREP as $E_{discharge} = E_{coef} (E_{RREQ} + E_{RREP})$. Energy remaining after each node forwards packets to its neighboring nodes is calculated as $E_{remaining} = E_{initial} - E_{discharge}$. Our simulation tests check energy discharge after node delivers its packets to other neighbor nodes at each instance. The energy remaining at an instance "i" for a node is calculated as

$$E(i)_{remaining} = \sum_{1}^{i} E(i-1)_{remaining} - E(i)_{discharge} \tag{1}$$

Substituting for the second term in the LHS of Eq. (1), we get

$$E(i)_{remaining} = \sum_{1}^{i} E(i-1)_{remaining} - \left(E(i)_{coef}\left(E(i)_{RREQ} + E(i)_{RREP}\right)\right) \tag{2}$$

The total energy remaining for all nodes at an instance "i" is

Table 1 Initial node configuration details

Nodes	Initial energy (mAh)	Energy discharge per minute (mAh)	Nodes in range	RREPs per RREQ
1	1000	1	1	1
2	1500	1	1	1
3	2000	1	2	1
4	2500	1	2	1
5	3000	1	3	1
5	3500	1	3	1
7	4000	1	4	1
8	4500	1	4	2
9	5000	1	4	2
10	5500	1	4	3

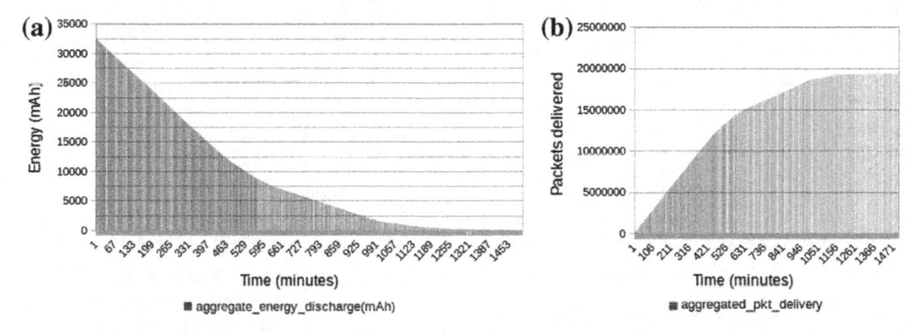

Fig. 2 **a** Aggregate energy discharge of 10 nodes lasting a time of 1500 min, **b** aggregate packets delivered by 10 nodes over period of 1500 min

$$E(n)_{remaining} = \sum_{1}^{n} \sum_{1}^{i} E(i-1)_{remaining} - \left(E(i)_{coef}\left(E(i)_{RREQ} + E(i)_{RREP}\right)\right) \quad (3)$$

Our simulation consists of 3 tests each with its initial node configuration details and energy left after packet delivery and simulated with modal cluster, with and without malicious node, and cluster using ADS with malicious node.

Table 1 gives the simulation results of 10 nodes of different initial energy values along with their discharge per minute (in mAh), nodes in the range 1–4 and RREPs per RREQ for each node is considered with threshold value of 1. Based on the above configuration parameters tests are conducted.

Test-1. Modal cluster without malicious node. When a malicious node is not present i.e. in the secured environment then the energy is least consumed with the highest throughput (packet delivery) as in Fig. 2.

Simulations indicate how nodes consume energy during the packet delivery. The aggregate energy discharge of 10 nodes lasting 1500 min is shown in Fig. 2a.

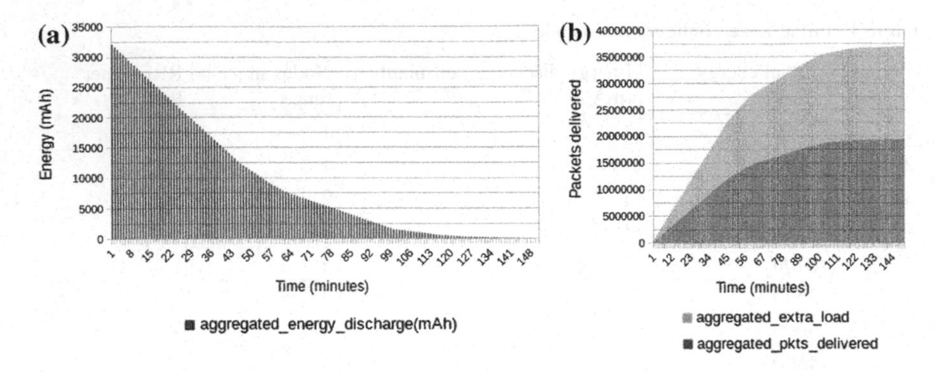

Fig. 3 **a** Aggregate energy discharge of 10 nodes lasting a time of 150 min, **b** aggregate packets delivered by 10 nodes over period of 150 min (color online)

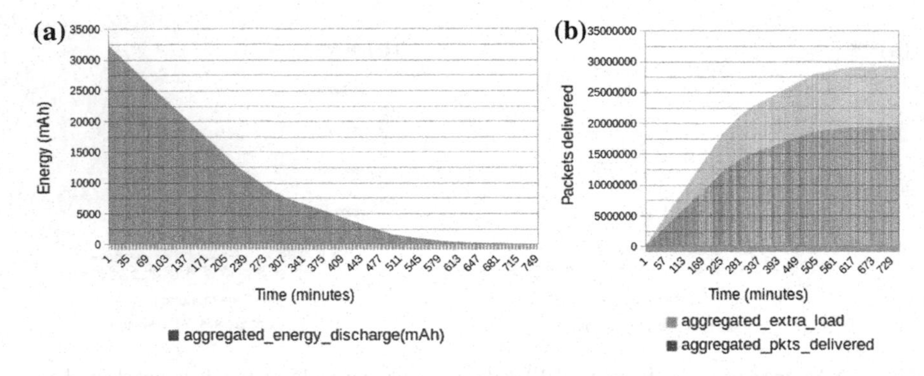

Fig. 4 **a** Aggregate energy discharge of 10 nodes lasting a time of 750 min, **b** aggregate packets delivered by 10 nodes over period of 750 min (color online)

The packets are delivered by 10 nodes over a period of 1500 min. The energy consumed during the packets delivery is shown as the aggregate in Fig. 2b.

Test-2. Cluster with malicious node. In real time when a malicious node is present in the unsecured environment with the unknown neighbors and no ADS then the energy is highly consumed with the lowest throughput (packet delivery). The aggregate energy discharge in a cluster is shown in Fig. 3a. Simulation results display the consumption of energy by the nodes in Fig. 3b, during the packet delivery in a cluster in the presence of malicious node.

Test-3. Cluster with malicious node along with ADS. In the real time situation involving a malicious node the results of energy consumed and throughput (packet delivery) obtained using the ADS are now discussed. The aggregate energy discharge of 10 nodes lasting for 750 min is shown in Fig. 4a.

The simulation results arising out of the ADS show how nodes consume their energy during the packet delivery in a cluster in the presence of a malicious node. The additional and aggregate packets delivered by 10 nodes over a period of 750 min is shown in Fig. 4b.

5.2 Summary of Results

The simulation results of all tests considering the cluster up-time, packets delivered and additional overhead are given in Table 2.

The energy discharged when a cluster node is in the normal mode (test-1) is low. Test-2 shows the energy consumption when a malicious node is present in a cluster. When the nodes are not authenticated the energy discharge is very high. Test-3 shows the real time situation around a cluster node in the presence of the ADS and malicious node. In this case the energy discharge is not so high. A comparison of all tests is shown in Fig. 5. In the past many detection techniques were proposed [4, 8, 16, 20, 21] relating the threshold and activity parameters. Though the statistical analysis of these parameters is made but the energy that plays a major role is ignored in the analysis. In the present work the effect of energy utilization during the conduction of three tests is shown to maintain the cluster energy and also the cluster up time.

Table 2 Simulation results for all tests

Simulation scenarios	Cluster up-time (min)	Packets delivered	Additional overhead
Test scenario-1	1500	19,494,000	0
Test scenario-2	150	19,392,000	17,452,300
Test Scenario-3	750	19,439,200	9,744,600

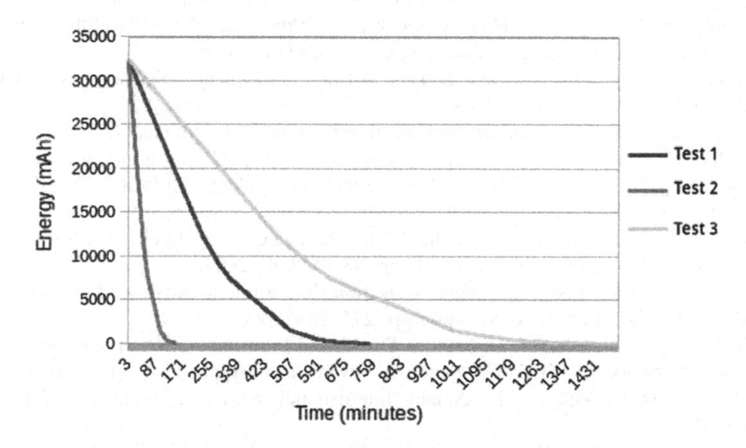

Fig. 5 Comparison of energy discharge for all three tests (color online)

6 Conclusions

The proposed Anomaly Detection System (ADS) is designed to disambiguate all kinds of attacks by detecting a malicious node. The attacks perpetrated one way or the other emanate due to the presence of malicious node. The activities of various attacks can be mirrored through the malicious node. The proposed ADS is demonstrated on a single cluster based topology with 10 nodes. Regarding the simulations, test-1 is concerned with a modal cluster and test-2 centers around a malicious node and test-3 represents real time situation comprising a malicious node and the ADS. The difference in the performance of test-2 and test-3 shows up as a huge difference in the energy discharge and throughput because of the activation of the ADS which dissuades the packets to a malicious node. By this the routing parameters of Manet such as PDR, routing overhead, throughput and end to end delay, are improved drastically. The extent of energy utilization and detection of malicious node during three tests help in keeping the steady discharge of the cluster energy and the cluster up time for a longer period. Further work is on to make the Manets intelligent so that ADS can function much more effectively.

References

1. Y. Hu, A. Perrig, and D. B. Johnson, "Packet leashes: a defense against wormhole attacks in wireless ad hoc networks", Proc. INFOCOM' 03, IEEE, San Francisco, CA, April 2003, pp. 1976–1986.
2. L. Zhou, Z. J. Haas, "Securing ad hoc networks", IEEE Network, Nov/Dec 1999, pp 24–30.
3. H. Deng, R. Xu, J. Li, F. Zhang, R. Levy, W. Lee, "Agent based Cooperative Anomaly Detection for Wireless Ad Hoc Networks," in Proc. the IEEE Twelfth International Conference on Parallel and Distributed Systems (ICPDS'06),2006.
4. C. Perkins, E. Belding-Royer, and S. Das, Ad hoc On-Demand Distance Vector (AODV) Routing, Jul. 2003. IETF RFC 3561.
5. Liy.weij "Guidelines on selecting intrusion detection method in manet" 2004.
6. M. Hollick, J. Schmitt, C. Seipl, and R. Steinmetz,"On the effect of node misbehavior in ad hoc networks," in Proc. IEEE Global Telecommun. Conf. GLOBECOM, Jun. 2004, pp. 3759–3763.
7. A. Mishra, K. Nadkarni, and A. Patcha, "Intrusion detection in wireless ad hoc networks," IEEE Wireless Commun., vol. 11, no. 1, pp. 48–60, Feb. 2004.
8. Y. Zhang and W. Lee, "Intrusion Detection in Wireless Ad-Hoc Networks", Proc. MOBICOM 2000, Boston, ACM press, pp: 275–283, 2000.
9. Fu, Yingfang; He, Jingsha; Li, Guorui, "A Distributed Intrusion Detection Scheme for Mobile Ad Hoc Networks", IEEE Computer Software and Applications Conference, 2007. COMPSAC 2007 - Vol. 2. 31st Annual International Volume 2, Issue, 24–27 July 2007 Page(s):75–80.
10. B. Sun, K. Wu and U. Pooch, "Alert Aggregation in Mobile Ad-Hoc Networks" ACM Wireless security (WISE.03), SanDeigo, CA, pp. 69–7.
11. ZaputeN "securing adhoc routing protocols" in ACM workshop on wiress security, USA 2002.
12. Eskin. E, Portray L "A Geometric framework for unsupervised anomaly detection intrusins in unlabelled data" in 2002.

13. Kimay sanzgiri, Daniel laflamne bridget dahil, clay shields and Elzabeth belding Royer "Authentication routing for adhoc networks" IEEE journal on selected areas in communications, vol 23, March 2005
14. C. Perkins, E. Belding-Royer, and S. Das, Ad hoc On-Demand Distance Vector (AODV) Routing, Jul. 2003. IETF RFC 3561 (Experimental).
15. M. Zapata, Secure ad hoc on-demand distance vector (SAODV) routing, Sep. 2006. IETF Internet Draft, draft-guerrero-manet-saodv-06.txt.
16. Hidehisa Nakayama, Satoshi Kurosawa, Abbas Jamalipour, Yoshiaki Nemoto, and Nei Kato, "A Dynamic Anomaly Detection Scheme for AODV-Based Mobile Ad Hoc Networks" IEEE Transaction on Vehicular Technology, vol. 58, NO. 5, June 2009.
17. Y. Huang, W. Fan, W. Lee, and P. Yu, "Cross-feature analysis for detecting ad-hoc routing anomalies," in Proc. 23rd ICDCS, May 2003, pp. 478–487.
18. Y. Huang and W. Lee, "Attack analysis and detection for ad hoc routing protocols," in Proc. 7th Int. Symp. RAID, Sep. 2004, pp. 125–145.
19. Hidehisa Nakayama, Satoshi Kurosawa, Abbas Jamalipour, Yoshiaki Nemoto "A Dynamic Anomaly Detection Scheme for AODV-Based Mobile Ad Hoc Networks" IEEE Trans. On Vehicluar Technology, Vol. 58, No. 5, June 2009.
20. P. Ning and K. Sun, "How to misuse AODV: A case study of insider attacks against mobile ad-hoc routing protocols," in Proc. 4th Annu. IEEE Inf. Assurance Workshop, Jun. 2003.
21. Y. Waizumi, Y. Sato, and Y. Nemoto, "A network-based anomaly detection system using multiple network features," in Proc. 3rd Int. Conf. WEBIST, Mar. 2007, pp. 410–413.

Temperature Data Transfer Using Visible Light Communication

Sindhubala Kadirvelu and Vijayalakshmi Baba

Abstract At present wireless communication is subjugated by radio frequency (RF) communication system. RF communication has drawbacks like limited bandwidth, radiation hazards, health concern, exposure to interception and interference and many more. Thus visible light communication (VLC) acts as an added supplement to RF communication by overcoming the above drawbacks. VLC provides both illumination and communication at the same time by using white LED at the transmitter. This chapter presents the measurement of temperature using LM35 temperature sensor and displays the same to the receiver using VLC. We tested the proposed prototype in indoor environment for different temperature monitoring applications like room temperature monitoring and core body temperature measurements of patient in hospitals and transmission of same to the receiver. The benefits of the proposed system are low cost, hazardless, accuracy, high range of temperature sensing, low power, high performance and communication distance of 0.60 m is achieved.

Keywords Light emitting diode · Radio frequency · Temperature monitoring · Visible light communication

1 Introduction

Currently in real time, temperature is measured and transmitted wirelessly using RF communication. RF radiation when used for long time will degrade the health of the human beings. Hence a supplement technology is needed for data transmission wirelessly. VLC can be used as alternate for radio frequency (RF) communication as it has got the benefits of bandwidth, less power consumption, visibility, free from

---

S. Kadirvelu (✉) · V. Baba
ECE Department, B.S. Abdur Rahman University, 600048, Chennai, Tamilnadu, India
e-mail: sindhubala_ece_phd_2014@bsauniv.ac.in

V. Baba
e-mail: vijayalakshmi.b@bsauniv.ac.in

S.C. Satapathy et al. (eds.), *Computer Communication, Networking and Internet Security*, Lecture Notes in Networks and Systems 5,
DOI 10.1007/978-981-10-3226-4_3

EMI and harmful radiation. VLC uses the spectrum of visible light from 380 to 780 nm. In the outlook, the indoor room illumination will be replaced by white LEDs instead of conventional fluorescent lamps due to the advantages like less energy consumption, reduced amount of health hazards and longer life time [1, 2]. VLC is standardized institute of electrical and electronics engineers (IEEE) [3]. Now Asia, Europe, Wireless World Research Forum are also working in VLC research [4]. VLC communication is believed to be energy saving solution in the future as the energy used to drive the LED will also be used to drive the wireless signal transmission. Thus it helps to save energy and will not consume extra amount of energy.

This research paper aimed to investigate, real time temperature measurement and data transmission of the same to the receiver using VLC system. The research paper is organized with the following sections. In Sect. 2, related work is described. In Sect. 3, proposed temperature monitoring using VLC system is explained. Section 4, includes experimental testing and results with the explanation. Finally, Sect. 5, discusses the conclusion of the chapter.

2 Related Works

Following paragraph illustrate the applications of visible light communication done by different researchers.

In [5], LED lights can serve as a strong candidate for not only illuminating the room but also for communicating the data is discussed. This technology is called visible light communication (VLC). Fundamental analysis of VLC system is also discussed.

In [6], the potential applications and future vision of VLC technology is described. A bidirectional VLC communication with the data rate of 500 Mb/s is experimented and their results are discussed.

In [7], novel VLC system for transmission of biomedical data such as information of patient as text and body signals was used for health care applications. This method reduced human health hazards caused due to EMI and radiation interference with the medical instruments. The distance achieved between the transmitter and receiver was about half meter.

In [8], VLC systems are used in mobile applications such as games, toys, location services, and augmented reality. When the cars are face to face and the cars are behind one another, they exchange messages. Distance of 1 m is achieved between the transmitter and receiver.

In [9], real time audio and video transmission using indoor VLC system is established for the distance of 3 m.

In [10], indoor navigation system for visually impaired person is developed. The location based information is transmitted via white LED to the visually impaired person connected to the headphones of the smart phone. VLC system also finds its applications in underwater communications, flight entertainment, and many more [11–13].

None of the applications discussed above used real time temperature data transmission using VLC in the indoor environment. This proposed method is very helpful for many applications like room temperature monitoring, body temperature monitoring and the temperature data is transmitted to the receiver wirelessly using VLC. The proposed VLC system uses receiver with the ambient light noise reduction technique [14]. This receiver helps to mitigate the ambient light noise produced by indirect sunlight and fluorescent lamp generated by conventional ballast.

3 Proposed Temperature Data Transfer Using VLC System

The proposed VLC system consists of three main blocks such as transmitter which uses LM35 temperature sensor, high power white LED which is used for both transmitting the data and illumination, air as the transmission medium and silicon PIN Photodiode at the receiver. Figure 1 shows the block diagram of temperature data transfer using VLC system.

3.1 Temperature Sensing and VLC Transmitting Procedure

This work considers, LM35 analog temperature sensor to measure the temperature data continuously in the transmitter side. The temperature data acquired is serially transmitted to laptop by NI USB-6009 and national instruments Lab VIEW software is used for data analysis. The temperature data sensed using the temperature sensor is interfaced to PIC18F45k22 microcontroller through an ADC channel to

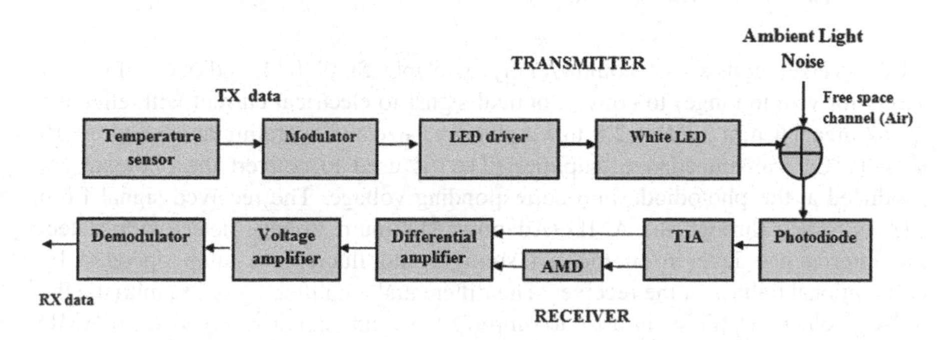

Fig. 1 Block diagram of proposed temperature data transfer using VLC system

convert the analog value to the digital value. It processes the data received from the sensor and sends to serial UART interface. On-off keying (OOK) modulation is used, to modulate the temperature data to the white LED. OOK is a simplest form of amplitude-shift keying (ASK) where the data signal is represented as binary one and absence of data is represented as binary zero. 1 W high brightness ultra white LED of forward current 0.35 A, forward voltage 3.2–3.6 V, viewing angle 90° and light intensity of 80 lm is used as an optical source in this experiment as it is of less cost and not complex compared to RGB LED.

3.2 LOS Channel Model

Visible light communication channel is considered as a linear additive white Gaussian noise (AWGN) channel and the noise model should also be included with the channel model [1, 2], the expression is given by Eq. (1)

$$I(t) = \eta\, p_i(t) \otimes h(t) + N(t) \quad (1)$$

where I(t) is the photo detector current, η is the photosensitivity of the Photo detector, $p_i(t)$ is the input optical power, ⊗ denotes the convolution (t) is the impulse response and N(t) denotes the AWGN noise.

This work considers, line of sight (LOS) channel model between the transmitter and receiver. The transmitter focuses the light beam directly in the direction of the receiver.

The received power at the photodiode is given by

$$P_{rLOS} = H_{LOS}(0)_{P_t} \quad (2)$$

where $H_{LOS}(0)$ the channel gain and P_t is the transmitted optical power.

3.3 VLC Receiving Procedure

VLC receiver consists of commercially available Si PIN Photodiode (S1787-04 series for visible range) to convert optical signal to electrical current with effective active area 6.6 mm^2, 2.4 × 2.8 mm active area size and operating temperature −10 to +60 °C. Transimpedance amplifier (TIA) is used to convert the photocurrent produced at the photodiode into corresponding voltage. The received signal from TIA is passed through the AMD (Adaptive Minimum Voltage detector) to detect the interference level from indirect sunlight and fluorescent lamp operated by conventional ballast, at the receiver. The differential amplifier is used to mitigate the noise. Voltage amplifier is used to amplify the weak signal received from AMD stage. The demodulator is used to recover back the original transmitted signal, which is serially transmitted to laptop by NI USB-6009 and national instruments Lab VIEW software is used for data analysis. IC 741 high gain operational amplifier is used in the experiment.

4 Experimental Testing and Results Discussion

The experimental test bench of the proposed real time temperature data transfer using VLC is shown in Fig. 2.

The experimental set up includes the temperature sensor to sense the temperature and the temperature data is transmitted using high power white LED. Electrical signal is converted into light waves, which is transmitted in the free space (air) and the receiver circuit using photodiode detects the optical signal from LED. TIA used to produce electrical voltage. Noise reduction circuitry using AMD and differential amplifier is used to mitigate the noise from ambient light noise sources. The transmitted signal is recovered back at the receiver. The maximum distance achieved between the transmitter and receiver is 0.60 m. The experimentation is performed at electronics laboratory of B.S. Abdur Rahman University which is considered as indoor room of VLC system. The prototype is tested for different applications like room temperature monitoring and body temperature monitoring of a patient in hospitals.

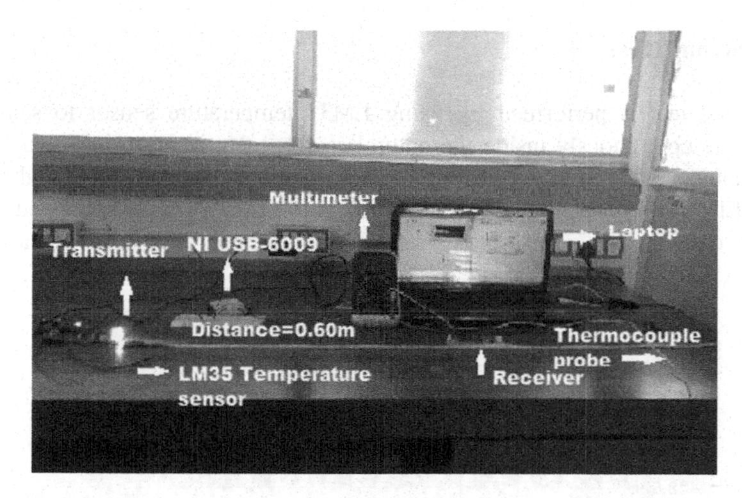

Fig. 2 Picture of real time temperature data transfer using VLC for the distance of 0.60 m

4.1 Room Temperature Data Transmission Using VLC

Temperature sensor (LM35) is used to measure the room temperature in the proposed design. NI USB-6009 is used to acquire the temperature data collected from the natural room environment and it is integrated with laptop using National Instruments Lab VIEW software for data analysis.

4.1.1 First Test

More than a few measurements are done to ensure that the sensor is sensing the correct temperature values or the change in temperature values are registered. To validate the LM35 temperature sensor performance, the same temperature measurements are performed using digital multimeter coupled with K-type thermocouple probe. The experiment is carried out on 6/4/2016 for different timings in electronics lab at B.S. Abdur Rahman University. Figure 3 shows the performance comparison between the temperature values acquired using LM35 and thermocouple.

The test concludes, LM35 temperature sensor performance in measuring the temperature values has only minor variations compared to thermocouple temperature measured values. Thus LM35 temperature is low cost and adequate for measuring the temperature of the natural environment and can be used for our proposed VLC system.

4.1.2 Second Test

The second test is performed by using LM35 temperature sensor to sense the temperature continuously inside the natural environment in electronics lab at B.S. Abdur Rahman University and the temperature data is transmitted using high power white LED. The optical signal is transmitted in the free space and received by the receiver circuit. The transmitted and received temperature data is acquired

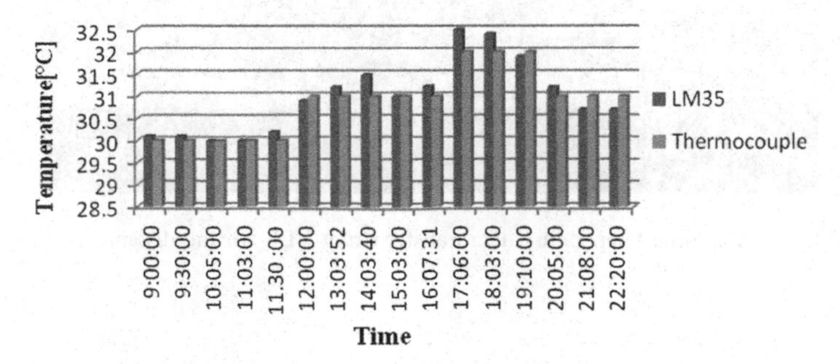

Fig. 3 Variation on the temperature values taken by the LM35 temperature sensor and thermocouple using multimeter (colour online)

and serially transmitted to laptop by NI USB-6009. National instruments Lab VIEW software is used for data analysis. The temperature is recorded for different timings on 6/4/2016 and the data is transmitted to the receiver at the distance of 0.60 m. The transmitted and the received room temperature data using VLC system is shown in Fig. 4.

Figure 4 infers, the transmitted temperature data is received back at the receiver with minimal loss. Sample measurement of temperature data at the transmitter and receiver is shown in Figs. 5 and 6 shows the transmitted and the received room temperature data using lab VIEW at 14:03:40 PM.

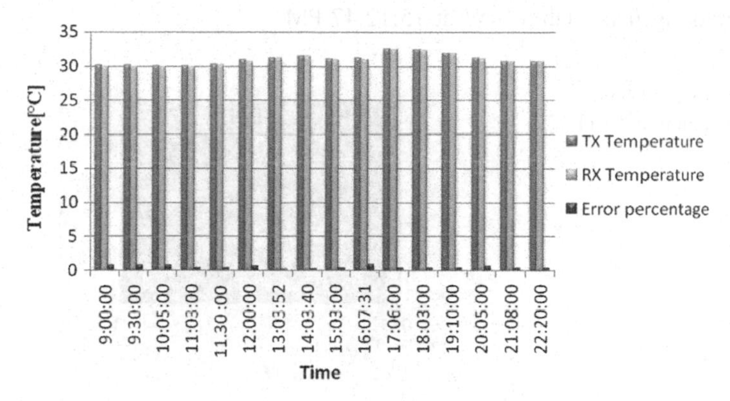

Fig. 4 Transmitted and received room temperature using proposed VLC system for different timings (colour online)

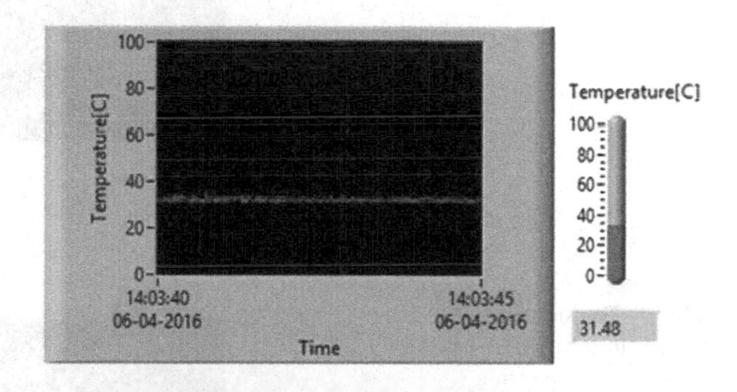

Fig. 5 Transmitted room temperature at 14:03:40 PM

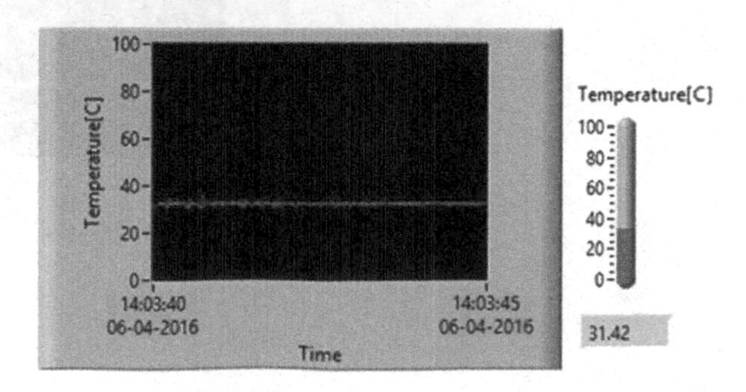

Fig. 6 Received room temperature at 14:03:40 PM

4.1.3 Third Test

The third test is performed to monitor the high temperature and low temperature inside the electronics lab at B.S. Abdur Rahman University on 6/4/2016. For high temperature monitoring, the test includes placing the LM35 temperature sensor to hot flame. High temperature monitoring is useful for emergency conditions like fire inside the indoor environment. The transmission distance of 0.60 m is kept between the transmitter and receiver. Figures 7 and 8 shows the transmitted and the received high room temperature data analysis using lab VIEW at 15:11:17 PM.

Figures 9 and 10 shows the transmitted and the received low room temperature data monitoring using lab VIEW at 15:12:47 PM.

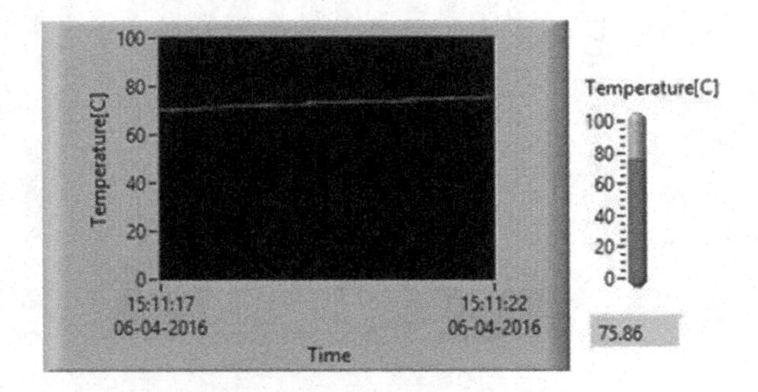

Fig. 7 Transmitted high room temperature at 15:11:17 PM

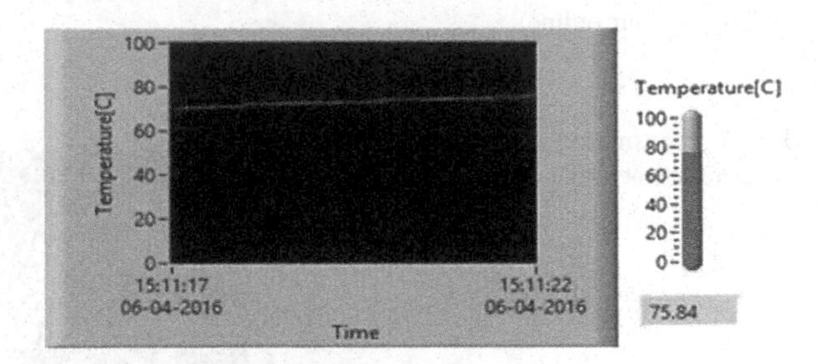

Fig. 8 Received high room temperature at 15:11:17 PM

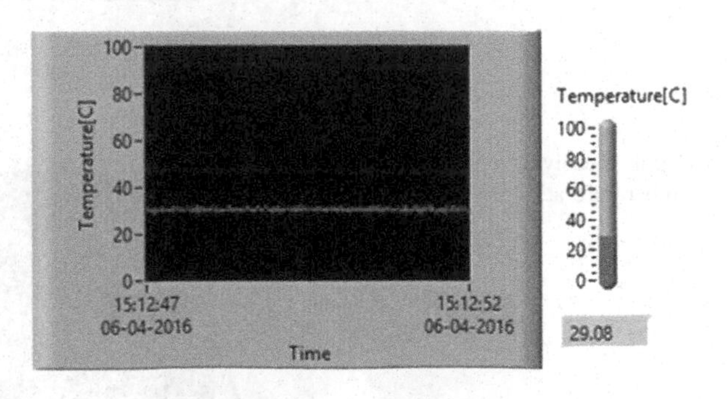

Fig. 9 Transmitted low temperature at 15:12:47 PM

4.2 Body Temperature Monitoring and Data Transmission Using VLC

In hospitals, the patient's body temperature needs to be constantly monitored for the diseases like viral fever and typhoid. Presently the temperature monitoring is done manually by the nurse or doctor. In our proposed system LM35 sensor reads the patient's body temperature, displays on the laptop using lab VIEW software and received by the doctor using VLC system. It helps the doctor to monitor the patient's body temperature wirelessly. The key advantage of using VLC in hospitals is free from electromagnetic interference (EMI). Hence it will not cause any adverse health effects to the patient's. For body temperature measurement, the test is carried out by placing the finger in contact with LM35 sensor. The other main areas to measure the body temperature are under arm, tip of the tongue and back of the ear lobe. The transmitted and the received body temperature data for different patients using VLC system is shown in Fig. 11.

Figures 12 and 13 shows the transmitted and the received body temperature data analysis using lab VIEW for patient2.

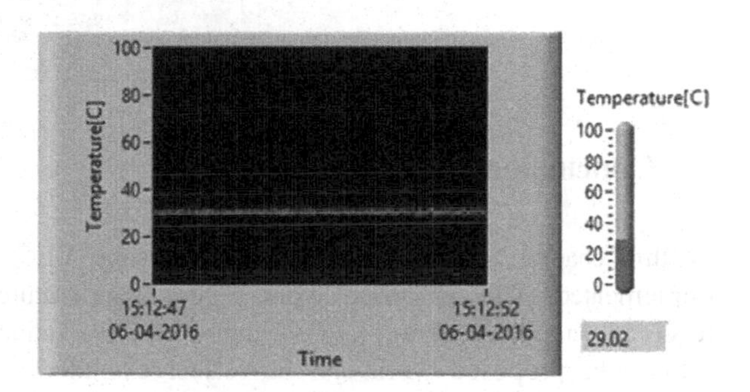

Fig. 10 Received low temperature at 15:12:47 PM

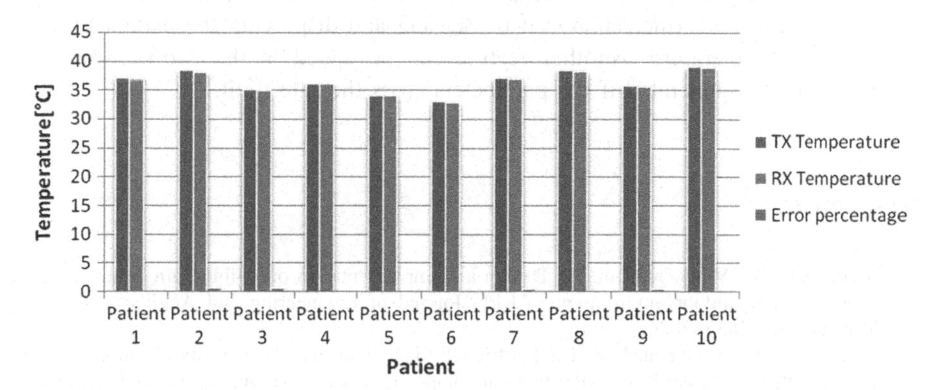

Fig. 11 Transmitted and received body temperature using proposed VLC system for different patients (colour online)

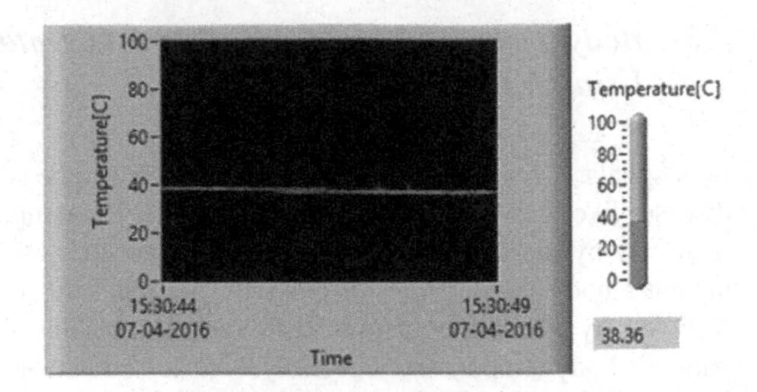

Fig. 12 Transmitted body temperature for patient2

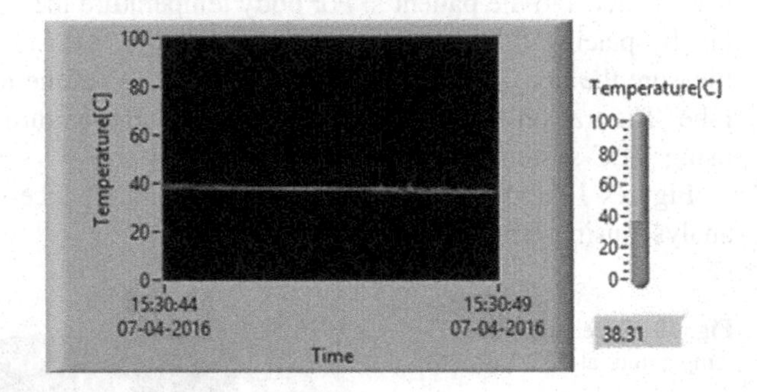

Fig. 13 Received body temperature for patient2

5 Conclusion and Future Work

In this chapter, temperature monitoring using VLC system is proposed and implemented. This proposed work reports temperature data is transmitted and received wirelessly using VLC technology with reduced noise interference. In future, the proposed work can be further extended with better communication distance of more than 0.60 m by using the LED of high wattage, better noise reduction circuitry and high gain amplifiers. For health care applications the proposed work can be further enhanced by sensing and displaying the patient information with temperature. Another feature can be added in this work, whereas warning signal is generated if the parameters cross the safe limit.

References

1. Sindhubala, K., Vijayalakshmi, B.: Design and implementation of visible light communication system in indoor environment. ARPN Journal of Engineering and Applied sciences, Vol.10 (2015)2882–2886.
2. Sindhubala, K., Vijayalakshmi, B.: Ecofriendly data transmission in visible light communication. IEEE International conference on computer, communication, control and information technology (2015).

3. IEEE Standards Association, IEEE Standard for Local and metropolitan area networks—Part 15.7: Short-Range Wireless Optical Communication Using Visible Light IEEE Std 802.15.7 (2011).
4. Wireless World Research Forum http://www.wireless-world-research.org.
5. Grobe, L., Anagnostics, P., Hilt., J, Schultz, F., Hartlieb., F, Christoph, K., Jungnickel, V., Langer, K.: High-speed Visible Light Communication Systems. IEEE communications magazine. Vol. 51(2013)60–66.
6. Komine, T., Nakagawa, M.: Fundamental Analysis for Visible Light Communication System using LED Lights. Vol.50 (2004)100–107.
7. Cheong, Y.K., Weing, X., YounChung, W.: Hazardless Biomedical sensing Data Transmission Using VLC. IEEE sensors journal, Vol.13 (2013) 3347–3348.
8. Corbellini, G., Gross, T., Mangold, Mkrtchyan, A., S., Schmid, S.: Demo LED to LED Visible Light Communication for Mobile Applications. IEEE Workshop on Optical Wireless Communications (2012).
9. Ding, L., Gong, Y., He, Y., Wang, W.: Real-time Audio & Video Transmission System Based on Visible Light Communication. Optics and Photonics Journal, Vol. 3 (2013)153–157.
10. Haruyama, S.: Advances in Visible Light Communication Technologies. ECOC Technical Digest (2012).
11. Asada, H.H., Rust, I.C.: A dual-use visible light approach to integrated communication and localization of underwater robots with application to non-destructive nuclear reactor inspection. IEEE international conference on Robotics and Automation (2012)2445–2450.
12. Endo, T., Fujji, T., Iwasaki, S., Kimura, Y., Premachandra, S., Tanimato, M.: Visible light road to vehicle communication using high speed camera.IEEE Intelligent Vehicles Symposium(2008)13–18.
13. Cheol Kim, H., Ho Yoo, J., Lee, R., Kyo Oh, J., WookSeo, H., YoonJung, S., YoungKim, J.: Demonstration of Vehicular Visible Light Communication based on LED Headlamp. International conference on Ubiquitous and Future Networks (2013)465–467.
14. Vongkulbhisal, J., Zhao, Y.: Design of Visible Light Communication Receiver for On-Off Keying Modulation by Adaptive Minimum-Voltage Cancellation, Vol.17 (2013).

A New Approach for Data Security in Cryptography and Steganography

Prashanti Guttikonda, Sandhya Rani Kaviti and Ashok Kumar Popuri

Abstract This work deals with the combination of cryptography and steganography to handle the problems of information security. A new encryption algorithm is proposed based on matrix scrambling which uses arithmetic operations, random functions, circular shifting and exchanging operations to get the encrypted message. Further, security is enhanced by steganography method where the encrypted message is hidden in the image based on the gini index value. For each R G B component of the pixel, gini index is calculated and the component with least value will have the encrypted message. The proposed method provides dual security since the encrypted message is hidden at different positions of 4 LSB of image. Avalanche effect is used as the metric for analyzing the performance of the encryption algorithm.

Keywords Cryptography · Steganography · Information security · Avalanche effect · Gini index

P. Guttikonda (✉) · S.R. Kaviti
Vignan's Lara Institute of Technology & Science, Vadlamudi, Guntur, Andhra Pradesh, India
e-mail: prashantiguttikonda77@gmail.com

S.R. Kaviti
e-mail: sandhyakaviti@gmail.com

A.K. Popuri
VFSTR University, Vadlamudi, Guntur, Andhra Pradesh, India
e-mail: ashok_kumar_popuri@yahoo.com

A.K. Popuri
S.V.U. College of Engineering, S.V. University, Chittoor, Tirupati, Andhra Pradesh, India

S.C. Satapathy et al. (eds.), *Computer Communication, Networking and Internet Security*, Lecture Notes in Networks and Systems 5,
DOI 10.1007/978-981-10-3226-4_4

1 Introduction

Security of information passed over a network has become a fundamental issue and hence, the confidentiality is required to protect against unauthorized access and use. Cryptography and steganography are the two popular methods available to provide security. Cryptography and steganography when used individually provides security but when combined provides higher level of security [1]. An algorithm based on magic rectangle proposed by Suli Wu et al., uses random number approach, especially to select the sub-matrix randomly, helps to avoid the regularity in the resulted scrambling matrix which is transformed from plaintext matrix, hence improves the difficulty for decryption [2]. P-Hill algorithm which was put forth by HuLihong where the data is completely disrupted rearrangement by using 16 iterations [3]. Gurpreet Singh et al. proposed a hybrid encryption algorithm which integrates MVEA, Base64 and AES algorithm to improve the security of data [4]. Dipesh Agrawal et al. discussed about various techniques based on random bit image steganography and proposed some new methods of hiding a secret data into an image and also analyzed these steganography methods with the help of various analysis tools [5]. Badrinath R et al. proposed an approach by embedding the data (text file) in Least Significant Bits of the planes based on the minimum value constraint of MSB using variable embedding technique [6]. Manoj Kumar Ramaiya et al. presented unique technique for image steganography based on the Data Encryption Standard (DES) using the strength of S-Box mapping and Secrete key [7].

2 Proposed Method

To enhance the security of message a variant matrix scrambling technique has been proposed along with a steganographic technique for hiding the encrypted message into image. The steganographic technique uses inequality measure called gini index. Here, a bit of encrypted message is hidden in either LSB of red, green or blue component of a specific pixel which is decided by the gini index value. So, it is difficult to make decision in which component of the pixel the information is hidden.

2.1 Encoding Message

This work proposes a novel encryption algorithm which scrambles the matrix by the addition, shifting and exchange operations on bi-column, bi-row circular queue with better avalanche effect. The algorithm proposed is the symmetric encryption algorithm, where encryption and decryption is done with the same key. Thus, the key file should be maintained secret and have to be shared with the receiver through any secure communication channel.

Encryption Algorithm

1. Read the plaintext on M*N matrix. Input an integer parameter w, which represents number of rounds of operations to perform, which means w transformations are performed on the matrix. Let I = Random(), where Random() is the random function which generates random positive integers. Convert I into the binary value and place it in vector b, b = binary sequence ().
2. Let K = digit(b_t), where 't' is the bit position from LSB to MSB. K is either 0 or 1. If K = 0 then row transformation is performed on the matrix. If K = 1 column transformation is performed on the matrix.
3. Record the transformations as sub key and write this key to the key file. The sub key is given as (T, op, α1, α2, β1, β2) where T takes values 'r' or 'c' i.e. either row transformation or column transformation; op is operation to be performed on row and columns and it has a value of either 0 or 1 or 2; α1, α2 are two random rows or columns selected depending on the transformation; β1, β2 are minimum and maximum values of range for two selected rows α1, α2. Repeat the steps for w number of times.

Row Transformation with Addition Operation

1. To determine the range of rows on which transformation has to be performed, four random integers are generated r1 = Random(m), r2 = Random(m) such that r1 ≠ r2 and c1 = Random(n), c2 = Random(n) where c1 ≠ c2. Let x1 = min(c1, c2) and x2 = max(c1, c2). Here x1 and x2 becomes lowest index and highest index of the sub array selected in the rows r1 and r2. Perform the operation (op), where op = Random() mod 3.
2. If op = 0, select the sub array of rows r1, r2 in range x1, x2 and then except the first element add the elements in the sub array with the element with its left and perform circular left shift operation. If op = 1, perform addition operation (except first element) and then circular right shift operation on sub array of rows r1, r2 in range x1, x2. If op = 2, perform addition operation (except first element) and then reverse operation on sub array of rows r1, r2 in range x1, x2. For each operation a sub key (r:op-c1-c2-x1-x2) is constructed and recorded in a key file.

Column Transformation

1. To determine the range of columns on which transformation has to be performed, r1 = Random(m), r2 = Random(m), are the random values for rows selected such that r1 ≠ r2, similarly two columns are selected randomly, c1 = Random(n), c2 = Random(n) where c1 ≠ c2. Let x1 = min(r1, r2) and x2 = max(r1, r2), Here x1 and x2 becomes smallest index and largest index of the sub array selected in the columns c1 and c2.
2. Perform the operation (op), where op = Random() mod 3. If op = 0, select the sub array of columns c1, c2 in range x1, x2 and perform the operations on the elements of sub array. Except the first element add the elements in the sub array with the element above it and circular shift upward the sub array. If op = 1, perform addition operation (except first element) and then circular shift downward sub array of columns c1, c2 in range x1, x2.
3. If op = 2, perform addition operation (except first element) and reverse the contents of sub array of columns c1, c2 in range x1, x2. For each operation a sub key (c:op-c1-c2-x1-x2) is constructed and recorded in a key file.

Decryption Algorithm

1. Read the cipher text into M × N matrix. Separate the key file by row operations and column operations and read these sub keys in reverse order. For each sub key(T), if T equals to r then inverse row transformation is performed otherwise column transformation.

   i. In inverse row transformation when op = 0, subtract each element in the sub array with the element to its left and then do circular right shift operations, for op = 1 subtract and do left shift operations and op = 2 subtract and then perform reverse operations on rows.
   ii. In inverse column transformation when op = 0, except first element, starting from second element subtract each element in sub array with the element to its above and downward shift operations for op = 1, subtract and upward shift operations for op = 2, subtract and do reverse operations on columns.

2. This process is done until the key file is completed and at the end of the process matrix contains the required original message which is plain text.

2.2 Hiding the Encoded Message into an Image

The cipher text obtained from encryption algorithm is hidden in the image to enhance the security. Gini index is used as metric to hide the encrypted text into image.

Gini (T) is defined as $1 - \sum_{j=1}^{n} p_j^2$

Encoding Function
Input: Encoded message, color image; Output: Stego image

1. Get the 4 MSBs of R, G, B component for each pixel of the cover image. Calculate the gini index for each component. Take the component which has the least gini index and hide the encoded message in the 4 LSB of it.
2. If the values of the 3 components are equal then hide the message in the 4 LSB of the three components. If the values of any 2 components are equal then hide the data in 4 LSB of those two components. Continue for each pixel in the cover image until the entire message is hidden.

Decoding Function
Input: Stego image; Output: Encoded message

1. Get the 4 MSBs of R, G, B component for each pixel of the stego image. Calculate the gini index for each component. Take the component which has the least gini index and extract the encoded message from the 4 LSBs.
2. If the values of the 3 components are equal then extract the message in the 4 LSB of the three components. If the values of any 2 components are equal then extract the data in 4 LSB of those two components. Continue the process for each pixel in the cover image until the entire message is extracted.

3 Results and Discussion

3.1 At Sender Side

Encryption of Message Let the data to be encrypted is 1, 2, 3, 4, 5, 6, 7, 8, 9, 10, 11, 12, 13, 14, 15, 16, 17, 18, 19, 20. Let m = 4, n = 5, w = 7, i = 14, b = 01110. Let the contents of the key file are R: 1/2/1/2/4, C: 0/4/1/0/3, C: 1/3/0/0/3, C: 2/1/2/1/2, R: 1/1/2/0/4, R: 2/3/1/0/2, C: 1/2/0/0/1. The scrambling of the matrix is

1	2	3	4	5
6	7	8	17	19
11	12	13	27	29
16	17	18	19	20

R: 1/2/1/2/4 (addition)

1	2	3	4	5
6	7	19	8	17
11	12	29	13	27
16	17	18	19	20

R: 1/2/1/2/4 (shifting)

1	2	3	4	5
6	9	19	8	22
11	19	29	13	44
16	29	18	19	47

C: 0/4/1/0/3 (addition)

1	9	3	4	22
6	19	19	8	44
11	29	29	13	47
16	2	18	19	5

C: 0/4/1/0/3 (shifting)

1	9	3	4	22
7	19	19	12	44
17	29	29	21	47
27	2	18	32	5

C: 1/3/0/0/3 (addition)

27	9	3	32	22
1	19	19	4	44
7	29	29	12	47
17	2	18	21	5

C: 1/3/0/0/3 (shifting)

27	9	3	32	22
1	19	19	4	44
7	48	48	12	47
17	2	18	21	5

C: 2/1/2/1/2 (addition)

27	9	3	32	22
1	48	48	4	44
7	19	19	12	47
17	2	18	21	5

C: 2/1/2/1/2 (shifting)

27	9	3	32	22
1	49	96	52	48
7	26	38	31	59
17	2	18	21	5

R: 1/1/2/0/4 (addition)

27	9	3	32	22
48	1	49	96	52
59	7	26	38	31
17	2	18	21	5

R: 1/1/2/0/4 (shifting)

27	9	3	32	22
48	1	97	96	52
59	7	26	38	31
17	2	35	21	5

R: 2/3/1/0/2 (addition)

27	9	3	32	22
97	1	48	96	52
59	7	26	38	31
35	2	17	21	5

R: 2/3/1/0/2 (shifting)

27	9	3	32	22
124	1	51	96	52
59	7	26	38	31
35	2	17	21	5

C: 1/2/0/0/1 (addition)

124	9	51	32	22
27	1	3	96	52
59	7	26	38	31
35	2	17	21	5

C: 1/2/0/0/1 (shifting)

So the encrypted text is 124, 9, 51, 32, 22, 27, 1, 3, 96, 52, 59, 7, 26, 38, 31, 35, 2, 17, 21, 5.

Hiding the Encrypted Text into an Image A 24 bit color image consists of red, green and blue components, each of the 8 bits as Pixel 1: **11110001 11001000 00000111.** Convert the cipher text obtained from Sect. 1 into binary: 124-0111 1100, 9-0000 1001, 51-0011 0011…. The first pixel of the image is **11110001 11001000 0000111**, take the 4 MSBs of the red component and calculate the gini index.

For **1111**

1's	4
0's	0

Gini Index = 1-[$(4/4)^2$+$(0/4)^2$]
= 0

For **1100**

1's	2
0's	2

Gini Index = 1-[$(2/4)^2$+$(2/4)^{2)}$]
= 0.5

For **0000**

1's	0
0's	4

Gini Index = 1-[$(0/4)^2$+$(4/4)^2$]
= 0

By comparing the 3 values, the red and blue components have the least value i.e. 0. So, cipher text is embedded in 4 LSBs of the red and blue components. The altered pixel 1 is **11110111 11001000 00001100.** So place the cipher text in the 4 LSB of the three components. Continue this process until entire cipher text is hidden.

3.2 *At Receiver Side*

Extracting the Encrypted Text from the Image To get the cipher text form the stego image start from the first pixel: **11110111 11001000 00001100**. The gini index for the 4 MSB of the RGB is 0, 0.5, 0. Since red and blue has the least values data is hidden in these components. So, extract the message from LSB of red and blue components and convert the bits into decimal values. Continue this for each pixel until the entire cipher text is extracted from the image. The cipher text 124, 9, 51, 32, 22, 27, 1, 3, 96, 52, 59, 7, 26, 38, 31, 35, 2, 17, 21, 5 is obtained. Now, decrypt this cipher text to get the plaintext by using decryption algorithm which is as follows:

Decryption of the Encrypted Text
The cipher text is 124, 9, 51, 32, 22, 27, 1, 3, 96, 52, 59, 7, 26, 38, 31, 35, 2, 17, 21, 5

27	9	3	32	22
124	1	51	96	52
59	7	26	38	31
35	2	17	21	5

C: 1/2/0/0/1 (shifting)

27	9	3	32	22
97	1	48	96	52
59	7	26	38	31
35	2	17	21	5

C: 1/2/0/0/1 (subtraction)

27	9	3	32	22
48	1	97	96	52
59	7	26	38	31
17	2	35	21	5

R: 2/3/1/0/2 (shifting)

27	9	3	32	22
48	1	49	96	52
59	7	26	38	31
17	2	18	21	5

R: 2/3/1/0/2 (subtraction)

27	9	3	32	22
1	49	96	52	48
7	26	38	31	59
17	2	18	21	5

R: 1/1/2/0/4 (shifting)

27	9	3	32	22
1	48	48	4	44
7	19	19	12	47
17	2	18	21	5

R: 1/1/2/0/4 (subtraction)

27	9	3	32	22
1	19	19	4	44
7	48	48	12	47
17	2	18	21	5

C: 2/1/2/1/2 (shifting)

27	9	3	32	22
1	19	19	4	44
7	29	29	12	47
17	2	18	21	5

C: 2/1/2/1/2 (subtraction)

1	9	3	4	22
7	19	19	12	44
17	29	29	21	47
27	2	18	32	5

C: 1/3/0/0/3 (shifting)

1	9	3	4	22
6	19	19	8	44
11	29	29	13	47
16	2	18	19	5

C: 1/3/0/0/3 (subtraction)

1	2	3	4	5
6	9	19	8	22
11	19	29	13	44
16	29	18	19	47

C: 0/4/1/0/3 (shifting)

1	2	3	4	5
6	7	19	8	17
11	12	29	13	27
16	17	18	19	20

C: 0/4/1/0/3 (subtraction)

1	2	3	4	5
6	7	8	17	19
11	12	13	27	29
16	17	18	19	20

R: 1/2/1/2/4 (shifting)

1	2	3	4	5
6	7	8	9	10
11	12	13	14	15
16	17	18	19	20

R: 1/2/1/2/4 (subtraction)

Decrypted text is 1, 2, 3, 4, 5, 6, 7, 8, 9, 10, 11, 12, 13, 14, 15, 16, 17, 18, 19, 20.

3.3 Avalanche Effect

It is a property of cryptographic algorithm where if an input is changed slightly (single bit) the output changes significantly.

$$\text{Avalanche Effect} = \frac{\text{Number of flipped bits in cipher text}}{\text{Number of bits in cipher text}} \times 100$$

Table 1 Cipher text comparison with original encrypted text

Text and effect	Matrix scrambling technique	Proposed technique
Text	1,2,3,4,5,6,7,**8**,9,10,11,12,13, 14,15,16,17,18,19,20	1,2,3,4,5,6,7,**8**,9,10,11,12, 13,14,15,16,17,18,19,20
Encrypted text	17,7,14,19,9,16,1,3,15,4,20,6, 12,10,8,18,2,11,13,5	124,9,51,32,22,27,1,3,96 ,52,59,7,26,38,31,35,2,17,21,5
Modified text	1,2,3,4,5,6,7,**136**,9,10,11,12, 13,14,15,16,17,18,19,20	1,2,3,4,5,6,7,**136**,9,10,11,12, 13,14,15,16,17,18,19,20
Modified encrypted text	17,7,14,19,9,16,1,3,15,4,20,6,12, 10,136,18,2,11,13,5	252,9,179,32,150,27,1,3, 96,52,187,7,26,38,159, 35,2,17,149,5
Number of bits fipped	1	6
Avalanche effect	0.625%	3.75%

Since the proposed algorithm is based on ascii (256) characters it is difficult for the cryptanalyst to break the algorithm because it requires more characters to search as compared to the existing technique (uses 20 characters).

As compared to matrix scrambling technique our proposed method shows better avalanche effect as shown in Table 1.

4 Conclusion

The proposed encryption algorithm is easy to understand and execute due to its simple operations like shifting, exchanging and arithmetic. When the same text is processed again and again with different sub keys the scrambling is not same. It shows the efficiency of the algorithm and strong encryption that is not vulnerable to cryptanalysis. To increase the level of security the cipher text obtained from encryption algorithm is hidden randomly in RGB components of the image using an inequality measure called gini index. Since data is embedded in LSB of the pixels, image distortion is less and image quality is maintained.

References

1. Kiran Kumar, M., Mukthyar Azam, S., Shaik Rasool: Efficient Digital Encryption Algorithm Based on Matrix Scrambling Technique. International Journal of Network Security & Its Applications. 2(4), 30–41 (2010).
2. Suli Wu, Yang Zhang, Xu Jing: A Novel Encryption Algorithm based on Shifting and Exchanging Rule of Bi-column Bi-row Circular Queue. In: IEEE International Conference on Computer Science and Software Engineering, vol.3, pp. 841–844 (2008).

3. HuLihong: Research on Improved Algorithm Based on Encryption Algorithm of Hill. In: IEEE International Conference on Computer Application and System Modeling, ICCASM, vol. 8, pp. 659–662 (2010).
4. Gurpreet Singh, Supriya: Modified Vigenere Encryption Algorithm and Its Hybrid Implementation with Base64 and AES. In: Proceedings of the 2nd International Conference on Advanced Computing, Networking and Security, IEEE Computer Society, pp. 232–237 (2013).
5. Dipesh Agrawal, Samidha Diwedi Sharma: Analysis of Random Bit Image Steganography Techniques. International Journal of Computer Applications. International Conference on Recent Trends in Engineering & Technology (2013).
6. Badrinath, R., Anand, P.S.: MSB Constrain Based Variable Embedding. In: Computing Communication & Networking Technologies (2012).
7. Manoj Kumar Ramaiya, Naveen Hemrajani, Anil Kishor Saxena: Security Improvisation in Image Steganography using DES. In: 3rd IEEE Trans. International Conference IACC, pp. 1094–1099 (2013).

Improvement of Toward Offering More Useful Data Reliably to Mobile Cloud from Wireless Sensor Network

Ankita Singla and Deepali

Abstract Cloud based Wireless sensor network (WSN) technology is used in real-time applications as users can access the data from sensor nodes through cloud at any time and at any place. This provides various features like easy data storage, maintenance, availability, sharing etc. Sometimes the data which is requested by users may not be received on time as multiple users can access data from cloud at same time or may be lost during transmission. Therefore this results an unreliable and inefficient approach. In this paper, we proposed a protocol to increase reliability of cloud based WSN technology.

Keywords WSN · MEAL · LEACH · E-LEACH · TPSDT · PSS

1 Introduction

Sensor nodes have limited amount of energy and memory and they may or may not be chargeable depending on different application areas. Sensor nodes in WSN sense their surroundings and transfer the collected data to its sink for further processing. To send data from sensor nodes to sink, various protocols has been proposed such as LEACH [1], I-LEACH [2], ERA [3], E-LEACH [4], VLEACH [5], MEAL [6] etc. As sensor nodes sense data continuously due to which large amount of data gather at sink which is difficult to handle. Therefore it needs huge space for data storing which can be possible by using cloud.

Cloud is a common space utilized by users to share data and resources with the help of internet connection. Cloud provides easy data sharing, data managing and data interoperability including security and privacy features to maintain trust of users.

A. Singla (✉) · Deepali
Computer Science Department, Guru Nanak College, Budhlada, India
e-mail: singlaankita92@gmail.com

Deepali
e-mail: goyaldeepali1@gmail.com

S.C. Satapathy et al. (eds.), *Computer Communication, Networking and Internet Security*, Lecture Notes in Networks and Systems 5,
DOI 10.1007/978-981-10-3226-4_5

WSN with cloud computing provides various useful features like real time data processing, data sharing at any time or any place. Due to availability of recent data, emergency detection is very easy in any application area.

Key Features of WSN with cloud computing are-

1) *Data handling*—Due to common server for all users it is easy to handle huge amount of data of WSN.
2) *Virtualization*—Users can request for data from WSN without knowing sensor node's location.
3) *Confidentiality*—Cloud handles the user's requests and sends sensor data only to the authorized users.

2 Related Work

Kantarci et al. [7] proposed a framework named as Trustworthy Sensing for Crowd Management (TSCM) to get data from trustworthy users having sensors with-in their mobile devices. Users provide accurate data to cloud and are paid according to their tasks of sending data to cloud. Sometimes malicious users provide data at low cost. But they provide insufficient data to cloud which reduce efficiency of cloud. To avoid this, TSCM stores the user's information on cloud. And it checks previous records of users before assigning new task to any user.

Phan et al. [8] proposed an architecture named as Sensor-Cloud Integration Platform as a Service (SC-iPaaS) to optimize the communication between edge and cloud. Edge is a collection of sink which receive data from sensor nodes and send data to cloud. A lot of bandwidth, energy and data are utilized during communication between edge and cloud. To reduce all these parameters Evolutionary Multi-objective optimization (EMOA) algorithm is used to select optimum solution for data transmission according to priorities of users.

Zhu et al. [9] proposed Authenticated Trust and Reputation Calculation and Management (ATRCM) system to avoid intruder's attacks during data transmission and maintain trust and reputation between cloud and WSN. It helps users to select trustable cloud and helps cloud service providers to select trustable WSN. Trust is calculated by Eq. (1).

$$t = \frac{s+1}{f+s+2} \tag{1}$$

where s represents positive feedback and f represents negative feedback. Reputation of cloud is calculated by Eq. (2).

$$Rc = \frac{CNc}{N'u} \tag{2}$$

where CN_c represents number of users select the cloud services and N'_u represents number of users need services of cloud. Reputation of WSN is calculated by Eq. (3).

$$Rk = \frac{\mathrm{CNk}}{\mathrm{N'c}} \tag{3}$$

where CN_k represents number of cloud service providers selects WSN and N'_c represents number of cloud service providers need WSN.

Zhang et al. [10] proposed an architecture to increase the efficiency of cloud based WSN. In this cloud acts as a virtual sink having collection of sinks of different areas to collect data from sensor nodes of their respective areas and then process the data with the help of Hadoop [11] system to reduce data movement.

Alam et al. [12] proposed cloud based mental state monitoring system for suicide risk reconnaissance (CBMSMS-SRR) system to monitor suicide risks for mentally retarded patients. In this patients are registered with cloud healthcare agent with unique ID. Cloud healthcare agent collects data from sensor nodes and then checks the history of patient which is stored at cloud healthcare service and updates the current state of patient to different users.

Chatterjee et al. [13] proposed data caching mechanism which introduces two types of cache- external and internal for caching the data. External cache presents between cloud and WSN which stores the data only when sensor nodes sense any change in physical environment and internal cache presents with-in cloud to provide data to users from external cache. This mechanism reduces redundancy of data transmission, hence increase energy efficiency.

Liang et al. [14] proposed a framework to enhance the efficiency of WSN based on cloud by introducing QOS (quality of service) approach using Service Oriented Middleware (SOM) and multi-dimensional cloud model to select the optimum methods like selection of nodes, resources, routing technique etc. for the execution of tasks.

3 Toward Offering More Useful Data Reliably to Mobile Cloud from Wireless Sensor Network

A. Problems in traditional WSN-mobile cloud architecture

1) WSN consists of sensor nodes which sense the surrounding and send the data to sink which further transmit it to mobile cloud. But as sensor nodes have limited energy so by continuous sensing and sending will reduce the lifetime of WSN.

2) WSN sends data without knowing user's demand. This will lead to use of extra bandwidth during transmission which will slow down the overall performance of network.
3) As continuous transmission of non-useful data may not satisfy user's demands on time.

B. Description of protocol

Zhu et al. [15] proposed protocol TPSS having two components- TPSDT (time and priority based selective data transmission), PSS (priority based sleep scheduling). Authors mainly focus on less energy depletion and getting more reliability by providing useful data to users. Two schemes that make this possible are described as follows:

1) *TPSDT* is a scheme in which data is sent to users according to their demands which will be managed at sink. A Point versus Time & Priority (PTP) table in cloud is organized on the basis of user's requests with in different time intervals to set priorities of sensor nodes.
2) *PSS*—It provides sleep-scheduling scheme to sensor nodes. PTP table is updated to set priorities of sensor nodes according to user's demands in each time intervals. In that specific period, sensor nodes that have high priority are active to transmit data to sink while rest of others is in sleep mode.

C. Work flow of protocol

1) Users send requests on cloud for getting data from sensor nodes.
2) Cloud updates PTP table and sends user's requests to sink.
3) Further sink sets the priorities of sensor nodes using PSS scheme.
4) Sensor nodes with higher priority are activated to send data to sink by multi-hop manner.
5) After receiving data from active nodes, sink transmit data to cloud in priority order using TPSDT scheme.
6) If any failure occurs while transmission of data from sensor nodes to sink then retransmission is performed until it get successful (Fig. 1).

4 Proposed Protocol

Improvement of Toward Offering More Useful Data Reliably to Mobile Cloud from Wireless Sensor Network

A. Problem definition

As discussed in Sect. 3, protocol [15] provides reliable and useful data to the users by implementing WSN with cloud computing in real time environment. In this, authors have taken assumption of large amount of data storage at sensor nodes

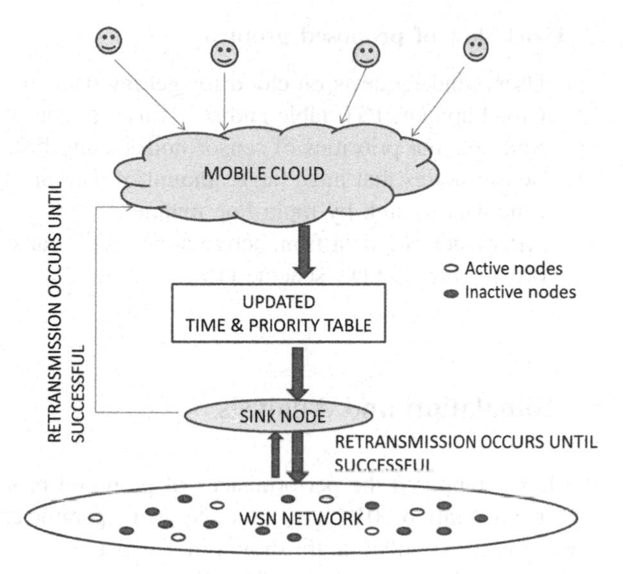

Fig. 1 Flow control

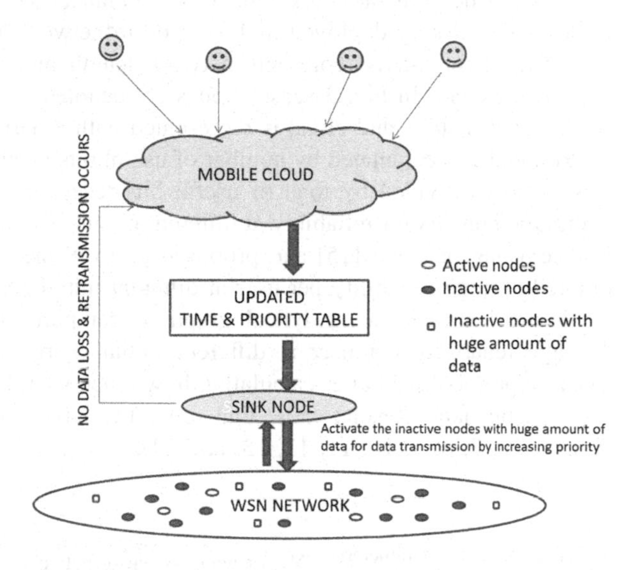

Fig. 2 Proposed flow control

which is not so practical, causes data loss or retransmission and more energy will be consumed by sensor node to retransmit the data.

B. Proposed work

The assumption made in existing protocol [15] cannot consider practically as storage capacity of sensor node is not unlimited. Sometimes, sensor nodes sense large amount of data. But due to lack of storage, useful data will be lost. So, we improve the reliability of protocol by increasing the priority of inactive nodes that have large amount of data in WSN to reduce data loss and retransmission (Fig. 2).

C. Work flow of proposed protocol

1) Users send requests on cloud for getting data from sensor nodes.
2) Cloud updates PTP table and sends user's requests to sink.
3) Sink sets the priorities of sensor nodes using PSS scheme [15].
4) Sensor nodes that have huge amount of data or high priority are activated to send data to sink by multi-hop manner.
5) After receiving data from active nodes, sink transmit data to cloud in priority order using TPSDT scheme [15].

5 Simulation and Analysis

We have compared the performances of proposed protocol with existing protocol [15] using Matlab-2010 simulator. Network parameters which are taken in consideration for simulation are shown in Table 1.

Figure 3 demonstrates the four WSN initialization where in each WSN 100 nodes are randomly deployed in 100×60 m^2network area and sink is located at (50, 30). Here, x-axis represents network length and y-axis represents network height in meters. In Fig. 3 sensor nodes are denoted with circle shape and sink is represented with $\times$ and cloud is represented with * and blue curves.

Reliability is calculated by number of useful bits received by user from different sensor nodes divided by total of useful bits received and not received by users. Simulation results for reliability at different nth users at different initial energies for both existing protocol [15] and proposed protocol are shown in Table 2 and calculated average reliability per user at different initial energy is shown in Table 3.

Figure 4 shows the analytical graph of comparison of simulation values of average reliability per user at different initial energy of proposed protocol and existing protocol [15] after calculating how much useful data has sent to user from total useful data. Results show that reliability has been increased for each case (Figs. 5, 6, 7, 8, 9, 10, 11, 12, 13, and 14).

Table 1 Network parameters

No. of wireless sensor networks	4
Network size of each WSN	100×60 m^2
Number of sensor nodes in each WSN	100
End users	10
EDA	5 nJ/bit/signal
Efs	10 pJ/bit/m^2
Emp	0.0013 pJ/bit/m^4
Message size	Assigned randomly
Initial energy	0.1–0.9 J

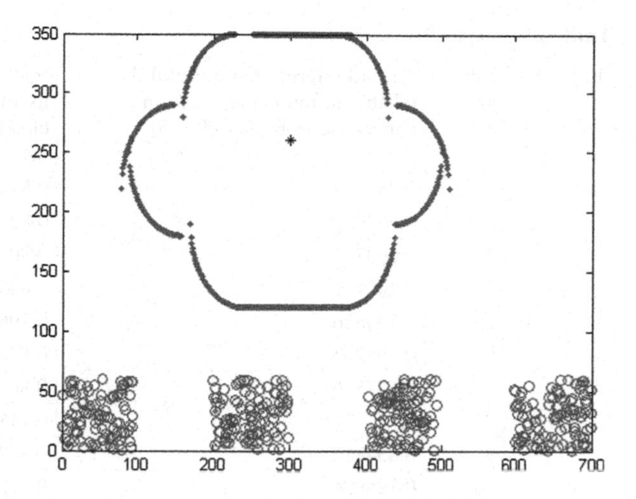

Fig. 3 Network area

Table 2 Reliability at different initial energy

Energy	Nth user	Toward offering more useful data reliably to mobile cloud from wireless sensor network [15]	Improved toward offering more useful data reliably to mobile cloud from wireless sensor network
0.1	1	0.673684	0.846939
	2	0.67474	0.84669
	3	0.717687	0.813725
	4	0.745387	0.818505
	5	0.719424	0.823729
	6	0.704918	0.805281
	7	0.691228	0.80427
	8	0.704082	0.856643
	9	0.664407	0.811189
	10	0.711409	0.829268
0.14	1	0.691756	0.846939
	2	0.644518	0.84669
	3	0.69863	0.813725
	4	0.761566	0.818505
	5	0.684028	0.823729
	6	0.715719	0.805281
	7	0.681159	0.80427
	8	0.686131	0.856643
	9	0.703971	0.811189
	10	0.680921	0.829268

(continued)

Table 2 (continued)

Energy	Nth user	Toward offering more useful data reliably to mobile cloud from wireless sensor network [15]	Improved toward offering more useful data reliably to mobile cloud from wireless sensor network
0.2	1	0.75	0.875433
	2	0.668966	0.829861
	3	0.690722	0.807818
	4	0.743421	0.830508
	5	0.704626	0.796364
	6	0.705263	0.84669
	7	0.677536	0.869128
	8	0.7	0.818182
	9	0.634812	0.850174
	10	0.725086	0.838488
0.3	1	0.674497	0.821549
	2	0.704467	0.797753
	3	0.697917	0.84375
	4	0.738983	0.808511
	5	0.696864	0.837545
	6	0.673759	0.838384
	7	0.657534	0.817544
	8	0.705882	0.801347
	9	0.676976	0.804196
	10	0.667832	0.852349
0.4	1	0.678445	0.837545
	2	0.741497	0.86014
	3	0.725694	0.825623
	4	0.722408	0.817204
	5	0.677852	0.877622
	6	0.721429	0.855705
	7	0.710247	0.849673
	8	0.749129	0.836237
	9	0.672474	0.828767
	10	0.701031	0.847458
0.5	1	0.697509	0.820513
	2	0.666667	0.830258
	3	0.718121	0.828467
	4	0.634483	0.845638
	5	0.7	0.821306
	6	0.702614	0.799363
	7	0.67	0.814433
	8	0.68	0.838596

(continued)

Table 2 (continued)

Energy	Nth user	Toward offering more useful data reliably to mobile cloud from wireless sensor network [15]	Improved toward offering more useful data reliably to mobile cloud from wireless sensor network
	9	0.697279	0.848921
	10	0.75082	0.820598
0.6	1	0.702509	0.846939
	2	0.685121	0.84669
	3	0.655629	0.813725
	4	0.707143	0.818505
	5	0.675958	0.823729
	6	0.682594	0.805281
	7	0.727599	0.80427
	8	0.707483	0.856643
	9	0.687273	0.811189
	10	0.693431	0.829268
0.7	1	0.674497	0.793478
	2	0.704467	0.814947
	3	0.697917	0.8125
	4	0.738983	0.886121
	5	0.696864	0.881119
	6	0.673759	0.821053
	7	0.657534	0.784247
	8	0.705882	0.83557
	9	0.676976	0.865517
	10	0.667832	0.796552
0.8	1	0.696751	0.798013
	2	0.689189	0.820339
	3	0.758389	0.866894
	4	0.675768	0.797753
	5	0.69258	0.849123
	6	0.715686	0.847222
	7	0.64094	0.800687
	8	0.684588	0.835526
	9	0.672241	0.848485
	10	0.730769	0.802721
0.9	1	0.647651	0.818841
	2	0.697183	0.84058
	3	0.701695	0.825503
	4	0.707317	0.864407
	5	0.729097	0.772727

(continued)

Table 2 (continued)

Energy	Nth user	Toward offering more useful data reliably to mobile cloud from wireless sensor network [15]	Improved toward offering more useful data reliably to mobile cloud from wireless sensor network
	6	0.735395	0.830619
	7	0.665529	0.848797
	8	0.685921	0.836735
	9	0.733788	0.836237
	10	0.670068	0.84589

Table 3 Average reliability per user at different initial energy

Energy	Toward offering more useful data reliably to mobile cloud from wireless sensor network [15]	Improved toward offering more useful data reliably to mobile cloud from wireless sensor network
0.1	0.7006966	0.8256239
0.14	0.6948399	0.8256239
0.2	0.7000432	0.8362646
0.3	0.6894711	0.8222928
0.4	0.7100206	0.8442699
0.5	0.6917493	0.8268093
0.6	0.692474	0.8256239
0.7	0.6894711	0.8291104
0.8	0.6956901	0.8266763
0.9	0.6973644	0.8320336

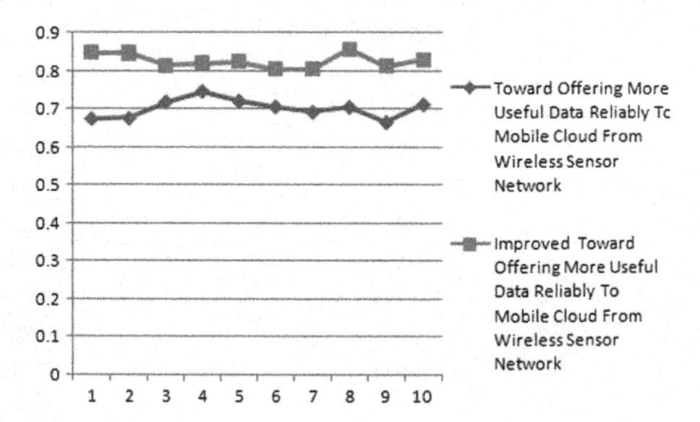

Fig. 4 Increase in reliability for each user at initial energy = 0.1

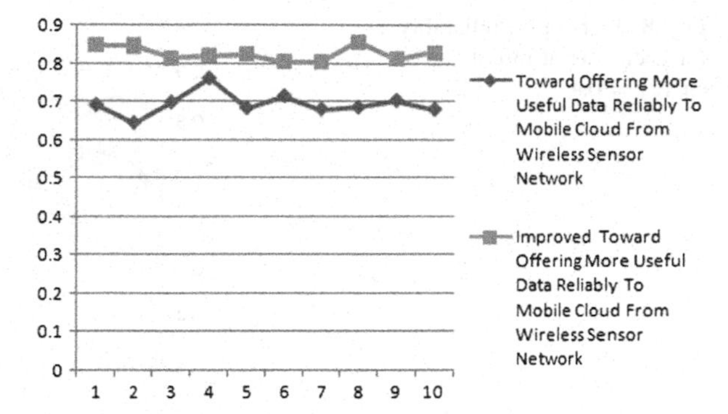

Fig. 5 Increase in reliability for each user at initial energy = 0.14

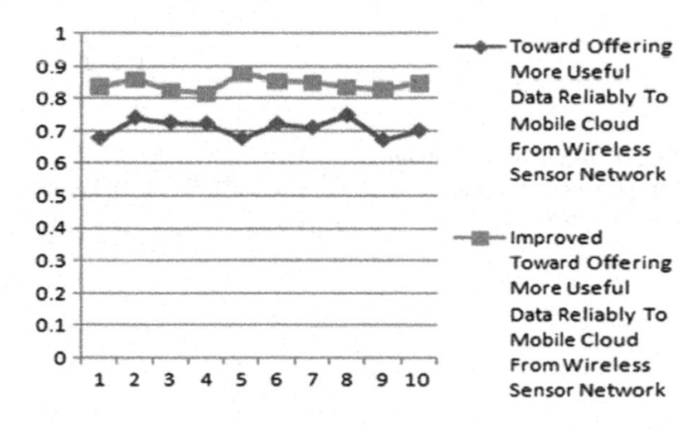

Fig. 6 Increase in reliability for each user at initial energy = 0.2

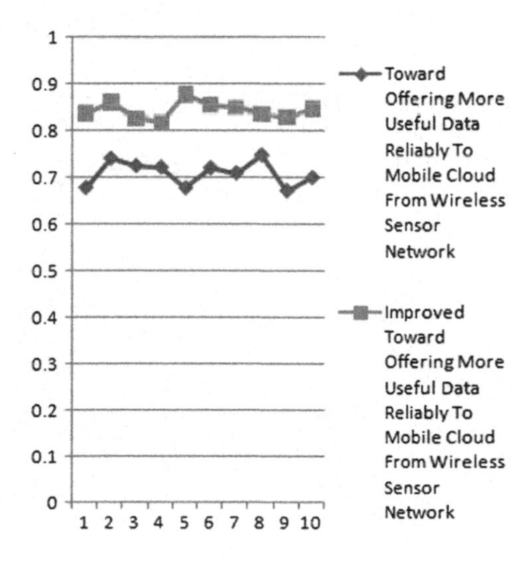

Fig. 7 Increase in reliability for each user at initial energy = 0.3

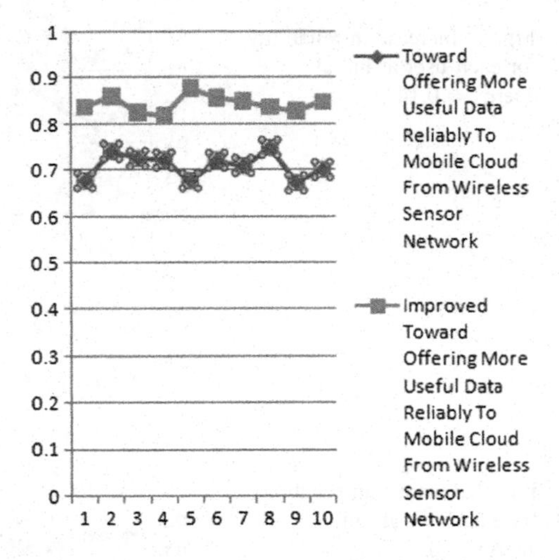

Fig. 8 Increase in reliability for each user at initial energy = 0.4

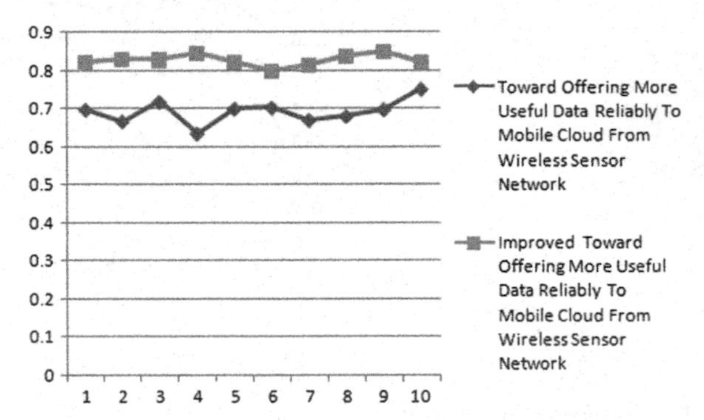

Fig. 9 Increase in reliability for each user at initial energy = 0.5

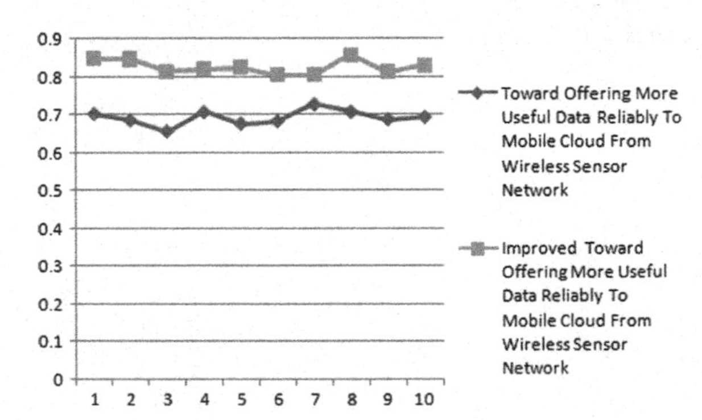

Fig. 10 Increase in reliability for each user at initial energy = 0.6

Fig. 11 Increase in reliability for each user at initial energy = 0.7

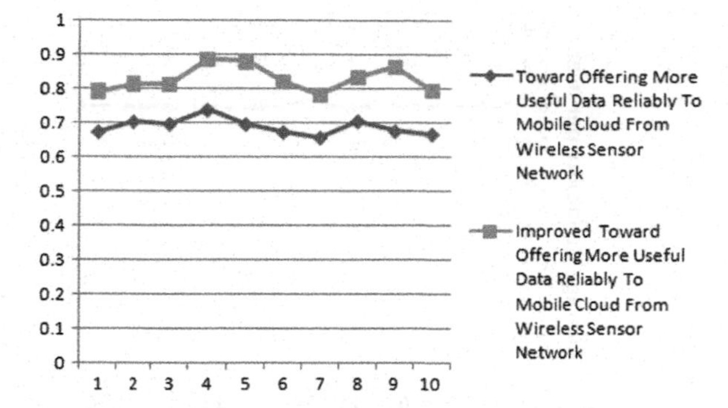

Fig. 12 Increase in reliability for each user at initial energy = 0.8

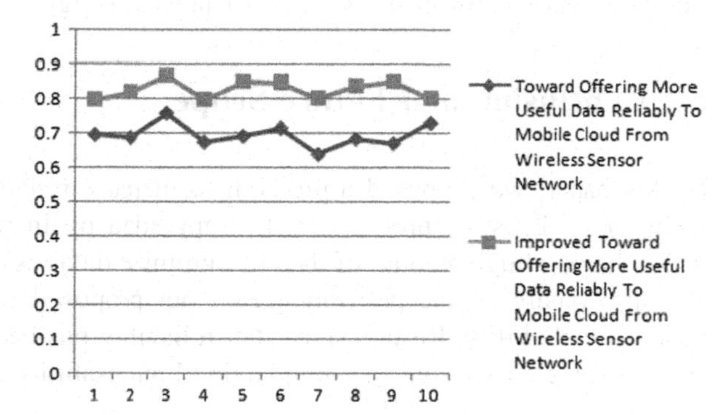

Fig. 13 Increase in reliability for each user at initial energy = 0.9

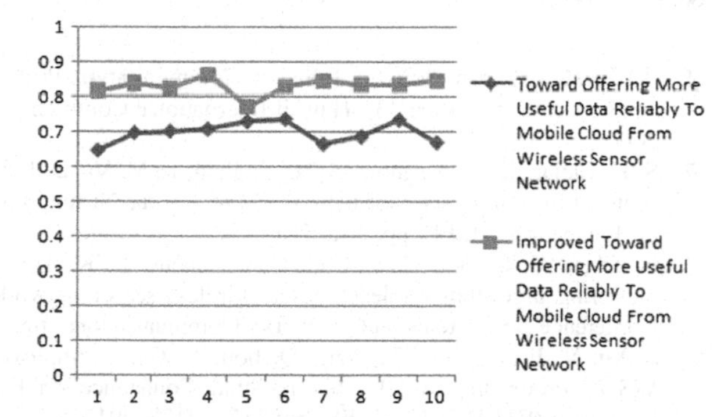

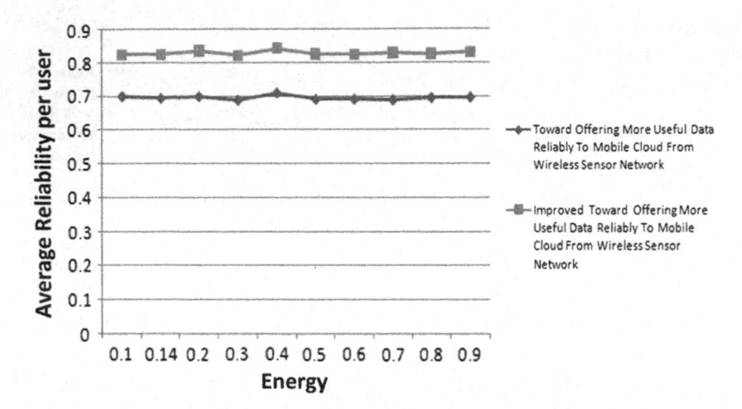

Fig. 14 Average reliability per user at different initial energy

6 Conclusion and Future Scope

In this paper, we proposed a protocol to increase reliability of cloud based WSN technology. Existing protocol [15] is upgraded by increasing priorities of sensor nodes having huge amount of data to minimize data loss and to increase reliability. We have analyzed the performance of our proposed work and existing protocol based on reliability. Results show that reliability has been increased for each case. In future, we will design a system having high usability and high energy efficiency.

References

1. W. R. Heinzelman, "Energy-Efficient Communication Protocol for Wireless Micro sensor Networks," Proc. of the 33rd Hawaii International Conference on System Sciences, pp. 1–10, 2000.
2. S. H. Gajjar, K. S. Dasgupta, S. N. Pradhan, K. M. Vala, "Lifetime improvement of LEACH protocol for wireless sensor network," Proc. of the Nirma university International conference on Engineering, IEEE, pp. 1–6, 2012.
3. H. Chen, C. S. Wu, Y. S. Chu, C. C. Cheng, L. K. Tsai, "Energy residue aware(ERA) clustering algorithm for leach-based wireless sensor networks," Proc. of 2nd International Conference on Systems and Networks Communications, pp. 40–45, 2007.
4. J. Xu, N. Jin, X. Lou, T. Peng, Q. hou, Y. Chen, "improvement of LEACH protocol for WSN," Proceedings of 9th International Conference on Fuzzy Systems and Knowledge Discovery (FSKD 2012), IEEE, pp. 2174–2177, 2012.
5. A. Ahlawat, V. Malik, "An Extended Vice-Cluster Selection approach to improve VLEACH in WSN," Proceedings of 3rd International conference on Advanced Computing & Communication Technologies, pp. 236–240, 2013.
6. MEAL Based Routing in WSN, Patalia, Tejas, Tada, Naren, Patel, Chirag, Satapathy, Chandra Suresh, Das, Swagatam Proceedings of First International Conference on Information and Communication Technology for Intelligent Systems: Volume 2 2016

Springer International Publishing Cham 978-3-319-30927-9 Patalia 2016 10.1007/978-3-319-30927-9_11 http://dx.doi.org/10.1007/978-3-319-30927-9_11 109–114.

7. B. Kantarci and H. T. Mouftah, "Trustworthy Sensing for Public Safety in Cloud-Centric Internet of Things," in *IEEE Internet of Things Journal*, vol. 1, no. 4, pp. 360–368, Aug. 2014. doi:10.1109/JIOT.2014.2337886.
8. D. H. Phan, J. Suzuki, S. Omura, K. Oba and A. Vasilakos, "Multiobjective Communication Optimization for Cloud-Integrated Body Sensor Networks," *Cluster, Cloud and Grid Computing (CCGrid), 2014 14th IEEE/ACM International Symposium on*, Chicago, IL, 2014, pp. 685–693. doi:10.1109/CCGrid.2014.48.
9. C. Zhu, H. Nicanfar, V. C. M. Leung and L. T. Yang, "An Authenticated Trust and Reputation Calculation and Management System for Cloud and Sensor Networks Integration," in *IEEE Transactions on Information Forensics and Security*, vol. 10, no. 1, pp. 118–131, Jan. 2015. doi:10.1109/TIFS.2014.2364679.
10. Zhang, Peng, Zheng Yan, and Hamlin Sun. "A novel architecture based on cloud computing for wireless sensor network." *Proceedings of the 2nd International Conference on Computer Science and Electronics Engineering. Atlantis Press.* 2013.
11. Marinelli, Eugene E. *Hyrax: cloud computing on mobile devices using MapReduce.* No. CMU-CS-09-164. Carnegie-mellon univ Pittsburgh PA school of computer science, 2009.
12. Md. Golam Rabiul Alam, Eung Jun Cho, Eui-Nam Huh, and Choong Seon Hong. 2014. Cloud based mental state monitoring system for suicide risk reconnaissance using wearable bio-sensors. In *Proceedings of the 8th International Conference on Ubiquitous Information Management and Communication* (ICUIMC '14). ACM, New York, NY, USA, Article 56, 6 pages. DOI:http://dx.doi.org/10.1145/2557977.2558020.
13. S. Chatterjee and S. Misra, "Dynamic and adaptive data caching mechanism for virtualization within sensor-cloud," *2014 IEEE International Conference on Advanced Networks and Telecommunications Systems (ANTS)*, New Delhi, 2014, pp. 1–6. doi:10.1109/ANTS.2014.7057243.
14. Jun-bin Liang, Ning-jiang Chen and Min-min Yu, "A cloud model based multi-dimension QoS evaluation mechanism for Wireless Sensor Networks," *Computer Science & Education, 2009. ICCSE '09. 4th International Conference on*, Nanning, 2009, pp. 348–352. doi:10.1109/ICCSE.2009.5228429.
15. C. Zhu, Z. Sheng, V. C. M. Leung, L. Shu and L. T. Yang, "Toward Offering More Useful Data Reliably to Mobile Cloud From Wireless Sensor Network," in *IEEE Transactions on Emerging Topics in Computing*, vol. 3, no. 1, pp. 84–94, March 2015. doi:10.1109/TETC.2014.2364921.

Computational and Emotional Linguistic Distance and Language Learning

Krishnaveer Abhishek Challa and Mounika Vurity

Abstract Computational linguistics is a fast evolving concept in the contemporary era of technological advancements. The methods of teaching language to the machine so as to reap better benefits from a linguistically knowledgeable machine is the aim of computational linguistics. However, the concept of linguistic distance has arisen in the spheres with the introduction of multiple languages to the e-world. This paper aims to study the problems of computational linguistic distance and the methods to mitigate it. The paper also delves into Computer Assisted Language Learning and its Emotional Connection and Aspects of Language Acquisition.

Keywords Linguistic distance · Speech recognition systems · Computational linguistic distance · Computational linguistics · Emotional Linguistic Distance · Emotional Language Teaching · Computer Assisted Language Learning

1 Introduction

Linguistic Distance is the distance between two languages which greatly impacts the social, economic and political relations between the countries, individuals and regions adhered to the respective languages. Computational Linguistics is a fast evolving concept in the multi-disciplinary fields of computer sciences and linguistics. It is the use of computer generated programs to study and effectively apply linguistics in the day-to-day lives of the individuals. In the technological driven lives of the contemporary era, computational linguistics is a must to enhance the mutual understanding between man and machine. To teach the machine the language of man is nothing but computational linguistics.

K.A. Challa (✉)
Department of Linguistics, Andhra University,
Visakhapatnam, Andhra Pradesh, India
e-mail: com2mass@gmail.com

M. Vurity
Department of Journalism and Mass Communications, Andhra University,
Visakhapatnam, Andhra Pradesh, India

S.C. Satapathy et al. (eds.), *Computer Communication, Networking and Internet Security*, Lecture Notes in Networks and Systems 5,
DOI 10.1007/978-981-10-3226-4_6

Computational Linguistic Distance is the use of computer generated programming to measure the linguistic distance and to find appropriate methods to mitigate the linguistic distance between languages. It also refers to the barriers of linguistic distance that are hurdling the progress of computational linguistics. In the once globalized system of computational linguistics, the new trends of innumerable languages have entered and the roots of linguistic distance have followed suit.

Computational linguistic distance, as a concept shall reap exceeding results in the field of linguistics by efficiently measuring the linguistic distance and systematically providing necessary solutions to overcome the barriers posed by linguistic distance in maintaining collateral relations among nations and their people. The study is still a long way to go in the sphere of multiple language learning and understanding. The field of CALL involves the use of a computer in the language learning process. CALL programs aim to teach aspects of the language learning process through the medium of the computer. CALL programs can be (and have been) developed for the many parts of the language learning process. Some of the factors that determine the characteristics of any CALL program include:

- the language taught,
- the language of instruction,
- the language writing system (both roman and non-roman character based),
- the level of the language to be taught (from absolute beginners to advanced),
- what is to be taught (grammar, informal conversation and pronunciation) and
- how it is to be taught.

2 Linguistic Distance

Linguistic Distance can be termed as the distance between two languages. In other words it is the measure of difficulty faced by the individuals of two different languages in learning and reproducing the language of their counterparts. Linguistic Distance impacts the bilateral relations between two nations. According to the Gravitational formula for Trade, the trade between two countries is inversely proportional to the distance between them. Here distance can be either geographical or linguistic distance. Thus, linguistic distance can be said to possess great impact on the trade relations between two nations. According to the formula, the greater the linguistic distance between the nations, the lesser is the trade relation and vice versa.

Linguistic distance depends on various factors like language vicinity, language origin and language scatter. Methods like Language trees and Automatic Similarity Judgement Program are used to measure linguistic distance between two languages. The effective measurement of linguistic distance helps scientists and linguists to actually trace out the possible solutions to overcome the distance [1].

Linguistic distance impacts the social, economic and political relations between the countries. To promote any cause on a global platform, linguistic distance acts as a huge barrier for communication. Linguistic distance also hinders the development of a nation by stalling the flow of potential immigrants and by confining the content of the nation to its own language boundaries [2].

To abridge the linguistic distance between the countries, a number of innovative methods are being tested to effectively mitigate linguistic distance. One of such yielding methods is Computational Linguistic Distance, where computer generated programmes are used to measure linguistic distance and to provide suitable solution to lessen the same.

Although there are more than five thousand languages spoken across the world, every person will have his own emotional language. Emotional language is a language to which we are attached to, right from childhood. It is the language which a person acquires and is acquainted with right from his childhood. Best example for an emotional language is our mother tongue. We speak in our mother right from our childhood. No one has taught us specifically how to speak in our mother tongue at childhood but listening to people around us, we acquired that language. In other words, we can say that emotional language is the language through which we:

- Think
- Communicate
- Express our feelings
- Acquire an intrusive understanding of grammar

We can say that a person may know many languages but he processes his thoughts only in the language to which he is emotionally connected with. Our mother tongue can be described as our emotional language because we acquire our mother tongue while we learn the other tongues. It is a well known fact that things once acquired can not be forgotten for life time. Hence, our emotional language is protected from the oblivion's curse.

Another trait of emotional language is that we express our emotions through it. For instance, as a child we express our first happiness, our first fears in our emotionally connected language that is our mother tongue. A child's first comprehension of the world him and his perception of existence starts with the language that is first taught to him, his mother tongue. Our personality, character and other traits become more evident through our mother tongue because the sound of mother tongue in our ears gives us a feeling of warmth and confidence. This is the impact of emotional language on any person. One of the important feature of cultural identity is the ability to speak one's mother tongue.

It is rightly quoted that "Our mother tongue is the language we use to think, dream and feel emotion".

3 Computational Linguistics

Linguistic Distance which for many years had been the sole concern of the linguists has in the recent years exposed itself to computer scientists who try to study it with the help of computer based programs and algorithms. Computational Linguistics is concerned with the computational perception of natural language. The concept of computational linguistics is said to have originated with the basic efforts in using computers to translate other languages to English. While the earlier assumption of the concept was that the measurement of linguistics and their reproduction is easier using computers, later research identified that the theory is more complicated than anticipated. Later on a new field of study called Computational Linguistics came into light bringing along with it a new set of professionals called Computational Linguists [3].

Computational Linguistics is used not just to translate texts but also to reconstruct earlier forms of modern languages and to sub group the modern languages based on their similarities. Thus, the concept which started as an experiment of translation finally took shape into a complex process of using computer algorithms to study the roots of languages and thus in turn group the languages into their respective spheres of origin. The process of language tree drawing has also augmented with the introduction of computational linguistics.

Computational linguistics as an applied science has lots of scope in near future. Research in the field of computational linguistics is being carried out to enhance human-machine interactions and speech recognition systems. The speech recognition systems, if accurately designed, can take the human-machine interaction to a new level. Robotic understanding of language commands is also a part of computational linguistic application. With the help of computational linguistics, the robots working at single sentence command pace can supersede their contemporaries with actual language understanding capabilities [2].

3.1 Approaches to Computational Linguistics

Computational linguistics is developed using various approaches like the developmental approach, structural approach, production approach, text-based interactive approach, speech-based interactive approach and comprehension approach.

Language learning, which is a continuous process in the human beings is studied to enhance the work. The way children learn to speak languages is carefully observed and tested on robots. The experiments have resulted in a significant number of findings in the field of linguistics thus augmenting the study in both the fields. Computational programs have been generated to understand the continuously evolving language patters and to study the possibilities of new patters of language in near future.

Structural study of the language is another approach in the field of computational linguistics. To carefully detail the computer of a language, an in-depth understanding of the language is quite essential. The basic structural guidelines of a language when fed to the computer can yield greater results in the study of the language patterns. While structural approach is quite essential, production approach needs to go hand in hand with the structural approach to help the system reproduce the learn language. According to Alan Turing, the day when a machine succeeds in responding to a human query without any disparities between man and machine, is the day when the machine can be termed as intelligent as man. This is known as Turing Test and the world is yet to witness a machine which has passed the test.

Text-based interactive approach is used to enhance human-computer interaction with the help pf typed texts and keyword recognitions. Speech recognition techniques are the new trend in computational linguistics where the machine is expected to recognize the speech of the individual and act accordingly. Though the speech based recognition systems have reached a level of efficiency in the reception of the commands, they are yet to reach a level of proficiency in the reproduction of the language.

Computers promote interactivity. Learners have to interact with the computer and cannot hide behind their classmates. If the learner does nothing, nothing happens. At the very least, learners have to start the CALL program. The program can only pass from one section to another with the "consent" of the learner. Thus learners have to drive the program. Usually they have to use the target language in exercises within the program. They have plenty of opportunities to practise the language in a one-on-one situation. They can practise the exercises as many times as they like, until they are satisfied with their results. CALL programs promote interactivity using many senses. Not only is text presented, sound can be heard and videos viewed. Sub-titles to videos can be switched on and off. Videos can be viewed in mute mode, so that learners can use various strategies to ascertain what is happening. Graphics can be used to demonstrate not just grammar items (for example, moving words around to form questions) but also for spatial related language topics (for example, the use of "in front of" and "behind").

3.2 Speech Recognition Systems

The speech recognition systems are one of the latest developing systems in terms of linguistics. Speech recognition is a process of feeding the computer with the required data to recognize the speech and to respond accurately. The system though might not understand the said words and their meanings. All it takes to processing is the keywords and an automated generated response is given back. Speech recognition helps a lot in the fields of stenography and other typing works where the sole purpose of the individual is to convert the speech to text. With this aspect accurately done by the system, the individual can concern oneself with better works that need a human brain to work on. The typing down of official documents, the automatic replies to frequently asked questions can all be assigned to the speech recognizing machines [4].

Speech Recognition is also in the recent years being used in the security departments to recognize the voice of the accused or the victim with the help of the number of unique usage of sounds by the individual. This process of identifying a person through his voice which is almost impossible to human calculation is being done by the speech recognition systems with compelling accuracy.

4 Computational Linguistic Distance

In the year 1966, Joseph Weizenbaum developed a program called ELIZA, the first of its kind designed to provide natural interaction between man and machine. Later on a number of programs have been developed to understand the commands given by the human to the computer and to cleverly answer them in their own language. For many years when English language was the sole language of the internet, the developed language response programs acted as a good connection between the people and their computers. However, in the recent years with the advent of multiple language usage in the computer world, the barriers of linguistic distance have once again started to play a key role in hindering the flow of communication. Be it a man or a machine, the presence of linguistic distance shall obstruct the communication process between the participants in communication.

Computational linguistics is an applied science which is slowly gathering momentum in the day-to-day lives of the individuals. It is slowly creeping into the spaces left vacant in the human acceptance of latest technology which is developing every single second in leaps and bounds. The search engines, online translators, automated response generation software are all a part of the fast evolving sphere of computational linguistics.

Linguistic distance is not only present between languages, but also within the individual languages. This concept is known as inter language linguistic distance and intra language linguistic distance. The inter language linguistic distance itself poses a great difficulty in enhancing the relations among the people. The intra language linguistic distance further complicate the scenario.

The presence of dialects for every language further catalyze the effect of linguistic distance. Dialectical distance within the language deters the individuals from understanding the same language. For a machine to understand the dialects within the languages and to respond accordingly is nearly impossible but for the speech recognition systems. The speech recognition systems are so trained to differentiate between every single voice based on the pitch, tone and intonation. It is the sounds made by the individual that more matters to the system than the language used.

In one distinctive method, the languages are all categorized based on their sounds and fed to the computers. The system classifies the languages based on the sounds produced in sequence and responds accordingly. However, the method could not yield necessary results as the sounds do not refer to the meanings of the words and thus clever responses are not possible.

The speech-to-text conversions are a considerable success yet the interaction between the man and machine still has a long way to go. The use of multiple languages in the computers has now distanced the languages of the virtual world too. For the linguistic distance to reduce and common relations to form between nations and people, the presence of a common language is a must. The globally accepted and largely spoke English language can fulfill this criteria. A study of the native languages and their introduction in the computational spheres is a welcome change. However, the use of a common language to enhance the bilateral ties between the nations and to connect the entire globe with a single language is also a necessity of the hour.

The disparities arising between the people because of the presence of multiple languages has to an extent come to a halt but is once again increasing its scope. The use of the native language in the computer generated programs is once again linking the linguistic groups together thus separating them from the global community. The love for one's own language is separating the individuals on the virtual platforms and even the global cloud is being filled with vernacular languages of the natives.

India alone has innumerable languages and dialects which are slowly trying to find a place in the global platform. The multimedia dimensions which extend to all languages are being hurdled by the computational linguistic distance. This distance to reduce and the global reach to once again embrace the both poles of the earth, the use of English language as the primary language for machines is advisable.

Computers increase interactivity due to multitudes of codes of transmission of the intended message. This appeals to all our senses equivocally and simultaneously. This makes us emotionally connect with the language which we wish to learn. Through CALL, Emotional Language Teaching takes place as we connect more though two-way communication and thereby acquire the English Language as our mother tongue.

References

1. Chsiwick, B. R., Miller, P. W.: Linguistic Distance: A Quantitative Measure of the Distance Between English and Other Languages. J. Social Science Research Network. 1246, 1–20 (2004).
2. Nerbonne, J., Hinrichs, E.: Linguistic Distances. In: Proceedings of the Workshop on Linguistic Distances, pp. 1–6. Association for Computational Linguistics, Sydney (2006).
3. Jurafsky, D.: Pragmatics and Computational Linguistics. In: Laurence R. Horn and Gregory Ward (eds.) Handbook of Pragmatics, pp. 1–36. Oxford, Blackwell.
4. Renals, S., Hain, T.: Speech Recognition. The handbook of computational linguistics and Natural Language Processing, United Kingdom (2010).

Collaborative Attack Effect Against Table-Driven Routing Protocols for WANETs: A Performance Analysis

E. Suresh Babu, Satuluri Naganjaneyulu, P.S.V. Srinivasa Rao and M. Srinivas Rao

Abstract Recently, wireless networks are rapidly gaining importance and generating more interest in wireless communication, due to remarkable improvements in wireless devices such as low-cost laptops, palmtops, PDA, etc. Revolutionary development of these widespread wireless networks has resulted in a number of new applications such as vehicle-to-vehicle communication in remote areas. This paper attempts to comprehend the performance issue by evaluating of two proposed two table-driven (DSDV, OLSR) routing protocols with varying traffic and mobility condition in a realistic manner. Moreover. We modelled a collaborative attack against these two routing protocols that can obstruct the normal operations of the ad hoc networks. Precisely, our experiential result shows the comparative analysis of OLSR, DSDV under hostile environments with varying mobility and traffic scenario.

Keywords OLSR · DSDV · Collaborative attack · Adhoc networks

E. Suresh Babu (✉)
Department of Computer Science and Engineering, K L University, Guntur, AP, India
e-mail: sureshbabu.erukala@gmail.com

S. Naganjaneyulu
Department of Information Technology, LBRCE, Mylavaram, AP, India
e-mail: svna2198@gmail.com

P.S.V.S. Rao
Department of Computer Science and Engineering, SVECW, Bhimavaram, AP, India
e-mail: parimirao@yahoo.com

M.S. Rao
Department of Computer Science and Engineering, SVIET, Machilipatnam, AP, India
e-mail: srinu.mekala@gmail.com

S.C. Satapathy et al. (eds.), *Computer Communication, Networking and Internet Security*, Lecture Notes in Networks and Systems 5,
DOI 10.1007/978-981-10-3226-4_7

1 Introduction

Recently, wireless networks are rapidly gaining importance and generating more interest in wireless communication, due to remarkable improvements in wireless devices such as low-cost laptops, palmtops, PDA, etc. Revolutionary development of these widespread wireless networks has resulted in a number of new applications such as vehicle-to-vehicle communication in remote areas. Existing cellular infrastructure networks may become infeasible to support these new applications. Hence, there is a growing need for the connectivity in situations where, there is no fixed infrastructure; for instance, there is no base station available or inconvenient to employ these typical applications. This resulted in the emergence of a new kind of large network called adhoc network, in which wireless user can easily communicate with each other on the fly without any fixed infrastructure. These networks are not dependent on any pre-existing fixed infrastructure for relaying the packets to the other nodes. However, these networks may provide unreliable wireless communications due to its dynamic nature. Mobile Adhoc Networks (MANETs) [1, 2] is a self-configuring multi-hop network, usually composed of a collection of highly autonomous mobile nodes that can dynamically set up anytime and anywhere to form a temporary network without the need of any centralized infrastructure. Typically, the mobile nodes in this network make use of wireless connectivity for exchanging the information. Hence, every node has often acted as wireless mobile host and wireless mobile router for forwarding the packets to the destination. Specifically, a mobile router allows relay the traffic within the network and a mobile host may access the information within the network domain. Moreover, these nodes have a flexibility to move freely and arbitrarily that may experience rapid and unpredictable topology changes. Nevertheless, these networks contain self-organizing and self-configuring capabilities for establishing the connections on the fly. Therefore, this communication technology truly supports pervasive computing environments. Some of the characteristics of MANET provide great opportunities for wide commercial deployment of the MANET. However, this technology poses several challenges such as Routing, Power Consumption, Internetworking, Quality of Service (QoS), Security, Reliability, Scalability and Energy Efficiency etc. Due to the inherent characteristics of MANETs and wide usage of applications, there is a need to design the efficient protocols for ad hoc networks. This paper attempts to comprehend the performance issue by evaluating of two proposed two table-driven (DSDV, OLSR) routing protocols with varying traffic and mobility condition in a realistic manner Moreover. We modelled a collaborative attack [3–6] against these two routing protocols that can obstruct the normal operations of the ad hoc network. We modelled a collaborative attack against these two routing protocols that can obstruct the normal operations of the ad hoc networks. Precisely, our experiential result shows the comparative analysis of OLSR, DSDV under hostile environments with varying mobility and traffic scenario. Rest of the paper is organized as follows, related work is discussed in Sects. 2 and 3 specifies the overview of routing protocols. In Sect. 4 presents Collaborative Attack against Adhoc Routing Protocols. In Sects. 5 and 6, we discussed the simulation work and conclusion.

2 Related Work

This section presents an overview of related work and background details required for this paper. Aleksandr Huhtonen [7] compared two well-known AODV and OLSR routing algorithms. In his comparison, The AODV protocol performs better than OLSR, because it takes fewer resources such as less bandwidth, reducing the computational power. Hence, it can be used in resource critical environments. While OLSR protocol is more efficient in networks with highly sporadic and traffic high density. However, OLSR continuously requires some bandwidth for topology updates. The limitation of this paper is security of the protocols is undone. In [8] Nadia Qasim et al. conducted the performance comparisons of MANETs with respective to QoS parameters. In their comparison is mainly performed between AODV, OLSR and TORA by considering various performance metrics of end-to-end delay, packet delivery ratio, throughput and media access delay. The author concluded that the OLSR outperforms than TORA in terms of routing packet delivery ratio with high node mobility. However, TORA perform well in small network size. While AODV performs better than OLSR and TORA. In [9] Mohamed Amnal et al. conducted a behavior study of OLSR protocol using two traffic types VBR and CBR with various mobility models as Mobgen Steady State, Random Direction and Random Way Point. Their experimental results illustrate the behavior of OLSR changes with used traffics and the model.

3 Overview of Routing Protocols

Most of the MANET applications are based upon unicast communication, which is used to transmit the data packets from one source to one destination. Generally, most of the unicast routing protocols are Topology-Based Routing Protocols. This routing protocol makes use of metrics of the network links for finding the route from a source to a destination. Moreover, these protocols forward the packets based on the address of the destination node. The topology-based routing protocols are divided into two categories. One is proactive routing protocols and other is reactive routing protocols. This paper discuss only Proactive Routing Protocols [10–12], while reactive proactive routing protocols [13, 14] are out of scope of this paper.

3.1 Proactive Routing Protocols

The Proactive Routing Protocols are also called table-driven routing protocols that work like a classical Internet routing protocol. In these protocols, each node's share the routing information maintains fresh lists of destinations and a constantly updated route by periodically distributing routing tables throughout the network in order to maintain a consistent network state. However, these protocols suffer from a constant overhead of routing traffic because of maintenance of large amounts of data and very slow reaction in handling the failures and the process of restructure. Nevertheless, these routing protocols have no initial delay in communication. Few of the existing proactive routing protocols, which are directly proposed by the researchers.

3.1.1 Destination-Sequenced Distance-Vector (DSDV) Routing Protocol

DSDV [15] is one of the antecedent proactive routing protocols for mobile ad hoc networks that extend the basic Bellman-Ford mechanism. It makes use of distance vector algorithm by attaching the sequence number that is originated and updated by the destination to each distance node. Subsequently, the update messages (sequence number) will be broadcasted throughout the network and advertise to every next hop to find a change in the routing information in its routing table, when a significant new information is available about destinations. However, these nodes only update its routing table, if any one of the two cases is satisfied. (a) The newer sequence number is larger than the old one that is already recorded. or (b) if the newer and recorded one have a same sequence number, then it is preferred with a better cost metric (shortest path). In other words, each node continuously maintains up-to-date routes (information) using its routing table to every next-hop in the network. Routing information such as a sequence number, number of hops, the destination's address is periodically transmitted throughout the network in order to maintain a consistent routing table. Eventually, all the above procedure is performed with two exchanges, one is event-driven incremental updating, and another is periodic full-table broadcast. Moreover, DSDV reduces control message overhead, increases convergence speed and eliminates route looping by associating each route entry with a sequence number to determine the "freshness" of a route. In case of route failure, the node immediately broadcasts this information to its intermediate nodes and instantaneously updates the sequence number of a route when it detects a link break on that route. When an intermediate node receives routing information with newer or updated sequence number, then it checks in its routing table. If it found an entry in its routing table, then it compares the sequence number with the received one and updates them accordingly, by following the same procedure, which is explained above. Otherwise updates the routing table, if the entry is not found.

Advantages of DSDV

- DSDV is suitable, when a small number of nodes are available for creating the ad hoc networks.
- DSDV is free from the routing loop problem.
- DSDV overcomes the Count to infinity problem.
- DSDV only maintains the optimal path instead of multiple paths to every destination.

Disadvantages of DSDV

- DSDV consumes more battery power for regular updating the routing tables.
- DSDV requires the smallest amount of bandwidth even though when the network is in standby mode.
- The new sequence number is necessary, whenever the topology of the network changes occurs,
- DSDV is only suitable for smaller networks. But, it is not suitable for highly dynamic networks. Therefore, it is abandoned.

3.1.2 Optimized Link State Routing (OLSR) Routing Protocol

OLSR [7] is also one of the predominant proactive routing protocols tailored for mobile ad hoc networks. It is an optimization over the classical link state protocols that have several advantages such as the immediate availability of routes, reduces the size of the control packet and less number of transmission and retransmission of control packet, etc. The key concept of this protocol makes use of multipoint relays (MPRs) that reduces the control traffic overhead significantly, by just flooding the control traffic only to the selected nodes. Particularly, MPR is a one-hop neighbor node, which can be used to retransmit its packets, instead of the flood a message to all nodes in the network. The selection of Multipoint Relay is chosen in such a way that all the two-hop neighbors can be reached through the MPR node. In general, OLSR uses two types of control packets-one is Topology Control (TC) packet and other is Hello packet. The main purpose of using TC packet is to broadcast the information to own advertised intermediate nodes, which includes at least the MPR Selector list. The MPR host only can forward these TC packets periodically, and the Hello packets are used to find the host's neighbors and link status information. These hello packets are sent to the one hop (one neighbor) only. Besides this, the MPR Selector set is built with the help of hello packets that describes which intermediate node has chosen to act as MPR host. Moreover, OLSR only makes use of partial link state to provide shortest path routes to the destination with the help of MPR. In order to provide the shortest path, OLSR considerably reduces the time interval for the control message transmission can bring more reactivity to topological changes. Additionally, the entire MPRs node in the network must announce the link information between itself and its MPR selectors (i.e. the nodes actually selected the MPR node). Eventually, OLSR complete works in a distributed fashion and maintains routes to all destinations in the network, which is particularly suitable for large and dense networks.

Advantages

- OLSR reduces the control information and thereby reduces the number of retransmissions of topological information
- OLSR efficiently minimizes bandwidth usage and broadcast traffic
- In OLSR, every route involves an MPR node for forwarding the information (may not be the shortest path) between source and destination.

Disadvantages

- OLSR suffers from more routing delays
- OLSR consumes more bandwidth overhead at the MPR nodes, because an MPR node acts as localized forwarding routers.

3.2 *Performance Assessment of Proactive Routing Protocols*

This section compares the performance of two proactive routing protocol for mobile ad hoc networks, because DSDV and OLSR are also two challenging table driven routing protocol. Although DSDV and OLSR routing protocols share the mechanism of maintaining current up-to-date routing information between the nodes in their routing tables. However, differs in the protocol working procedure that had led to significant performance differentials such as network load, network size, routing load and mobility. Moreover, this assessment gives the clear depiction of two typical proactive routing protocols for MANETs.

3.2.1 Assessment of DSDV and OLSR Routing Protocol

- DSDV maintains best shortest path to every destination. Hence, the amount of space in routing table for selecting the path is reduced. While OLSR will not necessary, maintain the shortest path, because every path selection will be performed or forwarded through a MPR node.
- DSDV avoids extra traffic with incremental updates. While OLSR produces, more bandwidth overhead and routing delays, because MPRs nodes are acts localized forwarding routers.
- Both DSDV and OLSR protocols provides routing information to all participated nodes in the network. Hence, both the routing protocols send the unnecessary updated topology information throughout the entire network, even

Table 1 Characteristic and comparison of DSDV and OLSR routing protocols

Mechanisms	OLSR	DSDV
Neighbors detection method	Periodic hello messages	Periodic as required
Routing information storage	Three routing table: routing, neighbor, topology control	Two routing table: incremental updating and periodic full-table broadcast
Frequency of update	Periodic	Periodic
Optimized broadcast	Multi-Point Relaying (MPR)	Periodic full-table broadcast
Route freshness	Up-to-Date	Up-to-Date
Storage complexity (Memory overhead)	O(n2)	O(n)
Time complexity (Convergence time)	O(D-1)	O(D-1)
Control overhead	O(n2)	O(n)

if there is no change in the network topology. This creates wastage of bandwidth or more usage of bandwidth However DSDV works better for smaller networks, while OLSR is more suitable for large and dense networks.

- OLSR protocol has low latency compared to DSDV, because OLSR make use of TC packet to avoid the stale route problem, which enables bandwidth efficiency. While DSDV suffers long delays whenever the topology changes occur and a new sequence number is necessary before the network reconverges. Hence OLSR is suitable for application that has short delays of transmission of the data packets.
- DSDV generates high routing load, because, each node should maintain two tables in which updates are transmitted to neighbors periodically that add large amount of routing overhead into the network. While OLSR reduces routing load, because, it makes use of multipoint relays (MPRs) that reduces the control traffic overhead significantly, by just flooding the control traffic only to the selected nodes.
- OLSR considerably reduces the time interval for the control messages transmission can bring more reactivity to topological changes. While DSDV determine a more time delay to react to the topological changes.
- In brief, the following Table 1 depicts the characteristic and comparison of DSDV and OLSR routing protocol

4 Modelling Collaborative Attack Against Adhoc Routing Protocols

This section gives about collaborative attack in which the attackers are tends to become more and more advanced with the combinations of blackhole attack, and wormhole attack. In this paper, we specifically modelled the collaborative attack in

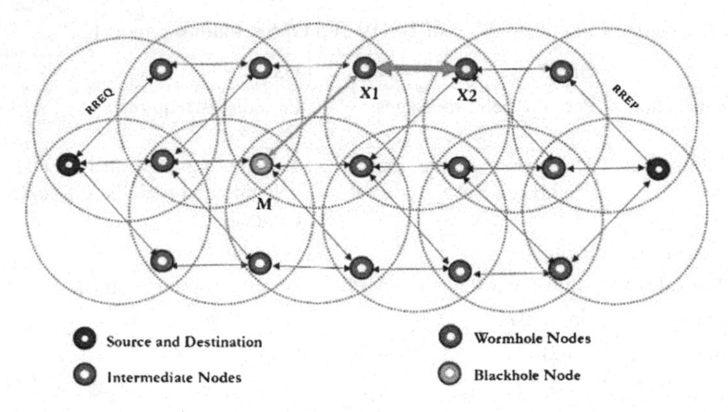

Fig. 1 Illustrates the collaborative attack

which cooperate blackhole nodes are coordinated with wormhole node. In other words the how the different attack such as black hole and wormhole are cooperating each other to form heterogeneous attack. As shown from the following Fig. 1. Suppose the sender the route request to the receiver via some intermediate relying nodes. One of the Neighbor node 'M' along the path performs a black hole attack. Correspondingly, the other two intermediate nodes 'X_1' and 'X_2' are worm hole nodes which cooperate each other. As depicted in Fig. 1, Let us assume the neighbor nodes both 'M' and 'X_1' are cooperated each other during the path discovery phase. The cleaver neighbor node 'M' will send the route reply packet back to the sender informing that has the optimal route to receiver nodes 'D'. Meanwhile, the neighbor node 'M' will establishes the path to other intermediate node 'X_1' which in sequence forwards the route request packet through MPR nodes that cooperate neighbor node X_2 through the tunnel which transmits the packet into the networks. After the path establishment, the compromised node 'M' will simply forward the data packets to its neighbor node X_1 even though it is black hole nodes. Further, the two malevolent nodes 'X_1' or 'X_2', which are wormhole nodes that selectively drops the data packets. We incorporated the collaborative attack in both DSDV and OLSR to evaluate its performance with various metrics such as routing overhead, packet delivery ratio and end-to-end delay.

5 Simulation Results and Performance Analysis

This section presents the simulation results of DSDV and OLSR routing protocols to know the practicability of our hypothetical work, The DSDV and OLSR routing protocol software is taken from authorized website. For this existing routing protocol, we added the collaborative attack [16, 17] module in both the DSDV and OLSR software in network simulator [NS2] which is a popular simulator running in

Ubuntu-14.04. We conducted a sequences of experiment to evaluate its performance.

Figure 2 depict the Average End-to-End Delay of DSDV and OLSR routing protocol with and without collaborative attack for 50 nodes with changeable pause time. OLSR has comparatively lower delay than DSDV. Moreover, OLSR and DSDV have almost identical delays. Moreover, the same figure also shows that both OLSR and DSDV considerably increases the latency with 10% black hole nodes (i.e. one blackhole node) and 10% wormhole nodes (i.e. one wormhole node), which are collaborative to each other to form collaborative attack

Figure 3 shows the packet delivery fraction for OLSR and DSDV for 50 nodes, the PDFs for OLSR and DSDV are almost similar with 3, 5, and 7 sources.

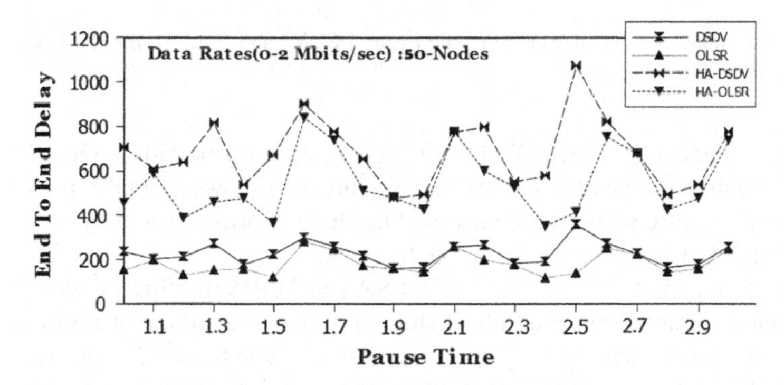

Fig. 2 Latency versus pause time between DSDV and OLSR for 50-Nodes under collaborative attack

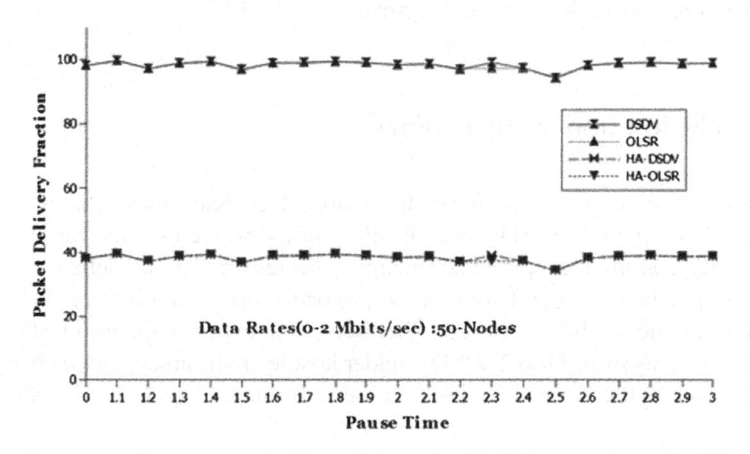

Fig. 3 PDF versus pause time between DSDV and OLSR for 50-Nodes under collaborative Attack

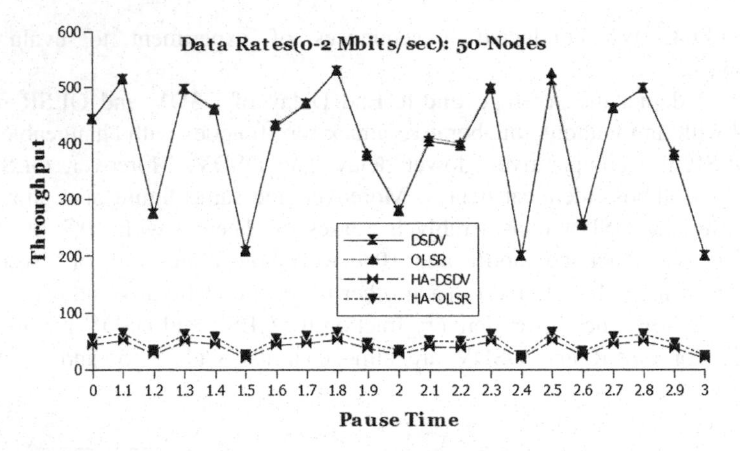

Fig. 4 Throughput versus pause time between DSDV and OLSR for 50-Nodes under collaborative attack

However, OLSR outperforms DSDV at higher pause times with 9 and 12 sources before the collaborative attacks. The same figure also shows that 60% packet drops in the presence of collaborative attacks (blackhole, wormhole nodes) for all pause times, which is more overwhelming DoS attack.

Figure 4 depicts the throughput of OLSR and DSDV routing protocol without the presence of heterogeneous attack for 50 nodes with different traffic sources. OLSR outperforms than DSDV for all pause times. The figure also shows both the OLSR and DSDV considerably that affects the overall throughput from 60 to 75% of the data packets However, OLSR provides the relatively better throughput than DSDV. The impact on DSDV observed was a decrease of approximately ten percent (10%) approximately under this attacks. Hence, In DSDV, the attacker is more operative in upsetting the routing information than OLSR.

6 Conclusion and Future Work

This paper attempts to comprehend the performance issue by evaluating of two proposed two table-driven (DSDV, OLSR) routing protocols with varying traffic and mobility condition in a realistic manner. Moreover. We modelled a collaborative attack against these two routing protocols that can obstruct the normal operations of the ad hoc networks. Precisely, our experiential result shows the comparative analysis of OLSR, DSDV under hostile environments with changeable mobility and traffic scenario. These routing protocols lack in cryptographic mechanism due to its cooperative and characteristics nature. Thus, deployment of security in such MANETs is one of the challenging issues that need specialized security solutions.

References

1. Babu, E. Suresh, C. Nagaraju, and MHM Krishna Prasad. "An Implementation and Performance Evaluation Study of AODV, MAODV, RAODV in Mobile Ad hoc Networks." vol 4 (2013): 691–695.
2. Abolhasan, T. Wysocki, and E. Dutkiewicz, "A review of routing protocols for mobile ad hoc networks," vol. 2, pp. 1–22, 2004.
3. Babu, E. Suresh, C. Nagaraju, and MHM Krishna Prasad. "Inspired Pseudo Biotic DNA based Cryptographic Mechanism against Adaptive Cryptographic Attacks." International Journal of Network Security 18, no. 2 (2016): 291–303.
4. Babu, E. Suresh, C. Nagaraju, and MHM Krishna Prasad. "Light-Weighted DNA-Based Cryptographic Mechanism Against Chosen Cipher Text Attacks." Advanced Computing and Systems for Security. Springer India, 2016. 123–144.
5. Babu, E. Suresh, C. Nagaraju, and MHM Krishna Prasad. "Analysis of Secure Routing Protocol for Wireless Adhoc Networks Using Efficient DNA Based Cryptographic Mechanism." Procedia Computer Science 70 (2015): 341–347.
6. Babu, E. Suresh, C. Nagaraju, and M. H. M. Prasad. "A Secure Routing Protocol against Heterogeneous Attacks in Wireless Adhoc Networks." In Proceedings of the Sixth International Conference on Computer and Communication Technology 2015, pp. 339–344. ACM, 2015.
7. Huhtonen, Aleksandr. "Comparing AODV and OLSR routing protocols." Seminar on internetworking, Sjkulla. 2004. Kullberg, "Performance of the Ad-hoc On-Demand Distance Vector Routing Protocol". 2004.
8. Qasim, F. Said, and H. Aghvami, "Mobile Ad Hoc Networks Simulations Using Routing Protocols for Performance Comparisons", 2008.
9. Amnai, Mohamed, Youssef Fakhri, and Jaafar Abouchabaka. "QOS routing and performance evaluation for Mobile Ad Hoc Networks using OLSR protocol." arXiv preprint arXiv:1107. 3656 (2011).
10. Babu, E. Suresh, and M. H. M. K. Prasad. "An Implementation Analysis and Evaluation Study of DSR with Inactive DoS Attack in Mobile Ad hoc Networks." vol 2 (2013): 501–507.
11. Kumar, S. Ashok, E. Suresh Babu, C. Nagaraju, and A. Peda Gopi. "An Empirical Critique of On-Demand Routing Protocols against Rushing Attack in MANET." International Journal of Electrical and Computer Engineering 5, no. 5 (2015).
12. Gopi, A. Peda, E. Suresh Babu, C. Naga Raju, and S. Ashok Kumar. "Designing an Adversarial Model Against Reactive and Proactive Routing Protocols in MANETS: A Comparative Performance Study." International Journal of Electrical and Computer Engineering 5, no. 5 (2015).
13. Babu, E. Suresh, C. Nagaraju, and MHM Krishna Prasad. "An Implementation and Performance Evaluation of Passive DoS Attack on AODV Routing Protocol in Mobile Ad hoc Networks." International Journal of Emerging Trends & Technology in Computer Science 2, no. 4 (2013): 124–129.
14. Swarna, Mahesh, Syed Umar, and E. Suresh Babu. "A Proposal for Packet Drop Attacks in MANETS." In Microelectronics, Electromagnetics and Telecommunications, pp. 377–386. Springer India, 2016.

15. Perkins and P. Bhagwat, “Highly dynamic destination-sequenced distance-vector routing (DSDV) for mobile computers”, Proc. ACM SIGCOMM Conf. Comms. Architectures, pp 234–244, 1994.
16. Babu, E. Suresh, C. Nagaraju, and MHM Krishna Prasad. “Light-Weighted DNA Based Hybrid Cryptographic Mechanism Against Chosen Cipher Text Attacks.” International Journal of Information Processing and Indexed With arXiv, Indian Citation Index-2015, ISSN-0973–821.
17. Babu, E. Suresh, C. Nagaraju, and MHM Krishna Prasad. “Efficient DNA-Based Cryptographic Mechanism to Defend and Detect Blackhole Attack in MANETs.” In Proceedings of International Conference on ICT for Sustainable Development, pp. 695–706. Springer Singapore, 2016.

Body Biased High Speed Full Adder to LNCS/LNAI/LNBI Proceedings

D. Khalandar Basha, B. Naresh, S. Rambabu and D. Nagaraju

Abstract In various domains like VLSI design, Embedded Systems, Signal processing, Image processing etc. combinational and logical circuit are the basic building blocks whereas addition is the fundamental step involved in any of it. Increasing the efficiency of these fundamental blocks is one of the major concerns in the application development. Moreover, power and delay are the major concern in the VLSI design so as to increase the efficiency of the circuit. In the combinational system performance and speed of the circuit is directly related with delay. In this paper hybrid adder circuit is designed using both Complementary Metal Oxide Semiconductor (CMOS) logic and transmission gates which performs addition at a low power and reduced delay. Further the speed of operation of the circuit is improved by introducing Gate Level Body Biasing (GLBB) in the design. The design was first implemented for full adder and then extended for 8 bit ripple carry adder. The circuit was implemented using Cadence Virtuoso tools in 180 nm technology. For 1.8 V supply at 180 nm technology, the average delay of the circuit is (114.5 ps), having moderate power consumption (27.52 mW) is found to have extremely low values than that resulted from the use of very weak CMOS inverters coupled with strong transmission gates. With additional GLBB circuit incorporated with the design proved useful in boosting the circuit. In comparison with the existing full adder the proposed Full adder found to offer significant improvement in terms of speed at the cost of power.

Keywords Delay · Hybrid full adder · High speed adder · Transmission gates · Body biasing

D. Khalandar Basha (✉) · B. Naresh · S. Rambabu
Institute of Aeronautical Engineering, Hyderabad, India
e-mail: bashavlsi@gmail.com

D. Nagaraju
Matrusri Engineering College, Hyderabad, India

S.C. Satapathy et al. (eds.), *Computer Communication, Networking and Internet Security*, Lecture Notes in Networks and Systems 5,
DOI 10.1007/978-981-10-3226-4_8

1 Introduction

In the advanced world, integration of large number of transistors on a single chip is achieved by VLSI designing. Increase in the usage of these portable devices like mobile phones, notebooks, laptops the demand for ultra large scale integration designs with high speed and low power characteristics has increasing enormously. Full adder design is one of the significant steps in designing the logic circuits which are the preliminary blocks of almost all the digital applications. There by, through increasing the performance of the full adder circuit, the overall performance of the circuit could be increases tremendously. Various design styles have been evolving for implementation the functionality of full adder, though they have their own pros and cons. To meet the present and future needs, the designs should be effective in terms of power and speed. From the basic Conventional CMOS (CCMOS) Full adder to the present day Hybrid full adder circuits as the best design in all the individual blocks of the full adder circuits. Further it was found that the performance of the circuit could be improved by GLBB technique.

With increase in demand for Ultra-low energy digital complementary CMOS is exponentially raising [1]. Now, the speed limit could be improved by introduction of biasing techniques. As this operates at sub-threshold region of the design, it speeds up the switching operation with preserving the power. The main source for power consumption in bulky designs is the supply voltage (VDD). Decreasing this voltage dynamically drops the power consumption but drastically raises the delay of the circuit. From [2, 3]. Before the introduction of the technique we have Forward Body Biasing (FBB) technique [4–7] where the speed factor cannot be overlooked. At the time of instigating various new techniques to limit the delay and power dissipation by decreasing the transistors count, various aspects have been enlightened that the RC delay in charging the body of the devices and the input switching without changing the logic gate status have no effect on the speed of the logic and that the body capacitance are no more uselessly charged/discharged [8, 9]. These inventions of the thesis have helped in introducing GLBB [19] technique to logical circuits. The expected output is supposed to me more speed efficient as operating at sub-threshold and near-threshold regions.

2 Literature Survey

There exist various designs in present era to implement Full adder circuit. Starting from the Conventional CMOS (CCMOS) [10] initiated the research work by proposing a full adder built by 28 transistors to the present design which uses hybrid circuits to enhance the performance of the circuit. There are different logic styles in favour of good performance aspect at expense of other parameters among them some of the styles are discussed and which are like standard static CMOS [11], Complementary Pass transistor Logic (CPL) [12, 13], Dynamic CMOS logic [14].

Due to the use of large number of transistors, these circuits incorporate high delay which in turn affects the overall performance of the system. It is obvious that using of many transistors result in large power consumption for operation that in turn lead to high power dissipation in the form of heat and these parameters have a huge effect on the lifetime of the system.

2.1 CCMOS 28 Transistor Full Adder

The conventional CMOS circuit utilized 28 transistors to design full adder circuit [10]. Since the proposed full adder circuit comprises of large number of transistors count, the power and delay characteristics are high.

2.2 GDI Full Adder

Gate Diffusion Input (GDI) is the popular technique to reduce the transistor count which in turn significantly decreases the power and delay characteristics of the circuit. A GDI cell has three inputs and one output pin. These three inputs are fed to sources of pFET and nFET transistors and the common gate of it. Similarly the output is taken at the common drain point. PMOS and NMOS substrates are connected to their respective source terminals shown in [15, 16].

2.3 MG-GDI Full Adder

The design of MG-GDI is similar to the GDI cell but the only difference is the voltage at substrate terminal. This technique in MG-GDI improves the performance of the circuit and it significantly reduces the power consumption and delay of the circuit [17, 18].

2.4 Low Power High Speed Hybrid Full Adder

The existing techniques does not meet the requirement of the future needs so still research are going on to improve full adder functionality. One such attempt is evolution of hybrid design of the full adder [19] in which weak inverter's are used in order to reduce the power consumptions.

2.5 High Speed GLBB Technique

Biasing voltage of a transistor plays an important role when the topic is focused on the power dissipation and delay minimization. As V_{DD} dropping down can resolve the above mentioned problem that is depended on V_{TH}, its controlling is of great importance that is achieved by using GLBB technique. In this, from the different blocks output, different biasing voltages (V_B) have been generated and are applied to those respective block transistors. Thus, the threshold voltage is indirectly in control of the supplied voltage V_{DD} as the output is directly depended on it. Every GLBB block contains two transistors namely PMOS and NMOS [18]. Thus further using this block increases the transistor count and thereby raising the area and power requirement. However, these are of cordial importance when the speed is the key factor at the cost of area size.

3 Proposed System

3.1 Modified XNOR Module

The XNOR module [19] compromises of 6 transistors with varied L/W ratio. In the proposed XNOR module the average delay of the XNOR circuit which is reported in [19] is further improved a lot by incorporating Body Biasing circuit with additional inclusion of two more transistors. We know that inverter circuit consumes more power when compared with the other modules in the circuit. In order to design XNOR module inverter circuit is mandatory which in turn leads to the increment of the power consumption. In all the existing full adder circuits, XNOR module is responsible for the high power consumption of the entire adder circuit. The basic inverter CMOS circuit which is used in design of the adder circuits consumes high power in order to eliminate this problem we are using hybrid circuits in our design. Figure 1 shows the modified XNOR module in which the power consumption is significantly reduced by using weak invertors (channel width of the transistors are small). The weak inverter is formed by using transistor n1 and p1 which is shown in Fig. 1. The default length and width values of a transistor in 180 nm technology are 180 nm, 2 um respectively are modified as transistor p1 (L = 180 nm, W = 800 nm) and transistor n1 (L = 180 nm, W = 400 nm). Voltage degradation problem is observed in the output due to the use of weak inverter and this problem is eliminated by using level restoring transistors p3, n3 which guarantee the full voltage swing in output. Further to increase the performance of the XNOR module we propose body biasing technique which is incorporated at the output stage of the XNOR module.

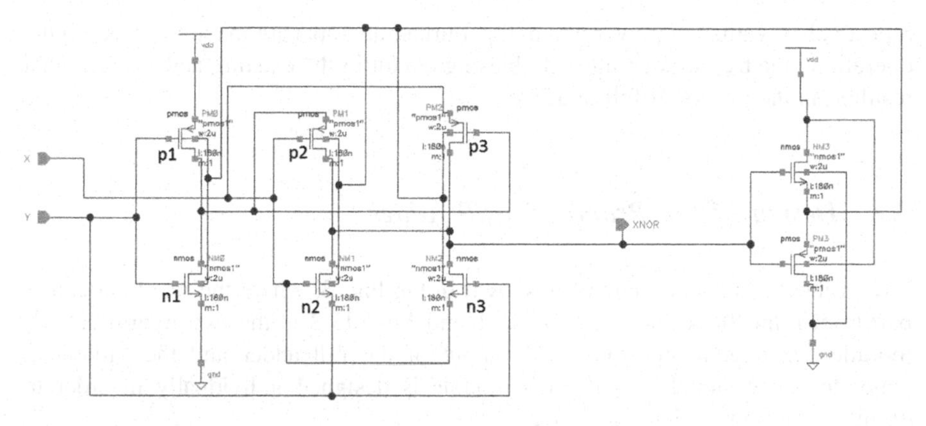

Fig. 1 Modified XNOR module

3.2 *Carry Generation Module*

The speed of operation of full adder is mainly depends on the carry signal which is minimized in the proposed carry module by reducing the path length of the carry input. Figure 2 shows the carry generation module. In the proposed circuit, the carry signal is generated using p7, p8, n7 and n8 transistors. The input carry signal (C_{in}) is transmitted through single transmission gate which reduces the delay in the carry module thereby reducing the overall delay of the full adder circuit. The output of the carry module is feed to the body biasing circuit which in turn boosts the performance of the carry module. In [19] hybrid full adder has been implemented with varied transistor sizes for efficient performance of the circuit. This design

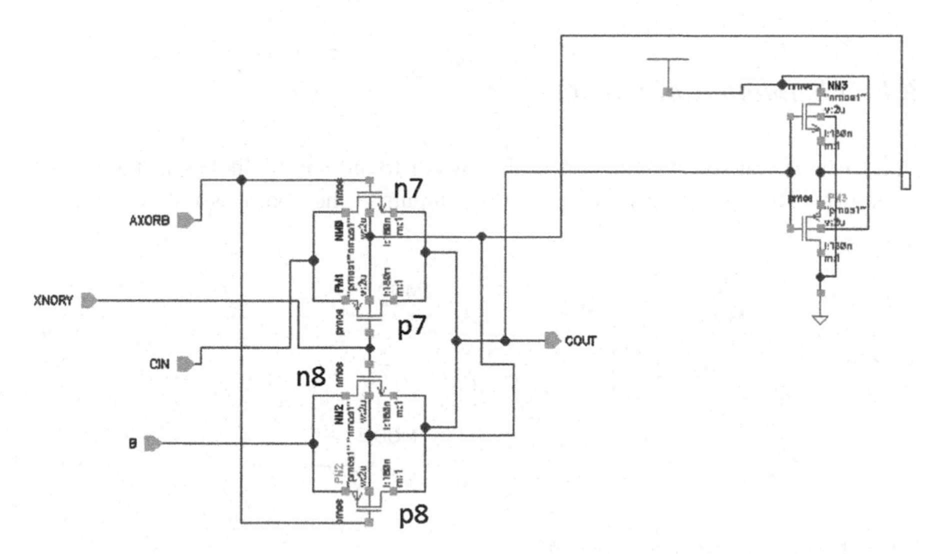

Fig. 2 Carry generation module

reported best values of power and delay. Further attempts are made to speed up the operation time by introducing body biasing circuit in the existing adder circuit that resulted in the proposed full adder.

3.3 Design of the Proposed Full Adder

The proposed full adder circuit is shown in the Fig. 3, where the whole circuit is partitioned into three blocks. Module 1 and Module 2 is the two hybrid XNOR module which generates the SUM output of the full adder and the Module 3 generates carry signal (C_{out}). Each module is designed individually in order to optimize the power, delay and area.

The proposed full adder uses more than one design technique therefore it is named as Hybrid full adder. It is designed using weak transistors to reduce the power consumption and strong transmission gates in order to reduce the carry path through which the delay of the circuit is reduced. Module 1 carries out XNOR operation between two inputs and its output is fed to the second XNOR module (Module 2) to generate the end SUM operation between three inputs. The first XNOR module is designed using transistors p1, n1, p2, n2, p3, n3. The second XNOR module (module 2) is designed using transistors n6, p6, n5, p5, n4, p4 among which n6, p6 represents inverter circuit and p5, n5 are level restoring transistors which is shown in Fig. 4. By analyzing the truth table of full adder we found some conditions for generation of carry signal (C_{out}). If both the inputs A, B are equal then C_{out} follows the B input which is implemented using the transmission gate p8 and n8, else the carry input (C_{in}) will be passed on to the output which is implemented using another transmission gate p7, n7.

3.4 Elimination of Glitches

Due to some transistor switching problems we found some glitches in the output waveform. The switching between the two inputs at the output is not accurate. In

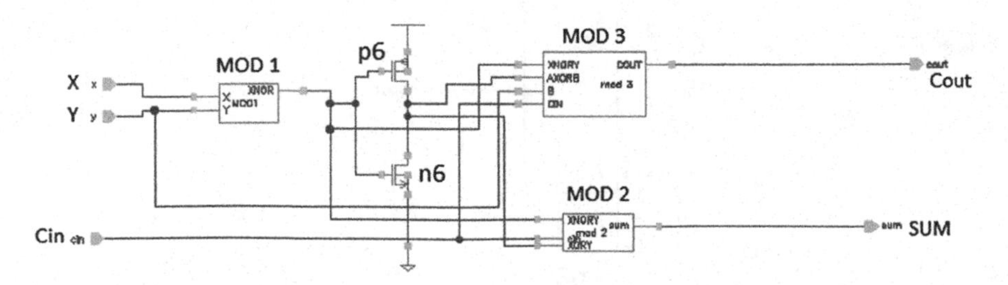

Fig. 3 Proposed full adder without buffers

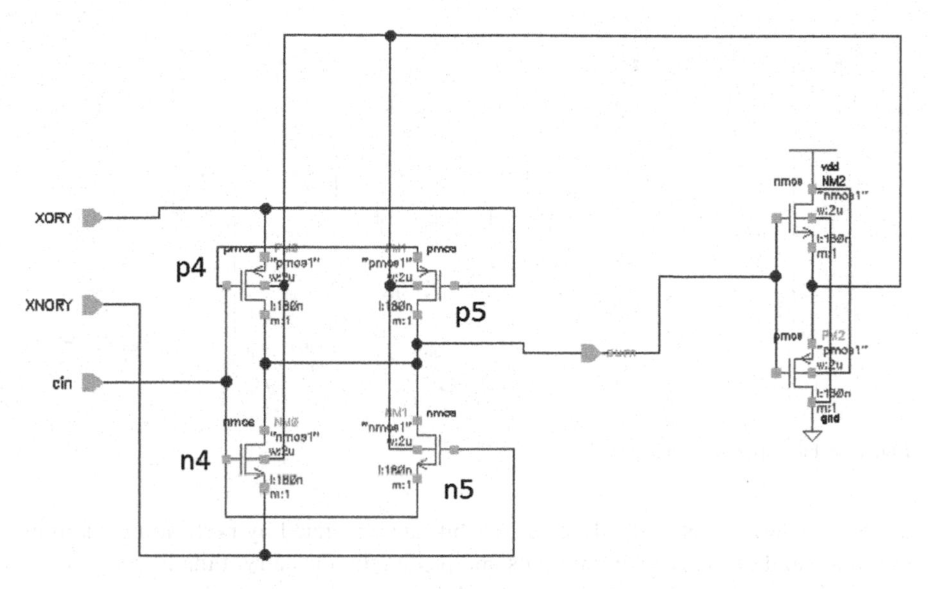

Fig. 4 Module 2 circuit

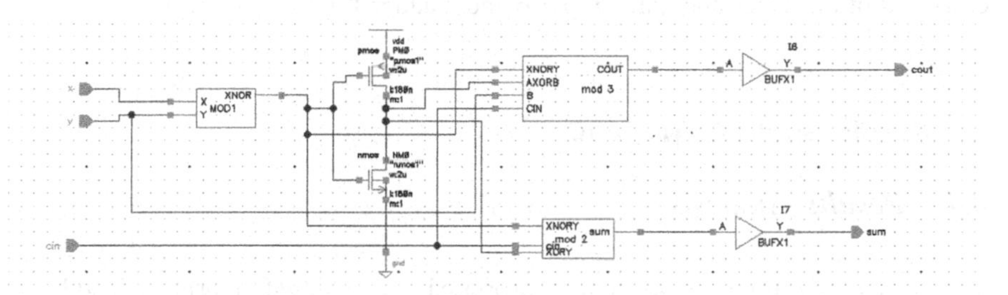

Fig. 5 Proposed full adder with buffers

order to overcome this problem, our proposed design as shown in Fig. 5 compromises of buffers at the output which reduces the glitches in output at the cost of increasing in average power. Driving problems also significantly reduced due to the presence of buffers at output stage.

3.5 Body Biased LPHS Full Adder with 8 Bit Input

After observing better result in full adder we have got motivated and we have extended it for 8 bit adder. The 8 bit adder was designed in ripple carry adder (RCA) technique Fig. 6. We need 8 such single bit adders in order to compute 8 bit addition between two inputs. First bits of two inputs a0 and b0 are added and results are passed on to s0 output pin and the carry generated is applied to the next stage

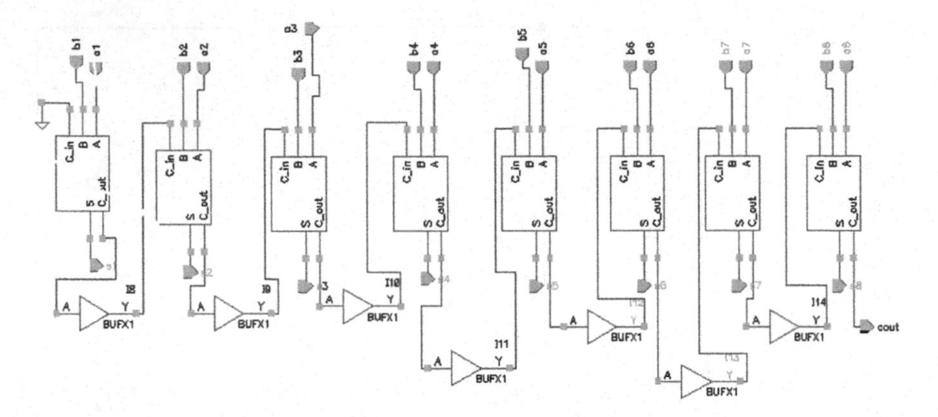

Fig. 6 8-bit ripple carry adder

adder. Similarly s1, s2, s3, s4, s5, s6, s7 bits are generated by performing addition operation to their respective input bits and passing on the carry. Finally we get only one final carry output from last bit adder. In order to eliminate the driving problems the proposed design consists of buffers in the cascaded path which lies between the carry out of an adder and carry input of next adder Fig. 6.

4 Results and Discussions

4.1 Results Analysis

The technology used for designing proposed body biased hybrid full adder is gpdk180. The simulation was performed in a common environment with supply voltage 1.8 V in 180 nm technology. Simulation is done using Cadence spectra tool and the results are depicted Table 1. The body biased hybrid full adder circuit has also compared with other full adder circuits [1, 15, 17]. The circuits of other full adders were re-simulated in cadence spectra tool in 180 nm technology. So, the paper can be authentically compared the proposed full adder design with other full adder circuits reported in [1, 15, 17].

The circuits of other full adders were re-simulated in cadence spectra tool in 180 nm technology. So, the paper can be authentically compared the proposed full adder design with other full adder circuits reported in [1, 15, 17]. The simulation results shows that the proposed adder offers improved Average Delay than the earlier techniques at the cost of power consumption and area.

Table 1 Simulation results of full adder in 180 nm technology with 1.8 V supply

Design	Average delay (ps)	Average power	Transistor count
C-CMOS [1]	292.1	15.1 uW	28
M-GDI [15]	250.9	52.56 mW	10
LPHS [17]	224	1.17 uW	16
Proposed	114	27.5 mW	22

4.2 Simulation Results

The proposed body biased hybrid full adder circuit is simulated using cadence Spectra tool and the functionality of the proposed system is verified. Results are shown in Fig. 7.

In the above results there are some glitches in output waveform due to the improper switching between the two outputs. This problem is eliminated by placing buffers at the output stage and the waveforms that are obtained after insertion of buffer are accurate as shown in Fig. 8.

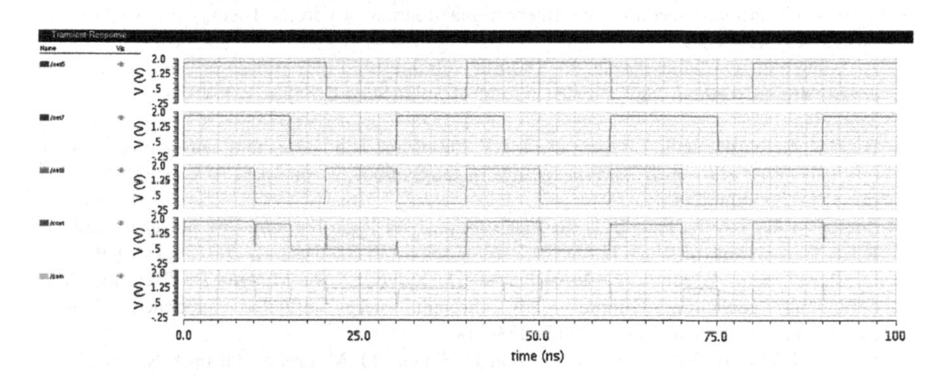

Fig. 7 Simulation results without buffers

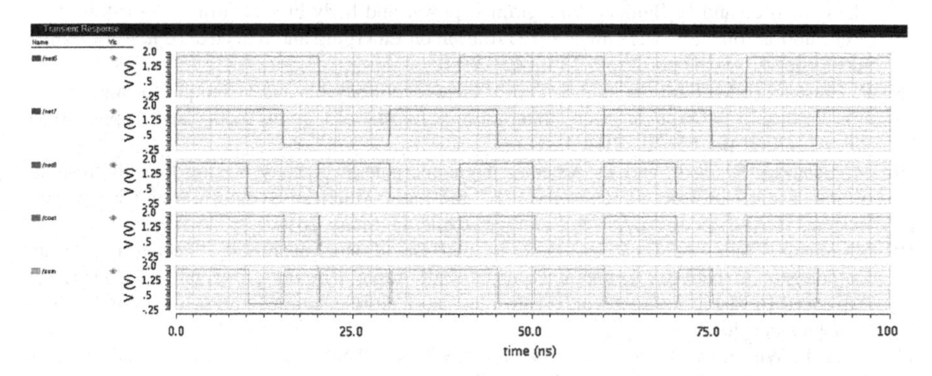

Fig. 8 Simulation results with buffers

5 Conclusion

In this paper, a body biased low power high speed 1 bit full adder has been proposed and has extended for 8 bit case also. The simulation results were carried out using cadence spectra tool with 180 nm technology and compared with the other full adder circuits like CCMOS, M-GDI, and LPHS. The simulation results shows that the proposed adder offers improved Average Delay with the earlier reports. The proposed adder circuit also eliminates glitches in the output by having buffers at the output stage.

Acknowledgements The author would like to thank the management, Director, Principal, Head of Department of Institute of aeronautical college, Hyderabad for their support and guidance in completion of this research paper.

References

1. Rjoub A, Al-Ajlouni M. “Efficient multi-threshold voltage techniques for minimum leakage current in nanoscale technology”, International Journal of Circuit Theory and Applications 2011; 39:1049–1066.
2. A. Wang and A. Chandrakasan. “A 180 mV sub threshold FFT processor using a minimum energy design methodology”, IEEE Journal of Solid-State Circuits, vol. 40, no. 1, pp. 310–319, 2005.
3. B. Zhai, L. Nazhandali, J. Olson et al. “A 2.60pJ/inst sub—threshold sensor processor for optimal energy efficiency,” in Proceedings of the Symposium on VLSI Circuits (VLSIC’06), pp. 154–155, June 2006.
4. Soeleman H, Roy K, Paul BC. “Robust subthreshold logic for ultra-low power operation”, IEEE Transactions on Very Large Scale Integration (VLSI) Systems 2001; 9(1):90–99.
5. Dreslinski RG, Wieckowski M, Blaauw D, Sylvester D, Mudge T. “Near-threshold computing: reclaiming Moore’s law through energy efficient integrated circuits” Proceedings of the IEEE 2010; 98(2):253–266.
6. Hanson S, Zhai B, Seok M, Cline B, Zhou K, Singhal M, Minuth M, Olson J, Nazhandali L, Austin T, Sylvester D, Blaauw D. “Exploring Variability and Performance in a Sub-200-mV Processor” IEEE Journal of Solid State Circuits (JSSC) 2008; 43(4):881–891.
7. M. R. Kakoee and L. Benini. “Fine-grained power and body-bias control for near-threshold deep sub-micron CMOS circuits,” IEEE Journal on Emerging and Selected Topics in Circuits and Systems, vol. 1, no. 2, pp. 131–140, 2011.
8. P. Corsonello, M. Lanuzza, and S. Perri. “Gate-level body biasing technique for high-speed sub-threshold CMOS logic gates,” International Journal of Circuit Theory and Applications, vol. 42, no. 1, pp. 65–70, 2014.
9. M. Lanuzza, R. Taco, and D. Albano. “Dynamic gate-level body biasing for subthreshold digital design,” in Proceedings of the IEEE 5th Latin American Symposium on Circuits and Systems (LASCAS’14), pp. 1–4, Santiago, Chile, February 2014.
10. Saradindu Panda, A. Banerjee, B. Maji and Dr. K.Mukhopadhyay. “Power and Delay Comparison in between Different types of Full Adder Circuits”, International Journal of Advanced Research in Electrical, Electronics and Instrumentation Engineering, Vol. 1, Issue 3, September 2012.
11. N. H. E. Weste, D. Harris, and A. Banerjee. CMOS VLSI Design: A Circuits and Systems Perspective, 3rd ed. Delhi, India: Pearson Education, 2006.

12. D. Radhakrishnan. "Low-voltage low-power CMOS full adder," IEE Proc.-Circuits Devices Syst., vol. 148, no. 1, pp. 19–24, Feb. 2001.
13. R. Zimmermann and W. Fichtner. "Low-power logic styles: CMOS versus pass-transistor logic," IEEE J. Solid-State Circuits, vol. 32, no. 7, pp. 1079–1090, Jul. 1997.
14. J. M. Rabaey, A. Chandrakasan, and B. Nikolic. Digital Integrated Circuits: A Design Perspective, 2nd ed. Delhi, India: Pearson Education, 2003.
15. J.-M. Wang, S.-C. Fang, and W.-S. Feng. "New efficient designs for XOR and XNOR functions on the transistor level," IEEE J. Solid-State Circuits, vol. 29, no. 7, pp. 780–786, Jul. 1994.
16. Neeraj Devararil, Shashank Sundriyal. "Efficient Design Methodologies and Power Comparison For Full Adder Circuits ISTP" Journal of Research in Electrical and Electronics Engineering (ISTP-JREEE) 1st International Conference on Research in Science, Engineering & Management (IOCRSEM 2014).
17. D. Khalandar Basha, S. Padmaja Reddy, P. Monica Raj. "Low power and speed M-GDI based full adder", International journal of advanced trends in computer science engineering, vol 5, no 1 pages 88–91 (2016) Special Issue of ICACEC 2016-Held during 23-24 January, 2016 in Institute of Aeronautical Engineering, Hyderabad.
18. Partha Bhattacharyya, Bijoy Kundu, Sovan Ghosh, Vinay Kumar and Anup Dandapat. "Performance Analysis of a Low-Power High-Speed Hybrid 1-bit Full Adder Circuit", IEEE TRANSACTIONS ON VERY LARGE SCALE INTEGRATION (VLSI) SYSTEMS.
19. Ramiro, Taco, Marco, Lanuzza, and Domenico Alban Ultra-Low-Voltage Self-Body Biasing Scheme and Its Application to Basic Arithmetic Circuits, Hindawi Publishing Corporation VLSI Design Volume 2015, Article ID 540482, 10 pages http://dx.doi.org/10.1155/2015/540482.

Optimization of Contiguous Link Scheduling

Hardik Gupta, Siddhesh Mhatre, M.M. Chandane, Akshita Shah and Shreeya Laad

Abstract Wireless Sensor Networks (WSNs) constitute battery operated devices. These devices stop functioning once the battery has completely drained out. It is not feasible to recharge or to replace the batteries of these devices either due to the hostile environments where these devices are deployed or due to the cost associated with the batteries. Hence energy consumption is crucial problems in WSNs. Researchers have tried to address this issue through various techniques. Here the attempt has been made to reduce the number of active time slots through a coloring scheme which will eventually reduce the energy consumption, while increasing the throughput. The proposed scheme is termed as Optimized Contiguous Link Scheduling. The proposed algorithm has been implemented using the EXata Network Simulator and the obtained results are compared with other scheduling algorithms like TDMA, Aloha, Contiguous Link Scheduling and 802.11e, with respect to energy consumption, throughput and jitter.

Keywords Wireless sensor networks · Energy optimization · Data aggregation · Sleep scheduling · Contiguous link scheduling

H. Gupta (✉) · S. Mhatre · M.M. Chandane · A. Shah · S. Laad
Department of Computer Engineering and I.T, Veermata Jijabai Technological Institute, Mumbai Maharashtra 400019, India
e-mail: hardikgupta@gmail.com

S. Mhatre
e-mail: sid1793@gmail.com

M.M. Chandane
e-mail: mmchandane@vjti.org.in

A. Shah
e-mail: akshita.s.shah@gmail.com

S. Laad
e-mail: shreeyalaad@gmail.com

S.C. Satapathy et al. (eds.), *Computer Communication, Networking and Internet Security*, Lecture Notes in Networks and Systems 5,
DOI 10.1007/978-981-10-3226-4_9

1 Introduction

Wireless Sensor Network consists of large number of small sensing, spatially distributed, autonomous, self-powered nodes which gather information or detect spatial events with the end goal of working collaboratively to perform a common application [1]. The key characteristic of sensor nodes is that they are battery operated and that it is not feasible to replace or recharge either because of the cost or the environment where these devices are placed. Energy consumption is therefore a crucial problem in WSNs. The energy consumption must be optimized while retaining the normal functions of WSNs. There are many available techniques for power saving and energy optimizations in WSNs discussed in [2].

TDMA MAC protocols provide an effective way for nodes to wake up to transmit or receive messages without interferences. They are inherently devoid of collisions and hence possess no overhead due to contentions. TDMA MAC protocols may be classified as broadcast scheduling and link scheduling. The time is slotted in TDMA MAC protocols, and in TDMA link scheduling the communication links are assigned time slots. Reducing the number of time slots by making use of concurrent transmissions, results in an enhancement of network throughput as well. In terms of energy optimization, TDMA broadcast scheduling has a degraded performance as compared to TDMA link scheduling because only the intended receiver needs to switch on from sleep mode in link scheduling [3].

Several approximation algorithms have been proposed for both broadcast scheduling and link scheduling. Centralized and distributed algorithms have been proposed as well. Traditional TDMA link scheduling algorithms don't take into account the energy utilized in state transitions, which makes the energy model separate from the practical scenario. If a node happens to start-up frequently and the packets are small, then it may so happen that the transmission time is less than the start-up time. Accordingly, the energy consumption in the start-up process can be higher than that actually utilized in transmission. Certain TDMA scheduling algorithms that have however taken into account the energy utilized in state transitions, have not formally defined the contiguous link scheduling.

Table 1 EXata simulation parameters

Parameter	Specification
Physical	802.11
Mac	Aloha, 802.11e, TDMA Auto, TDMA File
Routing protocol	AODV
Number of nodes	8, 9, 15
Topology	Tree
Simulation time	180 s
Grid	1500*1500
Tx current load	17.4 mAmp
Rx current load	19.7 mAmp

Table 2 Total energy saved in optimized versus contiguous link scheduling

Topology	Total Energy Saved (in mWh) in Optimized
1	0.002
2	0.001
3	0.002

Table 3 Transition energy saved in optimized versus contiguous link scheduling

Topology	Transition energy saved (in mWh) in optimized
1	0.000009
2	0.000018
3	0.000009

There have been several approaches to obtaining contiguous time slots including search heuristics. [4] formally defined the contiguous link scheduling problem and proposed centralized algorithms that make use of the interval vertex colouring in the merged conflict graph. It addresses the problem of assigning consecutive time slots to nodes such that a node needs to start-up only once to fulfil all its tasks. In a tree network topology, each intermediate node would need to wake up at most twice; once to receive data from its children nodes and once to transmit data to its parent nodes.

This chapter presents the optimized contiguous link scheduling algorithm which is the optimization of the original Centralized scheduling algorithm [4] proposed by I. W. L. M. I. Junchao Ma, and I. Xiang-Yang Li. The optimization has been done to reduce the number of time slots. The work contributions are summarized as follows:

i. Addressing the scheduling problem in WSNs by proposing an optimization of the contiguous link scheduling problem by considering both the conventional energy model and that with energy transitions,
ii. Presenting an optimized contiguous link scheduling algorithm which provides an interference free link scheduling with minimum number of time slots,
iii. Supporting the algorithm with theoretical performance bounds in homogeneous networks,
iv. Conducting simulations to validate our theoretical results, and display the efficiency of our algorithm in terms of total energy consumed, network throughput and network latency.

The proposed optimization considers only tree topologies used for the aggregation of data. The contents of the chapter are organized as follows; related work is presented in Sects. 2 and 3 proves the proposed optimizations, results are discussed in Sect. 4 and finally Sect. 5 presents the conclusion.

2 Related Work

TDMA link scheduling has been widely investigated and researched with respect to WSNs, due to its plusses which include interference-free channel access and energy efficiency. In link scheduling, the data transmitted by a node must be received by a particular neighboring node, without interferences. Numerous algorithms have been proposed in link scheduling to come up with possible approximations.

Broadcast scheduling and link scheduling are analyzed by Ramanathan and Llyod [5]. An optimal scheduling can be achieved in a tree network whereas in an arbitrary network, the performance depends on the breadth of the network. A two phase link scheduling algorithm was proposed in Gandham [6]. In the first phase, a valid edge coloring is obtained. Following this, the second phase involves assigning each unique color a time slot in order to obtain an interference free scheduling. The hidden terminal and exposed terminal problems are circumvented by assigning a direction of transmission to each edge. Djukic and Valee [7] performed research on the minimum delay round trip schedules in link scheduling. To solve the problem they proposed applying an efficient min-max delay scheduling to a conflict graph. A many nodes to one node scheduling problem in sensor networks was formulated by Ergen and Varaiya [3] to find the smallest length interference free scheduling such that the packets generated at each node end up at the sink. For general communication graphs they prove this problem to be NP-complete. However a general communication graph in some instances may not be a deployable graph. A deployable graph is one in which there exists a valid deployment of nodes for the given communication graph in WSNs.

Ma et al. [8] propose a robust link scheduling using only data which states whether a given pair of links interfere or not. Xu and Valee [9] show an effective interference free link scheduling with Signal to Interference-Plus-Noise Ratio constraints in WSNs. They propose two algorithms with an efficient delay by constructing a routing tree. To avoid the hidden terminal problem Alsulaiman and Djukic [10] frame the link scheduling problem in the form of a distance-2 edge colouring of a bi-directed graph. Their proposal is under the synchronous message passing and the asynchronous message passing communication models. [4] formally defined the contiguous link scheduling problem and proposed centralized algorithms that make use of the interval vertex coloring in the merged conflict graph. It addresses the problem of assigning each node consecutive time slots such that it needs to start-up only once to fulfil all its tasks. Each node in a network topology will have to wake up at most twice if it is in the form of a tree. Once to receive data from its children and once to transmit data to its parents. Asratian and

Table 4 Increased throughput in optimized versus contiguous link scheduling

Topology	Increased throughput in optimized (in bits/s)
1	NA
2	2143.44
3	2095.7

Kamalian [11] introduced interval edge coloring. In this the colors of the edges incident on a vertex should be contiguous. In other words the colors must be composed of an integer interval.

Work presented in this chapter talks about the optimization to centralized scheduling algorithm proposed in [4] to reduce the number of time slots. In the optimized algorithm, the interval edge coloring is applied to the conflict graph of a communication network.

3 Proposed Optimized Contiguous Link Scheduling Algorithm

The proposed Optimized Contiguous Link Scheduling algorithm is an optimization to [4]. It uses a coloring scheme to optimize the link scheduling. The objective of this scheme was to achieve high throughput, low jitter and low energy consumption. Each node *(i, j)* in this graph represents a link in the original graph and an edge between two nodes represents interference between the two edges. Then we compute the frequency of the occurrence of number *j* in the second index of the nodes and arrange these numbers in decreasing order in a list. After this we parse the entire list and assign the minimum number of consecutive time slots that do not conflict with any other node in that time slot.

3.1 Algorithm

Step 1: Construct a conflict graph *G*(*V, E*). An edge between a node (i_1, j_1) and (i_2, j_2) represents interference between the two links

Step 2: Sort the second index of each node in list *L*, in decreasing order of frequency of occurrence in graph

Step 3: For every item *k* in list *L*:

i. Arrange the nodes (*x*, *k*) for all x, in consecutive slots such that they do not conflict with any other nodes in that slot

ii. Create a new slot if no such slot is found

Sevastjanov [12] has proved that the determination of the existence of an interval edge colouring is an NP-complete problem. In [13], it was proved through experiment that an interval edge colouring exists for small and sparse graphs with a high probability. The Optimized Contiguous Link Scheduling algorithm is based on the vertex colouring of the conflict graph of the given network topology. The chromatic number of a graph is defined as the minimum number of colours that are required to colour all the vertices of a graph such that no two adjacent vertices possess the same colour [14]. In the context of the proposed algorithm, the chromatic number represents the minimum number of time slots required for an interference free contiguous link schedule.

The Brooks Theorem [15] gives an upper bound on the chromatic number of vertex colouring as,

$$X(G) \leq \Delta(G) \quad (1)$$

where X(*G*) is the chromatic number of the graph *G* and Δ(*G*) is highest degree of the graph *G*.

Hence the above equation gives the maximum number of time slots required for Optimized Contiguous Link Scheduling. The Contiguous Link Scheduling [16] algorithm gives the upper bound on the number of time slots as,

$$T < 2\left(\sum_{l=1}^{l=L} w_{il} + w_i\right) \quad (2)$$

where w_{il} denotes the weight of the vertices adjacent to node *i* and w_i denotes the weight of the node *i* in the merged conflict graph of graph G.

Comparing the above Eqs. (1) and (2), it is evident that the number of active time slots to obtain a feasible schedule are lesser in Optimized Contiguous Link Scheduling. The topologies used for experimentation are shown in Figs. 1, 2 and 3.

Table 5 Reduced jitter in optimized versus contiguous link scheduling

Topology	Reduced jitter in optimized (s)
1	NA
2	NA
3	0.07017

Fig. 1 Topology 1

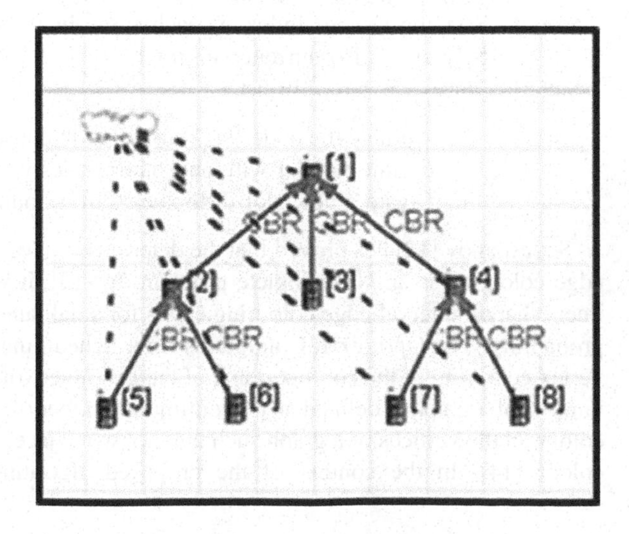

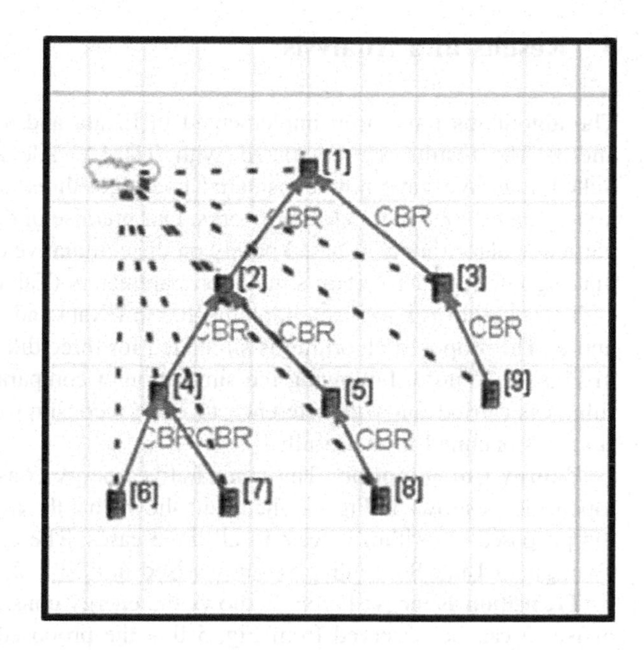

Fig. 2 Topology 2

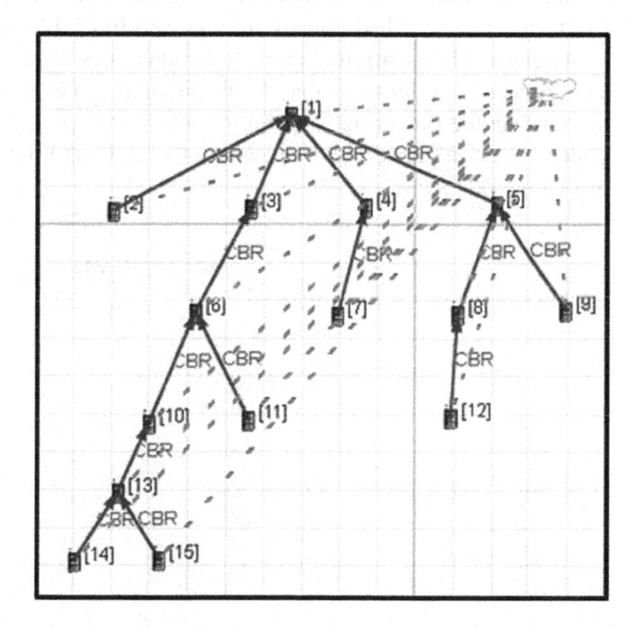

Fig. 3 Topology 3

4 Results and Analysis

The algorithms have been implemented in EXata and are executed on Figs. 1, 2 and 3. The results are compared with TDMA, Aloha, and Contiguous Link Scheduling. We have now considered five algorithms using which scheduling of nodes can be done in wireless networks. Our premise of determining how efficient a particular algorithm is, is based purely on the cumulative energy consumption using that algorithm under certain simulation parameters (Table 1).

The algorithms have been implemented in EXata and are executed on Figs. 1, 2 and 3. The proposed algorithm is simulated for three different topologies as shown in Figs. 1, 2 and 3. Following the simulation, a comparative analysis of the algorithms is carried out with reference to energy consumption, throughput and jitter statistics obtained in the results.

Energy Consumption: The comparative energy consumption for the different topologies is shown in Fig. 4. The result shows that the overall energy consumed by the proposed algorithm is least in all three cases. The comparison with respect to Contiguous Link Scheduling is summarised in (Table 2).

Transition Energy: Figure 5 shows the energy consumed during the transition phase. It can be observed from Fig. 5 that the proposed algorithm uses the least amount of energy as compared to others. The comparison with respect to Contiguous Link Scheduling is summarised in (Table 3).

Average Throughput: Figure 6 demonstrates the average throughput obtained. The obtained results show that the throughput achieved with proposed algorithm is higher than the others for Topology 2 and Topology 3, which have relatively more nodes. The comparison with respect to Contiguous Link Scheduling is summarised in (Table 4).

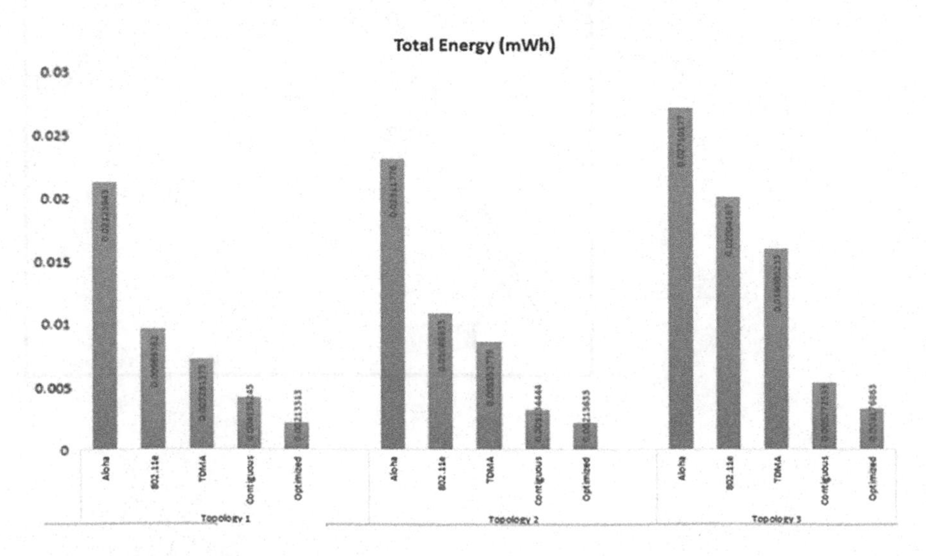

Fig. 4 Comparison of total energy consumption

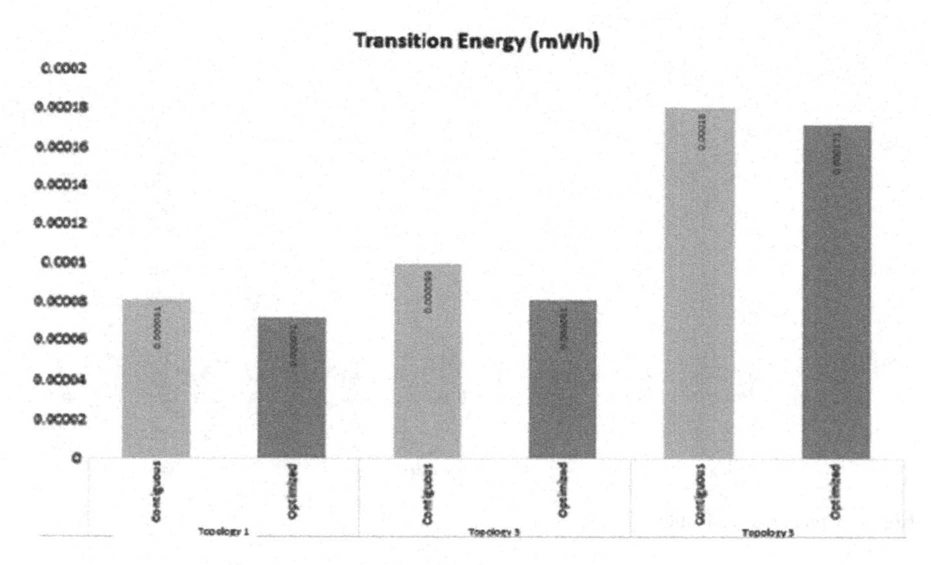

Fig. 5 Comparison of transition energy consumption

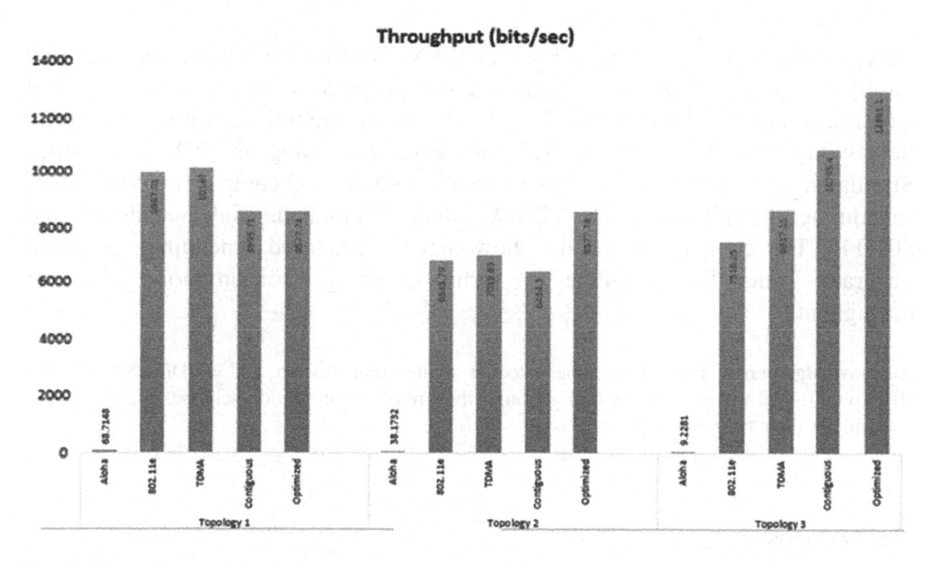

Fig. 6 Comparison of throughput

Jitter: Figure 7 shows the comparative analysis of jitter. The obtained results clearly illustrate that the proposed algorithm produces less jitter than most of the other algorithms. The jitter with the proposed algorithm is the lowest for the topology with the maximum number of nodes i.e. Topology 3. The comparison with respect to Contiguous Link Scheduling is summarised in (Table 5).

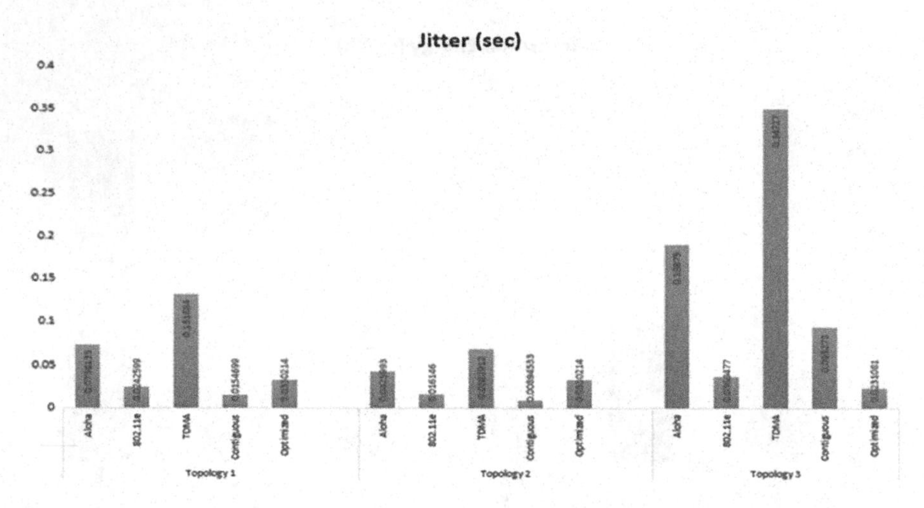

Fig. 7 Comparison of jitter

5 Conclusion

Energy consumption is a major issue in WSNs. The problem has been addressed through a graph coloring method and the proposal is termed as Optimized Contiguous Link Scheduling. The algorithm for Optimized Contiguous Link Scheduling has been implemented and simulated using the EXata Network Simulator. The results are tested for various topologies and compared with tradition scheduling algorithms, namely TDMA, Aloha, Contiguous Link Scheduling and 802.11e. The comparative results show that the proposed scheduling algorithm generates better link schedules in terms of energy consumption, jitter and throughput.

Acknowledgements This work is the outcome of the grant Ref No. 8023/RID/RPS-51/(Policy III)Govt/2011-12 provided by AICTE through their research promotion scheme. We are highly grateful for their financial support.

References

1. F. L. Lewis, “Wireless sensor networks,” Smart Environments: Technologies, Protocols and Applications, 2004.
2. M. G. Sandra Sendra, Jaime Lloret and J. F. Toledo, “Power saving and energy optimization techniques for wireless sensor networks,” Journal of Communications, vol. 6, no. 6, 2011.
3. P. V. Sinem Coleri Ergen, “Tdma scheduling algorithms for wireless sensor networks,” Wireless Networks, 2010.
4. I. W. L. M. I. Junchao Ma, Student Member and I. Xiang-Yang Li, Senior Member, “Contiguous link scheduling for data aggregation in wireless sensor networks,” IEEE Transactions on Parallel and Distributed Systems, vol. 25, no. 7, 2014.

5. Ramanathan and Llyod, "Broadcast scheduling and link scheduling," IEEE Transactions on Parallel and Distributed Systems, vol. 26, no. 8, 2011.
6. Gandham, "Link scheduling using edge coloring," IEEE, vol. 27, no. 15, 2010.
7. Djukic and Valee, "Link scheduling and minimum round trip delay," ACM, vol. 30, no. 1, 2013.
8. J. Ma and W. Lou, "Interference-freewakeup scheduling with consecutive constraints inwireless sensor networks," International Journal of Distributed Sensor Networks, vol. 2012, no. 525909, 2012.
9. Xu and Valee, "Energy efficient interference free link scheduling," ACM, vol. 30, no. 1, 2011.
10. Alsulaiman and Djukic, "Distance 2-edge coloring of a bi-directed graph," IEEE, vol. 31, no. 5, 2010.
11. A. S. Asratian and R. R. Kamalian, "Investigations on interval edgecolorings of graphs," Journal of Combinatorial Theory, vol. B 62, no. 5, 1994.
12. S. V. Sevastjanov, "On interval colorability of a bipartite graph," Metody Diskretnogo Analiza, vol. 50, 1990.
13. K. Giaro, "Compact task scheduling on dedicated processors with no waiting periods," Ph.D. dissertation, Technical University of Gdansk, ETI Faculty, Gdansk, Poland, 1999.
14. S. Pemmaraju and S. Skiena, Computational Discrete Mathematics: Combinatorics and Graph Theory with Mathematica, 1997.
15. K. Giaro, "Compact task scheduling on dedicated processors with no waiting periods," Ph.D. dissertation, Technical University of Gdansk, ETI Faculty, Gdansk, Poland, 1999.
16. W. L. Junchao Ma and X.-Y. Li, "Supplementary document: Contiguous link scheduling for data aggregation in wireless sensor networks.

[illegible] "Broadcast scheduling and link scheduling," IEEE Transactions on Parallel and Distributed Systems, vol. 2[illegible], no. 8, 2011.
[illegible] "Link scheduling [illegible] edge coloring," IEEE, vol. [illegible], no. 1[illegible], 201[illegible].
[illegible] "Link scheduling and [illegible] delay," ACM, vol. [illegible], no. [illegible], 201[illegible].
8. [illegible] "Multi-frequency [illegible] scheduling with consecutive constraints in wireless sensor networks," International Journal of Distributed Sensor Networks, vol. 201[illegible], no. [illegible], 2012.
9. [illegible] "Energy efficient [illegible] link scheduling," ACM, [illegible].
10. [illegible] and [illegible] "[illegible] edge coloring [illegible]" [illegible].
11. A. S. [illegible] and K. [illegible] "[illegible]" Journal of Combinatorial Theory, [illegible] B 52, no. [illegible].
12. [illegible] "[illegible] graph," [illegible] Mathematics, vol. [illegible] 19[illegible].
13. K. Giaro, "Compact task scheduling on dedicated processors with no waiting periods," Ph.D. dissertation, Technical University of Gdansk, EТI Faculty, Gdansk, Poland, 1999.
14. S. Pemmaraju and S. Skiena, Computational Discrete Mathematics: Combinatorics and Graph Theory with Mathematica, 2007.
15. K. Giaro, "Compact task scheduling on dedicated processors with no waiting periods," Ph.D. dissertation, Technical University of Gdansk, EТI Faculty, Gdansk, Poland, 1999.
16. W. [illegible] and [illegible] "[illegible] Contiguous link scheduling for data aggregation in wireless sensor networks," [illegible].

A Novel Reversible EX-NOR SV Gate and Its Application

D. Krishnaveni and M. Geetha Priya

Abstract Reversible logic in quantum computing and nanotechnology provides solution to optimize the power and this technology can be put into use in a variety of low power applications such as optical computing, and CMOS VLSI design. An EXNOR gate is a digital logic gate that can be used as an equivalence gate and also in variety of applications. A new reversible SV gate is proposed in this paper that functions as an EXNOR gate and whose quantum cost is 1. An equivalence checker also has been designed in this paper that can be used to check the equivalence of two n-bit binary numbers. The proposed SV gate has been validated by comparing its performance with some of the existing circuits in literature. The results prove to be more reliable, efficient and provide a basis for building more complex arithmetic systems using the concept of reversible logic that can be used in digital data transmission circuits.

Keywords Quantum computing · Reversible logic gates · Equivalence gate · Low power CMOS

D. Krishnaveni (✉)
Department of Telecommunication Engineering, A P S College of Engineering, Bangalore, Karnataka, India
e-mail: mailkveni@gmail.com

M. Geetha Priya
Centre for Incubation, Innovation Research and Consultancy, Jyothy Institute of Technology, Bangalore, Karnataka, India
e-mail: geetha.sri82@gmail.com

S.C. Satapathy et al. (eds.), *Computer Communication, Networking and Internet Security*, Lecture Notes in Networks and Systems 5,
DOI 10.1007/978-981-10-3226-4_10

1 Introduction

Major thrust is being given to the design, implementation, and analysis of logic circuits that are reversible, in research domain. The idea of reversible logic is increasingly employed in the areas of nanotechnology, quantum computing, low power VLSI design, and optical computing. As the complexity in CMOS VLSI circuits increases, the dissipation of power in the circuit becomes a major challenge in the design. Due to loss of the information, there will be dissipation of energy and consumption of power in irreversible logic circuits. It was demonstrated by Landauer in 1961 [1] that heat energy of kT*log2 joules is dissipated for every bit of information lost, where the absolute temperature is represented by T (Kelvin) and k is the Boltzmann's constant respectively. It was shown by Bennet [1973] that power dissipated in logic circuits that comprise of logic gates that are reversible is zero [2].

The primary requirement for an 'n' input, 'k' output logic expression to be reversible is that the values of 'n' and 'k' must be equal. Constant Inputs (CI) are the additional inputs added to fulfill this requirement. Similarly, the additional outputs that are not used further in the computation, in the reversible circuits are called the Garbage Outputs (GO).

In this paper the design of a Reversible EXNOR gate called SV gate with a quantum cost of 1 is proposed. This design forms the basis for arithmetic circuits that are reversible. A logic circuit of equivalence checker is designed using the proposed SV gate that checks whether any two n-bit numbers are equal or not. This is the first time an effort has been made to design a reversible gate that acts as only an EXNOR gate, as per the literature survey and information available with us. The parameters, constant inputs, quantum cost, hardware complexity, reversible gates used, computational delay, and garbage outputs, of the reversible EXNOR gates using existing gates are compared with that of the proposed SV gate. In Sects. 2 and 3, preview of reversible gates used and the design of proposed SV gate is discussed followed by simulation results, analysis, and conclusion in Sects. 4 and 5 respectively.

2 Preview of Reversible Logic Gates

A few of the numerous reversible gates that are available in literature are highlighted in this section with respect to quantum cost, input and output vectors and are listed in Table 1. Few other Reversible logic gates available are TG (Toffoli), FRG (Fredkin) [3], DKGP [4].

Table 1 Preview of reversible gates

Name of reversible gate	Inputs	Outputs	Quantum cost
Feynman (FG) [3]	Iv = (X, Y)	$O_V = (M, N : M = X, N = X \oplus Y)$	1
Peres (PG) [3]	Iv = (X, Y, Z)	$O_V = (I, J, K : I = X, J = X \oplus Y, K = XY \oplus Z)$	4
SRK [5]	Iv = (X, Y, Z)	$O_V = (I, J, K : I = X, J = X \oplus Y \oplus Z, K = \overline{X}Y \oplus XZ)$	4
M Gate [6] & DG [7]	Iv = (X, Y, Z)	$O_V = (I, J, K : I = X, J = \overline{(X \oplus Y)}, K = X\overline{Y} \oplus Z)$	5
L gate [6]	Iv = (X, Y, Z)	$O_V = (I, J, K : I = X, J = Y, K = \overline{(X \oplus Y)} \oplus Z$	Not applicable
R Gate [8]	Iv = (X, Y, Z)	$O_V = (I, J, K : I = X \oplus Y, J = X, K = XY \oplus \overline{Z}$	Not applicable
TS-3 Gate [9]	Iv = (X, Y, Z)	$O_V = (I, J, K : I = X, J = Y, K = X \oplus Y \oplus Z)$	2
TRG [10]	Iv = (X, Y, Z)	$O_V = (I, J, K : I = X, J = X \oplus Y, K = X\overline{Y} \oplus Z$	4
DG [7]	Iv = (X, Y, Z)	$O_V = (I, J, K : I = X, J = \overline{X \oplus Y}, K = X\overline{Y} \oplus Z$	5
HNG [11]	Iv = (W, X, Y, Z)	$O_V = (D, E, F, G : D = W, E = X, F = W \oplus X \oplus Y,$ $G = (W \oplus X)Y \oplus (WX \oplus Z)$	6
MRG [12]	Iv = (W, X, Y, Z)	$O_V = (D, E, F, G : D = W, E = W \oplus X,$ $F = W \oplus X \oplus Y, G = (WX \oplus Z) \oplus (W \oplus X) \oplus Y$	6

3 Proposed Design

3.1 2 × 2 Reversible EXNOR (SV) Gate

An Exclusive-NOR or EXNOR gate is a 2 input logic gate whose output is logic 1 when both the inputs are equal else logic 0. This gate implements logical equality. It is also called equivalence gate. The primary component in error detecting circuits that detect the odd or even parity of a given binary input is the EXNOR logic gate. The application of these parity checkers are in digital data transmission circuits. EXOR and EXNOR gates are mainly used in arithmetic and encryption circuits. Exclusive-NOR circuits can be used to compare two binary data words and determine if they are exactly the same.

The proposed reversible 2 × 2 SV gate functions as an EXNOR gate with high performance when compared to other existing EXNOR gate structures. The Fig. 1 gives the gate level schematic view of the proposed SV gate along with the symbol

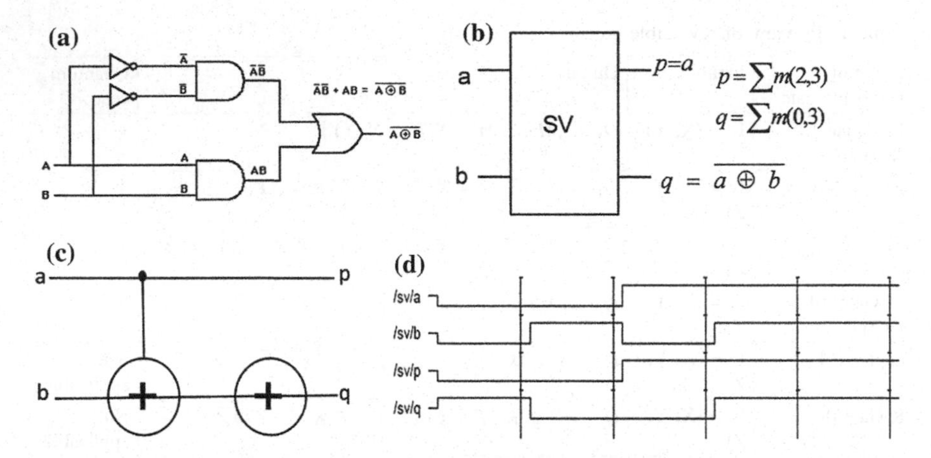

Fig. 1 **a** Gate level Schematic of proposed reversible SV gate **b** Proposed SV gate as an EXNOR gate **c** Quantum circuit of proposed SV gate **d** Simulation wave form

and quantum circuit obtained from Xilinx respectively. It can be proved that a particular input can uniquely be determined from corresponding output as applicable to reversible logic. The quantum circuit in Fig. 1 is used to calculate the quantum cost and is calculated to be 1. Figure 1d displays the input and output waveform results obtained during simulation of the reversible SV gate.

3.2 2 × 2 Reversible EXNOR (SV) Gate Based Equivalence Checker

As one of the important application of EXNOR gate is Equivalence checking, (checking whether the inputs given to it are equal or not) the same can be used in an array forming an n-bit Equivalence checker. In this paper an equivalence checker is also designed using an array of SV gates to compare two n-bit numbers. To get a single output that indicates whether the given inputs are equal or not, AND gates are used at the output of EXNOR gates.

The design of equivalence gate is shown in Fig. 2. Two n-bit inputs ($A_{N-1} - A_0$ & $B_{n-1} - B_0$-inputs) are given to the array of SV gates. When both the inputs of the array of SV gates are equal, the outputs X0 X1 X2 … Xn−1 would be 111… 111 i.e. all 1's. These are given as input to an array of PG gates [3] (Fig. 2 b) whose output is the signal K. PG gate acts as an AND gates when input c is '0'. K='1' indicates both the inputs are equal and K='0' indicates they are not equal.

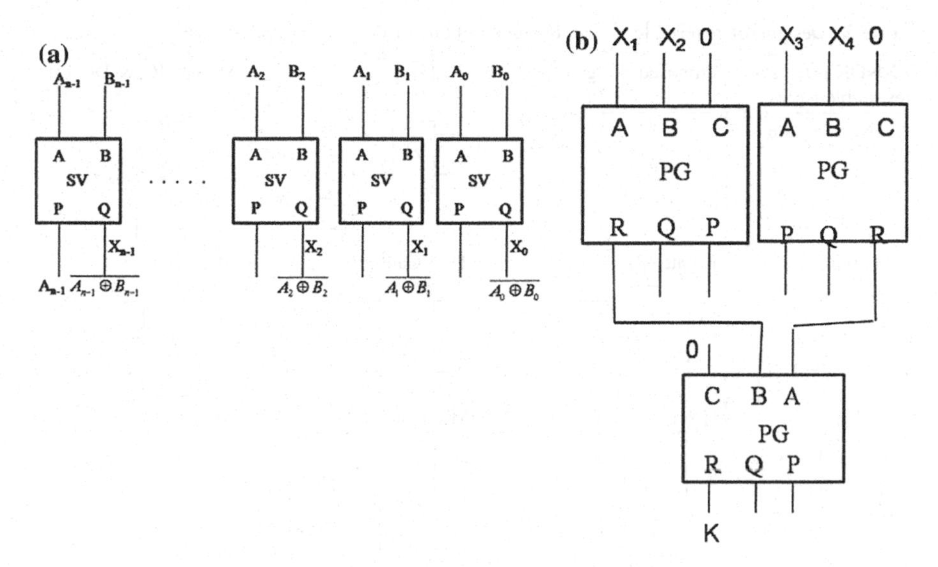

Fig. 2 **a** Array of Reversible SV gates **b** Generation of signal K='1' indicating 2 n-bit numbers are equal (here n = 4)

4 Results and Discussions

For the purpose of experimental evaluation for the performance of the proposed reversible SV gate, simulation has been carried out using Verilog language on Xilinx version 7.1, and ModelSim version 6.1 softwares. EXNOR gates can be realized using EXOR gate and a NOT gate or basic gates such as AND, OR and NOT gate. It can also be realized using Universal gates such as NAND or NOR gates. Thus for analysis purpose, different reversible gates are used to design EXNOR gate and are compared with the proposed reversible SV gate which works as an EXNOR gate.

To have a better analysis of the performance of proposed reversible SV gate, the results are compared with designs of reversible EXNOR gates obtained from the gates proposed in [6–12] as shown in Table 2. For the evaluation purpose the parameters taken for comparison are quantum cost, reversible gates used, garbage outputs, and constant inputs. The comparative analysis has been presented in Table 3.

Reversible EXNOR gate is constructed using reversible gates present in literature such as M Gate and L gate [6], R gate [8], TS-3 gate [9], TRG gate [10], DG gate [7], HNG gate [11] and MRG gate [12]. These designs are compared with the proposed reversible SV gate with respect to the parameters such as constant inputs (CI), reversible gates used (RG), quantum cost (QC) and garbage outputs (GO).

Table 2 Designs of reversible EXNOR gates obtained from the reversible gates

EXNOR gate using reversible gates	Proposed SV gate	FG [3]	M Gate [6] & DG [7]
A, B → Q	a, b; SV; $p=a$; $q = \overline{a \oplus b}$	X, Y; FG; A; $X \oplus Y$; 1; FG; $\overline{X \oplus Y}$	X, Y, Z; M Gate = DG; $I=X$; $J=\overline{X \oplus Y}$; $K=X\overline{Y} \oplus Z$
L Gate [6]	R gate [8]	TS-3 Gate [9]	TRG [10]
X, Y, 0; L Gate; $I=X$; $J=Y$; $K=\overline{X \oplus Y}$	X, Y, Z; R Gate; $I=X \oplus Y$; 1; FG; $\overline{X \oplus Y}$; $J=X$; $K=XY \oplus \overline{Z}$	X, Y, 1; TS-3; $I=X$; $J=Y$; $K=\overline{X \oplus Y}$	X, Y, Z; TRG; $I=X$; $J=X \oplus Y$; 1; FG; $\overline{X \oplus Y}$; $K=X\overline{Y} \oplus Z$
SRK [5]	HNG [11]	MRG [12]	
X, Y, 1; SRK; $I=X$; $J=\overline{X \oplus Y}$; $K=\overline{X} \oplus XY=\overline{X}$	W, X, 1, Z; HNG; $D=W$; $E=X$; $F=\overline{W \oplus X}$; $G=(W \oplus X) \oplus (WX \oplus Z)$	W, X, 1, Z; MRG; $D=W$; $E=W \oplus X$; $F=\overline{W \oplus X}$; $G=\overline{(WX \oplus Z) \oplus (W \oplus X)}$	

Table 3 Comparison of different parameters of proposed Reversible SV EXNOR Gate with existing reversible gates

Reversible EXNOR Gate	Quantum cost	Garbage outputs	Reversible Gates	Constant Inputs
Proposed SV gate	1	1	1	0
FG [3]	2	2	2	1
M Gate [6] & DG [7]	5	2	1	0
L Gate [6]	NA	2	1	1
R gate [8]	NA	3	2	1
TS-3 Gate [9]	2	2	1	1
TRG [10]	5	3	2	1
SRK [5]	4	2	1	1
HNG [11]	6	3	1	1
MRG [12]	6	3	1	1

Quantum cost of L gate and R gate has not been computed. The analysis of reversible EXNOR gate with respect to Table 3 is as follows.

- While constructing a reversible logic network, the usage of gates that are reversible need to be minimum for better performance. The design of the EXNOR gate using SRK gate [5], M Gate [6] or DG [7], L gate [6], TS-3 gate [9], HNG gate [11] and MRG gate [12] uses 1 reversible gate where as design of EXNOR gate using FG [3], R gate [8], and TRG gate [10], uses 2 gates. The proposed SV gate uses 1 gate.

- The constant inputs given as input to the reversible gates so that they can function as EXNOR gate is 1 for EXNOR gate using FG gate [3], SRK gate [5], L gate [6], R gate [8], TS-3 gate [9], TRG gate [10], HNG gate [11] and MRG gate [12]. The proposed SV gate and the design using M gate [6] or DG [7] uses 0 constant inputs.
- To increase the performance and minimize the complexity of the logic network, garbage outputs that are the unwanted or unused outputs generated from the reversible logic network should be minimized. Garbage outputs generated in the proposed design is 1 and that in designs using FG gate [3], SRK [5], M gate [6] or DG [7], L gate [6], and TS-3 gate [9], is 2. The garbage outputs generated in EXNOR gate design using R gate [8], TRG gate [10], HNG gate [11] and MRG gate [12] is 3. Thus there is a reduction in generation of unwanted outputs.
- Quantum cost of the reversible EXNOR gate designs using HNG [11] and MRG [12] is 6, using M gate [6] or DG gate [7], and TRG gate [10] is 5, using SRK gate [5] is 4, and using FG gate [3], and TS-3 gate [9] is 2. The quantum circuit for the proposed reversible SV gate is obtained, its the quantum cost is computed to be 1. This value is significantly lower than the other designs. Critical path delay is directly related to the quantum cost of the design and thus, the value of critical delay for the proposed SV gate is 1Δ.

An n-bit equivalence checker requires an array of 'n' EXNOR gates. For generation of final output that indicates the inputs are equal or not, AND gates are used. Two n-bit numbers are given as input to the EXNOR gates in the array and the output of these gates are as input to the AND gates which generate a signal indicating whether the given n-bit inputs are equal or not. The logical analysis of the proposed reversible equivalence checker has been obtained by conducting the simulation using Verilog language on Xilinx version 7.1, and ModelSim version 6.1 software. The Fig. 3 shows the obtained simulation results of 4 bit equivalence checker.

An array of 4 proposed SV gates would have 4 garbage outputs, a quantum cost of 4, and zero constant inputs. PG gate structure, generates signal K that tells whether the given inputs are equal or not (K = 0, unequal inputs; K = 1, equal inputs), has 6 garbage outputs, a quantum cost of 12, and 3 constant inputs. Together, 4-bit equivalence checker designed using proposed SV gate would have 10 garbage outputs, a quantum cost of 16, and 3 constant inputs. Total quantity of reversible gates that are used to implement the digital design of Equivalence checker is 7 (4SV + 3PG). The comparison with other circuits is as follows:

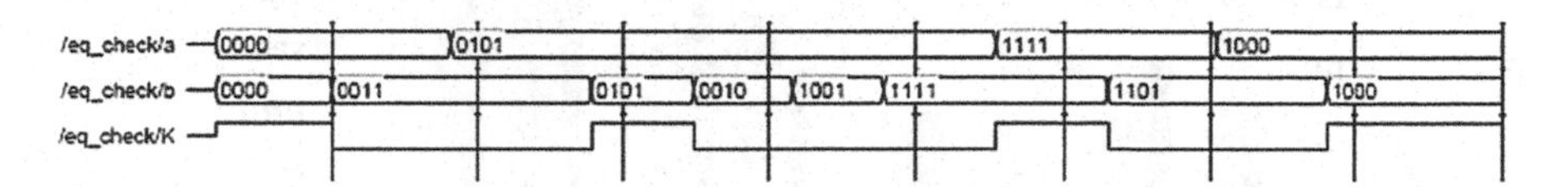

Fig. 3 Simulation waveform of Equivalence checker using proposed reversible SV gate

- 4-bit equivalence checker designed using FG gates [3] would have 14 garbage outputs, a quantum cost of 20, and 7 constant inputs. 11 reversible gates have been used in the design.
- The logic design of 4-bit equivalence checker using M gates [6] or DG gates [7] would have 14 garbage outputs, a quantum cost of 32, and 3 constant inputs. Total quantity of reversible logic gates used is 7.
- The logic circuit of 4-bit equivalence checker designed using L gates [6] would have 14 garbage outputs, and 7 constant inputs. 7 reversible logic gates are used. Quantum cost is not computed by the author.
- The network design of 4-bit equivalence checker designed using R gates [8] would have 18 garbage outputs, and 7 constant inputs. Total reversible logic gates used are 11. Quantum cost is not computed by the author.
- The logic network of 4-bit equivalence checker designed using TS-3 gates [9] would have 14 garbage outputs, a quantum cost of 20, and 7 constant inputs. Seven reversible logic gates are used.
- The circuit design of 4-bit equivalence checker designed using TRG gates [10] would have 18 garbage outputs, a quantum cost of 32, and 7 constant inputs. 11 reversible logic gates are used.
- The design of 4-bit equivalence checker designed using HNG gates [11] would have 18 garbage outputs, a quantum cost of 36, and 7 constant inputs. Total reversible gates used are 7.
- The logic design of 4-bit equivalence checker designed using MRG gates [12] would have 18 garbage outputs, a quantum cost of 36, and 7 constant inputs. The complete logic circuit of equivalence checker requires 7 reversible logic gates.
- The design of 4-bit equivalence checker using SRK gates [5] would have 14 garbage outputs, a quantum cost of 28, and 7 constant inputs. 7 reversible logic gates are used in the circuit.

The comparison of 4 bit equivalence checker designed using proposed SV gate and that designed with existing reversible logic gates is depicted in Fig. 4. Thus, the

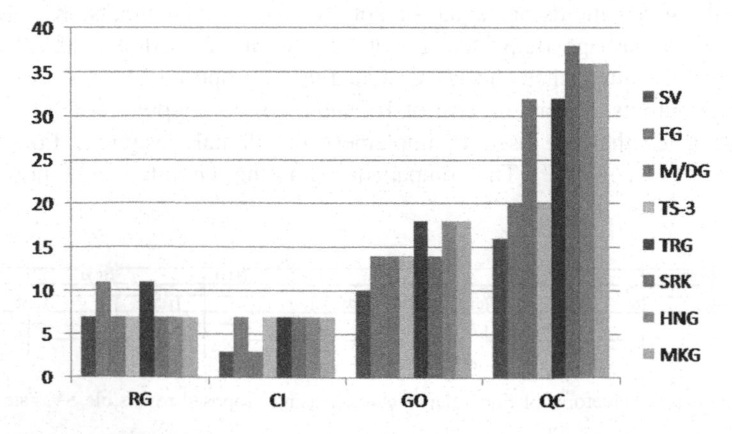

Fig. 4 Comparison of Equivalence check designs using proposed SV gate, FG, M/DG, TS-3, TRG, SRK, HNG, & MKG gates

values of quantum cost, garbage outputs, critical path delay, reversible gates used, and constant inputs, has been optimized for the proposed reversible SV gate that works as an EXNOR gate and in turn for Equivalence checker that uses the proposed SV gate.

5 Conclusion

A new 2 × 2 reversible gate (SV gate) that works as an EXNOR gate with quantum cost 1 and the logic network of a Reversible Equivalence checker is suggested in this paper. The proposed design of reversible EXNOR gate has reduced values of constant inputs, reversible gates utilized, quantum cost, and garbage outputs with respect to the designs already existing. The reversible SV gate can be utilized in building higher order reversible arithmetic network designs, error detecting circuits in digital data transmission and complex design of quantum computers. Thus this design forms the basis for a reversible arithmetic circuit design and thus contributing to reduced or zero power dissipation.

References

1. R. Landauer, "Irreversibility and Heat Generation in the Computing Process", IBM J. Research and Development, vol. 3, pp. 183–191, July 1961.
2. C. H. Bennett, "Logical Reversibility of Computation", IBM J. Research and Development, pp. 525–532, November 1973.
3. Thapliyal, H. and Ranganathan, N., "Design of Reversible Sequential Circuits optimizing Quantum Cost, Delay, and Garbage Outputs", 2010 ACM J. Emerging. Technology. Computing. System. 6, 4, Article 14, (December 2010), 31 pages.
4. D. Krishnaveni, M. Geetha Priya, K. Baskaran, "Design of an Efficient Reversible 8x8 Wallace Tree Multiplier" World Applied Sciences Journal 20 (8): 1159–1165, © IDOSI Publications, 2012.
5. D. Krishnaveni, M. Geetha Priya, "A Novel Design of Reversible Universal Shift Register with Reduced Delay and Quantum Cost", Journal of Computing, Volume 4, Issue 2, February 2012.
6. Madhina Basha, V. N. Lakshmana Kumar, "Transistor Implementation of Reversible Comparator Circuit Using Low Power Technique", International Journal of Computer Science and Information Technologies, Vol. 3 (3), 2012, 4447–4452.
7. Bahram Dehghan, Abdolreza Roozbeh, Jafar Zare, "Design of Low Power Comparator Using DG Gate", Circuits and Systems, 2014, 5, 7–12.
8. P. K. Lala, J. P. Parkerson, P. Chakraborty, "Adder Designs using Reversible Logic Gates", WSEAS Transactions On Circuits And Systems, Issue 6, Volume 9, June 2010, ISSN: 1109–2734.
9. Himanshu Thapliyal, Saurabh Kotiyal and M. B Srinivas, "Novel BCD Adders and Their Reversible Logic Implementation for IEEE 754r Format

10. Himanshu Thapliyal and Nagarajan Ranganathan, "A New Design of The Reversible Subtractor Circuit", 11th IEEE International Conference on Nanotechnology, August 15–18, 2011, Portland, Oregon, USA.
11. Haghparast M. and K. Navi, "A Novel Fault Tolerant Reversible Gate For Nanotechnology Based Systems". Am. J. Applied Sci., 5 (5): 519–523, 2008.
12. M. Morrison and N. Ranganathan, " Design of a Reversible ALU based on Novel Programmable Reversible Logic Gate Structures", IEEE Computer Society Annual Symposium on VLSI, 2011, pp. 126–131.

Internet of Things and Wireless Physical Layer Security: A Survey

Ankit Soni, Raksha Upadhyay and Anjana Jain

Abstract Internet of Things (IoT) has been a focus of research in the last decade with emphasis on the security aspects like wireless network security, communication security, sensor data security, integrity of physical signals and actuating devices. The existing security techniques are not suitable for IoT applications as the involved devices at the ground level have limited resources, low complexity, energy constraints etc. This survey analyzes various IoT concepts in terms of IoT elements, architecture and communication standards. We also analyze the existing wireless security techniques and security attacks at all the layers of the Open Systems Interconnection (OSI) model with special attention on applicability of wireless physical layer security (WPLS) techniques to achieve security for IoT devices.

Keywords Internet of Things (IoT) · Physical layer security (PLS) · Wireless security techniques · Communication standards · Security attacks and challenges

1 Introduction

One of the key technologies that conceptualize the Internet of Things (IoT) in the real world is the wireless communication. IoT is an integrated part of future internet in which "smart things/objects" are expected to become active participants in real time processes where they are enabled to interact and communicate among themselves. The basis of the security management of IoT is laid by exploring the security

A. Soni (✉) · R. Upadhyay
Institute of Engineering & Technology, DAVV, Indore, M.P, India
e-mail: soniankit15@gmail.com

R. Upadhyay
e-mail: raksha_upadhyay@yahoo.co.in

A. Jain
Shri Govindram Seksaria Institute of Technology and Science,
Indore, M.P, India
e-mail: jain.anjana@gmail.com

S.C. Satapathy et al. (eds.), *Computer Communication, Networking and Internet Security*, Lecture Notes in Networks and Systems 5,
DOI 10.1007/978-981-10-3226-4_11

performance of wireless systems [1]. Due to the open and heterogeneous nature of the wireless medium, data exchange may suffer from various attacks, resulting major threat to the security which is a critical concern in wireless network and so in IoT [2]. Physical layer security (PLS) is the primary security solution that focuses on utilizing the physical (PHY) layer properties of the wireless channels to safeguard the confidential information transmission against various attacks and is applicable for IoT [3].

The discussion proceeds with motivation in Sect. 2, in Sect. 3 we discuss the basic elements, architecture and communication standards for IoT. Section 4 flashes on the wireless network security at different layers of OSI model, WPLS and PLS methods for IoT. Section 5 concludes the survey with some areas identified for future work.

2 Motivation

Wireless network security is a very critical issue to solve for the IoT. There are various techniques in the literature for wireless security however not all existing techniques are suitable for IoT because the IoT communication devices have some unique characteristics compared to smart phones and tablets. They generally have low data rate requirements, periodic data traffic arrivals, limited hardware and signal processing capabilities, limited storage memory and significant energy constraints [2]. Achieving security at the physical layer overcomes the energy and the hardware constraints and is the motivation behind this discussion over IoT and wireless physical layer security (WPLS).

3 IoT Concepts

IoT can be considered as network of anything, where a variety of things (like sensors, mobile phones, gadgets, people) can interact with one another from any place in the world through an infrastructure like internet to serve specific application [4]. The IoT offers a great market opportunity for equipment manufacturers, internet service providers and application developers. The IoT smart objects are expected to reach 212 billion entities deployed globally by the end of 2020 [5].

3.1 IoT Elements

This section proceeds with the functional classification and discussion over the basic IoT elements with examples of each element [4] as shown in Fig. 1.

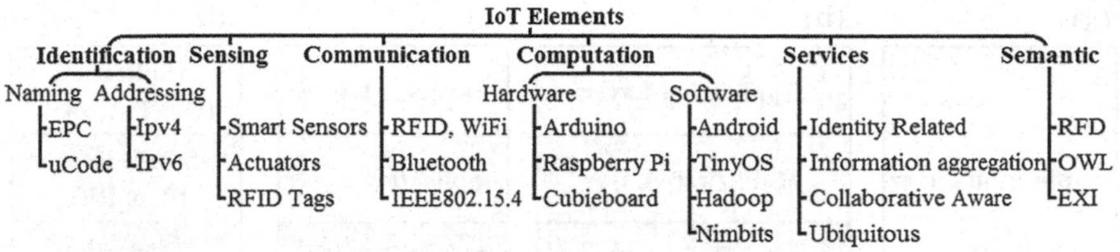

IoT Elements

Identification Naming: EPC, uCode; Addressing: Ipv4, IPv6
Sensing: Smart Sensors, Actuators, RFID Tags
Communication: RFID, WiFi, Bluetooth, IEEE802.15.4
Computation Hardware: Arduino, Raspberry Pi, Cubieboard; Software: Android, TinyOS, Hadoop, Nimbits
Services: Identity Related, Information aggregation, Collaborative Aware, Ubiquitous
Semantic: RFD, OWL, EXI

Fig. 1 Elements of IoT with their categories and examples

IoT end user devices are recognized within the network by identification. Identification broadly constitutes Naming and Addressing. Naming refers to object id (EPC: Electronic Product Code, uCode: Ubiquitous Codes) like "S1" for any sensor in the network while the addressing refers to its IP address (IPv4, IPv6) globally.

Sensing means collecting information from the objects such as smart and embedded sensors, actuators, RFID tags, wearable sensors etc. and sending it to the data storage units/services (memory/Cloud/Big data).

Communication involves the exchange of information between the heterogeneous nodes connected in the internet. Low power communication protocols such as Bluetooth, wifi, IEEE802.15 are applied over communication links.

Computation constitutes the hardware processing unit and the software counterpart and is considered as the "brain" of the IoT. The computation unit for IoT should be low complexity and low power consuming as compared to traditional smart devices.

The IoT services can be divided into four groups [6]: Identity-related Services: most basic services which supports other services, Information Aggregation Services: collects raw data from sensors and supplies it to the IoT applications, Collaborative-Aware Services: rely on Information Aggregation Services and take decision on the collected data, and Ubiquitous Services: provide Collaborative-Aware Services anytime they are needed to anyone who needs them anywhere.

Semantic refers to the capability of the system to extract compiled information from various available resources and provide it to the required services. It is supported by Resource Description Framework (RDF), the Web Ontology Language (OWL) and Efficient XML Interchange (EXI) format.

3.2 *IoT Architecture*

Billions of heterogeneous objects are interconnected in real time systems through internet in the IoT and so a robust and flexible layered architecture is required. The numerous proposed architecture in the literature has not yet converged to an authentic model [7]. In the traditional literature various models were proposed like

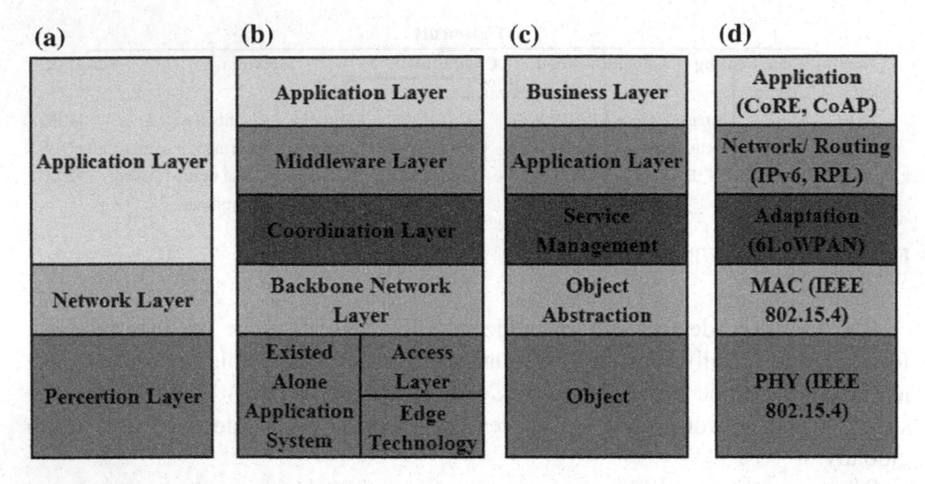

Fig. 2 Various IoT architecture models **a** 3-layer model, **b** 5-layer with enhanced perception layer model, **c** 5-layer model, **d** 5-layer low energy model

3-layer model [8], 5-layer model, 5-layer with enhanced perception layer [9, 10], low energy 5-layer model [11] as shown in Fig. 2. From a bottom-up approach, the following are the main characteristics of the various protocols in this low energy 5-layer model:

PHY Layer represents the physical sensors which collects and process information from ambient. Low-energy communications at the PHY and Medium Access Control (MAC) layers are supported by IEEE 802.15.4 [12]. IEEE 802.15.4 therefore sets the rules for communications at the lower layers of the stack and lays the ground for IoT communication protocols at higher layers.

MAC layer transfers data produced by the PHY layer to the adaptation layer. Low-energy communication environments using IEEE 802.15.4 requires much lesser bytes as compared to other counterparts.

Adaptation Layer pairs a service with its requester based on address and names. Routing over 6LoWPAN environments is supported by the Routing Protocol for Low-power and Lossy Networks (RPL) [13]. Constrained Application Protocol (CoAP) serves the end customers by supporting communication at application layer.

3.3 IoT Communication Standards

Many IoT standards are proposed to simplify the job of application developer and service providers. Basic communication standards like 6LoPAN are generally considered. Figure 3 shows IoT protocol stack with basic communication protocols.

The IoT protocols are classified into four broad categories [4], namely: application protocols, service discovery protocols, infrastructure protocols and other

S. No.	IoT Protocol Stack (Four broad categories)	Position of basic communication protocols in IoT Protocol Stack & Layered Architecture			IoT Layered Architecture (5-Layered Model)
1	Application Protocols	CoAP	MQTT-SN	AMQP	Application Layer
		DDS	HTTP	REST	
2	Service Discovery Protocols	mDNS	DNS-SD		Network Layer
3	Infrastructure Protocols	RPL			
		6LoWPAN	IPv4	IPv6	Adaptation Layer
		IEEE802.15.4			Data Link Layer
		LTE-A	EPC Global	IEEE802.15.4	Physical Layer
4	Influential Protocols	IEEE188.3	IPSec	IEEE1905.1	

Fig. 3 IoT protocol stack with basic communication protocol located in the IoT protocol stack and 5-layered low energy model for IoT

Wireless Security Methodologies

- Authentication
- Authorization
- Encryption
- Latency
- Complexity
- Intercept Probability
- Secracy Capacity
- Channel characterstics

Security Demands	Specific Objective
Authenticity	To differentiate authorized and unauthorized users.
Confidentiality	To limit the confidential data access to authorised user only.
Integrity	To guarantee the accuracy of the transmitted information.
Availability	To make sure that the authorised user can access the services any time on request.

Fig. 4 Wireless security methodologies and demands

influential protocols. However, not all of these protocols have to be bundled together to deliver a given IoT application. Moreover, based on the nature of the IoT application, some standards may not be required to be supported in an application. Basic communication protocols like Advance Message Queuing Protocol (AMQP), Message Queue Telemetry Transport (MQTT), Data Discovery Services (DDS), Representational State Transfer (REST), Routing Protocol (RPL), Hyper Text Transfer Protocol (HTTP), etc. are considered in the IoT protocol standard.

4 Wireless Network Security

Wireless networks usually follow the open systems interconnection (OSI) model constituting the basic seven layers from application to the physical layer considering the top down model. Security threats associated with these protocol layers are generally considered at individual layer level taking into account the integrity, authenticity, availability, and confidentiality [14] as summarized in Fig. 4.

OSI Layer	Basic Protocol		Attack		Security Approach
Application	HTTP SMTP	FTP	Malware Attack SMTP Attack	FTP Bounce Data Attack	Unique pairwise keys and cryptography approach.
Transport	TCP	UDP	Desynchronisation Flooding TCP Sequence Prediction Attack		Client puzzle authentication approach.
Network	IP	ICMP	Sel. Forwarding Sybil Attack	Hijacking Spoofing	Authentication, Monitoring, Probing, Redundancy.
MAC Layer	ALOHA CDMA	CSMA/CA OFDMA	MAC Spoofing Identity Theft	Collision Exhausion	Error correcting code, Rate limitation.
PHY Layer	WiFi 802.15.4	Ethernet Bluetooth	Eavesdropping Radio Interference	Jamming Temporing	Lower duty cycle, Priority message, Spread Spectrum

Fig. 5 Various wireless attacks at different layers of OSI model and the probable security approach with basic protocol applicable at each layer

4.1 OSI Model for Wireless Systems: Attacks and Security Approach

In wired networks, the communicating nodes are physically connected through cables. By contrast, wireless networks are extremely prone to the security threats due to the broadcast nature of the wireless medium. Figure 5 shows various security attacks at different layer of traditional OSI model and the probable security approach.

Since every layer in OSI model rely on different basic protocols so each of them have their own security issues [15–17]. Moreover, wireless networks are vulnerable to malicious attacks like eavesdropping attack, denial-of-service attack, etc.

4.2 Wireless Physical Layer Security

It is common to handle the issues like confidentiality, authentication and privacy in the upper layer of the protocol stack by using key based cryptosystems in the communication systems. The essential requirement of physical layer security is to perform the exchange of confidential information over a wireless medium in the presence of illegitimate user, without relying on higher-layer encryption techniques.

In the recent research many outcomes from the information theory, signal processing, and cryptography reveals that a higher degree of security can be achieved in designing the wireless networks by exploiting the inherent characteristics of physical layer. Physical layer contains definition of hardware specifications, encoding and signaling, data transmission and reception, topology and physical network design. Some key techniques in physical layer are: Multiple Input Multiple Output (MIMO), Code Division Multiple Access (CDMA), Orthogonal Frequency

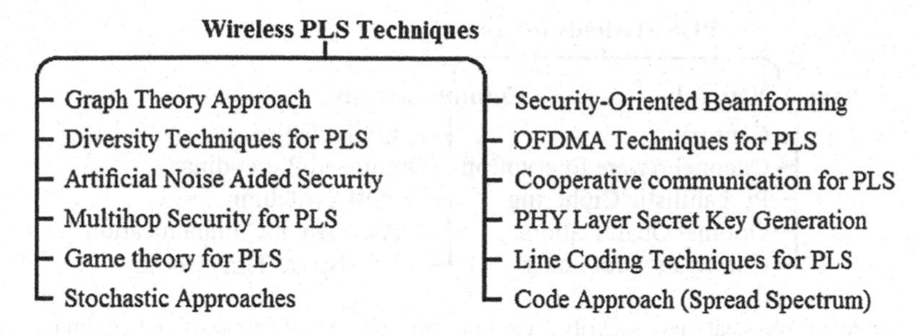

Fig. 6 Different physical layer security techniques applied to wireless networks

Division Multiplexing (OFDM), Algebraic Channel Decomposition Multiplexing (ACDM) etc. [1]. The physical layer security techniques are usually quantified in terms of complexity, secrecy rate, energy efficiency, and Channel State Information (CSI) requirements, relative SINR's, relative BER, relative MSE etc., of the legitimate and illegitimate users. Various wireless physical-layer security techniques are listed in Fig. 6.

4.3 *IoT and WPLS Methods*

There are some challenges to employ many of the traditional wireless physical layer security schemes in an IoT at different levels. The principal barrier is the problem of accurate CSIT acquisition that includes both channel amplitude and phase information. In the IoT, the acquisition of accurate legitimate CSIT is prevented by limited channel training opportunities and the lack of high-rate feedback channels. Transmitting frequent training signals for channel estimation is highly energy inefficient and wastes spectrum access occasions in dense IoT deployments [2]. Second, eavesdropper CSIT is also difficult to acquire when eavesdroppers are external to the IoT system and remain completely passive. Thirdly, the security techniques employed for IoT sensing applications should be of low-complexity and energy-efficient. Fourth, Considering the PLS for wireless network at the sensor level, various factors has to be considered like: multipath effects, fading, randomness, spatially distributed nature of the sensors, heterogeneity, etc. [18, 19].

Wireless physical layer security techniques such as physical layer signal processing can be applied at a gateway receiver to authenticate whether a transmission came from the expected IoT transmitter in the expected location. Investigating approaches such as ciphers or encoding for physical layer confidentiality that are efficient and have little to no message expansion is a promising direction for investigation that would greatly benefit IoT devices. There are many security methods specified in the literature for wireless sensor network security but keeping in mind the distributed nature of sensors and parallel channel access we suggest security methods like censoring, type based access, channel aware encryption

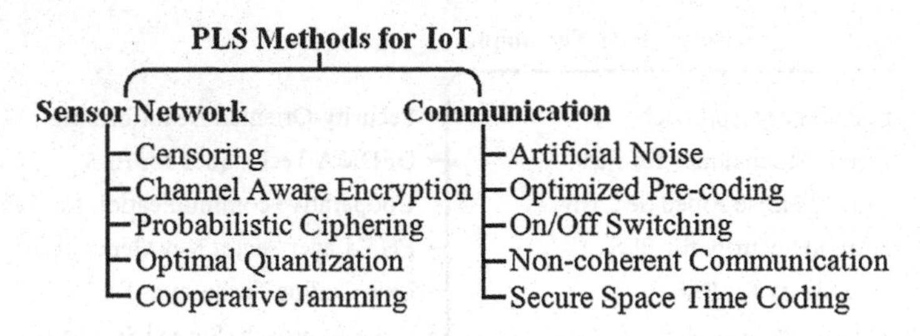

Fig. 7 Different physical layer security techniques for IoT sensor networks and communication

(CAE), optimal quantization, probabilistic ciphering for wireless network security for IoT at sensor level, among these techniques the CAE method offers best optimized combination in terms of low complexity, energy efficiency and CSI requirements [2].

Various signal processing and power approaches considering the artificial noise, Optimized precoding, RFID cryptographic techniques and coding approaches [20] including error correction coding, spread spectrum coding and secure channel analysis [21] can also be employed to achieve security at physical layer. Figure 7 lists out different physical layer security techniques suitable for IoT [2].

5 Conclusion

The IoT may represent a big step ahead for smart and efficient communication through deployment of embedded devices. It represents high degree of consideration towards their security aspects in terms of confidentiality, integrity, privacy and authenticity.

In this survey article we provide an overview of IoT concepts and wireless security with special attention on PLS techniques. It has been concluded that since the objects employed in IoT have limited resources, low complexity design, severe energy constraints so there is a need of designing low energy and low complexity architecture and the security techniques. Further we analyze some PLS techniques for IoT applications considering heterogeneity and distributed nature of the objects at the ground level and observe the appropriateness of CAE method.

We believe that this survey may provide the researchers an overview about the IoT and various security threats in wireless network communication with an approach of WPLS for IoT applications.

Acknowledgements This work is being supported by Department of Electronics and Information Technology (DeitY), Ministry of Communications and IT, Government of India, under Visvesvaraya PhD Scheme for Electronics and IT, www.phd.medialabasia.in.

References

1. Shiu, Y., S., et. al.: Physical Layer Security in Wireless Networks: A Tutorial, pp. 66–74, IEEE Wireless Communications (2011)
2. Mukherjee, A.: Physical-Layer Security in the IoT: Sensing and Communication Confidentiality Under Resource Constraints, vol. 103, pp. 1747–1764, Proc. IEEE (2015)
3. Trappe, W.: The Challenges Facing Physical Layer Security, pp. 6–10, IEEE Communication Magazine (2015)
4. Al-Fuqaha, A. et al,: Internet of Things: A Survey on Enabling Technologies, Protocols, and Applications, vol. 17, pp. 2347–2356, IEEE Comm. Surveys & Tutorials (2015)
5. Gantz, J., Reinsel, D.: The Digital Universe in 2020: Big Data, Bigger Digital Shadows, And Biggest Growth In The Far East, vol. 2007, pp. 1–16, IDC Anal. Future (2012)
6. Gigli, M., Koo, S.: Internet of Things: Services and Applications Categorization., vol. 1, pp. 27–31, Adv. Internet Things (2011)
7. Krco, S., Pokric, B., Carrez, F.: Designing IoT architecture(s): A European perspective. Proc, pp. 79–84, IEEE WF-IoT (2014)
8. Khan, R., Khan, S.U., Zaheer, R., Khan, S.: Future Internet: The IoT Architecture, Possible Applications And Key Challenges, pp. 257–260, in Proc. 10th Int. Conf. FIT (2012)
9. Yang, Z., et al.: Study and Application On The Architecture And Key Technologies for IoT.: In Proc., pp. 747–751, ICMT (2011)
10. Wu, M., Lu, T. J., Ling, F. Y., Sun, J., Du, H. Y.: Research on the Architecture of Internet of Things.: In pp. V5-484–V5-487 Proc. 3rd ICACTE (2010)
11. Granjal, J., Monteiro, E., Silva, J.: Security for the IoT: A Survey of Existing Protocols and Open Research Issues, vol. 17, pp. 1294–1312, IEEE Comm Surveys & Tut. (2015)
12. IEEE Standard for Local and Metropolitan Area Networks—Part 15.4: Low-Rate Wireless Personal Area Networks (LR-WPANs) Amendment 1: MAC Sublayer, IEEE Std. 802.15.4e-2012 (Amendment to IEEE Std. 802.15.4-2011), (2011)
13. Thubert, P. et al.: RPL: IPv6 Routing Protocol for Low-Power and Lossy Networks.: RFC 6550 (2012)
14. Kolias, C., Kambourakis, G., Gritzalis, S.: Attacks and countermeasures on 802.16: Analysis and assessment, vol. 15, pp. 487–514, IEEE Comm. Surveys & Tutorials (2013)
15. Bellovin, S.: Security problems in the TCP/IP protocol suite, vol. 19, pp. 32–48, ACM SIGCOMM Computer communications Review, (1989)
16. Zargar, G., Kabiri, P.: Identification of effective network features to detect Smurf attacks.: Proc. of IEEE Student Conference on Research and Development, UPM Serdang (2009)
17. Shon, T., Choi, W.: An analysis of mobile WiMAX security: Vulnerabilities and solutions, vol. 4658, pp. 88–97, Lecture Notes in Computer Science (2007)
18. Zou, Y., Wang, X., Hanzo, L.: A Survey on Wireless Security: Technical Challenges, Recent Advances and Future Trends, pp. 1–31, Proceedings of IEEE (2015)
19. Granjal, J., Monteiro, E., Silva, J.: Security in the Integration of Low-Power Wireless Sensor Networks with the Internet: A Survey, pp. 264–287, Ad Hoc Networks, Elsevier (2015)
20. Atzori, L., Iera, A., Morabito, G.: The Internet of Things: A survey, vol. 54, pp. 2787–2805, Computer Networks, Elsevier (2010)
21. Pecorella, T., Brilli, L., Mucchi, L.: The Role of Physical Layer Security in IoT: A Novel Perspective, vol. 7(3), Information, MDPI (2016)

Privacy Preservation in Cloud Computing with Double Encryption Method

K. Shivanna, S. Prabhu Deva and M. Santoshkumar

Abstract Cloud computing is advanced technology, which provides a road map to access the applications over the internet. It is a platform, where cloud consumer and data owner customize their applications through online. Due to storing vast amount of data on cloud, there may be many issues related to the security in cloud network. The major among them is the privacy preservation of the network that accesses the cloud network. So in order to address the problem, we proposed a double encryption method that increases privacy for storing and accessing resources on cloud platform. The proposed method provides authentication and privacy features to data owner, cloud service providers and cloud users.

Keywords Authentication · Cloud computing · Data center · Security · Privacy

1 Introduction

Basically cloud is a communication between networks. It provides the services at any time and any remote location. Example of cloud includes Amajan, flip cart etc. and these applications uses the cloud provider and data owner to exchange the information. Cloud computing is a process of accessing the software, applications and store vast amount of data through online. Cloud user need not install piece of software for specific application to run online. Hence, cloud computing making our

K. Shivanna · M. Santoshkumar (✉)
Department of Computer Science and Engineering, GM Institute of Technology, Davanagere, India
e-mail: santoshkumarm@gmit.ac.in

K. Shivanna
e-mail: shivannak@gmit.ac.in

S.P. Deva
Department of Information Science and Engineering, Jawaharalal Nehru College of Engineering, Shivamogga, India
e-mail: pdshirematt@gmail.com

S.C. Satapathy et al. (eds.), *Computer Communication, Networking and Internet Security*, Lecture Notes in Networks and Systems 5,
DOI 10.1007/978-981-10-3226-4_12

lives easier. Computing and configuring the applications through internet that utilizes three service models such as Infrastructure as a Service (IaaS), Platform as a Service (PaaS) and Software as a Service (SaaS). In Software as a Service model, an application, along with required software and operating system are utilized by end users. In PaaS, an end user can install and develop their own applications. IaaS concerned about basic resources such as physical machines, virtual machines and virtual storage.

Nowadays, cloud computing has gain interest from academic and industry sectors. It is a pool of resources that minimizes the cost and provides an on-demand service is satisfies the requirement of end user. As millions of people utilize the facility of cloud computing, providing security for vast data that has been stored and accessed is the major issue. One of the issues is privacy preservation for data upload and data access.

Related to the cloud security, no need to use physical server, rather software security tools are used to protect and observe the flow of data into inside and outside of cloud resources. In IT sectors, cloud security is a major area that consists of much functionality includes protecting sensitive information from theft, data leakage and deletion. Data centers are used to store data securely. Multiple data centers in cloud provider are required to store complex requirement such as credit card numbers or any other secured information. Make sure that cloud provider is responsible for controlling and coordinating all type of events.

There is an opportunity in cloud environment for storing the data with decreased cost rather purchasing the data from local data center. The multiple cloud data center is used to distribute the shared data throughout the network. To ensure the security measure that access data either illegally or accidentally by unauthorized users. By using early encryption tools, private information contained in files can be encrypted before storing it to the cloud network. The tool is limited to encrypt the file in transmit and then decrypt it when file is needed.

Cryptography is plays a major role in management of data storage and used in and around the cloud computing technology. Strong encryption and decryption is required for data whether it is stored and transmitted. Another challenge in cloud

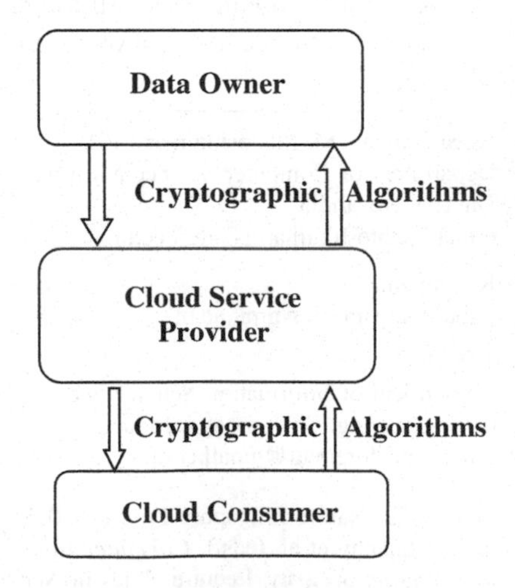

Fig. 1 High level view of cloud security

computing is authentication and authorization of data in distributed environments. The relationship between customer, data owner and service provider in Fig. 1 states that who has administrative rights to access sensitive information.

2 Related Work

Cloud computing is a revolution in information technology. The cloud consumer outsources their sensitive data and personal information to cloud provider's servers which is not within the same trusted domain. So, security is the key parameter to secure the customer's data. The cloud computing security issues are addressed in various standards and techniques which lacks in providing a complete solution.

In [1], authors have presented the concept that includes encryption of IP address in a network is the feasible solution for obtaining privacy in multi-tenant environment. In paper [2], the author have proposed a method to achieve fine grained security with combined approach of PGP and Kerberos in cloud computing. The proposed method provides authentication, confidentiality, integrity, and privacy features to Cloud Service Providers and Cloud Users. Kerberos performs secure verification of users and services based on the concept of trusted third party.

In [3], author have focused on several security issues such as data security, storage technologies and data management. These are the major concerns in present cloud computing and are starting to receive attention from the research community. In [4], the researcher does detailed survey on security and privacy issues in cloud computing. They suggested that authenticated user only access the private data, which secures the data from denial of service and check the existence of cloud resources also cloud environment repeatedly.

In [5], cloud computing architecture and security issues in cloud service models were proposed. Also, they have listed available platform to simulate the research idea. In [6], researchers have explained a storage, virtualization, and networks are the biggest security concerns in Cloud Computing. Virtualization allows multiple users to share a physical server is one of the major concerns for cloud users. Also, they listed some current solutions in order to deals with these threats.

The researchers have proposed a Cloud computing security taxonomy that includes Architecture, Privacy and Compliance [7]. It is strategic to develop new mechanisms that provide the required security level by isolating virtual machines and the associated resources while following best practices in terms of legal regulations and compliance to SLAs. In [8, 9], an overview on security issues in cloud computing and some solutions to deal with them. At the network level, firewall has filtered authorized traffic defined by security policies and intrusion detection systems to monitor the usage of systems and to detect insecure states. The researchers have explained that Mobile devices do not need to have large storage capacity and powerful CPU speed. Due to storing data on cloud there is an issue of data security. To ensure the correctness of users' data in the cloud, they proposed an effective mechanism with salient feature of data integrity and confidentiality [10].

3 Problem Formulation

Cloud consumers and providers faces various data protection risks regarding cloud computing. The major issues are leakage of sensitive data and also unavailability of data. In these situations, cloud consumer may not be check the data availability and accessibility. This may cause major problems in multiple transactions and leads to unreliable services.

4 Proposed Methodologies

The major issue related to security in cloud is the data protection. In most of the system this is handled with the help of secure authentication systems. To handle this threat, we will make use of an improved authentication system concentrates on unauthorized access and accidental loss or damage of the data. Here we proposed a double encryption method to obtain the data protection in cloud computing.

Initially, resource that provides data is given access to the cloud based on a secret key encryption. Later the cloud data are processed into different pools based on the characteristics of the data belongs to them. The secret key is generated based on the pool identifier, which is a unique access parameter to the pool. When a resource request for a data from the cloud, a secret key is generated to them based on their access credentials. That is, if the resource is requesting for a specific data, the secret key is generated for them based on pool ID that matches requirement of the resource.

The proposed method in Fig. 2 is for adding data to the cloud network. Here, the resource requests for adding data to the data center with data credentials. The cloud

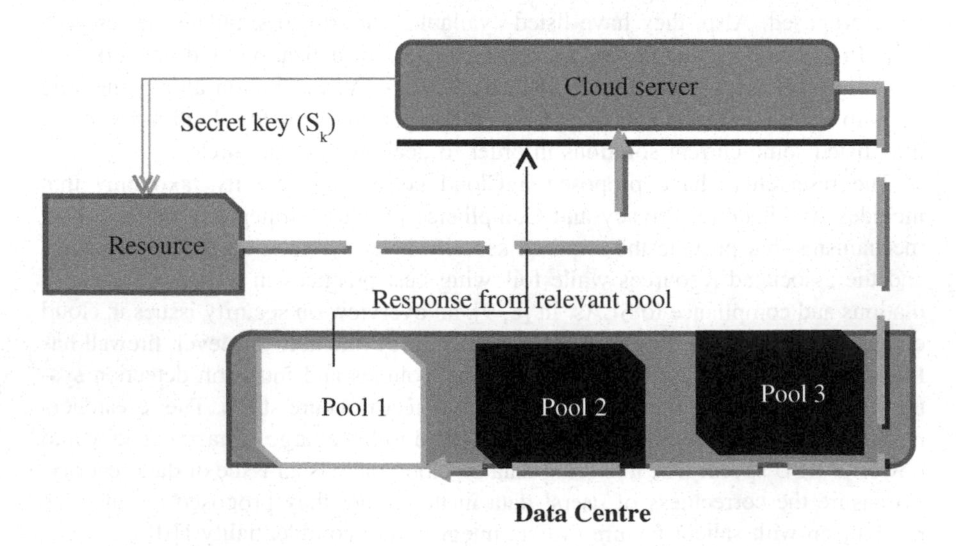

Fig. 2 Architecture for adding data to the data center

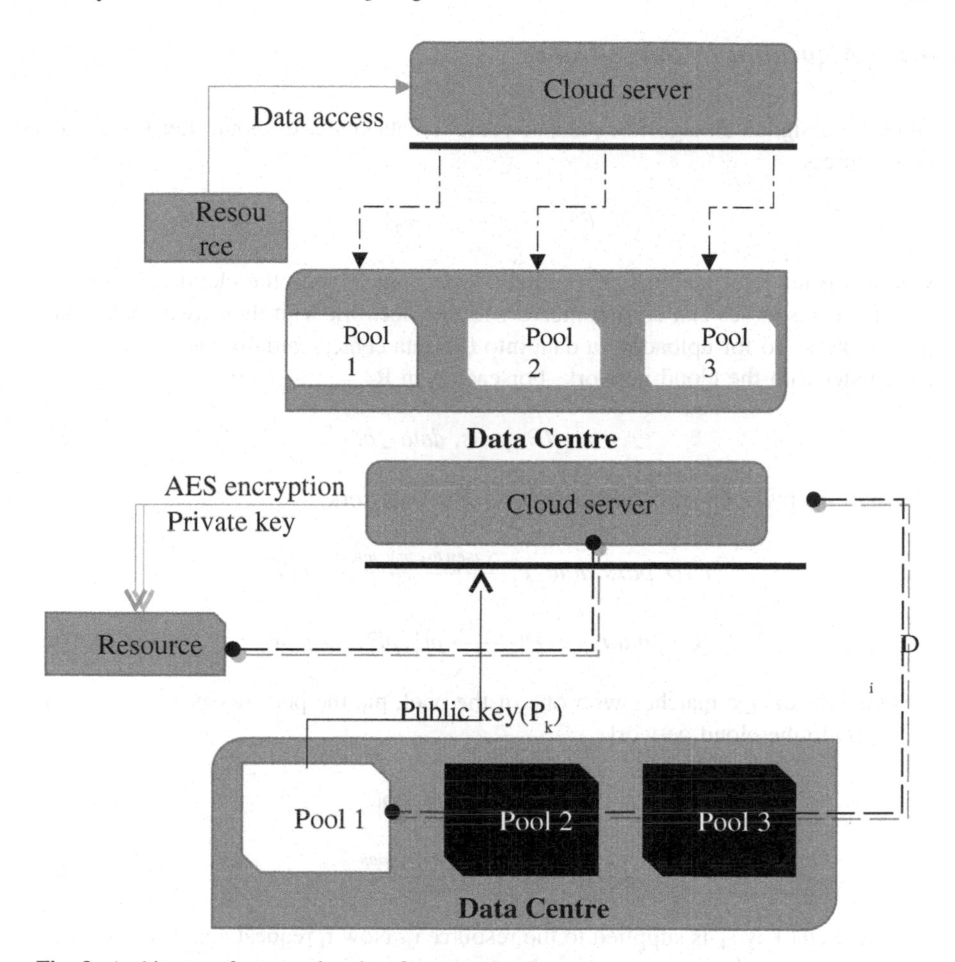

Fig. 3 Architecture for accessing data from the cloud

framework process the request by requesting the relevant pool to the data credentials defined by the resource. A secret key is generated for the resource once the relevant pool is identified. The resource can only give access to the cloud, once it provides the secret key. The secret key is generated by encrypting the resource ID and pool ID. Thus the identity of the resource is anonymized before accessing the cloud.

Now we consider the process of accessing a data from the cloud through resources in Fig. 3. The resource request an access to the cloud server with resource ID and data credentials to which it needs to access. The cloud server send a request to pools with resource's provided parameters and once the criteria are matched by the data center, it releases a public key for the resource via cloud server. The cloud server uses the AES encryption to encrypt the public key with resource ID and issues a private key to the resource. Thus the resource can access only the reserved pool. Through this way make the data secure.

4.1 Algorithm to Store Data

Here we designed an algorithm to store data to data center of cloud. Let R be a set of resources,

$$R = \{r_1, r_2 r_3 \ldots, r_n\} \quad (1)$$

where n is the total number of resources that connects with the cloud network.

All the resources can communicate with the network with their own secret and private keys. So for uploading a data into the data center, initially the resource has to register with the cloud network. For each r_i in R,

$$r = [rID, pass, data\_c] \quad (2)$$

Then the resource r_i request access to cloud network,

$$r[ID, pass, data\_c] \xrightarrow{\text{request to communicate}} [CN] \quad (3)$$

$$[CN][data\_c, rID] \xrightarrow{checks} [pl1, pl2, \ldots, p\ln] \quad (4)$$

Once the data_c matches with one of the pool, pl_i, the pool releases the pool id pl_ID to the cloud network.

$$[CN] \xleftarrow{issues} [pl_i, pl\_ID] \quad (5)$$

$$[CN][r\_ID, pl\_ID] \xrightarrow{encryption} s_k \quad (6)$$

The secret key s_k is supplied to the resource r_i. Now r_i request again with the s_k. Then the cloud grants access to the resource to write into the data.

$$r[ID, pass, data\_c, s_k] \xrightarrow{for\_data\_writing} [CN] \quad (7)$$

$$\text{Now}, r[ID, pass, data\_c] \xrightarrow{access} [CN][pl_i, pl\_ID] \quad (8)$$

Once the resource gets access to the pools, it can upload the relevant data to the pools.

4.2 Algorithm to Access Data

For accessing the resource from cloud, following steps are considered.

$$r[ID,\ pass,\ data\_r]\xrightarrow{request\ to\ communicate}[CN]] \tag{9}$$

Here, data_r is the data request to which the resource want to access.

$$[CN][data\_r, rID]\xrightarrow{checks}[pl1, pl2, \ldots., p\ln] \tag{10}$$

Once the data_r matches with one of the pool, pl_i, the pool releases the pool id pl_ID to the cloud network.

$$\begin{aligned} &[CN]\xleftarrow{issues}[pl_i, pl\_ID] \\ &[CN][r\_ID, pl\_ID]\xrightarrow{AES}p_k \end{aligned} \tag{11}$$

The private key p_k is supplied to the resource r_i. Now r_i request again with the p_k. Then the cloud grants access to the resource to read the data from the pool.

$$r[ID, pass, data\_r, p_k]\xrightarrow{for\_data\_reading}[CN] \tag{12}$$

5 Results and Discussion

In order to satisfy the privacy preservations, our architecture features a double encryption algorithm that minimizes the limitation of symmetric and asymmetric cryptography. In symmetric key encryption, one secret key shared between A and B for encryption and decryption. Here, security is failed if secret key falls on to the wrong hands. In asymmetric key encryption, public and private keys are used for encryption and decryption. Here, private key is kept in secret whereas public keys are stored in public repository. But in a large network, there is an ambiguity for finding public key belongs to specific user. Our proposed model has two approaches discussed below.

5.1 *Security Measures for Data Storing*

In order to measure the security for data storing, we proposed a model that requests for adding data to the data center with data credentials. The cloud provider generates the secret key once the relevant pool is identified. Here, secret key is obtained by encrypting the resource ID and pool ID. Once the data owner outsources the resource to data center by using secret key of the particular resource, ultimately privacy of data storing is increases.

Table 1 Comparative analysis

Challenges	Symmetric encryption	Asymmetric encryption	Proposed model
Secure data share	Medium	High	Very High
Key storage	Low	High	Medium
User authentication	No	No	Yes
Man in middle attack	Yes	Yes	No
Computational cost	Low	High	Medium

5.2 Security Measures for Data Access

To measure the security for data access, we designed architecture that request an access to the cloud server with resource ID and data credentials. To access the resource from the specific pool in data center, the cloud server uses the AES encryption to encrypt the public key with resource ID and issues a private key to the resource. The proposed technique uses combinations of keys that increase the security for accessing a resource from the cloud. Comparative study of earlier techniques with respect to proposed model is described in Table 1.

6 Conclusion and Future Scope

As all we know that cloud computing utilization is increased rapidly, privacy preservation on cloud is considered as a major challenge. To achieve privacy preservation in cloud, various services and mechanisms are proposed and presented. Our proposed mechanisms include double encryption methods for storing and accessing the data securely with data owner, cloud providers and cloud users. The proposed method performs the secure verifications of data between users and cloud providers. Further, this model is implemented in Java platform; in future this will be tested in cloud environment like CloudSim and GreenssCloud.

References

1. Sriram Natarajan, and Tilman Wolf, Network-Level Privacy for Hosted Cloud Services. IEEE Sponsored Second Int. Conference on Electronics and Communication Systems (ICECS 2015).
2. Subhash Chandra Patel, Ravi Shankar Singh and Sumit Jaiswal, Secure and Privacy Enhanced Authentication Framework for Cloud Computing. IEEE Sponsored Second Int. Conference on Electronics and Communication Systems (ICECS), 2015.
3. Qi Zhang, Lu Cheng, Raouf Boutaba, Cloud computing: state-of-the-art and research challenges. Springer, J Internet Serv Appl (2010) 1: 7–18.

4. IC. Saravanakumar, and C. Arun, Survey on Interoperability, Security, Trust, Privacy Standardization of Cloud Computing. IEEE International Conference on Contemporary Computing and Informatics (IC3I), 2014.
5. Deepak Puthal, Sambit Mishra, and Satyabrata Swain, Cloud Computing Features, Issues and Challenges: A Big Picture. IEEE International Conference on Computational Intelligence & Networks, 2015.
6. Keiko Hashizume, David G Rosado, Eduardo Fernández-Medina and Eduardo B Fernandez, An Analysis of Security Issues for Cloud Computing. Hashizume et al. Journal of Internet Services and Applications, Springer, 2013.
7. Nelson Gonzalez, Charles Miers, Fernando Redigolo, Marcos Simplicio, Tereza Carvalho, Mats Naslund and Makan Pourzandi, A Quantitative Analysis of Current Security Concerns and Solutions for Cloud Computing. Gonzalez et al. Journal of Cloud Computing: Advances, Systems and Applications,Springer,2012.
8. Dijiang Huang, Zhibin Zhou, Le Xu, Tianyi Xing and Yunji Zhong, Secure Data Processing Framework for Mobile Cloud Computing. IEEE INFOCOM Workshop on Cloud Computing, 2011.
9. Balasaraswathi V.R and Manikandan S, Enhanced Security for Multi-Cloud Storage using Cryptographic Data Splitting with Dynamic Approach. IEEE International Conference on Advanced Communication Control and Computing Technologies (ICACCCT), 2014.
10. Preeti Garg, and Dr. Vineet Shanna, An Efficient and Secure Data Storage in Mobile Cloud Computing through RSA and Hash Function. IEEE International Conference on Issues and Challenges in Intelligent Computing Techniques (ICICT), 2014.
11. Tumpe Moyo, and Jagdev Bhogal, Investigating Security Issues in Cloud Computing. IEEE Eighth International Conference on Complex, Intelligent and Software Intensive Systems, 2014.

A Machine Learning Based Approach for Opinion Mining on Social Network Data

Fayeza Arif and Uma N. Dulhare

Abstract Micro-blogging has been widely used for voicing out opinions in the public domain. One such website, Twitter is a point of attraction for researchers in the areas such as prediction of electoral events, movie box office, stock market, consumer brands etc. In our paper, we focus on using Twitter, for the task of opinion mining. We explore how combining the different parameters affect the accuracy of the machine-learning algorithms with respect to the consumer products. In this paper, we have combined the methods of feature extraction with a parameter known as negation handling. Negation words can awfully change the meaning of a sentence and hence the sentiment expressed in them. We experimented with supervised learning methods like Naïve Bayes (NB) Classifier and Maximum Entropy (MaxEnt) Classifier along with optimization iteration algorithms i.e., Generalized Iterative Scaling (GIS) and Improved Iterative Scaling (IIS). Experimental evaluations show that our proposed technique is better. We have obtained a 99.29% of specificity measure using the MaxEnt-IIS Classifier.

Keywords Opinion mining · Sentiment analysis · Negation detection · Supervised learning

1 Introduction

Big data is an evolving term used to describe large structured, semi-structured and unstructured data that has the potential to be mined for useful information. It is difficult to process big data through traditional data processing techniques. Big data analytics is implemented in various domains like network simulation, cloud computing, statistical analysis, user behavioral study etc.

F. Arif (✉) · U.N. Dulhare
Muffakham Jah College of Engineering and Technology, Hyderabad, Telangana, India
e-mail: arif.fayeza@gmail.com

U.N. Dulhare
e-mail: prof.umadulhare@gmail.com

S.C. Satapathy et al. (eds.), *Computer Communication, Networking and Internet Security*, Lecture Notes in Networks and Systems 5,
DOI 10.1007/978-981-10-3226-4_13

With the dramatic increase of social network, millions of users broadcast their thoughts and opinions on various subjects and topics. Huge amount of data from these text messages are appearing everyday on this mainstream micro blogging sites such as Tumbler, Twitter, Facebook etc. The users get the freedom to voice out their opinion on various topics [1]. Micro-blogging services provide easier accessibility options and a restriction less message format. Because of these advantages, internet users are inclined more to social media than the traditional blogging and mailing list. Twitter is a popular social-network website that lets its user to post short messages: less than 140 characters i.e. around 11 words per message. The tweets coming from all around the world reflects a mass real-time information system with more than one million messages per hour.

Users express their views in various ways which becomes difficult to process using the traditional text processing techniques. Many established methods use literal meaning of the sentence to process it as positive or negative irrespective of its hidden meaning. In this paper, we try to introduce a new parameter i.e., negation detection and scope of negation in the feature extraction. The need of negation handling can be explained by the difference in the meaning of the two phrases, "This is good" with "This is not good". The meaning of the sentence changes with the presence of negative word. Such words are said to be in the scope of negation.

The rest of the paper is organized as follows: Sect. 2 describes the prior works on opinion mining and their application for micro-blogging. Section 3 deals with Methodology and system flow while Sect. 4 gives the Results and Evaluation; Conclusions are detailed in the Sect. 5.

2 Background and Related Work

With the rise of blogs and social networks, opinion mining and sentiment analysis has become a field of interest for many researchers. Opinion mining focuses on using information processing techniques to gain valuable information from the large user generated content.

Malhar Anjaria et al., presents an approach for the sentiment extraction using the direct and extended features of the twitter data [2]. The authors have made use of the supervised learning algorithms like Support Vector Machines (SVM), Naïve Bayes, Maximum Entropy and Artificial Neural Networks. The SVM is combined with principal component analysis (PCA) in or to perform dimensionality reduction. They have come to a conclusion that twitter can pass or fail in predicting the electoral event outcome. The authors propose an approach for user influence factor for electoral event prediction. Pak and Paroubek [3] presents a method for an automatic collection of dataset that is used to train the sentiment classifier, which determines the sentiments of documents. Pang and Lee [4] proposed a one to five star ratings instead of stating whether a review is positive or negative. The ratio of positive words to total words was considered to estimate the opinion.

Sidorov et al. [5] examined the working of different classifier for opinion mining over Spanish twitter data. They used different settings for feature extractions, corpus size and machine learning algorithms focusing on precision. They also provided a resource for analysis of emotions based on tweets. Walaa Medhat et al., provided an in-depth survey on different sentiment analysis techniques [6]. They have focused on the basic data preprocessing, feature extraction and feature selection techniques and emphasized on the machine learning algorithms. Finally it shows the evaluations of the techniques in terms of precision, recall and f1-score.

Boutet et al. [7] provided an algorithm to identify the political leaning of users using messages related to political parties. It made prediction with the political party characteristics with the use of user behavior analysis and influence factor. Naïve Bayes (NB), Support Vector Machines (SVM) and Maximum Entropy (MaxEnt) classifier are well discussed in many literatures such as Pang and Lee [4]. Majority of the researchers have carried out various feature extraction methods but ignored negation handling as this can add value to improve the prediction. Das and Chen [8] proposed a method by adding negation "NOT" to words that are close to negation terms, so that in the sentence "I don't like to work", the word "like" is converted into the new word "like-NOT". Meeyoung et al. presents a semantic based approach in order to identify the subject in a tweet, like a person, organization etc. [9]. They also determine that removing stop words is not necessary and may have a bad effect while training the classifier.

3 Methodology

In our proposed work, we used preprocessing techniques on tweets, feature extraction method to extract feature vector and then used two machine learning classifiers to predicate the best of the results for sentiment analysis of twitter. We have utilized the corpus provided by Stanford as our training data. The process as follows

- For data pre-processing, we have applied data cleaning, normalization and feature reduction technique. In data cleaning, we remove the redundant data whereas stemming algorithms are applied to normalize the data.
- For feature extraction, we propose a new parameter that should be considered to combine with the basic feature extraction models like unigram, bigram and trigram. The proposed parameter is Negation detection, which deals with explicit negation cues.
- We finally train our classifier using two machine-learning algorithms—Naive Bayes (NB) and Maximum Entropy (MaxEnt). We have also classified the optimization iterative parameters for MaxEnt i.e., Generalized Iterative Scaling (GIS) and Improved Iterative Scaling (IIS).
- Thus, we obtain the prediction accuracy of these two classifiers.

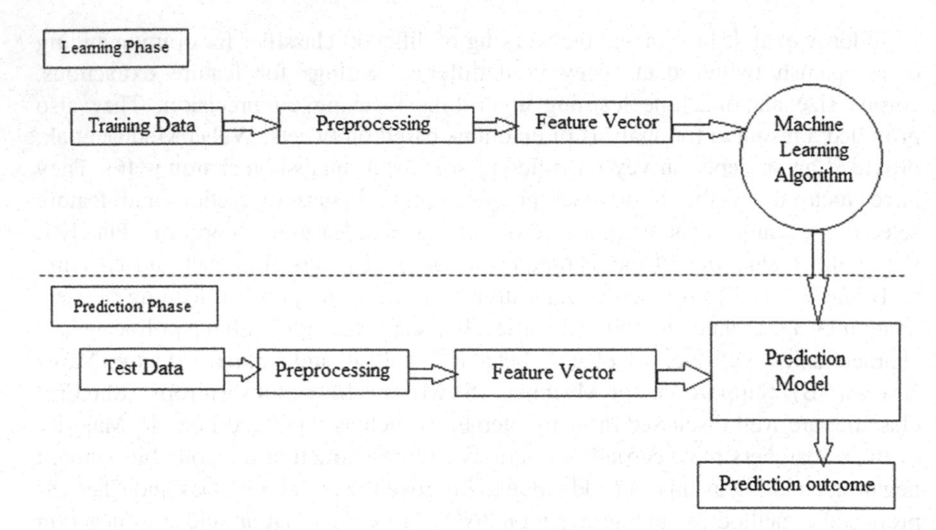

Fig. 1 System architecture

Figure 1 depicts our basic workflow for Twitter sentiment analysis utilizing features of machine learning classifiers. The algorithm-1 shows the basic system flow of the opinion mining with respect to consumer brands. Following subsections explains the system architecture (Fig. 1) in detail.

3.1 Data Collection

The corpus provided by Stanford Natural Language processing research group which is publicly available. The Twitter API is queried for obtaining training set by happy emoticons like ":)" and sad emotions like ": (" and labeled them to positive or negative. This corpus also contains 500 tweets manually collected and labeled for testing purposes. We randomly sampled and used 1000–20,000 tweets from the corpus.

3.2 Preprocessing and Stemming

User-generated content on the web is useful for learning. It is very important to normalize the data by a series of pre-processing steps. We have applied an extensive set of pre-processing steps in order to reduce the feature set so as to make it more suitable for learning algorithms.

Before preprocessing

1. @PrincessSuperC Hey Cici sweetheart! Just wanted to let u know I luv u! OH! and will the mixtape drop soon? FANTASY RIDE MAY 5TH!!!!
2. @Msdebramaye I heard about that contest! Congrats girl!!
3. UNC!!! NCAA Champs!! Franklin St.: I WAS THERE!! WILD AND CRAZY!!!!!! Nothing like it...EVER http://tinyurl.com/49955t3

After preprocessing

1. __HANDL hey cici sweetheart! just wanted to let u know i luv u! oh! and will the mixtape drop soon? fantasy ride may 5th!
2. __HANDL i heard about that contest! congrats girl!!
3. unc!!! ncaa champs!! franklin st.: i was there!! wild and crazy!!!!!! nothing like it...ever URL

Fig. 2 Before and after preprocessing

Stemming is the process of reducing the derived word to the root word. There are different algorithms associated with stemming. The stemming algorithm used here is the porter stemmer as it provides an excellent speed, readability, and accuracy. An important feature to note is that it doesn't involve recursion. We have also used wordnet lemmatizer for stemming and compared both resulting in negligible difference in accuracy (Fig. 2).

The data cleaning technique used in this approach is transformation data approach which involves the replacement of hashtags, handles, URLs, emoticons, repeated characters, and punctuation marks. Since we are using the twitter data, the hashtags, handles and URLs etc. are to be defined accordingly hence it has to be manually done. With this technique, we make sure that important information is not overlooked or deleted.

1. *Hashtags*: A hashtag is a word prefixed with the hash symbol (#). These are used to subject the current trends in Twitter. Regular Expression: # (\w+), Replace Expression: HASH_\1
2. *Handles*: Every twitter username is unique. Anything directed towards that user can be indicated as writing their username preceded by @ symbol. Regular Expression: @ (\w+)—Replace Expression: HNDL_\1
3. *URLs*: Many users share hyperlinks/URLs in their tweets. In text mining, usually a particular URL is not important, but presence of a URL can be useful feature.
   Regular Expression: `(http|https|ftp)://[a-zA-Z0-9\./]+`, Replace Expression: URL
4. *Emoticons*: Many users use emoticons and it is prevalent throughout web, more on micro-blogging sites. We identify the emoticons and replace them with a single word, the emoticons such as :), :-), (: are replaced by EMOT_SMILEY

where as emoticons like :,(, :"(, : ((are replaced by EMOT_CRY and other emoticons are ignored.

5. *Punctuations*: From the point of view of text classification, only some of the punctuation marks hold value like question mark and exclamation mark as they may provide information about sentiment of the text. Punctuation marks like ? are replaced by PUNC_QUES, ! by PUNC_EXCL etc.
6. *Repeating Characters*: People usually have the habit of using repeating characters while using colloquial language, like "chillllll" instead of "chill". We replace characters repeating more than twice as two characters.

3.3 Feature Vector

We use the unigram, bigram and trigram feature extraction methods. We have proposed and collaborated negation handling with unigram, bigram and trigram such as Unigram + Negation, Bigram + Negation, Trigram + Negation. As we get the normalized tweets from preprocessing tweets, which is given as input to extract positive or negative words, then we calculate the word frequency to obtain the feature vector [10]. Negation handling involves a two step process i.e., detecting the explicit negation words and the ranging the negation of these words.

1. *Detecting Explicit Negation Word*
   In order to detect explicit negation words, the explicit negation words like nothing, no, never, nowhere, no one, not, havnt, hasn't, hadn't, couldn't, cant, shouldn't, wont, wouldn't, dont, doesnt, didn't, isn't, aren't, aint and any word that ends with "n't are considered. The search is carried out using regular expressions.
2. *Ranging the Negation*
   Words that come immediately before and after the negation words or cues are the most negative. The words that are far away from negative cues do not belong to the range of negation shown from Fig. 3. The sentence that has a negative word changes the meaning of it. We have defined the left and right negativity of a word as there are chances that meaning of that word could be actually the opposite of the sentence. Left negativity depends on the closest negation word or cue on the left and similarly is done for calculating the right negativity.

```
Words: ['HASH_Skype', 'crash', 'too', 'much', 'PUNC_EXCL',
'not', 'expect', 'this', 'from', 'HASH_MICROSOFT']
Neg_l: [0.0, 0.0, 0.0, 0.0, 0.0, 1.0, 0.9, 0.8, 0.7, 0.6]
Neg_r: [0.5, 0.6, 0.7, 0.8, 0.9, 1.0, 0.0, 0.0, 0.0, 0.0]
```

Fig. 3 Range of negation

Algorithm 1—Performing Negation handling
Input: tweet, neg_regex (list of negation words)
Output: detecting and ranging the negation
def get_negtn_feat(words) {
//Search for negation words from the list
negtn = [bool(neg_regex.search(w)) for w in words]
//Assign values for left and negation words
return dict(zip(['negtn_l('+w+')' for w in words] + ['negtn_r('+w+')' for w in words], left + right)) }

The author analyzes the problem of polarity detection when negative words *not* and *hardly* appear in a sentence. They also show that there is no difference in accuracy with respect to different window sizes (negation scopes) [11]. Thus in the proposed system, instead of using only *not* and *hardly*, we make use of more negation cues defined in a list. Analysis show that there is an increase in accuracy with the proposed work as shown in the graph (Fig. 4).

The author addresses different ways a negation can be present in a sentence i.e., it can be a local negation, longer-distance negation dependencies or the negation of the subject [12]. It gives the understanding of the approach using bag-of-words approach later resolving it using dependency tree [13]. The proposed approach starts with calculating the number of words in a sentence then allotting values to the words based the left negativity (neg_l) words that immediately preceds the negation cue and words that follow the negation cure to be right negativity (neg_r). The dependency of left negativity is on closest negation words that lie on the left of the sentence and similarly for right negativity.

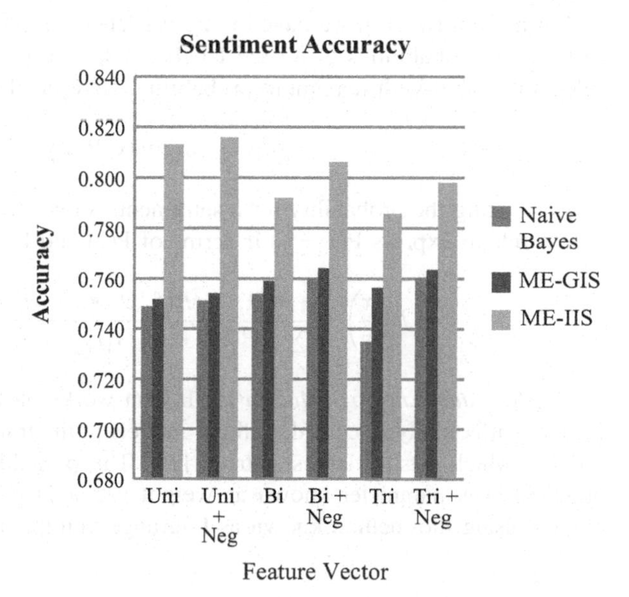

Fig. 4 Sentiment accuracy

Algorithm 2—Performing opinion mining on consumer products.
Input: Collection of the corpus data
Output: Outcome in terms of analyzing the metric parameters

1. Tweet = { {tweeted, tweetdata, userId, reTweet, tweetDate} }
   Store the tweet information in the .csv file.
2. normTweet = normalization(Tweet). Perform data cleaning and normalization
   Store the normalized tweets in .csv file
3. FeatVect = word_features(normTweet). For every word set create feature vector
4. Perform feature extraction and generate graphs for unigram, bigram, trigram
5. Generate the .csv files for respective n-grams after feature extraction.
6. Apply incremental machine-learning training with featVect for output class.
7. Measure the outcome in terms of accuracy, f1-score, precision, recall, specificity.

3.4 Machine Learning Techniques

In text classification and opinion mining domain, machine learning algorithms are widely used. We use two machine learning algorithms like Naïve Bayes and Maximum Entropy Classifiers.

i. Naïve Bayes (NB): NB classifier is the simplest and the fastest probabilistic classifier. It is based on Bayes rule with independent feature selection, which works well on text categorization [14]. It is the fastest learning algorithm especially for the training phase.

For a given tweet, if we have to find the label i.e., sentiment for it, we will have to find the probabilities of all the labels, as we are given that feature. Finally, it selects the label with maximum probability. Here label = x, features = y.

$$\mathrm{x_{NB}} := \mathrm{argmax_x\, P(x|y)} \tag{1}$$

For finding the probability for a sentiment, Naïve Bayes algorithm first uses the Bayes rule to express P(x − y) in terms of P(x) and P(y − x) as,

$$P\left(\frac{x}{y}\right) = \frac{P(x)\ *\ P(f_1|x)\ *\ \cdots\ *\ P(f_n|x)}{\sum_1 (P(1)\ *\ P(f1|1)\ *\ \cdots\ P(f_n|1))} \tag{2}$$

ii. Maximum Entropy (MaxEnt): MaxEnt works on the principle of Maximum Entropy, it is likely to consider all the models in the training data, and then selects the data which gets the largest entropy [15]. This probability distribution function is attached by a parameter known as weight vector [16]. The optimal value is calculated using the method known as Lagrange multipliers.

Table 1 Statistical significance of F-test variance

F-test two-sample for negation		
	Negation = true	Negation = false
Mean	0.80295	0.799311
Variance	9.18767	0.000105
Observations	4	4
Df	3	3
F	0.877678905	
P(F<=f) one-tail	0.45855686	
F critical one-tail	0.107797789	

$$P\left(\frac{x}{y}\right) = \frac{\sum_i w_{if_i(x)}}{\sum_{1 \in x} \sum_i w_{if_i(l)}} \tag{3}$$

MaxEnt classifier makes use of iterative scaling algorithms i.e., Generalized Iterative Scaling (GIS) and Improved Iterative Scaling (IIS). These are simple algorithms that help in estimating the parameters of the MaxEnt Classifier. GIS is significantly a simpler algorithm even though it requires more number of iterations than IIS. GIS is much faster than IIS.

NB and MaxEnt are supervised learning algorithms which work with labeled data and are known for applications in text classification [17]. These algorithms show different approach of estimating parameter for common log-linear models. The results show inclination towards MaxEnt than NB. NB is known to perform with independent assumptions where as MaxEnt performs well with dependent features using estimation parameters like GIS and IIS. As we are dealing with labeled data, we make use of supervised learning techniques like NB and MaxEnt (Table 1).

From previous works, applying machine learning techniques based on unigram models can achieve accuracy over 80%. The f-test is calculated by dividing variance of variable 1 by variance of variable 2. Thus, we can say that our classifier obtained 87.7% and performs at par with the state of the around result.

4 Results and Discussion

We use the Stanford corpus provided by Stanford natural language processing research group that has tweets in English language with tweet id, tweet id number, subject, time and date and their respective sentiment. We have used Python programming language. Data cleaning and normalization gives us 54.34% of the original data. Table 2 shows the result after cleaning and normalization process.

We used unigram, unigram + negation, bigram, bigram + negation, trigram, trigram + negation in order to achieve the sentiment values for each value. We follow tenfold accuracy matrix where the datasets were taken at random. Among

Table 2 Effects of data cleaning and normalization

Preprocessing	Words	Percentage (%)
Hashtags	15,550	97.74
Handles	13,245	83.25
URLs	15,335	96.39
Emoticons	15,541	97.68
Punctuations	11,225	70.55
Repeatings	15,276	96.02
All	8646	54.34

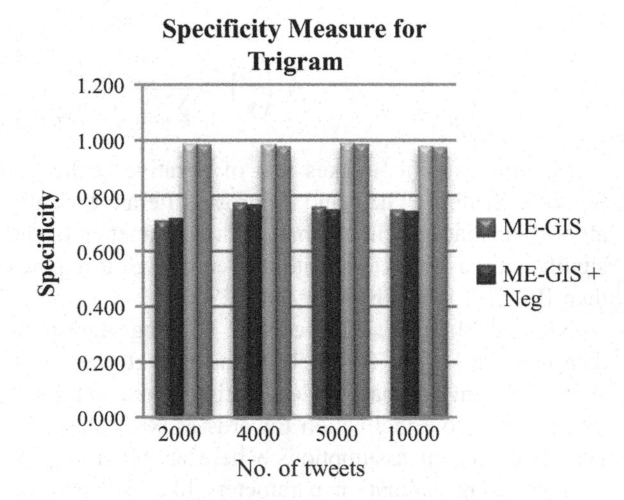

Fig. 5 Specificity versus number of tweets

feature extraction model, the trigram + Negation provide better accuracy results. From Fig. 4 shows that the highest accuracy achieved is 81.6% for unigram. From the graph, we can also analyze that basic feature extractors are giving lesser accuracy than that which is collaborated with negation in all cases. Hence, this approach yields good results. Even though NB processes the training phase fast, but it lacks few points of percentages when compared to MaxEnt Classifier.

Specificity is also known as True Negative Rate which correctly identifies the negatives. For example, no healthy persons are identified as sick. The results of specificity are exceptionally well compared to other metrics. The MaxEnt-IIS obtains 99.29% of specificity yielding the highest result for 5000 tweets as shown in Fig. 5.

F1-score is calculated by the harmonic mean of precision and recall. Figure 6, shows that unigram extractor combined with negation outperforms in all cases and the highest f-score measure is obtained by MaxEnt-IIS classifier with 0.77 f-score.

$$\mathrm{F1-score} = \frac{2 * (\mathrm{precision} * \mathrm{recall})}{\mathrm{precision} + \mathrm{recall}}$$

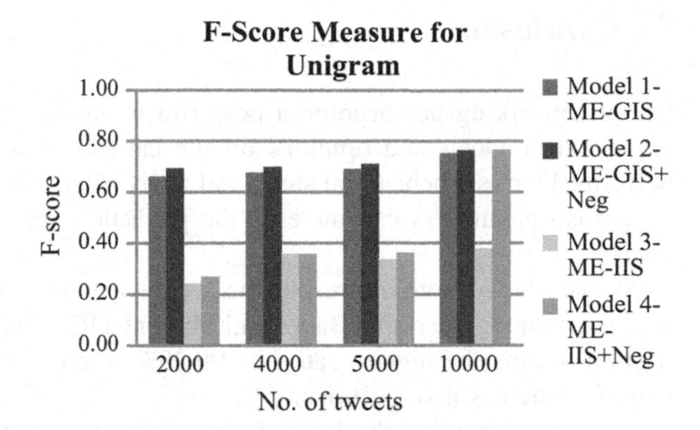

Fig. 6 F1-Score measure for unigram

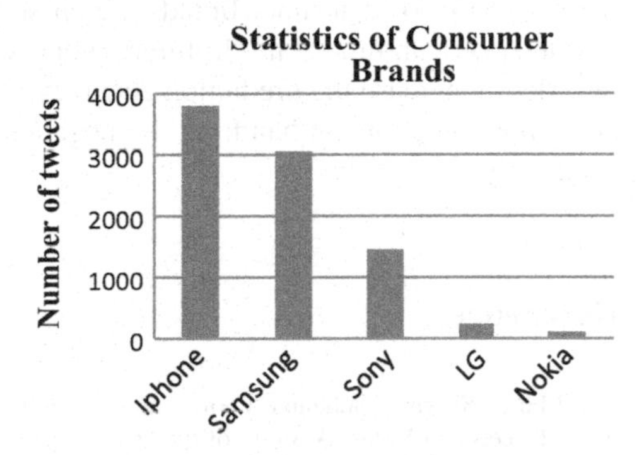

Fig. 7 Graph showing statistics of consumer brands

Figure 7 depicts the statistics of consumer brands which here is specifying the mobile phone brands. We can see that the number of tweets posted by users in an hour shows drastic inclinations towards iPhone rather than Samsung. This graph has been generated based on the American/Chicago timezone. Experimental evaluations show that implementing Maximum Entropy with Improved Iterative Scaling (IIS) yields increase in accuracy when compared to Generalized Iterative Scaling (GIS). But IIS takes more running time than GIS. The evaluations also show that the addition of feature from n-gram only to n-gram + negation gives an increase in accuracy in every case i.e., unigram, bi-gram, trigram. The stemming algorithms have been tried in this project; it shows that there is negligible difference between Porter stemmer and Wordnet Lemmatizer in accuracy when trained on both the classifiers. Experimental analysis shows that the higher accuracy can be acquired by tuning the different parameters with respect to feature extraction methods and classification models. We have evaluated the graphs for different measures like F1-score, specificity and sentiment accuracy. The highest results obtained using MaxEnt-IIS classifier is from the specificity for 99.29%.

5 Conclusion

Social networking has become a powerful means of communication for users to express their views and opinions on the current trend. The big data analysis is performed for user behavioral study and many other important domains. These user behavioral parameters can increase the prediction accuracy along with sentiment analysis.

We concluded from our results the accuracy of MaxEnt-IIS yields 81.6% which is highest among the Naïve Bayes and MaxEnt–GIS classifier. The specificity of the trigram obtains the highest rate i.e., 99.29% where it shows that it has correctly identified the negative portions.

Also we compared the basic feature extractors and extractors with negation handling, the results of our proposed approach yields higher accuracy. We analyze the popularity of consumer brands which shows inclination towards Iphone based on the American time zone. In future work we train the data on different classifiers in order to increase the prediction accuracy along with sentiment analysis. As there are some exceptions in handling the negation, we should attempt to model it more accurately.

References

1. Marko Skoric, Nathaniel Poor, Palakorn Achananuparp, Ee-Peng Lim, Jing Jiang et al. "Tweets and Votes: A Study of the 2011 Singapore General Election" In proceedings at 2012 45th Hawaii International Conference on System Sciences., (2012)
2. Malhar Anjaria, Ram Mahana Reddy Guddeti, "Influence Factor Based Opinion Mining of Twitter Data Using Supervised Learning", Proceeding of the Sixth International Conference on Communication System and Networks, (2014)
3. Alexander Pak and Patrick Paroubek. "Twitter as a corpus for sentiment analysis and opinion mining", Proceedings of the Seventh International Conference on Language Resources and Evaluation (LREC' 10), (2010)
4. Bo Pang. Lilliam Lee, "Seeing Stars: Exploiting class relationships for sentiment categorization with respect to rating scales", (2002)
5. Grigori Sidorov, Juan Gordon et al., Empirical Study of Machine Learning Based Approach for Opinion Mining in Tweets. In 11th Mexican International Conference on Artificial Intelligence, MICAI 2012, San Luis Potosí, Mexico, (2012)
6. Walaa Medhat, Ahmed Hassan, Hoda korashy, Sentiment analysis algorithms and applications: A survey, In Ain Shams Engineering Journal, www.sciencedirect.com, (2014)
7. Antonie Boutet et al., What's in your tweet: I know Who You Supported in the UK 2010 general elections, Association for the Advancement of Artificial Intelligence, (2012)
8. Sanjiv Das and Mike Chen. Yahoo! for Amazon: Extracting market sentiment rom stock message boards. In Proceedings of the Asia Pacific Finance Association Annual Conference (APFA), (2001)
9. Meeyoung, C. et al., Measuring User Influence in Twitter: The Million Follower Fallacy. In Fourth International AAAI Conference on Weblogs and Social Media, (2010)

10. Johan Bollen, Alberto Pepe, and Huina Mao. "Modeling public mood and emotion: Twitter sentiment and socioeconomic phenomena". In Proceedings of the Fifth International AAAI Conference on Weblogs and Social Media (ICWSM), Barcelona, Spain, (2011)
11. Dadvar, Maral and Hauff, Claudia and Jong, Franciska de, Scope of negation detection in sentiment analysis. In Dutch-Belgian Information Retrieval Workshop, Amsterdam, the Netherlands (pp 16–20) (2011).
12. Luciano Barbosa and Junlan Feng. 2010. Robust sentiment detection on Twitter from biased and noisy data. In Proc. of CO LING, (2010)
13. Amna Asmi and Tanko Ishaya. Negation Identification and Calculation in Sentiment Analysis, In 2nd International Conference on Advances in Information Mining and Management. IMMM, (2012).
14. Alec Go et al, Twitter sentiment classification using distant supervision, Stanford University, (2009)
15. Kamal Nigam, John Lafferty, Andrew McCallum, Using Maximum Entropy for Text Classification, In IJCAI-99 Workshop on Machine Learning for Information Filtering, pages 61–67, (1999)
16. Brendon O'Connor and Balasubramanyan et al,From Tweets to Polls: Linking Text Sentiment to Public Opinion Time Series, Proceedings of the International AAAI Conference on Weblogs and Social Media, Washington, DC, (2010)
17. Isaac G Councill, Ryan McDonald, and Leonid Velikovich. What's great and what's not: learning to classify the scope of negation for improved sentiment analysis. In Proceedings of the workshop on negation and speculation in natural language processing, pages 51–59. Association for Computational Linguistics, (2010)
18. Pang and Lee, 2002, Sentiment Classification using Machine Learning Techniques, Proceedings of the Conference on Empirical Methods in Natural Language Processing (EMNLP), Philadelphia, (2002)

Congestion Control Mechanism for Real Time Traffic in Mobile Adhoc Networks

Mamata Rath, Umesh Prasad Rout, Niharika Pujari, Surendra Kumar Nanda and Sambhu Prasad Panda

Abstract Maximum real time solicitations require Quality of Service (QoS) during data communication. Therefore many routing protocols for Mobile Adhoc Network (MANET) associating Real Time Applications have been established that uses improved Real time structure for optimization of delay and energy efficiency, basic objective being great deployment of resource in resource limited environment. Congestion Control is another important issue while directing towards QoS achievement specifically with highly transferable mobile stations. This paper highlights on congestion control issues in real time environment as well as proposes an upgraded traffic shaping mechanism in TCP/IP protocol suite of network model for real time applications with basic concept using the token bucket traffic shaping mechanism during packet routing at the intermediate nodes. Simulation findings illustrates that our proposed method performs better in highly congested traffic scenario with reduced queuing delay and improved packet delivery ratio.

Keywords MANET · AODV · PDR · Throughput · QoS

M. Rath (✉)
Department of IT, C.V.Raman College of Engineering, Bhubaneswar, Odisha, India
e-mail: mamata.rath200@gmail.com

U.P. Rout
Department of IST, Ravenshaw University, Cuttack, Odisha, India
e-mail: umesh.upr@gmail.com

N. Pujari · S.K. Nanda · S.P. Panda
C.V.Raman Computer Academy, Bhubaneshwar, Odisha, India
e-mail: niharika.pujari@cvrgi.edu.ac.in

S.K. Nanda
e-mail: situnanda@gmail.com

S.P. Panda
e-mail: sambhu.prasad.panda@gmail.com

S.C. Satapathy et al. (eds.), *Computer Communication, Networking and Internet Security*, Lecture Notes in Networks and Systems 5,
DOI 10.1007/978-981-10-3226-4_14

1 Introduction

During real time data transmission in MANET, the most promising issues include control of congestion at the network devices such as routers which have limited buffer capacity to hold incoming packets and forward them towards their respective destinations [1]. As per specific quality of service requirement, the traffic has to be created with necessary configuration. Therefore, before defining congestion control and QoS policies, it is essential to deliberate the data traffic. Generally three types of traffic profiles work in the network, Constant Bit Rate (CBR), Variable Bit Rate (VBR) and Bursty. In case of highly loaded data traffic, when the capability of the receiving network at the router level, becomes lesser than the amount of data incoming data packets, then the receiving network can not handle such large amount of incoming data [2] as a result of which there is more data loss leading to retransmission and hence the overall delay increases with decrease in throughput. Packet Delay and Throughput are affected greatly as they act as function of network load. The delay increases when the load increases and the throughput keeps on increasing till the load reaches the network capacity.

2 Related Work

Based on congestion control approach many researchers have contributed their improved approaches for better performance of the network. The important findings have been described in this section. A superior end to end congestion control platform has been proposed in [3] as per the parameters available in the channel in a non-uniform manner. The congestion control mechanism in this approach takes care and prevents the congestion as well as solves the related challenges from physical layer to transport layer. Considering the complexity of traffic, congestion status are introduced in paper [4] which allocates the portion of traffic across many sub-routes and simulation result obtained using Qualnet proves validity of this approach. Voice Over Internet Protocol (VOIP) performance and its analysis about traffic congestion has been presented in [5] with a concluding remark regarding relation between channel quality and VOIP performance. Chief functionality MAC (Medium Access Protocol) in Mobile Adhoc Network is to dispense the channel among numerous users uniformly who need to transmit [6]. A QoS related MAC protocol has been proposed in [7] which talks about fairness of channel distribution in multi-hop environment. Comprehensive survey and argument on several real time protocols have been presented in [8, 9]. Paper [10, 11] offers detailed issues and research carried out in real time scenario, power management, security aspects in MANET. An improved AODV Protocol with energy efficiency issues has been presented in [12] and performance analysis has been carried out in [13]. A detailed comparative analysis regarding various network parameters has been presented with various improved MANET routing protocols in [14]. For VANET a novel multi channel error recovery video streaming scheme has been presented [15]. Another wireless multi-hop protocol for real time applications [16] has been presented in [17] with performance analysis and an improved real time video streaming approach has been depicted [18] over MANETs.

3 Analysis of Congestion Control Issues in Real Time Scenario

To check congestion, many practices have been proposed and developed by researchers, some mechanisms are used for removing congestion and some avoids it. Depending on their operational activity, this controlling mechanism can be open loop congestion control for prevention of congestion or close loop congestion control [19] for elimination of congestion. Under open loop congestion control policies the re-transmission policy is important which is used when the sender feels that a sent packet is lost or its bits have been corrupted, and then it re-transmits the packets. This is one of the conventional ways which faces the issue of delaying in data transmission. In window policy. The selective repeat window flow control policy is used than the go-back n windows policy for congestion control. Then there is an acknowledgement policy, in which a receiver may send an acknowledgement only if it has packet to be sent or a specially programmed timer gets expired, other cases it will not send an acknowledgement. It prevents congestion by not sending too many acknowledgement packets which create congestion in the traffic. In discarding method, the routers discard some packets randomly without any basic criteria simply to avoid congestion that can be retransmitted at a later stage. In admission method, there are different switches that regulate the flow by checking the resource requirements before admitting it to the network. Hence this way avoiding chance of congestion in future. The provision of back pressure technique is a controlling mechanism of congestion in which a congested node does not receive any data from its immediate upstream node. The provision of back pressure technique is a controlling mechanism of congestion in which a congested node does not receive any data from its immediate upstream node. Another Choke Packet method is used in which a package sent by a node to the source regarding the event of congestion, so that the source pauses for some time without continuously sending the packets, and it starts when the congestion at that particular node is under control. Due to absence of any type of signal passing between the starting node and the congested station, a victim node which is affected with congestion, the initiator node believes there may be an event of congestion in the said node, so o it prevents sending more packets in this path and does not wait for receiving acknowledgement. This method is called implicit signaling. Similarly, in explicit signaling concept, the congested node explicitly sends a signal to the source computer. Congestion control also influences the Quality of Service during transmission which is a set of services that should be supported by a network, while transporting a packet from source to destination. By controlling and maintaining the flow characteristics of traffic, QoS can be achieved. The basic characteristics of a flow are Reliability, Delay, Jitter and Bandwidth. Bandwidth is an important parameter which is considered during video and multimedia data transmission.

4 The Proposed Approach

The proposed approach is an extension of our previous research work presented in [12, 13]. In our previous research work we have developed a basic platform [20] for Quality of Service [21] in MANET and in the current research work we have proposed a new approach for traffic shaping for congestion control in real time traffic. The cross layer design architecture that supports inter communication among layers includes Optimized Power and Delay AODV (PDOAODV) protocol [12] at network layer and a superior channel access method [21] for real time applications at the MAC layer. In the current work the congestion control technique used in the said architecture has been discussed with prominence on shaping the traffic with improved token bucket based technique for real time applications.

Figure 1 shows the basic functional diagram of the proposed RTB (Real time Token Bucket) method. The improved RTB method shapes the real time traffic in an organized manner using the proposed improved algorithm. This method checks the incoming traffic flow and streamlines it before inflowing to the network. The proposed method controls the rate of sending flow of traffic by introducing a set up time (Δt) decided by negotiation among the sending station and the carrier which decide a traffic pattern. In Real time Token Bucket (RTB) algorithm, there are two types of tokens generated by the system and the bucket holds all the tokens. To transfer a packet, a station has to seize one token of its application type and delete it. The tokens are produced by the system at a rate of one token in every Δt second.

Figure 2 shows the block diagram of the proposed RTB method. According to this method, in every alternative unit time, real time token and best effort tokens are produced. Stations which currently do not have any thing to send, can capture the tokens and preserve for future transmission purpose when they have higher bursts of flow. But it is limited to extreme size of the bucket. Packets are not discarded in our proposed method of RTB. Using RTB, a packet can only be communicated only if there is sufficient tokens to shelter its length. Improved RTB method permits the bigger bursts to be sent quicker by more rapidly sending up the output. RTB allows redeeming the tokens (authorizations) to send bulky bursts, which are not allowed in Leaky Bucket Algorithm [22].

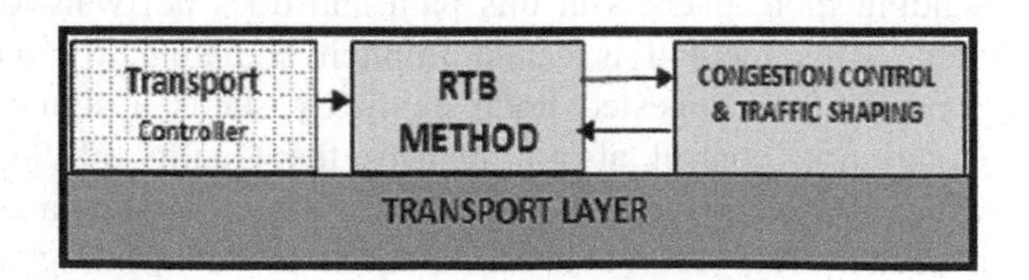

Fig. 1 Functional block diagram

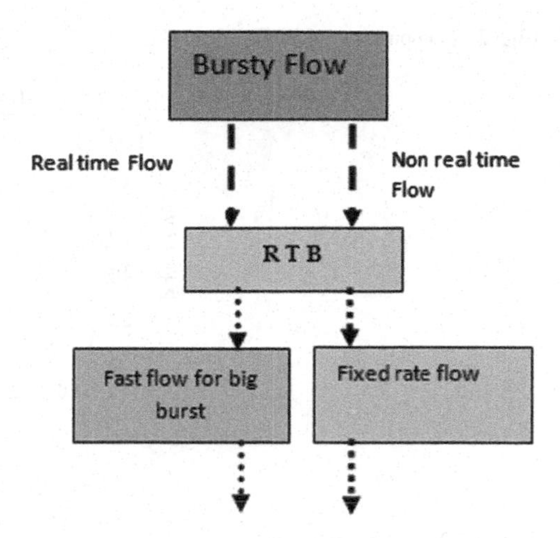

Fig. 2 Block diagram of RTB method

Table 1 Simulation parameters

Parameter name	Parameter value
Channel type	Wireless channel
Radio propagation model	Two ray ground
Network interface type	Wireless phy
Type of traffic	VBR
Simulation time	5 min
MAC type	Mac/802_11
Max speed	40–50 ms
Network size	1600 × 1600
Mobile nodes	120
Packet size	512 Kb
Interface queue type	Queue/Droptail
Simulator	Ns2.35

5 Simulation and Results

Network Simulator Ns2.35 is used for simulation and the and the simulation parameters are as given in Table 1. We have simulated our approach with Optimized AODV protocol [12] as part of the QoS architecture. Table 1 shows the network parameters considered during the simulation.

Figure 3 shows the comparative analysis graph between the Token Bucket Congestion Control with Traffic Shaping [19] and Our Proposed Improved RTB in terms of Packet Delivery Ratio (PDR). It clearly shows that our approach has a higher PDR percentage in a highly mobility scenario.

Figure 4 shows that due to improved congestion control mechanism in our approach, the average delay in packet transmission reduces up to a greater extent in comparison to other similar approach.

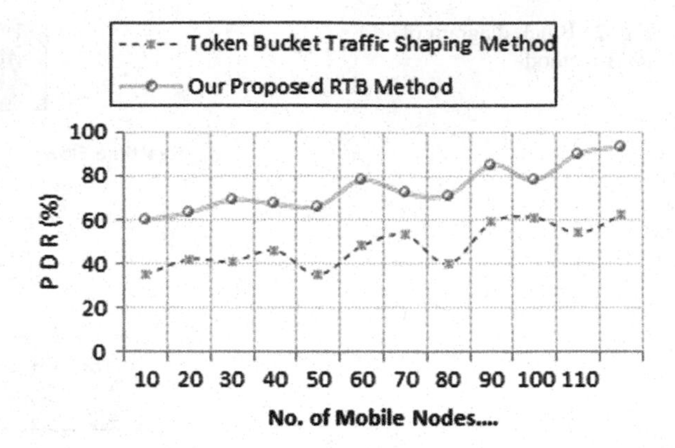

Fig. 3 Comparison of PDR

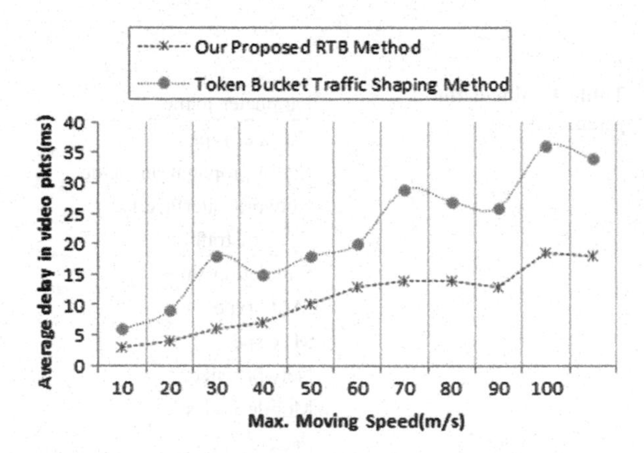

Fig. 4 Delay Comparison

6 Conclusion and Future Work

This research work concentrates on one of the dynamic concerns in real time data transmission in Mobile Adhoc Network. It resolves the stimulating difficulty of congestion due to queuing delay at the router interface during routing by proposing an enriched process called Real Time Token Bucket for fascinating control of inward as well as outbound traffic at the access point of the stations. This mechanism performs fantacstically when embedded as a part of our original Multi-Layer Communication based QoS architecture. In our future work we have scheduled to device this procedure in real podium and to focus on the other related technical issues including performance appraisal under variable parameters.

References

1. Wireless Adhoc and Sensor Networks, Houda Lablod, Wiley Publication, U.S.A, (2008).
2. GhulamYasin, Syed Fakhar Abbas, S. R. Chaudhry, "MANET Routing Protocols for Real-Time Multimedia Applications," *WSEAS Transactions on Communications,* Vol 12, Issue 8, Pages 386–395 August (2013).
3. Yi Song, Jiang Xie, "End-to-end congestion control in multi-hop cognitive radio ad hoc networks: To timeout or not to timeout?," Global *Communications Conference (GLOBECOM), 2013 IEEE,* Pages 4390–4395, 9–13 Dec. (2013).
4. Shrivastava, L., Tomar, G.S., Bhadoria, S.S., "A Load-Balancing Approach for Congestion Adaptivity in MANET," *Computational Intelligence and Communication Networks (CICN), 2011 International Conference on,* Pages 32–36, 7–9 Oct. (2011).
5. Al Alawi, K.; Al-Aqrabi, H., "Quality of service evaluation of VoIP over wireless networks," in GCC Conference and Exhibition (GCCCE), 2015 IEEE 8th, vol., no., pp. 1–6, 1–4, Feb. (2015).
6. Ahmad, Iftikhar, Samreen Ayaz, Syed Yasser Arafat, Faisal Riaz, and HumairaJabeen. "QoS routing for real time traffic in mobile ad hoc network", Proceedings of the 7th International Conference on Ubiquitous Information Management and Communication - ICUIMC 13, (2013).
7. Seth, D.D.; Patnaik, S.; Pal, S., "A Quality of Service assured & faired MAC protocol for Mobile Adhoc Network," in Communications and Signal Processing (ICCSP), (2011).
8. Rath M; & Pattanayak, B.K., "A methodical survey on real time applications in MANETS: Focussing on key issues," High Performance Computing and Applications (ICHPCA), 2014 International Conference on, Pages 1–5, 22–24 Dec. (2014).
9. Rath M, Pattanayak, B.K. & U.P.Rout " Study of Challenges and Survey on Protocols Based on Multiple Issues in Mobile Adhoc Network", International Journal of Applied Engineering Research,, 2015, Volume 10, pp 36042–36045, (2015).
10. Pattanayak B.K. & M.Rath,"A Mobile Agent Based Intrusion Detection System Architecture For Mobile Ad Hoc Networks", Journal of Computer Science, 2014, 10 (6): 970–975, (2014).
11. Rath M,B.K., Pattanayak & Bibudhendu Pati," A Contemporary Survey and Analysis of Delay and Power Based Routing Protocols in MANET", ARPN Journal of Engineering and Applied Sciences, 2016, Vol 11, No 1, Jan, 2016.
12. M.Rath, B.K.Pattanayak, B.Pati, " Energy Efficient MANET Protocol Using Cross Layer Design for Military Applications", Defense Science Journal Vol. 66, No. 2, March (2016).
13. Mamata Rath, Binod Kumar Pattanayak, " Energy Competent Routing Protocol Design in MANET with Real time Application Provision", International Journal of Business Data Communications and Networking, 11(1), 50–60, January–March (2015).
14. Rath M, Pattanayak B, Pati B., " Comparative analysis of AODV routing protocols based on network performance parameters in Mobile Adhoc Networks, Foundations and Frontiers in Computer, Communication and Electrical Engineering, pages 461–466; ISBN: 978-1-138-02877-7.May, 2016, CRC Press, Taylor & Francis Group, (2016).
15. Xie, H., Boukerche, A., Loureiro, A., "MERVS: A Novel Multi-channel Error Recovery Video Streaming Scheme for Vehicle Ad-hoc Networks," *Vehicular Technology, IEEE Transactions on,* Vol 99, Feb, (2015).
16. Rajib Mall, "Real-Time Systems: Theory and Practice", Pearson Education, Pearson India Education Services Pvt. Ltd, (2007).
17. DaniloTardioli, DomenicoSicignano, José Luis Villarroel," A wireless multi-hop protocol for real-time applications", *Computer Communications,* Vol 55, Pages 4–21, (2015).
18. Sondi, P., Gantsou, D., "Improving real-time video streaming delivery over dense multi-hop wireless ad hoc networks," *Wireless Days (WD), 2014 IFIP,* Pages 1–4, 12–14 Nov. (2014).
19. Behrouz A. Forouzan," Computer Networks",4th edition, McGraw-Hill, (2006).

20. Mamata Rath Bibudhendu Pati Binod Kumar Pattanayak, " Cross Layer Based QoS Platform for Multimedia Transmission in MANET", 3rd International Conference on Electronics and Communication Systems, (IEEE ICECS), Coimbatore, India, pp 3089–3093, (2016).
21. Mamata Rath, Binod Kumar Pattanayak, Bibudhendu Pati, "Inter-Layer Communication Based QoS Platform for Real Time Multimedia Applications in MANET", In the proceedings of IEEE International Conference on Wireless Communications, Signal Processing and Networking (IEEE WiSPNET 2016) pp 613–617, Chennai, India. (2016).
22. Sensor and Ad Hoc Networks, Theoretical and Algorithmic Aspects, Springer, (2008). DOI:10.1007/978-0-387-77320-9 [Last Accessed – 8th September,2016].
23. http://www.eecs.yorku.ca [last accessed on 13th Jan, 2016].

Correction of Ocular Artifacts from EEG by DWT with an Improved Thresholding

Vijayasankar Anumala and Rajesh Kumar Pullakura

Abstract Electroencephalogram (EEG) signals are widely being used for analyzing the activities of brain. It is extensively used for diagnosing different central nervous system disorders such as Alzheimer's, Parkinson's, seizures, epilepsy, etc. Ocular activity creates significant artifacts in EEG recordings. Analysis of the EEG and obtaining clinical information is difficult because of these noise sources. This paper proposes discrete wavelet transform (DWT) based denoising method with new statistical thresholding for single channel EEG signal. This method is evaluated on EEG signals taken from polysomnographic records, eegmmidb database. The effectiveness of the proposed method was measured using parameters such as signal to noise ratio (SNR), artifact rejection ratio (ARR) and comparing with the existing threshold method. Result of this study reveals that DWT with proposed thresholding method has shown superior performance in terms of SNR and ARR and effectively eliminates ocular artifacts.

Keywords EEG · Ocular artifacts · DWT · Statistical thresholding · SNR · ARR

1 Introduction

Physicians generally use EEG signals for analyzing the activity of brain, to diagnose certain neurophysiological states and disorders. Amplitude of EEG signals typically range between 10 and 100 μV mostly below 50 μV. These signals are more often contaminated with physiological artifacts such as electromyogram (EMG)

V. Anumala (✉)
Department of ECE, V R Siddhartha Engineering College, Vijayawada
520007, India
e-mail: vijayasankar.anumala@vrsiddhartha.ac.in

R.K. Pullakura
Department of ECE, Andhra University College of Engineering, Visakhapatnam
530022, India
e-mail: rajeshauce@gmail.com

S.C. Satapathy et al. (eds.), *Computer Communication, Networking and Internet Security*, Lecture Notes in Networks and Systems 5,
DOI 10.1007/978-981-10-3226-4_15

and electrooculogram (EOG). The EOG signals come from the eye movement and eye blinking that creates significant artifacts in the order of milli-volts. The frequency range of EEG signals is in the range from 0 to 64 Hz and the ocular artifacts (OA) occur within the range of 0–16 Hz. Overlapping of these artifacts over the desired signal causes a significant loss of valuable background EEG activity. Hence denoising of EEG signal is necessary to detect the neurophysiological disorders.

Numerous methods were in use for correcting the ocular artifacts [1–8]. Regression in time and frequency domain techniques is extensively used for removing ocular artifacts [1]. There are blind source separation (BSS) methods like principle component analysis (PCA) and independent component analysis (ICA), that identify the artifacts by decomposing the signals and reconstructing without artifacts. Principal component analysis cannot differentiate OAs from the EEG signal, when they have practically identical amplitudes [2]. Independent component analysis is an alternative approach to PCA that is not automated and requires visual classification of the components [3]. Except wavelet transform techniques most of these methods require multiple channels of EEG information or recording of EOG signal. Hence, for denoising EEG signals, Wavelet transforms with statistical thresholding methods are extensively used.

Tatjana Zikov et al. applied stationary wavelet transform (SWT) with coiflet 3 mother wavelet for denoising the EEG signal [4]. Furthermore, many investigators had removed EOG artifacts from EEG signals employing different combinations of wavelet transform techniques and mother wavelet to identify the artifact zones and apply thresholding to the identified zones to keep the background information [5–8]. SWT is not capable to correct ocular artifacts of overlapping spectrum [9]. Mantosh Biswas et al. suggested a new soft-thresholding image denoising method using wavelets which led to momentous improvement in peak signal to noise ratio (PSNR) [10]. In this paper we propose a new level dependent soft thresholding function for reduction of ocular artifacts. Comparative analysis has been performed using the combination of various wavelet functions (Daubechies, Coiflet and Symlet) and thresholds (Universal and proposed) to determine the suitable combination to effectively clean ocular artifacts.

2 Methodology

A typical EEG signal measured from the scalp will have amplitude of about 10–100 μV and a frequency in the range of 0.1 Hz to about 64 Hz. The EEG signals are further divided into five major sub-bands based on the frequency ranges. These bands form low to high frequencies respectively are called delta (δ) (Range 0.1–4 Hz), theta (θ) (Range 4–8 Hz), alpha (α) (Range 8–12 Hz), beta (β) (Range 13–30 Hz) and gamma (γ) (Range 30–45 Hz). Generally ocular artifacts (OA) lay in the low frequency bands i.e. Theta and Alpha bands [5, 7]. Decompose the EEG signal to a level of 'm' by means of discrete wavelet transform to correct the blink artifacts. Where $m = \log_2 N$ and N is the number of sample points of EEG signals taken into consideration [11]. Four channels of EEG data with 1600 sample point each was taken for this work. Decompose the original EEG signal using discrete wavelet transform to a level of 8, which result in a set of approximate (a_j) and detail (d_j) coefficients. Thresholding is done for the detail coefficients ranging from levels 8 to 5. Apply the inverse discrete wavelet transform to reconstruct the signal [12]. The wavelet decomposition, thresholding and reconstruction are carried out using MATLAB. The process of artifact removal using wavelet transforms method as shown in Fig. 1.

2.1 Data Acquisition

For this work four single channel EEG segments of 10 s duration each as shown in Fig. 2 are taken from polysomnographic records (https://physionet.org/cgi-bin/atm/ATM). The EEG signals are found dominant in the frontal and fronto-polar channels like FP1,

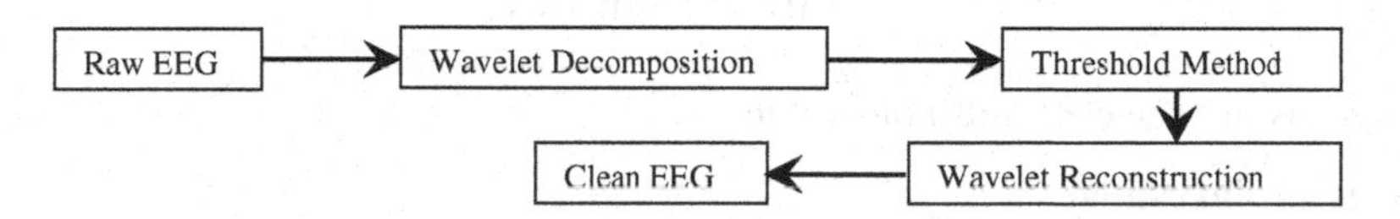

Fig. 1 Wavelet denoising method

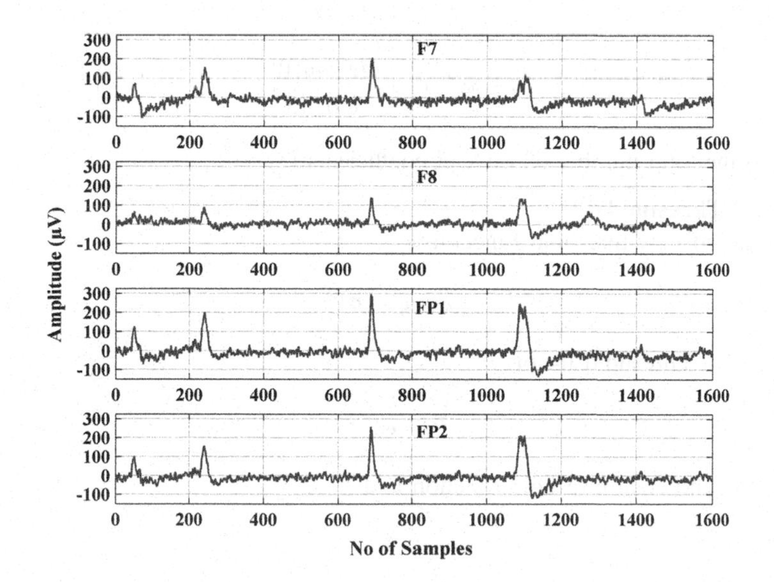

Fig. 2 EEG signals of electrodes F7, F8, FP1 and FP2

FP2, F7 and F8. Hence it is reasonable to take FP1, FP2, F7 and F8 as contaminated/corrupted EEG signals.

2.2 Existing and Proposed Thresholding Method

Wavelet thresholding depends on the choice of wavelet, level of decomposition and threshold estimation to a larger extent [11]. Choice of thresholding is critical step in the denoising process. It should not remove the original coefficients leading to loss of significant information in the analyzed data. The denoised signal remains noisy if the threshold is too small or too large. So, optimum threshold technique is to be found. There are two thresholding methods—hard and soft thresholding. In hard thresholding at each level, the wavelet coefficients above the threshold are unchanged and below the threshold reduced to zero where as in soft thresholding, the wavelet coefficients are reduced towards zero by threshold λ. Method of soft thresholding for the detail wavelet coefficients W_i is given below.

$$W_i = \begin{cases} W_i - \lambda & W_i \geq \lambda \\ W_i + \lambda & W_i \leq -\lambda \\ 0 & |W_i| < \lambda \end{cases} \tag{1}$$

where λ is the selected threshold value.

Universal Threshold
This is a global threshold function and it is computed as

$$\lambda = \sigma\sqrt{2\log N} \tag{2}$$

$$where \quad \sigma = \frac{Median|W|}{0.6745} \tag{3}$$

and N is the total number of wavelet coefficients [8].

Proposed Threshold
The proposed new threshold function is

$$\lambda_{NEW} = \sigma P \tag{4}$$

where P is a Threshold factor

$$P = e^{(\lambda - S)} \tag{5}$$

λ is the universal threshold function and

$$S = \frac{2^K \sum_{i=1}^{N} W_i}{N} \quad (6)$$

where K denotes the level of decomposition i.e. $K = 1, 2, 3\ldots J$.

2.3 Performance Evaluation

Different statistical performance metrics have been used to quantifying the effectiveness of the method like SNR and ARR. SNR is termed as the ratio of signal power to the noise power, expressed in decibels (dB). ARR is the ratio of the power of the removed artifacts to the power of the clean EEG. The comparisons were performed both qualitatively and quantitatively by simulation carried out in MATLAB environment.

$$SNR = 10\log_{10}\left(\frac{\sigma_x^2}{\sigma_y^2}\right). \quad (7)$$

$$ARR = \frac{\sum_{n=1}^{N} (x(n) - y(n))^2}{\sum_{n=1}^{N} y(n)^2}. \quad (8)$$

where $x(n)$ denotes the contaminated EEG signal, $y(n)$ means the clean EEG signal, σ_x^2 and σ_y^2 is the variance of the contaminated and clean EEG signals respectively.

3 Results and Discussions

EEG signals from electrodes F7, F8, FP1 and FP2 are considered for this work. Calculated the Signal to noise ratio (SNR) and Artifact rejection ratio (ARR) using Eqs. (7) and (8) respectively for each of the refined signals and tabulated. Tables 1 and 2 delineate the values of SNR and ARR for various mother wavelets with different thresholding functions respectively. The method adopted is completely data dependent that yields the variations in SNR and ARR for EEG signals of different electrodes. The values of SNR for different wavelet functions vary from 5.86 to 7.84 dB and 1.81 to 3.68 dB for the proposed and universal threshold methods respectively for the signal on electrode FP1. On the other hand, for the same signal the values of ARR ranging from 2.54 to 4.01 and 0.15 to 0.8 for the

Table 1 Comparison of SNR for different wavelet functions

Channels	F7		F8		FP1		FP2	
Method	Universal	Proposed	Universal	Proposed	Universal	Proposed	Universal	Proposed
db3	4.50	6.50	4.15	6.61	3.64	7.49	3.34	7.05
db4	4.60	5.96	3.69	5.69	3.35	5.86	3.29	5.73
db5	4.94	6.03	4.16	6.19	3.68	6.59	3.45	6.25
coif3	3.53	5.93	2.45	6.06	2.40	6.18	2.18	6.16
coif4	2.98	6.14	1.73	6.42	1.92	6.49	1.87	6.46
coif5	2.27	6.64	1.75	6.69	1.81	7.35	1.63	7.14
sym8	3.62	6.54	3.12	6.62	2.32	7.06	2.26	6.91
sym9	2.63	6.47	2.11	6.40	2.38	7.84	2.41	7.36
sym10	2.77	6.39	2.41	6.28	2.54	6.77	2.34	6.73

Table 2 Comparison of ARR for different wavelet functions

Channels	F7		F8		FP1		FP2	
Method	Universal	Proposed	Universal	Proposed	Universal	Proposed	Universal	Proposed
db3	0.35	2.08	0.84	2.93	0.69	3.73	0.54	3.69
db4	0.95	1.72	0.73	2.27	0.54	2.54	0.54	2.41
db5	1.06	1.80	0.96	2.62	0.80	2.96	0.58	2.94
coif3	0.49	1.75	0.27	2.53	0.25	2.91	0.22	2.63
coif4	0.33	1.81	0.13	2.77	0.16	3.14	0.15	2.86
coif5	0.18	1.99	0.13	2.97	0.15	3.76	0.12	3.53
sym8	0.49	1.98	0.42	2.72	0.22	3.56	0.21	3.30
sym9	0.23	1.99	0.19	2.76	0.25	4.01	0.26	3.99
sym10	0.27	1.92	0.25	2.68	0.27	3.39	0.24	3.08

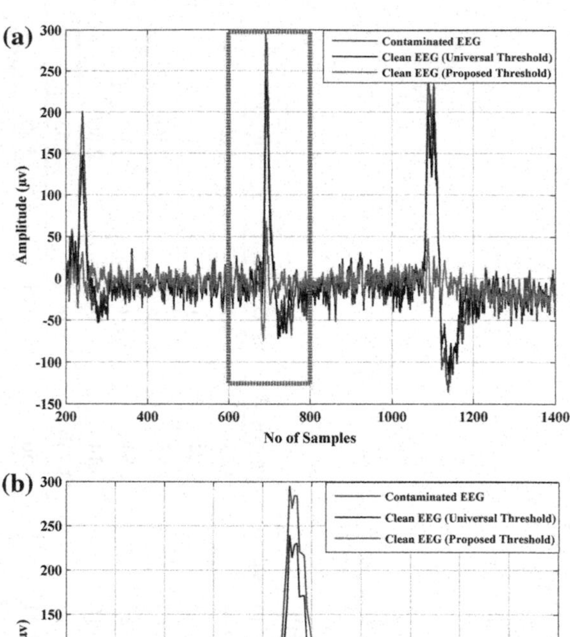

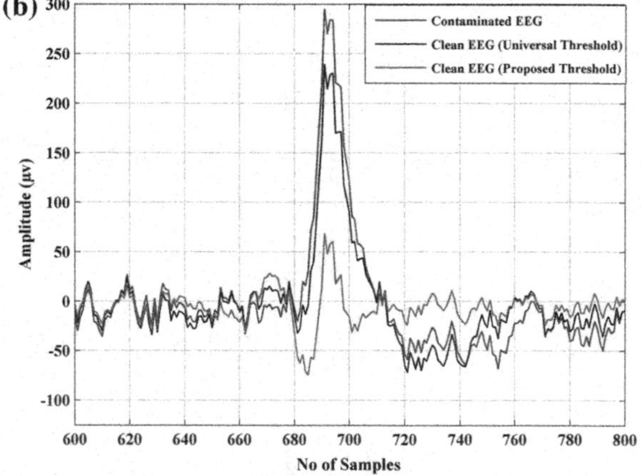

Fig. 3 **a** EEG signal of electrode FP1 recording before (*Blue*) and after (*Red*) denoising. **b** The zoomed portion of above plot in the interval 600–800

proposed and universal thresholding methods respectively. Indicate that, the proposed thresholding method has momentous improvement in SNR and ARR over universal thresholding method for all the wavelet functions. Wavelet functions db3, coif5 and sym9 have shown improved performance with proposed thresholding method where as universal threshold has shown better performance using db5.

Figure 3 compares the contaminated and clean EEG signals by universal and proposed thresholding methods for the signal on electrode FP1 using coif5 wavelet function. Careful perception demonstrates that wavelet transforms with proposed thresholding has better performance than universal thresholding method. Figure 4 illustrated the power spectrum of contaminated and clean EEG signals for the above mentioned combination, The clean EEG signal by DWT with proposed threshold contains lesser power in Alpha and Theta bands, indicate the effective removal of ocular artifacts in that region. Figures 5 and 6 addresses the SNR and ARR comparison of clean EEG signals of the channel F7, F8, FP1 and FP2 by DWT with

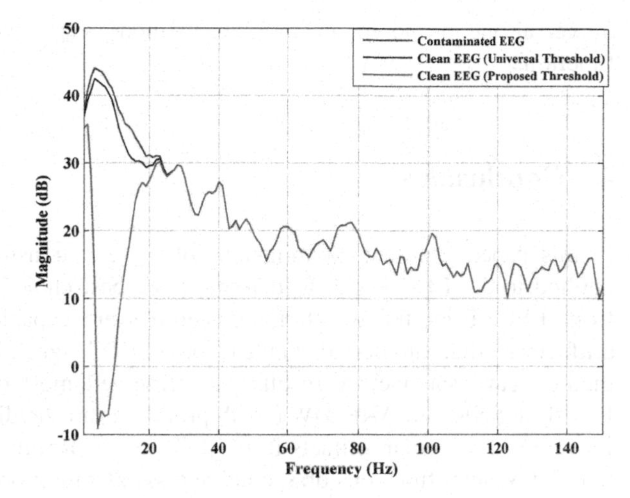

Fig. 4 PSD plot of contaminated and clean EEG signals

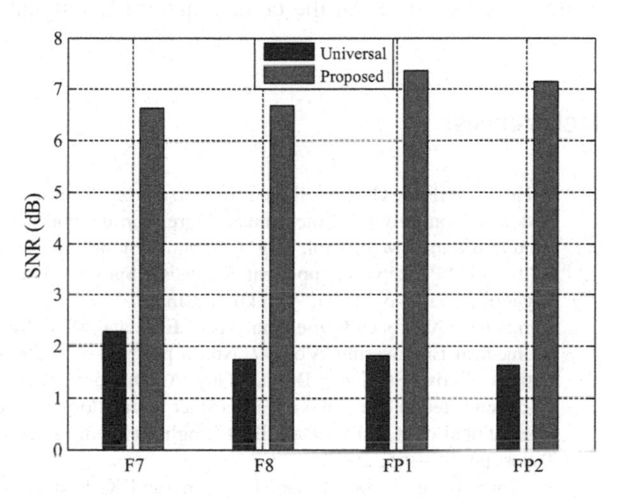

Fig. 5 SNR comparison of universal and proposed thresholding methods

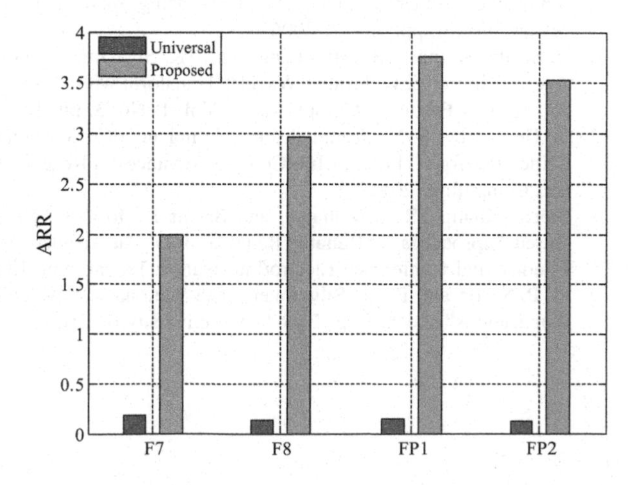

Fig. 6 ARR comparison of universal and proposed thresholding methods

universal and proposed thresholding methods respectively for the coif5 wavelet function.

4 Conclusions

In this paper, different combinations of wavelet transforms and thresholds were applied to the EEG signal to determine the best choice to remove ocular artifact from EEG. Both the thresholding methods are capable to correct the multiple artifacts at different intervals in the recorded EEG signal, but the proposed threshold method gave satisfactory results and preserved most of the signal information. Based on SNR and ARR DWT with proposed thresholding is a good combination for removal of ocular artifacts from EEG signals. Results of this study demonstrate that the wavelet functions db3, coif5 and sym9 with proposed threshold correct the artifacts effectively from the contaminated EEG signals.

References

1. P. He, G. Wilson, C. Russell, and M. Gerschutz, "Removal of ocular artifacts from the EEG: A comparison between time-domain regression method and adaptive filtering method using simulated data," Med. Biol. Eng. Comput., vol. 45, no. 5, pp. 495–503, 2007.
2. Joliffe I T, "Principal Component Analysis", Springer Verlag, New York, 1986.
3. Vigario R, Jaakko Sarela, Veikko Jousmaki, Matti Hamalainen, Erkki Oja, "Independent Component Approach to the Analysis of EEG and MEG Recordings", IEEE Transactions on Biomedical Engineering, Vol. 47, No. 5, pp. 589–593, 2000.
4. Tatjana Zikov, Stephane Bibian, Guy A. Dumont, Mihai Huzmezan, "A wavelet based de-noising technique for ocular artifact correction of the Electroencephalogram", 24th International conference of the IEEE Engineering in Medicine and Biology Society, Huston, Texas, pp. 98–105, 2002.
5. V. Krishnaveni, S. Jayaraman, L. Anitha and K. Ramadoss, "Automatic identification and removal of ocular artifacts from EEG using wavelet Transform," Measurement science review, Vol. 6, pp. 45–57, 2006.
6. P. Senthil Kumar1, R. Arumuganathan1, K. Sivakumar, and C. Vimal, "Removal of Ocular Artifacts in the EEG through Wavelet Transform without using an EOG Reference Channel" Int. J. Open Problems Compt. Math., Vol. 1, No. 3, pp. 188–200, 2008.
7. G. Geetha, Dr. S. N. Geethalakshmi "de-noising of EEG signals using Bayes shrink based on coiflet transform" Froc. of Int. Conf on Advances in Recent Technologies in Communication and Computing, 2011.
8. Saleha Khatun, Ruhi Mahajan, and Bashir I. Morshed "Comparative Analysis of Wavelet Based Approaches for Reliable Removal of Ocular Artifacts from Single Channel EEG" IEEE International Conference Electro/Information Technology (EIT), pp. 335–340, 2015.
9. G. P. Nason and B. W. Silverman, "The Stationary Wavelet Transform and some Statistical Applications", Tech. Rep. BS8 1Tw, University of Bristol, 1995.

10. Mantosh Biswas, Hari Om "A New Soft-Thresholding Image Denoising Method" 2nd International Conference on Communication, Computing & Security, pp. 10–15, 2012.
11. Lei, Chao Wang, and Xin Liu "Discrete Wavelet Transform Decomposition Level Determination Exploiting Sparseness Measurement" International Journal of Electrical, Computer, Energetic, Electronic and Communication Engineering Vol: 7, pp. 1182–1185, 2013.
12. K. P. Soman. N, K. I. Ramachandran, N. G. Resmi. "Insight into wavelets from theory to practice", 3rd edition, PHI Learning Private limited, Delhi, 2013.

EGRP: Enhanced Geographical Routing Protocol for Vehicular Adhoc Networks

N.V. Dharani Kumari and B.S. Shylaja

Abstract Vehicular Ad hoc Networks (VANETs) is an advanced wireless ad hoc network to communicate between the vehicular nodes. The unique characteristic of Vehicular Adhoc Networks leads to frequent network fragmentation and route reconstruction which cause an increase in packet drop ratio and control overhead. Thus, it brings challenges to establish an optimized routing path with high reliability and low latency. This paper presents an improved geographical forwarding strategy to select the next hop based on the mobility metrics such as distance, speed and moving direction of the nodes. These routing metrics have an impact on the performance of the routing protocol for Vehicular Ad hoc Networks (VANETs). Extensive simulations carried out based on the proposed solution have proved to outperform the existing GPSR approaches in terms of reliability, scalability and path latency.

Keywords VANETs · Geographical forwarding · Multi-metric node selection · Mobility model

1 Introduction

Vehicular Ad hoc Networks (VANETs) is a subclass of Mobile Ad hoc Network (MANETs) which allows communication among vehicular nodes without any pre-deployed infrastructure. Modern vehicles are equipped with IEEE802.11p for Wireless Access in Vehicular Network (WAVE), for the communication between nodes [1]. This wireless technology used in vehicles are considered for short-range

N.V. Dharani Kumari (✉)
Department of Computer Applications, Dr. Ambedkar Institute of Technology, Bengaluru 560056, India
e-mail: dharani.drait@gmail.com

B.S. Shylaja
Department of Information Science and Engineering, Dr. Ambedkar Institute of Technology, Bengaluru 560056, India
e-mail: shyla.au@gmail.com

S.C. Satapathy et al. (eds.), *Computer Communication, Networking and Internet Security*, Lecture Notes in Networks and Systems 5,
DOI 10.1007/978-981-10-3226-4_16

communications (DSRC), which enables vehicles to cover shorter range unless they have multi-hop communication. There are wide range of applications designed for VANETs which requires short distance to large area coverage and also require the reliable and an efficient routing protocol to forward data packets to the destination.

One of the challenges in VANET is the establishment of an optimized routing path among vehicular nodes to forward data packets from source to the destination. Routing in VANETs poses different challenges due to dynamic network topology, highly scalable network and frequent network fragmentation. The routing protocols of VANETs are broadly classified into topology based and geographical routing protocols. As the vehicles are embedded with navigation systems the geographical routing protocols are more acceptable for VANETs as it depends on the geographic position information [2].

The geographical routing protocol which adopts the basic greedy method selects the forwarder node using the position information to send the data packets to the destination. The source node in this method forwards the data packet to one of its neighbors, who gives maximum progress towards the destination node. This kind of routing under realistic VANET scenario may suffer from poor performance. Further, the longer forwarding distance can cause transmission errors due to high signal attenuation in the wireless link and also the chances of forwarding node moving out of communication range before receiving the data packets is also higher. Hence it may results into low reliability of routing path with increased path latency. Considering the single constrain metric such as position of nodes to make the routing decision in highly dynamic network like VANETs may not obtain an optimized routing path. There are several interrelated factors which may impact the quality of the routes are the location, speed and the moving direction of vehicles etc. Researcher in different work have proposed the multi-metric routing protocol by considering the different characteristics of vehicular ad hoc networks to obtain an efficient routing path [3–5]. The vehicles mobility is one of the important issues to be considered while designing an efficient routing protocol for VANETs. Since the movement of vehicles is constrained to the predefined path their mobility metric can be easily determined through different sensors equipped with the vehicles.

In the present study, it is aimed to optimize the routing decision of GPSR [6] a well-known position based routing protocol of MANETs to elect the best next hop around the smart vehicles. The simulation study of the proposed routing protocol called EGRP shows an enhanced performance in terms of successful packet delivery ratio and the reduced end-to-end delay.

2 Related Work

Authors in [7] proposed novel clustering scheme using node mobility metrics of speed, moving direction and distance to elect the best cluster head to route the data packets within a cluster for scalable routing. RBVT [8] leverage real-time traffic information of vehicular nodes to create road-based routing paths either reactively or proactive. Authors in [9] proposed the route lifetime based routing protocol by calculating the remaining time for which the link can be used for efficient communication among vehicular node in VANETS. The expected lifetime of the route is calculated by exploiting the vehicles location and its velocity vector. The local maximum problem is resolved by Tsiachris et al. in [10] by obtaining the most directed next hop at a recovery mode. Xue et al. [11] predicted the trajectories of the moving vehicle by leveraging the real traces of Vehicular Mobility Pattern (VMP) and road topology map to design an enhanced routing protocol for urban scenario. Taleb et al. [12] utilized the mobility metric of vehicular nodes and the digital map of the road to forecast the possible link destruction event prior to its occurrence to establish the stable communication link between the vehicles. This protocol had grouped the vehicles according to its velocity vector to enhance its performance in terms of throughput and thus reducing the number of link breakages.

The above discussed different approaches of routing protocols in Vehicular Adhoc networks motivated to further investigate the impact of different mobility metric on geographical forwarding decision and proposes an enhanced next hop selection strategy for the well-known position-based routing protocol called GPSR in urban scenarios of VANETs.

The greedy forwarding strategy of GPSR selects the neighbor node who is having the minimum geographical distance to the destination node. This approach if adopted directly to the highly dynamic networks like VANETs it may not provide the optimal solution due to the stale neighbors in its neighbor table. Thus it may fail to identify some of the efficient next hop candidate nodes to route the data packets. To overcome from such problems the enhanced next hop selection method is proposed to find the reliable routing path within the realistic urban scenario.

3 Proposed Work

In vehicular ad hoc networks, the availability of navigation systems makes it possible to exploit the mobility metrics of vehicular nodes such as geographical coordinates, direction and speed information, which can be later used to control the forwarding process of routing in VANETs. The proposed EGRP routing protocol selects the next hop during the process of data packet forwarding based on the calculation of the minimum angle and distance between all the neighboring nodes. We assume that each vehicles are furnished with GPS to get its geographical coordinates and speed information. The optimized next hop selection algorithm of EGRP is shown in Fig. 1. The following sections explains the steps carried out in the design of proposed routing solution.

1) The enhanced beacon packet is used to obtain the speed of each vehicular nodes along with its position information. Each vehicle stores its one hop neighbor's location and speed information in the neighbor table.

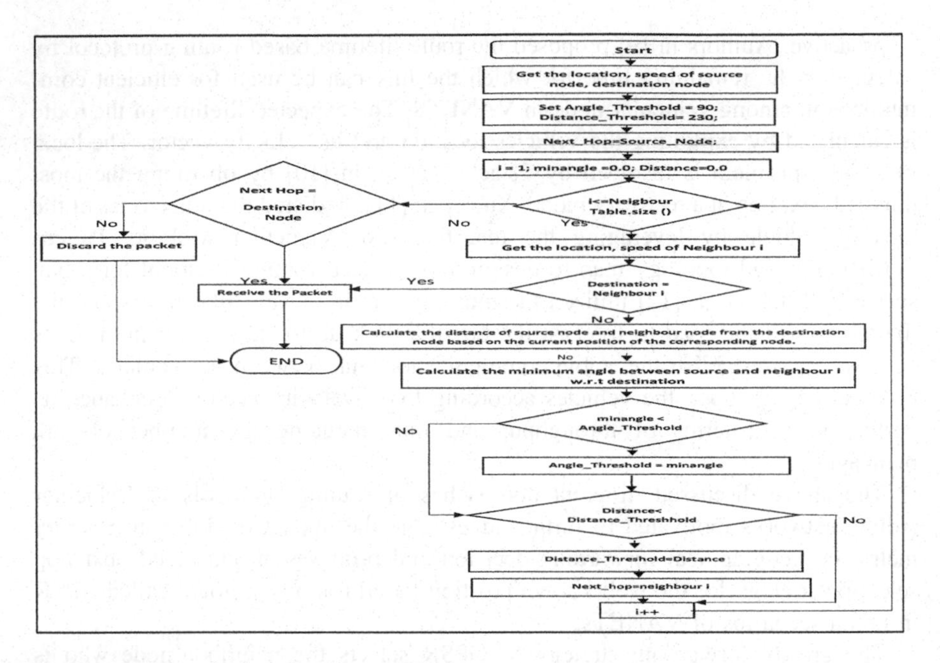

Fig. 1 Next hop selection algorithm of EGRP

2) The distance between source and each of its neighbor with respect to destination is calculating by using the Eq. (1). The source node calculates its position and each of its neighbor position using the Eqs. (3) and (4) while forwarding the data packets.
3) Further, the angle between source and next hop candidate nodes with respect to destination is obtained by using the formula (2).
4) When there is more than one neighbor with the same angle (i.e. minimum angle) then the node which is providing maximum progress towards the destination is selected as the optimal next hop node.
5) Finally, the optimal next hop node is selected with minimum angle and maximum progress towards the destination.

3.1 Determining Distance

The packet carrier node in the greedy method transmit the data packet to the node nearest to the destination. There are two possibilities when the selection of the next hop is done based on the distance. If the distance between the source and the relay node is very short, then it may leads to signal interference. On the other hand, if the distance between the nodes are equal to the maximum transmission range then due to the unreliable wireless link the probability of the vehicles moving out of the communication range before receiving the data packet is very high. Both of these situations may leads to the performance degradation of the protocol. In order to handle above mentioned issues, the present work set the transmission range to 250 m and *distance_threshold* to 230 m. If $distance\_threshold \geq DIST$ then the node is more favorable to forward the data packets. Suppose the source node s is at (x_1, y_1) and destination d is at (x_2, y_2) position then the $distance(DIST)$ between the nodes can be calculated using the following formula (1) [7]

$$\mathrm{DIST}_{s-d} = \sqrt{(x_2 - x_1)^2 + (y_2 - y_1)^2} \tag{1}$$

3.2 Determining the Moving Direction

In VANETs the movement of the vehicles are constrained by the road layout. The more reliable routing path can be established if more number of vehicles are moving in a particular direction. Therefore the minimum angle between the vehicles are considered to obtain the optimized next hop. Suppose the source vehicle s is at (x_0, y_0), and the destination vehicle d is at (x_1, y_1), and the neighbor is at (x_k, y_k), then the angle between source and neighbor vehicles with respect to destination can be obtained using the following formula (2).

$$\emptyset = arccos\frac{(x_l - x_0)(x_k - x_0) + (y_l - y_0)(y_k - y_0)}{\sqrt{(x_l - x_0)^2 + (y_l - y_0)^2}\sqrt{(x_k - x_0)^2 + (y_k - y_0)^2}} \tag{2}$$

3.3 Predicting the Current Position of the Vehicles

The vehicles mobility can be predicted based on the speed, location and moving direction. The source vehicle can obtain the current position of its one-hop neighbors at the time of transmitting the data packets to the destination. Suppose the source vehicle $\boldsymbol{s}$ is at $(\boldsymbol{x_0}, \boldsymbol{y_0})$, with the moving speed of $\boldsymbol{V_i}$ and the moving direction is $\boldsymbol{\theta}$ at time stamp $\boldsymbol{T_1}$ and its position at time stamp $\boldsymbol{T_2}$ can be obtained using the following formula. The time stamp $\boldsymbol{\delta T}$ is $(\mathbf{T}_2 - \mathbf{T}_1)$.

$$x_0' = x_0 + V_i * \partial T * \cos(\theta) \tag{3}$$

$$y_0' = y_0 + V_i * \partial T * \sin(\theta) \tag{4}$$

The proposed EGRP routing protocol adopts the above mentioned procedure to obtain an efficient next hop to forward the data packets.

4 Performance Evaluation

4.1 Simulation Setup

Simulations are conducted in two phases: In the first phase, the vehicular movement pattern trace files (that is used as input to the network simulator OMNET++ together with INET framework) are obtained from the SUMO, a microscopic road traffic simulator. The vehicles mobility is managed by the Stefan Kraus mobility model by using the 3 * 3 grid based two lane realistic street map of urban area with traffic lights in intersections. Figure 2 demonstrate the map used in the simulation. The mobility model parameters are summarized in Table 1.

In the second phase, the traffic generator SUMO's output is given as input to the network simulator OMNET++ together with its INET Framework extension for simulating EGRP. The random 15 source destination UDPBasicBurst traffic flows are chosen. Every source sends 2 UDPBasicBurst traffic per second, and the size of UDPBasicBurst packets is 512 bytes. Vehicles communicate with each other using the IEEE 802.11p DCF MAC layer. The radio transmission range is set to 250 m and the simulations is run for 600 s. The network simulation parameters are given in Table 2.

Fig. 2 Realistic urban vehicular scenario

Table 1 Vehicular mobility model

Parameters	Description	Values
N	No. of. nodes	(50–200) step of 25
A	Max acceleration	0.8
D	Max deceleration	0.3
Min-speed	Min velocity	5 m/s
Max-speed	Max velocity	(5–25) step 5 m/s

Table 2 Network simulation parameters

Parameter(s)	Value(s)
Topology size	2000 m * 1500 m
Simulation time	600 s
Number of traffic sources	15
Data packet length	512 byte
Vehicle beacon interval	1 s
Carrier frequency	5.8 GHz
Transmission range	250 m
Physical layer	IEEE802.11p
Transmission power	10 mW
Traffic type	UDP

4.2 Performance Metrics

The routing protocols performance were evaluated by varying the network density and the vehicles speed. The performance of the routing protocols were assess for the Packet Delivery Ratio (PDR), Average End-to-End delay (EED), Normalized Routing Load (NRL) and Average Hop Count (AHC).

4.3 Experimental Results

In VANET scenarios the vehicular traffic density and the mobility metric of speed are the dominant factor that impact the performance of geographical routing protocol. To analyze their impact on the performance of the EGRP compared to GPSR routing protocol is evaluate for different parameters. In the experimental setup to study the effect of node density we fix the maximum mobility speed to 15 m/s but vary the number of nodes from 50 to 200 in steps of 25 to represent the varying node densities and the obtained numerical results are plotted in Figs. 3, 4, 5, 6. To study the impact of vehicles velocity on network performance metrics we fix the number of nodes to 125 and vary the maximum speed from 5 to 25 m/s in steps of 5 m/s and the obtained results are shown in the same Figs. 3, 4, 5, 6.

The PDR of EGRP and GPSR is plotted in Fig. 3 clearly shows that as the node density increases the successful data packet delivery ratio also increases. The reason is that as the network becomes more connected the chances of getting into void problems are less, hence the PDR increases consistently with an increased node density. However, the EGRP has a better PDR than the existing GPSR because the proposed protocol is more scalable than the existing one and it performs well in dense network. Whereas the PDR drops with an increase in the

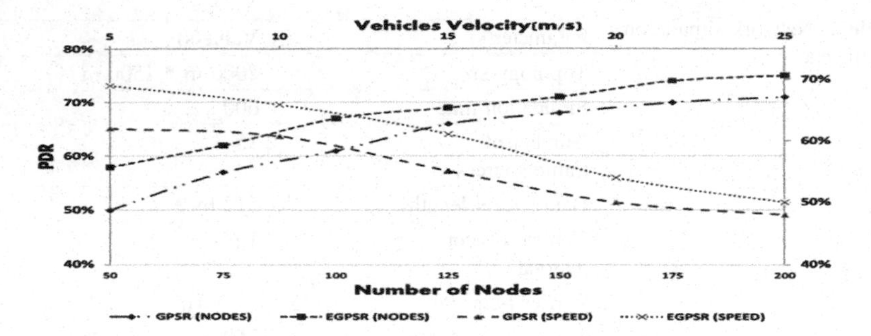

Fig. 3 The impact of node density and vehicle speed on PDR

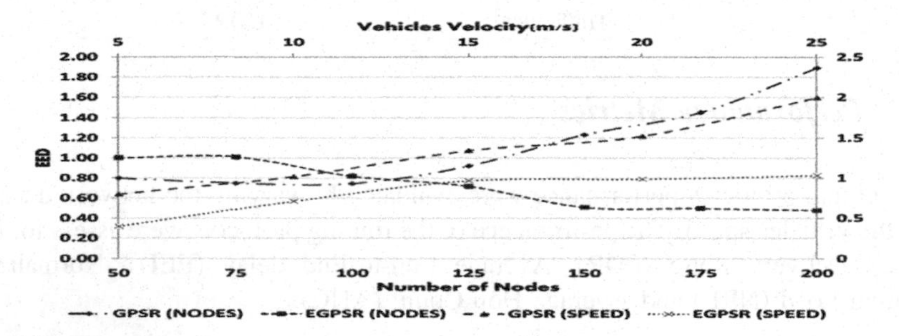

Fig. 4 Shows the impact of node density and vehicle speed on end to end delay

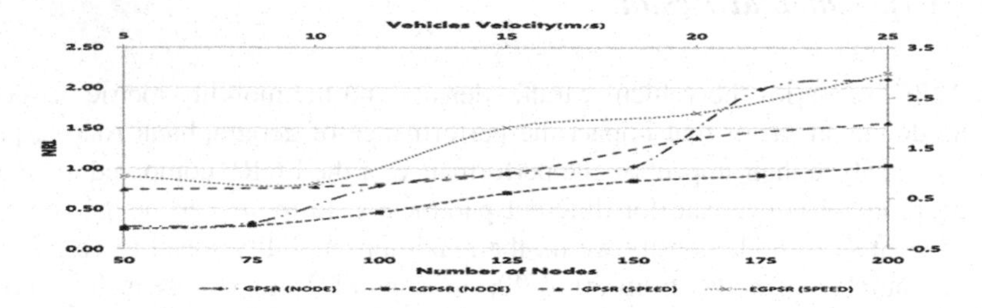

Fig. 5 Shows the impact of node density and vehicle speed on control overhead

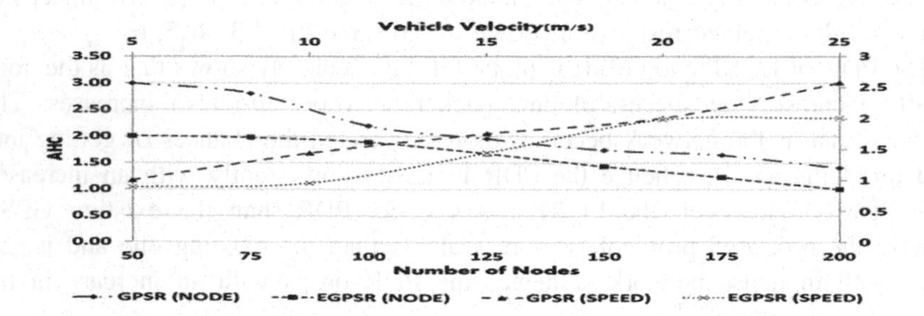

Fig. 6 Shows the impact of node density and vehicle speed on hop count

vehicles' speed. Since the high speed vehicles will remain in the communication for a very short period of time which may not be sufficient to forward all the data packets.

Figure 4 shows the average delay of data packets that have been received at the destination with different node densities and for the varying vehicles' speed. The EGRP has a lower delay when the network is more connected whereas the delay increases in GPSR due to higher level of contention.

The control overhead generated by the EGRP and GPSR is shown in Fig. 5. The EGRP produces less control overhead when compared to GPSR due to the enhanced next hop forwarding strategy. The increase in vehicles' speed might lead to frequent route failure and inturn to increased control overhead.

The graph in Fig. 6 plots the AHC of GPSR and EGRP for varying node density and velocity of vehicular nodes. The average number of hops required to deliver the data packets from source to destination is decreasing when the network is more connected. The increase in vehicular node velocity creates unstable routing path and thereby leads to increase in hop count. The proposed EGRP take less number of average hops to reach the destination since it considers the mobility metrics of vehicular nodes for the routing decisions.

5 Conclusion

In the present study an optimized next hop selection method for an existing GPSR routing protocol is proposed. The proposed Enhanced Geographical Routing Protocol (EGRP) considers the mobility metrics of vehicular node such as geographical coordinates, speed and moving direction to control the forwarding process of routing in urban scenarios of VANETs. The extensive simulation is performed to show the outperformance of EGRP routing protocol for different network performance metrics when compared with GPSR. The high speed vehicular nodes degrades the performance of EGRP and so does the GPSR in terms of PDR and control overhead. The EGRP has been shown to be reliable and scalable in urban environments.

Acknowledgements The authors would like to thank the web sources for OMNET++/INET (http://www.omnetpp.org/) and SUMO (http://sourceforge.net/projects/sumo/) software's.

References

1. Karagiannis, G., Altintas, O., Ekici, E.: Vehicular networking: a survey and tutorial on requirements, architectures, challenges, standards and solutions. In: IEEE Communication Survey, pp. 584–616. (2011).
2. Zhu, R., He, Y., Liu, T., Y., Ni, L. M.: Exploiting trajectory-based coverage for geocast in vehicular networks, IEEE Trans. Parallel Distrib. Syst, vol. 25, no. 12, pp. 3177–3189. (2014).

3. Wu, C., Ji, Y., Liu, F., Ohzahata, S., Kato, T.: Toward Practical and Intelligent Routing in Vehicular Ad Hoc Networks. IEEE Transactions on Vehicular Technology. Vol. 64 No.12. pp. 5503.19. (2015).
4. Salim Bitam, Abdelhamid Mellouk.: Bio-Inspired Routing Algorithms Survey for Vehicular Ad Hoc Networks. In: IEEE Communication Surveys & Tutorials. Vol. 17, No.2. (2015).
5. Mahmoud Hashem, Eiza, Thomas Owens.: Situation-Aware QoS Routing Algorithm for Vehicular Ad Hoc Networks. IEEE Transactions on Vehicular Technology. Vol. 64, No. 12. (2015).
6. Karp, B., Kung, H. T.: GPSR: greedy perimeter stateless routing for wireless networks. In: Proceedings of the 6th annual international conference on Mobile computing and networking, pp. 243–254, (2000).
7. Goonewardene, R. T., Ali, F. H., Stipidis, E. L. I. A. S.: Robust mobility adaptive clustering scheme with support for geographic routing for vehicular ad hoc networks. Intelligent Transport Systems. pp. 148–158. (2009).
8. Josiane Nzouonta, Neeraj Rajgure, Guiling Wang, Cristian Borcea.: VANET Routing on City Roads Using Real-Time Vehicular Traffic Information. IEEE Transactions on Vehicular Technology. Vol. 58. No.7. pp. 3609–26. (2009).
9. Nikoletta Sofra, Athanasios Gkelias, Leung, K. Kin.: Route Construction for Long Lifetime in VANETs, IEEE Transactions on Vehicular Technology. Vol. 60. No.7. (2011).
10. Tsiachris, S. Koltsidas, G. Pavlidou, FN.: Junction-based geographic routing algorithm for vehicular ad hoc networks, In: Wireless personal communications, Vol 71. N0.2. pp. 955–973. (2013).
11. Xue, G., Luo, Y., Yu, J., Li, M.: A novel vehicular location prediction based on mobility patterns for routing in urban VANET. EURASIP Journal on Wireless Communications and Networking. Vol 1. pp. 1–4. (2012).
12. Taleb, T., Sakhaee, E., Jamalipour, A., Hashimoto, K., Kato, N., Nemoto, Y.: A stable routing protocol to support ITS services in VANET networks. IEEE Transactions on Vehicular Technology, pp. 3337–3347. (2007).

PAPR Performance Analysis of Unitary Transforms in SLM-OFDM for WLAN 802.11a Mobile Terminals

Sukanya Kulkarni and B.K. Mishra

Abstract A high Peak to average power is a major disadvantage in Orthogonal frequency division multiplexing (OFDM) systems since it leads to presence of severe non-linearity in the final power amplifier stage thereby reducing the power efficiency of the transmitter. Precoded Selective level mapping (SLM) OFDM using high merit factor or low aperiodic autocorrelation based phase sequence set is proposed for PAPR reduction while ensuring satisfactory modulation accuracy in WLAN mobile terminals. Further, a novel unitary transform kernel based precoding method is applied as preprocessing stage to SLM and the performance is analyzed for a 64 sub-carrier OFDM system. Simulation results show that proposed discrete cosine transform (DCT) precoded SLM achieves a PAPR of 7.3 dB at 10^{-3} clipping probability and error vector magnitude of 8.9%.

Keywords Unitary transform · Selective level mapping (SLM) · Precoding · Discrete cosine transform (DCT) · Peak to average power ratio (PAPR)

1 Introduction

A high peak to average power ratio (PAPR) places a constraint on the design of the transmitter as it dictates the power efficiency of the HPA and the overall performance of the Orthogonal frequency division multiplexing (OFDM) system. Selective level mapping (SLM), is a popular distortion less method to reduce PAPR in which primarily many statistically independent sequences which represent similar information are generated and one with least PAPR is selected for transmission. However, the PAPR reduction of a conventional SLM-OFDM system depends on

S. Kulkarni (✉)
Sardar Patel Institute of Technology, Mumbai University, Mumbai, India
e-mail: sukanya_kulkarni@spit.ac.in

B.K. Mishra
Thakur College of Engineering and Technology, Mumbai University, Mumbai, India
e-mail: tcet.principal@thakureducation.org

S.C. Satapathy et al. (eds.), *Computer Communication, Networking and Internet Security*, Lecture Notes in Networks and Systems 5,
DOI 10.1007/978-981-10-3226-4_17

the choice of the phase sequence set. Hadamard phase sequence set is commonly used in SLM-OFDM [1, 2]. The elements of Hadamard (H) phase sequence set, no doubt is orthogonal but also exhibits a high aperiodic autocorrelation which results in high PAPR.

Precoded SLM-OFDM is considered as another suitable alternative to reduce PAPR of the transmit signal. Various precoding methods [3, 4] exists in literature which reduce the PAPR of transmitted signal such as the Vandermonde like matrix (VLM) based precoded SLM [5], DCT precoded [6] and Zadoff-Chu matrix transform (ZCMT) precoding [7]. However all these methods use Hadamard phase sequence set to generate multiple copies of transmit signal.

Hence to improve PAPR of conventional SLM-OFDM we may either choose a phase sequence set of low aperiodic autocorrelation or apply an unitary transform based precoding or both. In this paper, we propose a precoded SLM-OFDM using a new phase sequence set with low aperiodic autocorrelation based on Golay Complementary Sequences (GCS) [8] for optimal performance with reduced PAPR. We apply DCT precoding proposed in [9] and analyze the PAPR improvement. Further we analyze imperfections such as nonlinearity in the high power amplifier (HPA) that results in signal to noise degradation in terms of error vector magnitude (EVM).

2 Precoded SLM-OFDM

In Precoded SLM, the discrete complex valued OFDM symbol of length N is firstly partitioned into smaller V disjoint sub blocks using the adjacent sub block partitioning scheme. This is represented by the vectors $\{\mathbf{X}^{(v)}, v = 0, 1, 2\ldots, V-1\}$. Every partitioned subblock data block of length K is firstly transformed into U equivalent SLM candidate signals by multiplying the input symbol sequence with distinct phase sequences $B_k^v = [B_1^v \quad B_2^v \quad B_3^v \quad \ldots \quad B_k^v]$ chosen from the set $[-1\ 1]$.

$$U_k^{(u)} = B_k^{(v)} X^{(v)} \quad 1 \le k \le N/V \tag{1}$$

where $u \in \{0, 1, 2\ldots, U-1\}$. Each subblock phase rotated data is precoded by a unitary precoding matrix (P) which consists of multiplying the modulated data of each subblock by a precoding matrix before OFDM modulation. The predefined precoding matrix can be represented as:

$$P = \begin{bmatrix} P_{00} & & & P_{0,k-1} \\ P_{10} & & & P_{1,k-1} \\ \cdots & \cdots & \cdots & \cdots \\ P_{k-1,0} & & & P_{k-1,k-1} \end{bmatrix} \tag{2}$$

The precoded output signal of the every sub block is represented by the vectors $\{Y^{(v)} = PU_k^{(u)}, u = 0, 1,\ldots, U-1\}$. Each of the subblock sequence is then modulated by K orthogonal subcarriers $\{e^{j2\pi f_0 t}, e^{j2\pi f_1 t}, \ldots e^{j2\pi f_{k-1} t}\}$ and transformed into a time domain selective level signal.

$$x_n = \mathrm{IFFT}\left(\sum_{v=0}^{V-1} Y^{(v)}\right) = \mathrm{IFFT}\left(\sum_{v=0}^{V-1} PU_k^u\right) \quad 0 \le n \le N-1 \tag{3}$$

The PAPR of these U vectors are calculated separately after transforming the signal into the time domain by IFFT. Eventually, the sequence with the smallest PAPR will be selected for final serial transmission which is given by the equation:

$$\mathrm{PAPR}\{x_n\} = 10\log_{10}\frac{\max_{0\le n\le N-1}|x_n|^2}{E\left[|x_n|^2\right]} \tag{4}$$

We evaluate the PAPR statistically using the complementary cumulative distribution function (CCDF) which is the probability of PAPR exceeding a threshold (PAPR0).

$$P_r(\mathrm{PAPR}\{x_n\} > \mathrm{PAPR0}) = \left(1 - (1 - e^{-\mathrm{PAPR0}})^N\right)^U \tag{5}$$

3 Phase Sequence Set

The choice of phase sequence set selection is based on whether the members of the set are orthogonal to each other and the component wise product of any two members of a phase sequence set should not be periodic or similar to periodic sequences.

We analyze the second criteria on the basis of merit factor which is defined as follows.

Let $a = (a_0, a_1, \ldots a_{L-1})$ be a sequence of length L such that a € (+1, −1). Then we define the Aperiodic Auto-Correlation Function (AACF) of a sequence by

$$\rho_a(j) = \sum_{i=0}^{L-j-1} a_i a_{i+j} = a_0 a_j + a_1 a_{j+1} + a_2 a_{j+2} \quad 0 \le j \le L-1 \tag{6}$$

Then, the merit factor of a sequence is defined as

$$M(a) = \frac{L^2}{2\sum_{j=1}^{L-1} |\rho_a(j)|^2} \tag{7}$$

If b be defined similarly to a, then pair (a, b) is called a Golay Complementary Pair (GCP) for $j \neq 0$ if $\rho_a(j) + \rho_b(j) = 0$.

Each member of a GCP is a Golay complementary sequence (GCS). High merit factor GCS sequences have inherent property of low PAPR.

The first phase sequence set considered is the rows of well known Hadamard matrix (H). The other phase sequence set is also a Hadamard matrix generated recursively based on the GCP referred to as Golay-Hadamard (GH) is constructed as follows. Let (a, b) be Golay complementary pair of length L. Let A and B be L * L circulant matrices constructed with a and b as first rows respectively. Then resulting 2L * 2L GH matrix [10, 11]:

$$\begin{bmatrix} A & B \\ B^T & -A^T \end{bmatrix} \quad \text{or} \quad \begin{bmatrix} A & -B \\ B^T & A^T \end{bmatrix} \tag{8}$$

4 Unitary Precoding Transform

Orthogonal precoding transforms decorrelates or randomizes the OFDM data sequence further so as to reduce the PAPR. In this section, we briefly review DCT-II, proposed DCT and COSHAD kernels. Precoding kernels in OFDM systems expressed as matrix is characterized by two properties that helps to reduce the peak power of OFDM signal.

1. The square precoder matrix P is orthogonal i.e. its inverse equals it's transpose.
2. The matrix is unitary, then $P^{-1} = P^{*T}$.

4.1 Discrete Cosine Transforms

There are four common forms of DCT of which in SLM generally DCT-II [Zhou] is used as precoding kernel. Its precoding matrix of size M by M is given as:

$$P = \begin{cases} \dfrac{1}{\sqrt{M}} & k = 0, \qquad 0 \leq n \leq M-1 \\ \sqrt{\dfrac{2}{M}} \cos\left(\dfrac{\pi(2n+1)k}{2M}\right) & 1 \leq n \leq M-1 \\ & 0 \leq k \leq M-1 \end{cases} \tag{9}$$

We define for SLM-OFDM, a new form of orthogonal discrete cosine transform kernel proposed by Jianqin Zhou [9] which is variation of DCT-IV expressed in matrix form is as follows

$$P = S\begin{bmatrix} \cos\left(\frac{\pi}{2M-1}\right) & \cos\left(\frac{3\pi}{2M-1}\right) & \cdots & \cos\left(\frac{(2M-3)\pi}{2M-1}\right) & -\frac{1}{\sqrt{2}} \\ \cos\left(\frac{3\pi}{2M-1}\right) & \cos\left(\frac{9\pi}{2M-1}\right) & \cdots & \cos\left(\frac{3(2M-3)\pi}{2M-1}\right) & -\frac{1}{\sqrt{2}} \\ \cdot & & & & \\ \cos\left(\frac{(2M-3)\pi}{2M-1}\right) & \cos\left(\frac{3(2M-3)\pi}{2M-1}\right) & \cdots & \cos\left(\frac{(2M-3)(2M-3)\pi}{2M-1}\right) & -\frac{1}{\sqrt{2}} \\ -\frac{1}{\sqrt{2}} & -\frac{1}{\sqrt{2}} & \cdots & -\frac{1}{\sqrt{2}} & -\frac{1}{2} \end{bmatrix}$$

$$\text{where} \quad S = \sqrt{\frac{4}{2M-1}} \tag{10}$$

4.2 *COSHAD Transform*

COSHAD [12] kernel function first introduced by Merchant and Rao, is a hybrid version of discrete Cosine and Hadamard transform derived from the Kronecker product of the two kernel functions. The discrete orthogonal COSHAD kernel function ranges from the Hadamard kernel function (m = 0) to the DCT kernel function (m = n).

$$\begin{aligned} &P = [\text{COSHAD}_m(n)] = [\text{DCT}(m)] \otimes [\text{HAD}(n-m)] \\ &\text{DCT} : 2^m * 2^m \text{ matrix} \\ &\text{HAD} : 2^{n-m} * 2^{n-m} \text{ matrix.} \end{aligned} \tag{11}$$

5 Simulation Results

In this section, computer simulations are used to evaluate the peak-to-average ratio reduction capability with proposed scheme. In simulation, we assume an OFDM system with 64 subcarriers (N = 64), 24 Mbps data rate system with 1/2 rate convolutional encoding, 16-QAM data symbols and the phase factors are chosen from set [1 −1]. An oversampling factor of 4 is employed to approximate the continuous time OFDM signal and adjacent subblock partitioning is applied before precoding.

In the simulation results, SLM refers to conventional SLM of literature with H phase sequence set. The results of our work analyzes various precoded SLM methods such as Walsh Hadamard precoding (HAD), DCT precoding (DCT), proposed DCT precoding and COSHAD precoding with GH phase sequence set.

The two parameters analyzed are complement cumulative distribution function (CCDF) and error vector magnitude (EVM). Error vector magnitude (EVM) is a direct measurement of modulation accuracy and nonlinear behavior of transmitter high power amplifier (HPA) expressed as a percentage.

The merit factor evaluated using Eq. (7) is normalized and plotted in Fig. 1 for Hadamard (H) and Golay-Hadamard (GH) sequences. GH sequences exhibit high merit factor or low aperiodic autocorrelation compared to H sequences. The candidate signal copies generated in SLM, by component wise product of high merit factor orthogonal GCS and data sequence is also a GCS and hence PAPR reduction is observed.

In Fig. 2 it is observed that proposed DCT and COSHAD exhibit 0.8 dB and 1 dB PAPR reduction compared to DCT-II and Walsh-Hadamard precoding respectively. The two novel precoding methods with GH sequence set shows about 1.6 dB PAPR improvement compared to conventional SLM with H sequence set.

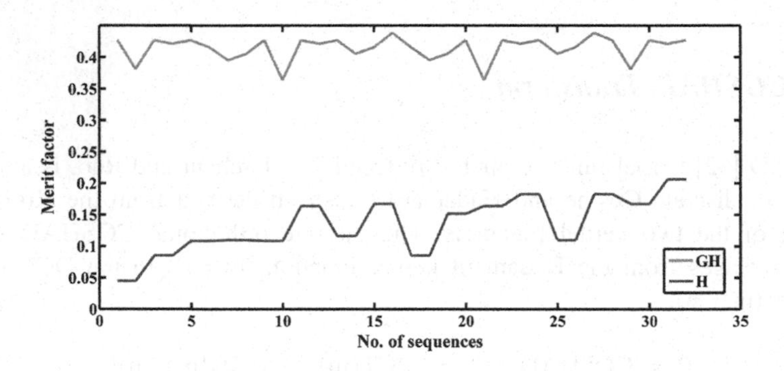

Fig. 1 Merit factor of phase sequences

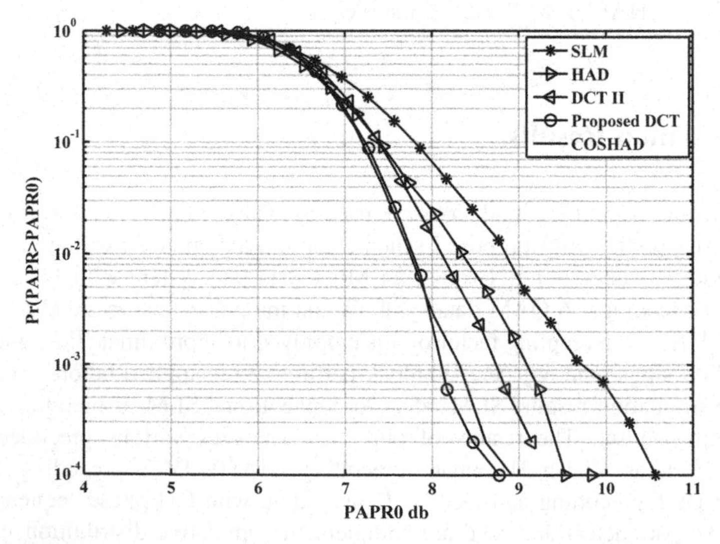

Fig. 2 Comparison of various precoded SLM-OFDM for 64 sub carrier system, 2 sub blocks and 8 candidate signals

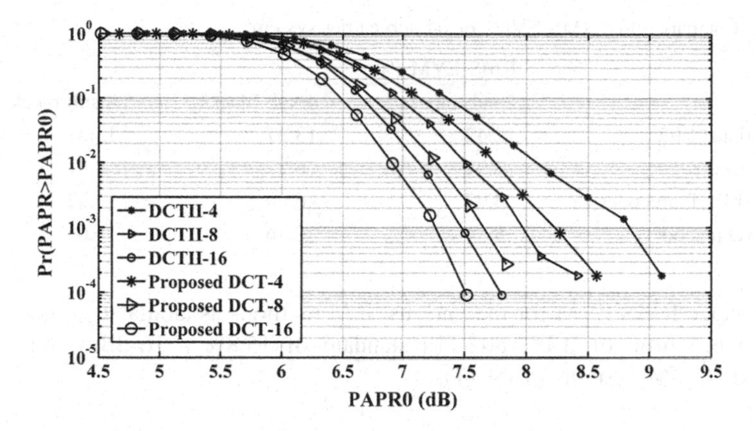

Fig. 3 Comparison of proposed DCT and DCT II for 4, 8, 16 candidate signals

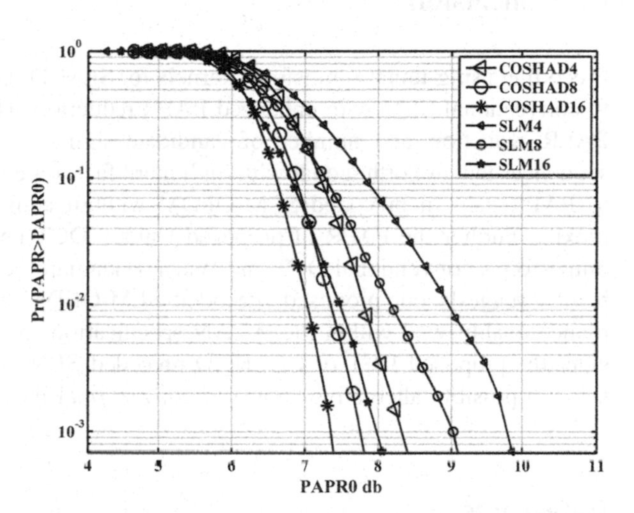

Fig. 4 Analysis of COSHAD precoded SLM-OFDM and conventional SLM-OFDM for 4, 8, 16 candidate signals

Precoded SLM OFDM with GH sequence set exhibits a improvement in PAPR reduction with increase in number of candidate signal mappings. It is observed from Fig. 3 that proposed DCT performs better than DCT-II for a given number of candidate signals since energy compaction is more as a result of higher decorrelation.

Figure 4 shows comparison of COSHAD precoded SLM and conventional SLM. As the number of candidate signals is increased, there is a better PAPR reduction. For 64 sub carrier system with two subblock partitioning, proposed precoded SLM shows 1.6 dB improvement for 4 candidate signals, 1.2 dB improvement for 8 candidate signals at a clipping probability of 10^{-3}.

The modulation fidelity of Power amplifier stage is simulated using Rapp Model and Error vector magnitude (EVM), a measure of inband distortion is evaluated to compare the various schemes proposed in an AWGN channel scenario. The values

Table 1 Comparison of RMS EVM for 64 sub carrier system

	RMS EVM (%)		
Method	2 sub blocks	4 sub blocks	8 sub blocks
Conventional SLM	12.42	13.27	15.2
DCT precoder	8.94	6.97	4.81
Proposed DCT precoder	8.97	6.68	4.72
COSHAD precoder	8.84	6.76	4.65

of percentage RMS EVM for the various SLM methods as tabulated is well within the tolerance limit of IEEE 802.11a standard of 15.8% (−16 dBm) for ½ rate 16-QAM OFDM systems (Table 1).

6 Conclusion

Precoding using unitary transform kernels for SLM-OFDM is a suitable alternative to conventional SLM with enhanced PAPR reduction. There is a trade off between PAPR reduction and number of candidate signal mappings considered. Phase sequences that are orthogonal with high merit factor are a better choice which leads to PAPR gain in precoded SLM-OFDM without additional computational complexity. Further the PAPR of proposed Unitary DCT kernel and COSHAD kernel outperforms conventional DCT and Walsh Hadamard kernels. The simulated EVM results suggests our proposed precoded SLM-OFDM exhibits transmitter performance quality well within the WLAN specifications his provides grounds to consider the proposed DCT or COSHAD precoded SLM with Golay phase sequence set as a possible alternative choice to conventional methods.

References

1. A. Kangappaden, A. R. Daniel, V. P. Peeyusha, M. P. Raja, P. Sneha and A. M. V. Das. Comparision between SLM-companding and precoding-companding techniques in OFDM systems. In: International Conference on Circuit, Power and Computing Technologies (ICCPCT), Nagercoil, pp. 1–5(2016).
2. Sghaier, Mouna, Fatma Abdelkefi, and Mohamed Siala.: New SLM-Hadamard PAPR reduction scheme for blind detection of precoding sequence in OFDM systems. In: IEEE Wireless Communications and Networking Conference (WCNC)(2014).
3. P. R. Lasya and M. S. Kumar.: PAPR and out-of-band power reduction in OFDM-based cognitive radios. In: International Conference on Signal Processing And Communication Engineering Systems (SPACES), Guntur, pp. 473–476(2015).
4. Wang, Sen-Hung,: A novel low-complexity precoded OFDM system with reduced PAPR. In: IEEE Transactions on Signal Processing vol.63,no.6, pp. 1366–1376 (2015).
5. Md. Mahmudul Hasan.: VLM Precoded SLM Technique for PAPR Reduction In OFDM Systems. In: Wireless Personal Communication (2013).

6. Baig, Imran, Varun Jeoti, and Micheal Drieberg.: On the PAPR Reduction in LTE-Advanced: Precoding-Based SLM–LOFDMA Uplink Systems. Arabian Journal for Science and Engineering vol.38, no.5, pp. 1075–1086(2013).
7. Baig, Imran, Varun Jeoti.: A ZCMT precoding based multicarrier OFDM system to minimize the high PAPR. In: Wireless Personal Communications vol.68, no.3, pp. 1135–1145(2013).
8. Davis JA, Jedwab J.: Peak-to-mean power control in OFDM, Golay complementary sequences, and Reed-Muller codes. In: IEEE Transactions on Information Theory., vol.45(7), pp. 2397–417(1999).
9. Zhou, Jianqin, Ping Chen.: Generalized discrete cosine transform. In: Pacific-Asia Conference on. C.Circuits, Communications and Systems, IEEE, (2009).
10. C.H. Yang,: On Hadamard matrices constructible by circulant submatrices. Math.Comp., vol.25, pp. 181–186, (1971).
11. Seberry J, Wysocki B, A Wysocki T.: On some applications of Hadamard matrices. In: Metrika. Nov 1, vol.62(2–3), pp. 221–39, (2005).
12. Zhu, H., Gui, Z., Zhu, Y. and Chen, Z.: Discrete Fractional COSHAD Transform and Its Application. In: Mathematical Problems in Engineering (2014).

Optimal Sensing Time Allocation for Energy Efficient Data Transmission in Amplify-Forward Cognitive Relay Assisted Network

Sutanu Ghosh, Aditya Chaudhuri and Sayantani Ghosh

Abstract In this chapter, the main goal is to enhance the energy efficient (EE) data transmission for an optimum value of sensing time in Amplify-and-forward (AF) relay assisted cognitive radio network. Performance of the system has been studied in standard optimization framework under the constraints imposed on probability of detection threshold and the limitation of total power budget. Based on the mathematical model, analytical output is shown for various system parameters like, false alarm probability, throughput and energy efficiency. Furthermore, such evaluation illustrates the existence of an optimal sensing time, which improves the energy efficiency of the system.

Keywords Cognitive radio network · AF relay · Spectrum sensing · Energy efficiency

1 Introduction

An insatiable demand has grown for radio frequency spectrum due to the increase in various wireless applications. So, there is an immediate requirement to combat the spectrum scarcity problem by its rational and optimal usage. The Federal Communication Commission (FCC) [1] inferred that the primary reason behind recent spectrum dearth was its inefficient utilisation rather than physical deficiency. A huge section of the ascribed spectrum either lingers idle or is scarcely utilised for

S. Ghosh (✉)
Indian Institute of Engineering Science and Technology, Shibpur, Howrah, India
e-mail: sutanu99@gmail.com

A. Chaudhuri
Dr. Sudhir Chandra Sur Degree Engineering College, Kolkata, India
e-mail: in.aditya.c@gmail.com

S. Ghosh
Jadavpur University, Kolkata, India
e-mail: qwerty123.hw@gmail.com

S.C. Satapathy et al. (eds.), *Computer Communication, Networking and Internet Security*, Lecture Notes in Networks and Systems 5,
DOI 10.1007/978-981-10-3226-4_18

a significant period of time. Hence, researchers had needed to develop a new technology that involved the sharing of unexploited radio spectrum by means of dynamic and opportunistic spectrum access (DOSA) [2]. This technology is referred to as the cognitive radio Network (CRN) [3]. It is a progression of software defined system (SDR). This technology enables the unregistered secondary users (SU) to dynamically and opportunistically utilize the registered spectrum without any adverse impinge on the quality of service for each coexisting incumbent registered primary users (PU).

The major objective of CRN is accomplished by spectrum sensing (SS) [4]. There are different forms of SS technique which include energy detection [5], waveform based sensing [6], wavelet based sensing [7], matched filter detection [8], cyclo-stationarity based sensing [9], multi-taper method [10] etc. However, energy efficient (EE) SS have recently gained importance in research area as sensing devices mostly work with limited energy.

Previously, Huang et al. [11] explored EE type of spectrum sensing under sensing reliability constraints. Eryigit et al. [12] investigated EE spectrum sensing scheduling problem to minimize the energy consumption and to maximize data transmission time with maintaining a perfect PU protection for multichannel cooperative network. Li et al. [13] dealt with joint optimization between the medium access layer (MAC) and physical layer to enhance energy efficiency.

This work is based on a previously proposed relay involving optimal power allocation to assist joint SS and data transmission [14]. The objective of this paper is to determine the optimal sensing time to enhance the energy efficiency of relay assisted transmission in cognitive radio architecture with maintaining proper PU protection along with the effective power budget of secondary source of the network.

The rest of the paper has been divided as follows—Sect. 2 describes the model of system structural design, Sect. 3 includes the optimal sensing time analysis, Sect. 4 provides the simulation results and the conclusion have been discussed in Sect. 5.

2 Description of System Exemplary

The system exemplary as shown in Fig. 1. consists of a PU, cognitive source (CS), cognitive destination (CD) and cognitive relay (CR) nodes. This CRN strategy provides two advantages—(a) enhances sensing performance of secondary network and (b) ensures higher rate of secondary transmission. It relies on a frame-by-frame basis operation as represented in Fig. 2. Each frame is divided into two time slots T_g and $T-T_g$. Altogether the period of total time frame is T_s. The CRN inspects the existence of PU at the beginning of each time frame which leads to the following two situations: (i) If PU is not found in the preceding frame (M') then CR receives data from CS along with supervision of existence of PU in the present frame. (ii) If PU is

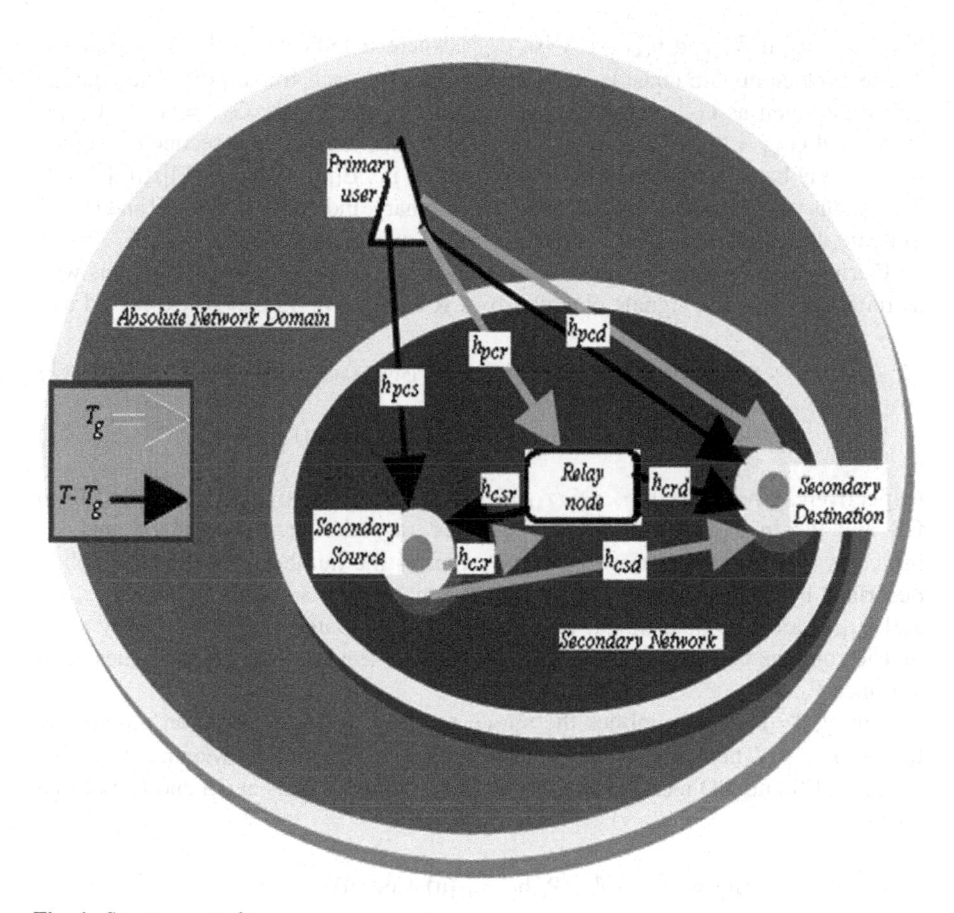

Fig. 1 System exemplary

Fig. 2 Frame format for current frame M

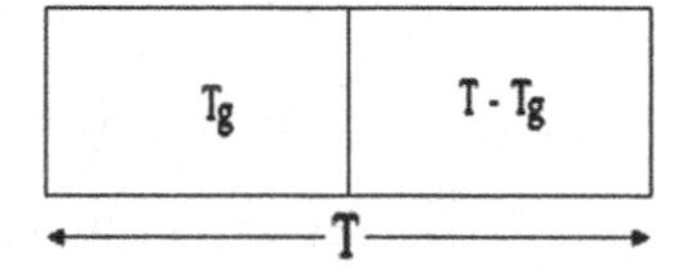

found to be present in the preceding frame (M′), then no data transmission takes place in secondary network and PU is monitored in the present frame (M).

Rayleigh flat fading channel is considered as channel of various links used in this exemplary. The coefficients of links between PU → CS, PU → CR, PU → CD, CS → CR (as well as CR → CS), CR → CD and CS → CD are represented by h_{pcs}, h_{pcr}, h_{pcd}, h_{csr}, h_{crd} and h_{csd}, respectively. Their respective distances include d_{pcs}, d_{pcr}, d_{pcd}, d_{csr}, d_{crd} and d_{csd}. Statistics of the fading coefficients are given as: $h_{pcs} \sim \mathcal{CN}\left(0,\, d_{pcs}^{-\alpha}\right)$, $h_{pcr} \sim \mathcal{CN}\left(0,\, d_{pcr}^{-\alpha}\right)$, $h_{pcd} \sim \mathcal{CN}\left(0,\, d_{pcd}^{-\alpha}\right)$, $h_{csr} \sim \mathcal{CN}\left(0,\, d_{csr}^{-\alpha}\right)$,

$h_{crd} \sim \mathcal{CN}(0, d_{crd}^{-\alpha})$ and $h_{csd} \sim \mathcal{CN}(0, d_{csd}^{-\alpha})$ where α indicates path loss exponent. These coefficients are considered to be constant for each frame [15]. The channel gains are given as $G_{pcs} = |h_{pcs}|^2$, $G_{pcr} = |h_{pcr}|^2$, $G_{pcd} = |h_{pcd}|^2$, $G_{csr} = |h_{csr}|^2$, $G_{crd} = |h_{crd}|^2$ and $G_{csd} = |h_{csd}|^2$. Let $z_p(n)$ and $z_s(n)$ be the primary and secondary signal, respectively having zero mean and unit variance i.e. $E[|z_p(n)|^2] = 1$, $E[|z_s(n)|^2] = 1$. The status of PU signal is described by a binary indicator ψ ($\psi = 0$ and $\psi = 1$ indicate the nonexistence and existence of PU, respectively).

During the first phase of time frame, CR and CD receive primary signal as well as data from CS. So, signals obtained by CR and CD in T_g can be expressed as:

$$w_{r1}(n) = \sqrt{P_{ts}} h_{csr} z_{s1}(n) + \psi \sqrt{P_{tp}} h_{pcr} z_{p1}(n) + e_{r1}(n). \tag{1}$$

$$w_{d1}(n) = \sqrt{P_{ts}} h_{csd} z_{s1}(n) + \psi \sqrt{P_{tp}} h_{pcd} z_{p1}(n) + e_{d1}(n). \tag{2}$$

respectively, where n = 1, …, N, $z_{s1}(n)$ and $z_{p1}(n)$ define signals transmitted from CS and PU in T_g, respectively, P_{ts} denotes transmission power of CS and P_{tp} indicates transmission power of PU. N indicates the number of samples which is described by the product of $T_g f_s$, where f_s denotes the sampling frequency. $e_{r1}(n)$ and $e_{d1}(n)$ are the independent and identically distributed (i.i.d.), circularly symmetric complex Gaussian (CSCG) form of random sequences having zero mean and variance P_v.

During $T-T_g$, CR amplifies the signal received during T_g with an amplifying factor of $\sqrt{\beta}$. Then this signal is forwarded to CS and CD. Moreover, at that instance, CS and CD receives PU signal. So, obtained signals at CS and CD can be written as:

$$\begin{aligned} w_{s2}(n) &= \sqrt{\beta} h_{csr} w_{r1}(n) + \psi \sqrt{P_{tp}} h_{pcs} z_{p2}(n) + e_{s2}(n) \\ &= \sqrt{\beta}\sqrt{P_{ts}} h_{csr} h_{csr} z_{s1}(n) + \psi \sqrt{P_{tp}} \left(\sqrt{\beta} h_{csr} h_{pcr} z_{p1}(n) + h_{pcs} z_{p2}(n) \right) \\ &\quad + \sqrt{\beta} h_{csr} e_{r1}(n) + e_{s2}(n). \end{aligned} \tag{3}$$

$$\begin{aligned} w_{d2}(n) &= \sqrt{\beta} h_{crd} w_{r1}(n) + \psi \sqrt{P_{tp}} h_{pcd} z_{p2}(n) + e_{d2}(n) \\ &= \sqrt{\beta}\sqrt{P_{ts}} h_{crd} h_{csr} z_{s1}(n) + \psi \sqrt{P_{tp}} \left(\sqrt{\beta} h_{crd} h_{pcr} z_{p1}(n) + h_{pcd} z_{p2}(n) \right) \\ &\quad + \sqrt{\beta} h_{crd}(n) e_{r1}(n) + e_{d2}(n) . \end{aligned} \tag{4}$$

respectively, where n = N + 1, …, 2 N, $z_{p2}(n)$ indicates the PU signal with the mean of zero and unity variance $E[|z_{p2}(n)|^2] = 1$ at T_g. Noise components $e_{s2}(n)$ and $e_{d2}(n)$ are i.i.d., CSCG random sequences having zero mean and variance of P_v.

Self-interference at CS is described by the first mathematical term in (3). Hence, by applying self-interference cancellation (cancelling $z_{s1}(n)$ which originates from CS), (3) can be re-written as

$$\widetilde{w_{s2}(n)} = \psi\sqrt{P_{tp}}\left(\sqrt{\beta}h_{csr}h_{pcr}z_{p1}(n) + h_{pcs}z_{p2}(n)\right) + \sqrt{\beta}h_{csr}e_{r1}(n) + e_{s2}(n). \quad (5)$$

The signal described in (5) aids in energy detection at CS for SS by using $\widetilde{w_{s2}(n)}$. Test statistic W_s, can be mathematically described as $w_s = \sum_{n=N+1}^{2N} |\widetilde{w_{s2}(n)}^2|$ and follows chi-square distribution. Based on the central limit theorem, for an abundant number of samples N, the distribution of test statistic Z_s nearly pursues a Gaussian pattern under both the hypothesis H_0 (ψ = 0) and H_1 (ψ = 1). For ease of analysis $z_{p1}(n)$ and $z_{p2}(n)$ are to be considered as CSCG form of random sequences and $z_{p1}(n)$, $z_{p2}(n)$, $e_{r1}(n)$ and $e_{s2}(n)$ are pair-wise mutually independent. The mean and variance of W_s under H_1 can be written as $E(W_{s1}) = N\mu_1$ and $\mathrm{var}(W_{s1}) = N\mu_1^2$, respectively, where $\mu_1 = G_{pcs}P_{tp} + \beta G_{csr}(G_{pcr}P_{tp} + P_v)$. Under the hypothesis H_0, W_s can be expressed as $E(W_{s0}) = N\ \mu_0$ and $\mathrm{var}(W_{s0}) = N\ \mu_0^2$, respectively, where $\mu_0 = \beta G_{csr}P_v + P_v$.

Let, ξ denote the decision threshold for detection probability p_d and false alarm probability p_f. These probabilities can be expressed as follows:

$$p_d = Q\left(\frac{(\xi - E(W_{s1}))}{(\sqrt{(\mathrm{var}(W_{s1}))})}\right) = Q\left(\frac{(\xi - (N\mu_1))}{\sqrt{N}\mu_1}\right). \quad (6)$$

$$p_f = Q\left(\frac{(\xi - E(W_{s0}))}{(\sqrt{(\mathrm{var}(W_{s0}))})}\right) = Q\left(\frac{(\xi - (N\mu_0))}{\sqrt{N}\mu_0}\right). \quad (7)$$

where, Q(.) denotes the complementary cumulative distribution function of standard form of Gaussian random variable.

Now, to meet the target threshold of detection p_d^{thr}, from (6) and (7), by cancelling the decision threshold ξ, false alarm probability p_f can be expressed in the following way:

$$\begin{aligned} p_f &= Q\left(\frac{Q^{-1}(p_d^{thr})\mu_1 + \sqrt{N}(\mu_1 - \mu_0)}{\mu_0}\right) \\ &= Q\left(\left(Q^{-1}(p_d^{thr}) + \sqrt{N}\right)\gamma_{p1} + Q^{-1}(p_d^{thr})\right). \end{aligned} \quad (8)$$

where,

$$\gamma_{p1} = \frac{(\mu_1 - \mu_0)}{(\mu_0)} = \frac{\gamma_{pcs} + \beta G_{csr}\gamma_{pcr}}{1 + \beta G_{csr}}$$
$$\gamma_{pcs} = \frac{G_{pcs}P_{tp}}{P_v} \& \gamma_{pcr} = \frac{G_{pcr}P_{tp}}{P_v}.$$

γ_{pcs} and γ_{pcr} are transmitted signal SNR along the link of PU to CS and PU to CR, respectively.

3 Optimal Sensing Time (OST) Analysis

The secondary relay throughput in CRN can be mathematically expressed as:

$$R' = \left(\frac{\mathrm{T}-\mathrm{T_g}}{\mathrm{T}}\right)(1-\sigma)(1-p_f)(\log_2(1+\gamma')). \tag{9}$$

where, γ' indicates the cognitive signal-to-noise ratio (CSNR) received at CD during the inactive state of PU ($\psi = 0$) and σ describes the probability when PU is in active state i.e., $\sigma = \mathrm{P}(\psi = 1) = \mathrm{P}(\mathrm{H_1})$. Received signal at CD from CR during $\mathrm{T}-\mathrm{T_g}$ phase, can be written using CSNR of (4) as:

$$\gamma' = \frac{\beta G_{crd} G_{csr} P_{ts}}{\beta G_{crd} \mathrm{P_v} + \mathrm{P_v}} \tag{10}$$

From (1), transmission power of CR can be calculated as:

$$P_R = \left(\beta G_{csr} P_{ts} + \beta\psi G_{pcr} P_{tp} + \mathrm{P_v}\beta\right) \tag{11}$$

The average transmission power of CR can be written as:

$$\begin{aligned}\overline{P_R} &= \sigma\left(\beta G_{csr} P_{ts} + \beta G_{pcr} P_{tp} + \beta \mathrm{P_v}\right) + (1-\sigma)(\beta G_{csr} P_{ts} + \beta \mathrm{P_v}) \\ &= \left(\beta G_{csr} P_{ts} + \beta\sigma G_{pcr} P_{tp} + \beta \mathrm{P_v}\right).\end{aligned} \tag{12}$$

On the basis of system exemplary we have found two different estimation of energy efficiency of system.

$$(EE') = \frac{\left(\frac{\mathrm{T}-\mathrm{T_g}}{\mathrm{T}}\right)(1-\alpha)(1-p_f)\left(\log_2\left(1+\frac{\beta G_{crd} G_{csr} P_{ts}}{\beta G_{crd} P_V + P_V}\right)\right)}{\left(\beta G_{csr} P_{ts} + \beta\sigma G_{pcr} P_{tp} + \beta P_V\right)}. \tag{13}$$

The objective of this work is to maximize EE satisfying sensing constraint to protect PU and maintaining a total transmission power budget P_m of secondary source alone. Then optimization problem can be formulated in two different ways as follows:

Prob.

$$\max_{\mathrm{T_g}}(EE') \tag{14a}$$

under the constraints:

$$\begin{aligned}&\mathrm{c_1} : p_d \geq p_d^{thr} \\ &\mathrm{c_2} : (\beta G_{csr} P_{ts} + P_{ts}) \leq P_m.\end{aligned} \tag{14b}$$

where, p_d^{thr} denotes the probability detection threshold and P_m indicates the total transmission power budget.

On the basis of (14b), power equality constraint can be reconstructed as:

$$(\beta G_{csr} P_{ts} + P_{ts} + \alpha) = P_m. \tag{15}$$

where, α is the slack variable. Now, P_{ts} can be mathematically defined as:

$$P_{ts} = \frac{P_m - \alpha}{(\beta G_{csr} + 1)}. \quad (\alpha \geq 0) \tag{16}$$

Therefore by putting (8) and (16) into (14a) the constrained optimization problem is converted into unconstrained optimization problem. Optimal value of T_g can be found using efficient searching technique.

4 Result Analysis

This section describes the performance analysis of this work. The parameter settings used to show the graphical plots to validate the mathematical analysis. The values of those parameters include: $\sigma = 0.3$, $p_d^{thr} = 0.95$, $P_v = 0$ dBW, $P_{tp} = 0$ dBW, $G_{csr} = G_{crd} = G_{pcr} = -4$ dB, $G_{csd} = G_{pcs} = -10$ dB, $\beta = 2$, $T = 10 \times 10^{-2}$ s and $f_s = 1000$.

Figure 3 represents the graphical plot of false alarm probability p_f versus sensing time T_g. It is observed that as T_g increases, the false alarm probability decreases. Figure 4. shows the secondary relay assisted throughput versus T_g. It is seen that initially throughput augments with an escalation of T_g and then after reaching an optimal value at $T_g = 0.4 \times 10^{-2}$ s, there is a gradual fall. This enhancement of secondary relay throughput occurs due to the fact of $(1 - p_f)$ factor increment.

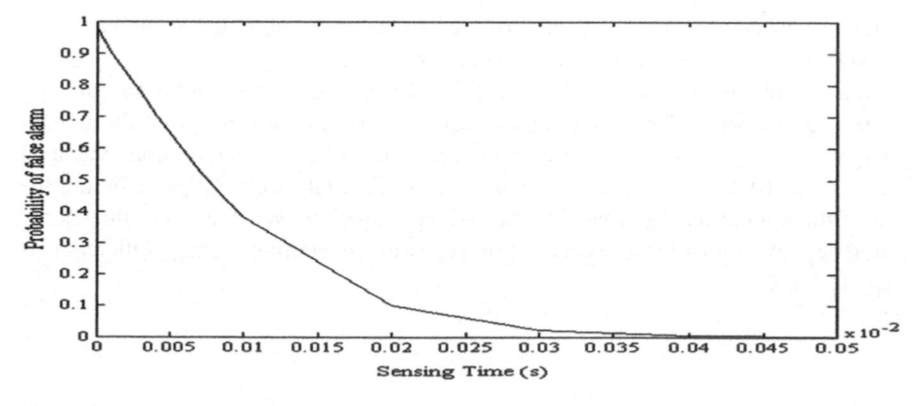

Fig. 3 False alarm probability versus sensing time

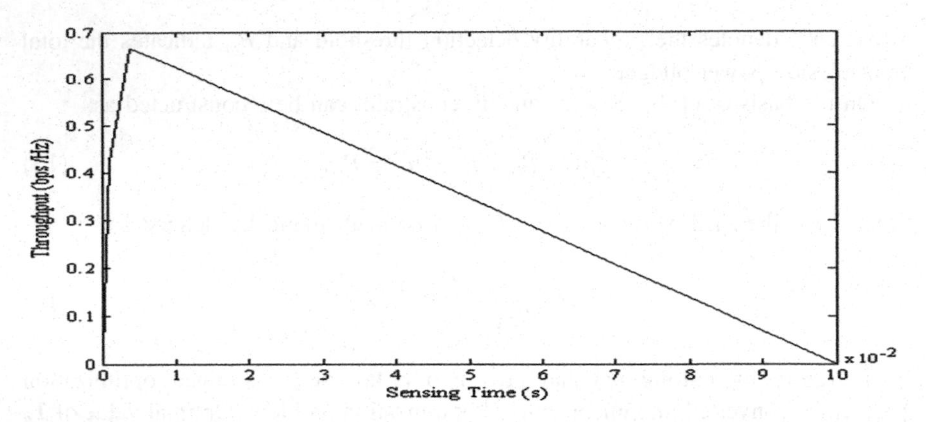

Fig. 4 Throughput at P_m more than 9 dBW (taken fixed 10 dBW) versus sensing time

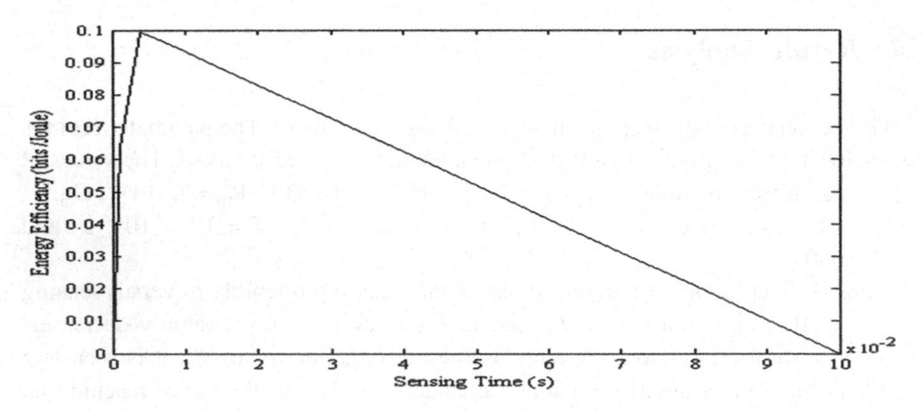

Fig. 5 Energy efficiency of secondary relay at P_m more than 9 dBW (taken fixed 10 dBW) versus sensing time

However, when $(1 - p_f)$ factor almost reaches its peak value, the secondary relay throughput decreases and reduces to nearly zero.

Energy efficiency is inversely proportional to power consumption and directly proportional to secondary relay throughput. Hence, with an upsurge of throughput energy efficiency also increases. However, after reaching an optimal value at $T_g = 0.4 \times 10^{-2}$ s energy efficiency decreases. This fall occurs slightly before the fall of throughput as after reaching the optimum level, power dominates the energy efficiency of the network. This entire scenario concerning energy efficiency is depicted in Fig. 5.

5 Conclusion

Hence, through the mathematical analysis and graphical results we can infer that a maximum value of energy efficiency is found for an optimum sensing time of cognitive relay assisted network. This sensing time is found to be less for more efficient data transmission. Moreover, in future this work can be extended for selection combining based cooperative scenario.

References

1. FCC, ET Docket No. 03-222 Notice of proposed rule making and order, 2003 Dec.
2. Akyildiz I F, Lee W Y, Vuran M C, Mohanty S.: NeXt generation/dynamic spectrum access/cognitive radio wireless networks: a survey. Computer networks. 2006 Sep; 50 (13):2127–2159.
3. Mitola III J, Maguire Jr G Q.: Cognitive radio: making software radios more personal. IEEE Personal Communications. 1999 Aug; 6(4):13–18.
4. Tang L, Chen Y, Hines E L, Alouini M S.: Effect of primary user traffic on sensing-throughput tradeoff for cognitive radios. IEEE Transactions on Wireless Communications. 2011 Apr; 10(4):1063–1068.
5. Atapattu S, Tellambura C, Jiang H.: Energy detection based cooperative spectrum sensing in cognitive radio networks. IEEE Transactions on Wireless Communications. 2011 Apr; 10 (4):1232–1241.
6. Tang H.: Some physical layer issues of wide-band cognitive radio systems. In New frontiers in dynamic spectrum access networks. First IEEE international symposium on DySPAN 2005. Nov (pp. 151–159).
7. Li X, Hu F, Zhang H, Shi C.: Two-branch wavelet denoising for accurate spectrum sensing in cognitive radios. Telecommunication Systems. 2014 Sep; 57(1):81–90.
8. Cabric D B.: Cognitive radios: System design perspective. ProQuest; 2007.
9. Dobre O A, Rajan S, Inkol R.: Joint signal detection and classification based on first-order cyclostationarity for cognitive radios. EURASIP Journal on Advances in Signal Processing. 2009 Mar: 1–12.
10. Haykin S, Thomson D J, Reed J H.: Spectrum sensing for cognitive radio. Proceedings of the IEEE. 2009 May; 97(5):849–877.
11. Huang S, Chen H, Zhang Y, Zhao F.: Energy-efficient cooperative spectrum sensing with amplify-and-forward relaying. IEEE Communications Letters. 2012 Apr; 16(4):450–453.
12. Eryigit S, Bayhan S, Tugcu T.: Energy-efficient multichannel cooperative sensing scheduling with heterogeneous channel conditions for cognitive radio networks. IEEE Transactions on Vehicular Technology. 2013 Feb; 62(6): 2690–2699.
13. Li L, Zhou X, Xu H, Li G Y, Wang D, Soong A.: Energy-Efficient Transmission in Cognitive Radio Networks. IEEE Consumer Communications and Networking Conference (CCNC). 2010 Jan: pp. 1–5.
14. Huang S, Chen H, Zhang Y.: Optimal power allocation for spectrum sensing and data transmission in cognitive relay networks. IEEE Wireless Communications Letters. 2012 Feb; 1(1):26–29.
15. Li L, Zhou X, Xu H, Li G Y, Wang D, Soong A.: Simplified relay selection and power allocation in cooperative cognitive radio systems. Wireless Communications, IEEE Transactions on. 2011 Jan; 10(1):33–36.

Data Privacy in Online Shopping

Shashidhar Virupaksha, Divya Gavini and D. Venkatesulu

Abstract The Online Shopping experience has provided the new ways of business and shopping. Now the traditional way of shopping has changed into easy and convenience manner according to customer shopping behavior and preferences. Extracting shopping patterns from increasing data is not a trivial task. This paper will help to understand the importance of data mining techniques i.e., Association rule mining is to get relationships between different items in the dataset, and frequent item set mining aims to find the regularities in the shopping behavior of customers, clustering and concept hierarchy to provide business intelligence to improve sales, marketing and consumers satisfaction. In this paper while using data mining techniques there is data susceptibility, which is influenced by attacks like membership disclosure protection and homogeneity attack. These attacks deal with reveal of information based on quasi identifier value in the data set. In this paper, protecting sensitive information is an important problem. Detailed analysis of these both attacks are given and proposed a privacy definition called L-Diversity, which can be implemented and experimental evaluation is also shown.

Keywords Homogeneity attack · K-Anonymity · L-Diversity · Dempster's rule · Apriori algorithm · FP growth · Quantitative association rule

S. Virupaksha (✉)
IT Department, VRSEC, Vijayawada, India
e-mail: shashidhar.virupaksha@gmail.com

S. Virupaksha
CSE Department, Vignan University, Guntur, India

D. Gavini
Department of Computer Science and Engineering,
Vignan's Lara Institute of Technology and Science, Guntur, India
e-mail: divyagavini.gd@gmail.com

D. Venkatesulu
Department of Computer Science and Engineering,
Vignan University, Vadlamudi, Guntur, India
e-mail: drv_cse@vignanuniversity.org

S.C. Satapathy et al. (eds.), *Computer Communication, Networking and Internet Security*, Lecture Notes in Networks and Systems 5,
DOI 10.1007/978-981-10-3226-4_19

1 Introduction

Online shopping means providing services to customers via internet. It is convenient to buy various items. The main advantage is to purchase item without need to wait in long lines and search from store to store for a required item. It is a suitable method of shopping and allows for a vast array of products to be at your fingertips. It is easy to retrieve the product quickly what is required. Online shopping is ruling the business world by making people to choose want they want from anywhere at any time. There are no limits for time and location in an online shopping. They provide attractive prices to customers which facilitates them to sell more items for highly discounted prices. This can lead to major cost benefits for shoppers. Today the leading position in business is Online Shopping. Every company needs to maintain a huge database to store list of items, customer details and handle transactions. To analyze all the data manually it is wastage of time and may not be accurate. Data mining is a powerful technology which is used in various applications. Data mining is the crucial for extracting and also identifying useful information from a large amount of data. To increase the investments every company need to focus on data mining techniques and capability. Company will be able to understand their customer by evaluating into behavior, profiles, profitability. It enables the company to be competitive and reduce risks. There are many data mining techniques like association rules, market basket analysis, clustering, and classification to extract the frequent shopping patterns of a customer.

Data mining techniques are used to design more enhanced goods transportation and items purchased ratio. These are also used to discover the hidden relationships in sales from the various applications. Data mining is also concerned with new developing methods to discover knowledge from online data stores. There is a need to maintain privacy for those databases since data is a gold mine for an attacker. The attacker may unlock hidden profitability and customer details. For any company one of its biggest challenges is to secure customer details that include record of purchased items, purchasing patterns, orders, area, and zip code. The accuracy of predicting customer shopping behavior like purchases and preferences not only depends upon the individual but also on the building process of database in the company.

The major problems are protection of behavior and security which will affect details of the population who buy online. Some of the customers are reluctant to buy in online because they consider that the process of online purchases are insecure due to some reasons like use of debit or credit cards, passwords, hacking information, untrustworthy, third party, social risks. First, online shoppers face fraud and security problems Hackers will still try to find a way to obtain the data even though many latest security facilities are available. Online shoppers should first familiarize by themselves with online stores and how they will protect their data before going on an online shopping. Customers trust, satisfaction and convenience are major factors to improve Online shopping business.

1.1 Literature Survey

Prediction uses only some part of information in shopping cart. Using prediction can reduce the rule mining cost by using a fastest algorithm [1] which uses Boolean vector with relational AND operation. It can be used to discover frequent item sets and generate the association rule. Association rules mining [2] helps to identify the various relationships among a set of items in database. Discovery of associations such as what items are frequently purchased by the customers and which items brings them better profit helps the retailers to improve their marketing strategies. The two types to find association among the products present in a large database are Boolean [3] and Quantitative. Boolean association rule mining used to find association for the entire dataset. Quantitative association rule mining used to find association for the clusters which are obtained from the dataset. To generate association rules for each of the item separately would create many rules with two obvious consequences. First, these rules occupy memory space which is larger than the original database. Second, the identifying the most relevant type of rules and combining leads to conflicting predictions may easily incurring prohibitive computational costs. Both of these problems are solved by developing a technique which is acceptable in terms of accuracy, time and space complexity. Then convert the data in raw database [3] into Boolean values [4] and form a Boolean matrix. The association rules are generated based on prediction [1] from already generated frequent item sets. Finally Dempster's rule of combination is used to get the predictions by combining the rules which are then suggested to the user. Here the memory occupied and execution time is very less since it does not generate any candidates item sets.

Data mining helps to extract useful information from large volume of data. It can be done by using several techniques. Among all those techniques, classification [5] is most popular technique and is being intensively used in many real business applications now-a-days. Consider classification using theorem of reasoning. The naive Bayesian classifier is one of the best data mining techniques for classifying the large dataset. Naive Bayesian classifier also efficiently applied in feature selection [6] and web classification [7].

To increase customer satisfaction, more efficient business-customer relationships were developed by CRM (Customer Relationship Management) to improve customer loyalty and retention. CRM identifies the customers who are most profitable and provide them the highest level of service [8]. Clustering analysis is a data mining technique which is used to map data objects into unknown groups of different objects with of high similarity. Clustering is the task of segmenting the total population into same type of clusters [9]. The customer details are segmented using K-means since it is not possible to discover theoretically the optimal number of clusters [10]. Then the association rules are to be used for the analysis of customer data. One of the most well-known association rule mining algorithm is the Apriori algorithm.

2 Problem Statement

Many organizations are releasing micro data which contain information about the individuals. If this information the micro data helps to reveal details about a particular individual, it shows that private information is leaked. This is not acceptable by the customers. To avoid this, important details should be kept secret. However this also does not seems to be privacy. The attributes are linked to the external data in order to identify the details of a particular customer which is quasi-identifier. To overcome this attack, a privacy technique called L-Diversity is used (Table 1).

Table 2 Shows online shopping records among different countries. Note that clearly that it does not contain any attributes like name, address, phone number which can be used to identify individual easily. Now consider attributes into two types, they are sensitive attributes and non-sensitive attributes. Sensitive attributes are to be kept secret and non-sensitive attributes are kept public.

Table 2 Now consider attributes into two types, they are sensitive attributes and non-sensitive attributes. Sensitive attributes are to be kept secret and non-sensitive attributes are kept public. The sensitive attribute must not be allowed to get identified personal information of any customer. Let the set of attributes {zip code, state} be quasi-identifier for this data-set. Table 2 describes 4-anonymous table which is derived from Table 3.

Homogeneity Attack. Alice and Bob are neighbors. One day Bob purchases Product in online. Having known about the purchasing of product, Alice thought to discover to which city that product is to be getting delivered. Alice will search for records that are published and will look up at 4-anonymous table in Table 2. As Alice is Bob's neighbour he knows that in which state he lives. If all products in that state are purchased from same city the Alice can easily guess to which place it is delivered.

Table 1 Customer micro data

S. No	Customer micro data	
	Non-sensitive (Zip Code)	Sensitive (City)
1	2903	Providence
2	2904	Providence
3	2906	Providence
4	2909	Providence
5	8002	Cherry Hll
6	8003	Cherry Hill
7	8014	Bridge Port
8	8077	Riverton
9	10002	New York
10	10003	New York
11	10009	New York
12	10011	New York

Table 2 4-anonymous customer micro data

S. No	4-anonymous customer micro data	
	Non-sensitive (Zip Code)	Sensitive (City)
1	29**	Providence
2	29**	Providence
3	29**	Providence
4	29**	Providence
5	80**	Cherry Hill
6	80**	Cherry Hill
7	80**	Bridge Port
8	80**	Riverton
9	100**	New York
10	100**	New York
11	100**	New York
12	100**	New York

3 Proposed Methodology

These days, many networked society places have great demand on the person specific data for many new uses. It can happen when there is more availability of public information. The data obtained from these are linked together so as to identify the particular person even when the information contains no explicit identifiers. The wide availability of personal data leads to the problem of privacy preservation. Lot of methods is proposed recently to handle privacy preservation for multi-dimensional data records. Among them two of the methods are K-Anonymity and L-Diversity which are used for privacy preservation of data.

3.1 K Anonymity

The main task is to provide security for the data and preventing identification of the individuals. This can be achieved using technique called K-Anonymity [11]. The released data is said to adhere to k-Anonymity if each and every released record contains at least (k − 1) other records also visible in the release whose values are same over a special set of fields called quasi Identifiers (QI) or pseudo identifier. Therefore k-Anonymity provides privacy protection through guarantee that each record which relates to at least k individuals even if released records are linked to external information.

A table which satisfies K-Anonymity is called K-Anonymous table. Here attributes are divided into two groups which are sensitive attributes and non-sensitive attributes. An attribute is said to be sensitive that must not be allowed to discover by an adversary and kept secret. Attributes that are not sensitive are treated as non-sensitive attributes. The K-Anonymity can leave the information by creating groups which can be done due to lack of Diversity in the sensitive attribute. This is called homogeneity attack. This attack leverages where all the values for the sensitive attributes within a set of records are identical. Another attack is membership disclosure attack this can be done based on QI value, it is easy to retrieve information. Therefore, the sensitive values for the records are easily predicted. Since K-Anonymity is susceptible to these both attacks there is need to focus on stronger definition of privacy.

3.2 L Diversity

L-Diversity is best in terms of practical and can be implemented. In a class if there are at least l well shown value for the sensitive attributes then it is said to have an L-Diversity. A table is said to be L-Diversity if every class of the table has L-Diversity. L-Diversity does not depend upon what kind of knowledge is possessed by adversary. The main idea of this technique is to maintain values of the sensitive attributes in a well-represented manner in each group. L-Diversity defends against the attacks occurred due to K-Anonymity. It can overcome instance level knowledge. It does not require the data publisher to have the information as the adversary contains. It also protects against highly knowledgeable adversaries.

4 Results

The dataset considered in this experiment is online shopping customer dataset which is taken from an online shopping directory. This data set consists of 12 raw attributes which also includes predictable attribute. Out of all these we now consider only seven attributes. In this dataset city is considered as sensitive attribute which refers to the place in the state to where the product has been delivered. The names and phone numbers are replaced with some dummy values to prevent reveal of personal information.

Privacy Preservation. The Privacy preservation can be evaluated using some of the following characteristics.

4.1 Membership Disclosure Protection

The adversary will obtain information about the membership of any particular customer based upon the QI value that is present in the dataset. If the QI values in not present in the data, then the desired customer is not present in original dataset. In Fig. 1 it shows protection against membership disclosure. The existing works without privacy, it provides protection up to 93.2%. The protection increases up to

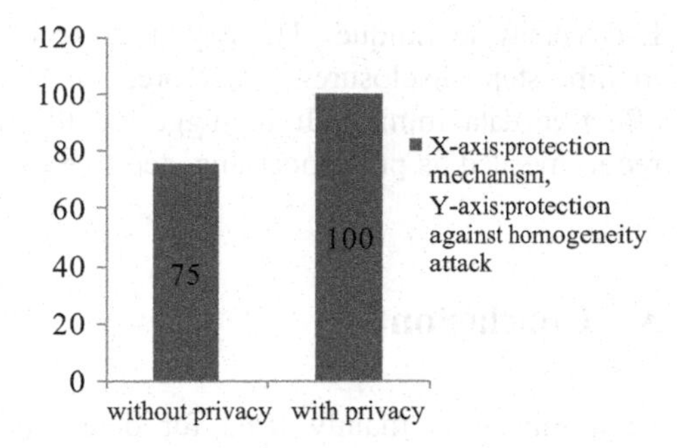

Fig. 1 Membership disclosure protection

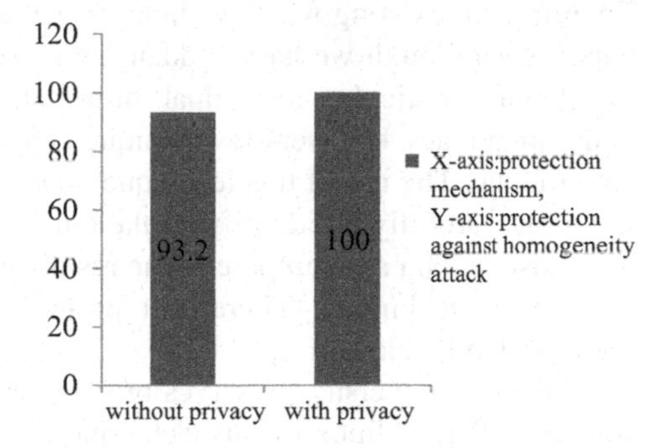

Fig. 2 Homogeneity attack

100% if L-Diversity technique is used along with the membership protection. This happens because membership protection changes upon privacy technique involved.

4.2 Homogeneity Attack

For any QI attribute (q1, q2, q3,.., qn) if any value of that attribute match with any other values then adversary can obtain the sensitive information of a particular customer. There is more chance of matching among values of QI attributes. Figure 2 represents the privacy protection provided by present work is only up to 75% whereas in addition of L-Diversity technique it can provide protection against homogeneity up to 100% for the same algorithms.

Data mining is mainly used for detection of individual details who purchase products online. The main risk is to protect sensitive information from the attacker. Therefore, in existing work we hide the personal details to provide privacy.

In this paper, even though we have hid the personal identifiable information it is easy to identify particular individual information or the sensitive information. To maintain privacy L-Diversity technique is applied to data which is then allowed for data mining. The use of this technique does not affect the characteristics of the data, it is done carefully by identifying the clusters of data. The results of modified data and existing data are compared. The results prove that the privacy is more in use of

L-Diversity technique. Thereby it protect against homogeneity attack and also membership disclosure. Therefore, Diversity preserves privacy and provides effective data mining. It is suggested that addition of this technique to data is recommended as pre-processing step to existing work in order to maintain privacy.

5 Conclusion

Data mining is mainly used for detection of individual details who purchase products online. The main risk is to protect sensitive information from the attacker. Therefore, in existing work we hide the personal details to provide privacy. In this paper, even though we have hidden the personal identifiable information it is easy to identify particular individual information or the sensitive information. To maintain privacy L-Diversity technique is applied to data which is then allowed for data mining. The use of this technique does not affect the characteristics of the data, it is done carefully by identifying the clusters of data. The results of modified data and existing data are compared. The results prove that the privacy is more in use of L-Diversity technique. Thereby it protect against homogeneity attack and also membership disclosure.

Therefore, Diversity preserves privacy and provides effective data mining. It is suggested that addition of this technique to data is recommended as pre-processing step to existing work in order to maintain privacy.

References

1. H. H. Aly, A. A. Amr, and Y. Taha, "Fast Mining of Association Rules in Large-Scale Problems," Proc. IEEE Symp. Computers and Comm. (ISCC "01), pp. 107–113, 2001.
2. Kasun Wickramaratna, Miroslav Kubat and Kamal Premaratne, "Predicting Missing Items in Shopping Carts", IEEE Trans. Knowledge and Data Eng., vol. 21, no. 7, july 2009.
3. P. Bollmann-Sdorra, A. Hafez, and V. V. Raghavan, "A Theoretical Framework for Association Mining Based on the Boolean Retrieval Model," Data Warehousing and Knowledge Discovery: Proc. Third Int l Conf. (DaWaK 01), pp. 21–30, Sept. 2001.
4. W. Li, J. Han, and J. Pei, "CMAR: Accurate and Efficient Classification Based on Multiple Class-Association Rules," Proc. IEEE Int l Conf. Data Mining (ICDM 01), pp. 369–376, Nov./Dec. 2001.
5. Quinlan, J. (1986), "Induction of Decision Trees," Machine Learning, vol. 1, pp. 81–106.
6. XHEMALI, Daniela and J. HINDE, Christopher and G. STONE, Naive Bayes vs. Decision Trees vs. Neural Networks in the Classification of Training Web Pages, IJCSI International Journal of Computer Science Issues, Vol. 4, No. 1, 2009.
7. A. Sopharak, K. Thet Nwe, Y. Aye Moe, Matthew N. Dailey and B. Uyyanonvara, "Automatic Exudates Detection with a Naive Bayes Classifier", Proceedings of the International Conference on Embedded Systems and Intelligent Technology, pp. 139–142, February 27–29, 2008.

8. E. W. T. Ngai, Li Xiu and D. C. K. Chau, 2009, Application of data mining techniques in customer relationship management: A literature review and classification, Expert Systems with Applications, Vol 36, Issue 2, Part 2, pp 2592–2602.
9. Magdalena G, T. Lasota, B. Trawiński "Comparative Analysis of Premises Valuation Models Using KEEL, RapidMiner, and WEKA" First International Conference, ICCCI, Wrocław, Poland, Oct 5–7, 2009. Proc., pp 800–812.
10. Tibshirani, R., Walther, G., and Hastie, T, 2001, estimating the number of clusters in a data set via the gap statistic. Journal of the Royal Statistical Society. Series B, Statistical methodology, 63:411–423.
11. G L. Sweeney. k-anonymity: a model for protecting privacy. International Journal on Uncertainty, Fuzziness and Knowledge-based Systems, 10(5):557–570, 2002.

Design and Performance of Resonant Spacing Linear Patch Array with Quarter Wave Transformer Feed Network for Wireless Applications

D. Prabhakar, P. Mallikarjuna Rao and M. Satyanarayana

Abstract This paper elucidates the simulated and analyzed designs of 4 element microstrip patch antenna arrays along with Quarter wave transformer feed which is to be used in Wireless applications in the ISM frequency bands. The paper focuses on different microstrip array antennas such as corporate feed, Quarter wave transformer feed, resonant ($\lambda/2$) spacing and element spacing ($\lambda/2$) which are simulated, analyzed and designed using HFSS 13.0 and fabricated by photolithography technique using the FR4 substrate. Patch is tested using Vector Network Analyzer E5071C. The various antenna parameters decide the optimum feeding system. The designed antennas are 1×4 arrays. Quarter wave transformer feed is used for better characteristic feature and the usage of which declines the reflection coefficient up to −16.767 dB. There is a direct improvement in gain up to 6.0643 dB and directivity up to 10.340 dB. With the help of Quarter wave transformer feed the bandwidth of an antenna can extend up to 116.6 MHz. We get the optimum return loss data with good matching and efficiency, with high gain, band width and reject bands by using Quarter wave transformer feed. The experimental results prove that the proposed antenna is a good candidate for applications in WLAN communication system.

Keywords Microstrip antenna · The quarter-wave transformer feed · The corporate feed · Reflection coefficient · WLAN · HFSS

D. Prabhakar (✉)
Department of ECE, DVR & Dr. HS MIC College of Technology, Kanchikacherla, India
e-mail: prabhakar.dudla@gmail.com

P. Mallikarjuna Rao
Department of ECE, AUCE (A), Andhra University, Visakhapatnam, India
e-mail: pmraoauece@yahoo.com

M. Satyanarayana
Department of ECE, M V G R College of Engineering (A), Vizianagaram, India
e-mail: profmsn26@gmail.com

S.C. Satapathy et al. (eds.), *Computer Communication, Networking and Internet Security*, Lecture Notes in Networks and Systems 5,
DOI 10.1007/978-981-10-3226-4_20

1 Introduction

The Microstrip antenna has an edge over for maintaining surface or developing a thin protrusion from the surface that provides microstrip with all the advantages of printed circuit technology. Medical application, satellite, radar communication etc. are the applications possessed by microstrip antenna. Narrow strip frequency band and its lack of ability to operate at high power levels of wave guide are said to be the limitation of microstrip antenna. Hence, microstrip antenna has brought forth a great challenge in its design to enhance its bandwidth and gain [1, 2].

Wide band width, improved efficiency together with high gain can be produced by different array configuration of microstrip antenna. The voltages are distributed among the elements of an array depends on the feeding network. The apt method of feeding network, when adopted, collates the incited voltages to feed into one single point. The right impedance matching across the corporate and series feeding array configuration supply with high efficiency micro strip antenna power distribution among the antenna elements can be remodelled by the corporate feed network, which can control the direction of the beam by recommending desired phase [3, 4].

2 Design of Feed Network for an Array

In the farthest communications, antennas with high directivity are regularly required. Single element antenna is not suitable for high gain or high directivity. High gain can be achieved by an assemblage of antennas, called an array. In the construction of an array, feed network design is essential. Feed network is used in an array to regulate the amplitude and phase of the radiating elements in order to control the beam scanning properties [5, 6]. Thus, in selecting and optimizing the feed network the design of an array is crucial. Different types of feed networks are parallel feed, T-Split power divider, Quarter wave transformer, and Metered bend feed.

Quarter Wave Transformer

The transmission lines such as coaxial cables, strip lines and microstrip lines are used in making most feed networks. Impedance matching is vital for ensuring efficient power transfer through feed network.

The use of quarter wave transformer is the best solution for achieving impedance matching. The reflection coefficient between two impedances Z_1 and Z_3 is cut down by a matching circuit. The quarter wave transformer shown in Fig. 1 is one of the most commonly used matching circuits. At the centre frequency, a section of transmission line $\lambda/4$ (the quarter wave transformer) long placed between the two transmission lines eliminating the reflection coefficient and its impedance is

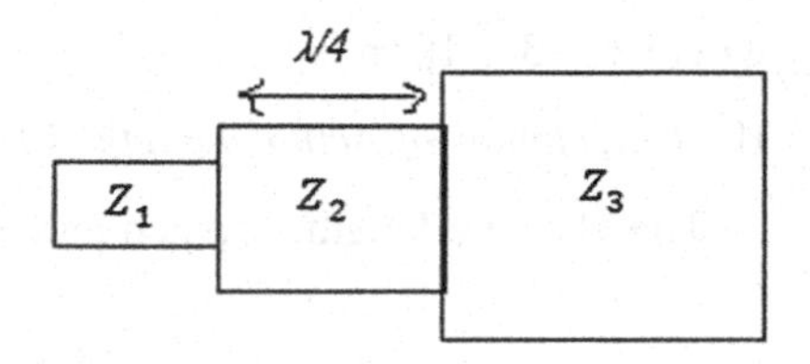

Fig. 1 Quarter-wave transformer

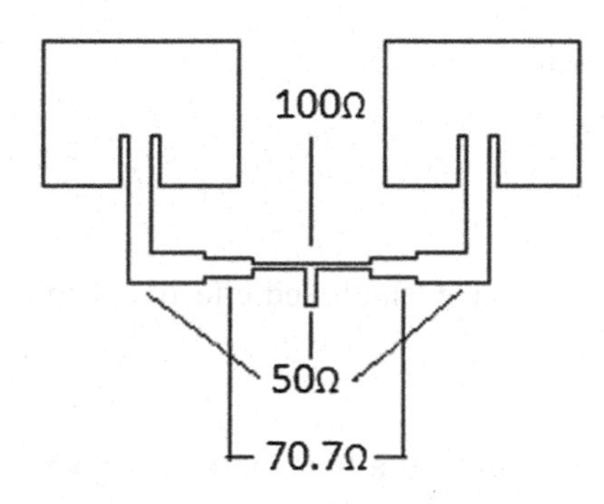

Fig. 2 2-element corporate-fed microstrip array with quarter wave transformer

$$Z_2 = \sqrt{Z_1 Z_3}$$

A 2-element corporate-fed micro strip uniform array is illustrated in the Fig. 2. The tree-like structure of the feed appropriately combines/distributes the signals from/to the elements. With a view to matching the lines of different impedances, a quarter-wave transformer appears at the splits. The 50 Ω input line splits into two 100 Ω lines. If the micro strip line continues to split like this, then the lines feeding the elements would be 200, 400 Ω and so on. Then the micro strip line will be very thin. Besides this, the element impedance should be very high for matching. But the element is fed with inset feed hence the transmission line will have an impedance of 50 Ω. Thus, the 100 Ω line is converted back to 50 Ω using a quarter-wave transformer of 70.7 Ω.

3 Design Consideration

A. *Calculation of Width* (W):

Width of the patch antenna is calculated by using

$$W = \frac{c}{2f_0\sqrt{(\varepsilon_r + \frac{1}{2})}} \tag{1}$$

where $c = 3 * 10^8$ m/s

B. *Calculation of Actual Length (L):*

The effective length of patch antenna depends on the resonant frequency (f_0).

$$L_{eff} = \frac{c}{2f_0\sqrt{\varepsilon_{reff}}} \tag{2a}$$

where

$$\varepsilon_{reff} = \frac{\varepsilon_r + 1}{2} + \frac{\varepsilon_r - 1}{2}\left[1 + 12\frac{h}{W}\right]^{-\frac{1}{2}} \tag{2b}$$

Actual length and effective length of a patch antenna can be related as

$$L = L_{eff} - 2\Delta L \tag{3}$$

where ΔL is a function of effective dielectric constant ε_{reff} and the width to height ratio $\left(\frac{W}{h}\right)$

$$\frac{\Delta L}{h} = 0.412\frac{(\varepsilon_{reff} + 0.3)\left(\frac{W}{h} + 0.264\right)}{(\varepsilon_{reff} - 0.258)\left(\frac{W}{h} + 0.8\right)} \tag{4}$$

C. *Calculation of Inset feed Depth* (y_0):

$$y_0 = 10^{-4}\{0.016922\varepsilon_r^7 + 0.13761\varepsilon_r^6 - 6.1783\varepsilon_r^5 + 93.187\varepsilon_r^4 - 682.69\varepsilon_r^3 + 2.561.9\varepsilon_r^2 - 4043\varepsilon_r + 6697\}\frac{L}{2} \tag{5}$$

D. *Calculation of feed width* (W_f):

To achieve 50 Ω characteristic impedance, the required feed width to height ratio $\left(\frac{W_f}{h}\right)$ is computed as

$$\frac{W_f}{h} = \begin{cases} \frac{8e^A}{e^{2A}-2}\frac{W_0}{h} \le 2 \\ \frac{2}{\pi}\left\{B - 1 - \ln(2B - 1) + \frac{\varepsilon_r - 1}{2\varepsilon_r}\left[\ln(B - 1) + 0.39 - \frac{0.61}{\varepsilon_r}\right]\right\}\frac{W_0}{h} \ge 2 \end{cases} \tag{6a}$$

where

$$A = \frac{Z_0}{60}\sqrt{\frac{\varepsilon_r + 1}{2}} + \frac{\varepsilon_r + 1}{\varepsilon_r - 1}\left(0.23 + \frac{0.11}{\varepsilon_r}\right) \tag{6b}$$

$$B = \frac{377\pi}{2Z_0\sqrt{\varepsilon_r}} \tag{6c}$$

[7–10].

4 Results and Discussion

4.1 Results of 1 × 4 RMPA Array

A 4 element antenna array of Quarter wave transformer feed with resonant ($\lambda/2$) spacing of elements is designed and fabricated with the design equations and is shown in Figs. 3 and 4 respectively. Simulated results such as reflection coefficient, VSWR, gain and directivity are observed in Figs. 5, 6, 7 and 8 respectively.

It is observed from Fig. 5 that a reflection coefficient (S_{11}) of 4 element array antenna is −16.7647 dB. Figure 6 VSWR of 4 element array antenna noticed to be

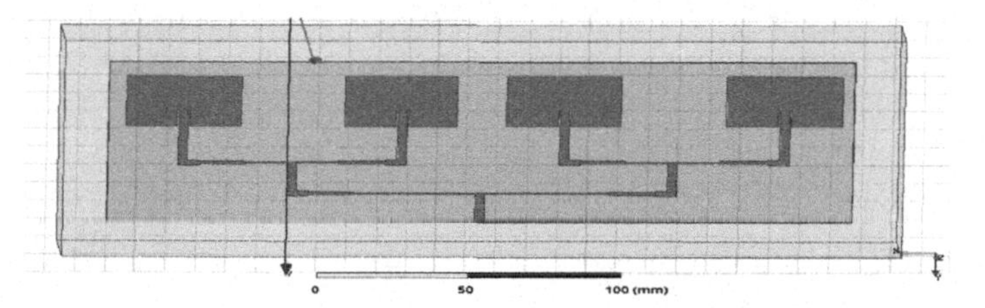

Fig. 3 Geometry for 4 element antenna array of Quarter wave transformer feed with resonant ($\lambda/2$) spacing

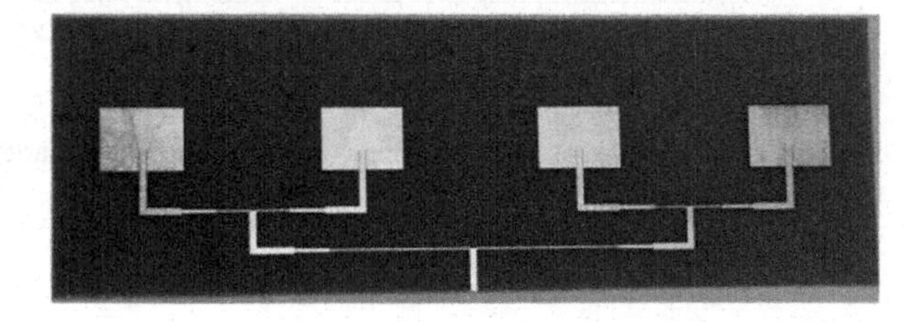

Fig. 4 Fabricated patch of 4 element antenna array of Quarter wave transformer feed with resonant ($\lambda/2$) spacing

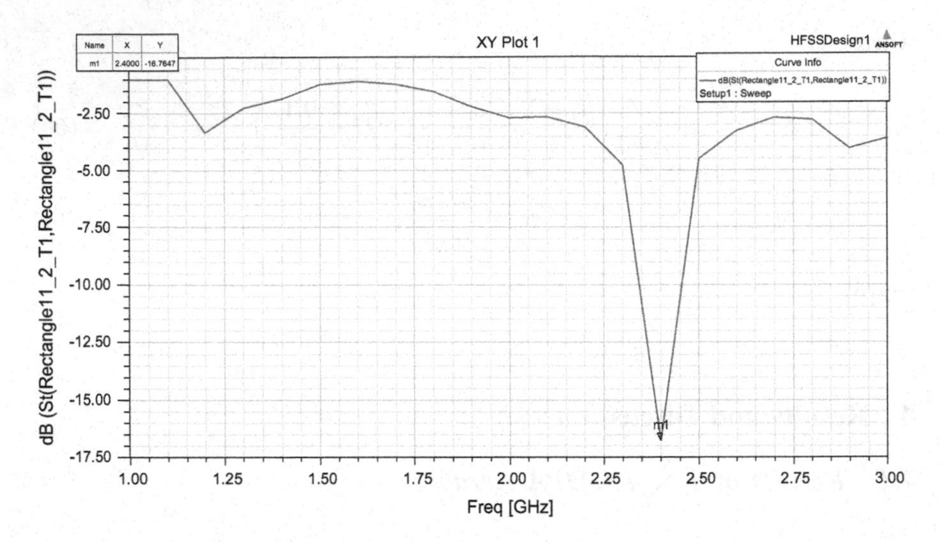

Fig. 5 Reflection coefficient curve of the 4 element antenna array of Quarter wave transformer feed with resonant ($\lambda/2$) spacing

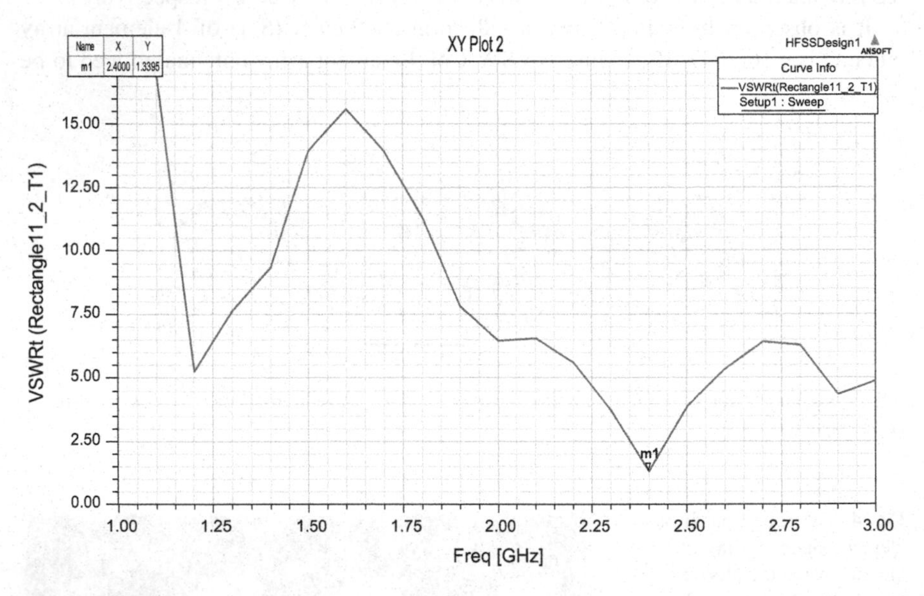

Fig. 6 VSWR curve of the 4 element antenna array of Quarter wave transformer feed with resonant ($\lambda/2$) spacing

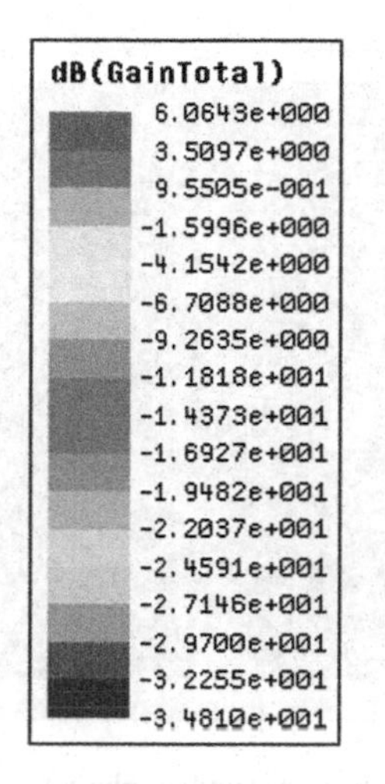

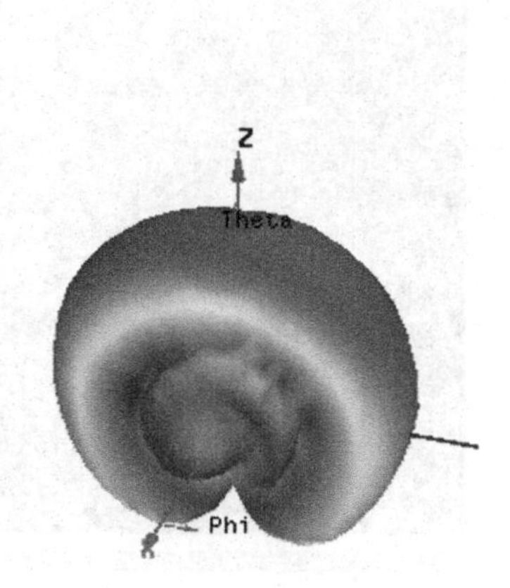

Fig. 7 Gain plot of the 4 element antenna array of Quarter wave transformer feed with resonant $(\lambda/2)$ spacing

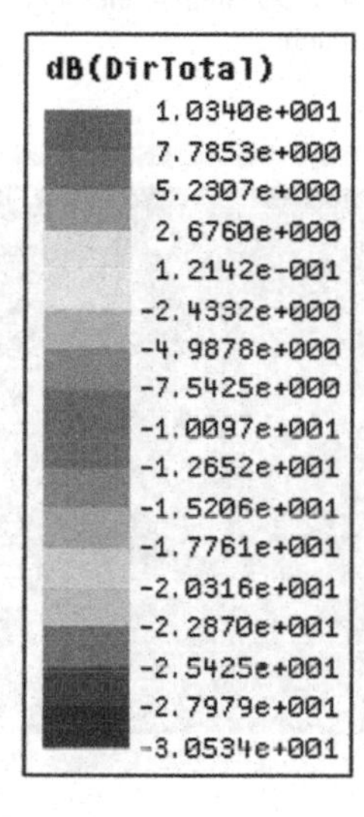

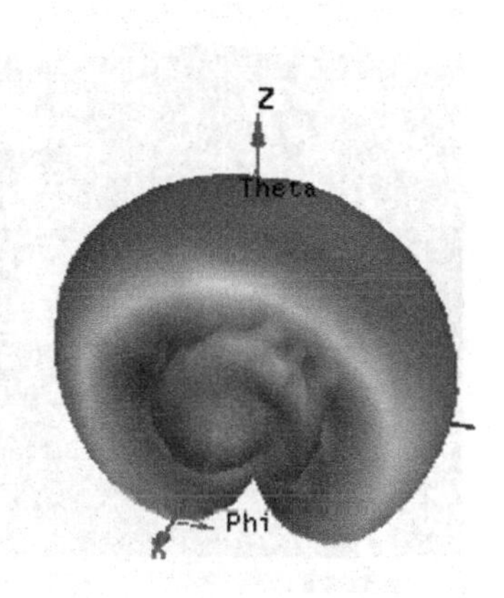

Fig. 8 Directivity plot of the 4 element antenna array of Quarter wave transformer feed with resonant $(\lambda/2)$ spacing

1.3395 at 2.4 GHz. The gain observed from Fig. 7 is 6.0643 dB and directivity as 10.340 dB at 2.4 GHz shown in Fig. 8. The obtained results of fabricated 4 element antenna array are shown in Figs. 9 and 10.

The Reflection coefficient is −16.187 dB and at 2.4358 GHz and VSWR of 1.3849 are observed from Figs. 9 and 10 respectively.

Results of 4 element patch antenna array of Quarter wave transformer feed with resonant $(\lambda/2)$ Spacing.

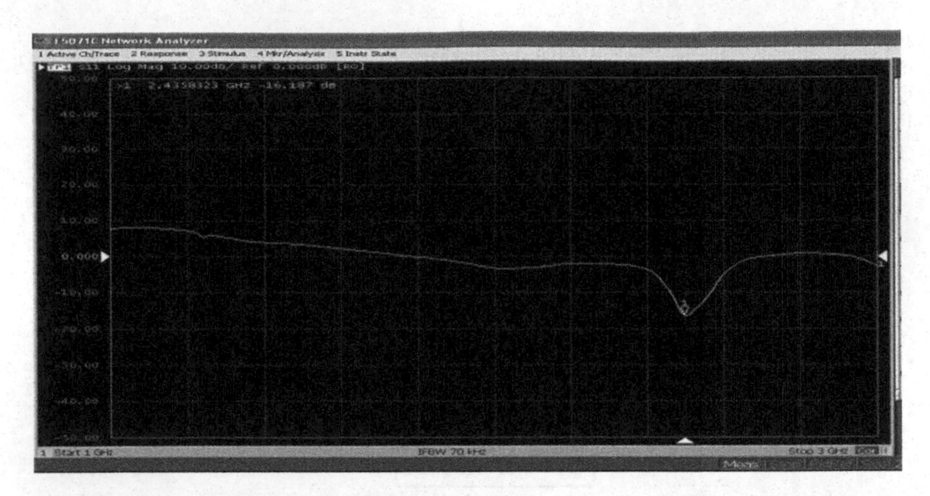

Fig. 9 Reflection coefficient curve of fabricated 4 element antenna array of Quarter wave transformer feed with resonant ($\lambda/2$) spacing

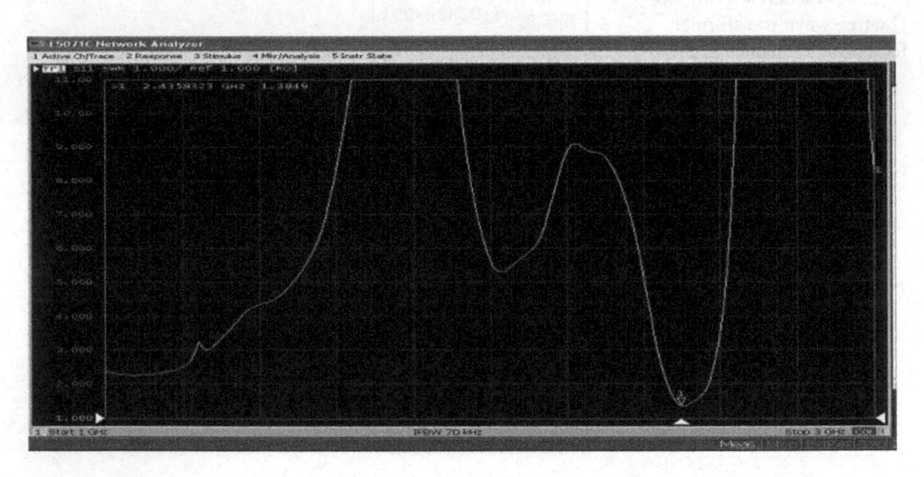

Fig. 10 VSWR of fabricated 4 element antenna array of Quarter wave transformer feed with resonant ($\lambda/2$) spacing

Parallel feed with quarter wave transformer feed along with resonant ($\lambda/2$) spacing		
Resonant frequency in GHz	Simulated	2.4
	Measured	2.438
Reflection coefficient (S_{11}) in dB	Simulated	−16.767
	Measured	−16.187
VSWR	Simulated	1.3395
	Measured	1.3849
Gain in dB	Simulated	6.0643

(continued)

(continued)

Parallel feed with quarter wave transformer feed along with resonant ($\lambda/2$) spacing		
Directivity in dB	Simulated	10.340
Band width in MHz	Simulated	116.6
Efficiency (%)	Simulated	59.05

5 Conclusion

A number of elements in an array is increased for better improvement in gain and directivity with the help of a Quarter Wave Transformer Feed. Then there is an observed improvement in gain from 3.0524 dB (single element) to 6.0643 dB (4 elements). Also the directivity of antenna is enhanced from 6.6480 dB to 10.340 dB. The VSWR decreased and band width enhanced up to 116.6 MHz. This leads to the increase in the number elements with suitable dimensions along with the patch design which is important for optimization of parameters.

References

1. M. T. I. Huque, et al., "Design and Simulation of a Low-cost and High Gain Microstrip Patch Antenna Arrays for the X-band Applications," in International Conference on Network Communication and Computer—ICNCC 2011, New Delhi, India., March 21–23, 2011.
2. Mohammed Moulay and Mehadji Abri, "Bowtie Antennas Design for Bluetooth/Wimax/Wifi Applications," International Journal of Microwave and Optical Technology,Vol.9, No.4, July 2014.
3. R. Mailloux, et al., "Microstrip array technology," Antennas and Propagation, IEEE Transactions on, vol. 29, pp. 25–37, 1981.
4. H. Cheng-Chi, et al., "An aperture-coupled linear microstrip leaky-wave antenna array with two-dimensional dual-beam scanning capability," Antennas and Propagation, IEEE Transactions on, vol. 48, pp. 909–913, 2000.
5. K. Gi-Cho, et al., "Ku-band high efficiency antenna with corporate series-fed microstrip array," in Antennas and Propagation Society International Symposium, 2003. IEEE, 2003, pp. 690–693 vol.4.
6. A. Abbaspour-Tamijani and K. Sarabandi, "An affordable millimeter wave beam-steerable antenna using interleaved planar subarrays," Antennas and Propagation, IEEE Transactions on, vol. 51, pp. 2193–2202, 2003.
7. C. A. Balanis. Antenna theory : analysis and design (3rd ed.), 2005.
8. R. Garg, Microstrip antenna design handbook. Boston, Mass. [u.a.]: Artech House, 2001.
9. D. Prabhakar, Dr. P. Mallikarjuna Rao, Dr. M.Satyanarayana "Design and Performance analysis of Micro strip Antenna using different ground plane techniques for WLAN application," International Journal of Wireless and Microwave Technologies, Vol.06, Issue.04, pp: 48–58, July 2016.
10. D. Prabhakar, Dr. P. Mallikarjuna Rao, Dr. M.Satyanarayana, "Characteristics of Patch Antenna with Notch gap variation for Wi-Fi Application," International journal Applied Engineering Research, Vol.11, Issue.0 8, pp: 5741–5746, April 2016,

Performance Analysis of PUEA and SSDF Attacks in Cognitive Radio Networks

D.L. Chaitanya and K. Manjunatha Chari

Abstract Cognitive radio is the promising technology for serving the needs of increasing demand for the wireless communications nowadays. Employing cooperative spectrum sensing gives better results as it involves the fusion of sensing reports from more than one secondary user for making the final decision regarding the presence or absence of primary users. Cooperative spectrum sensing is susceptible to security issues such as primary user emulation attack (PUEA) and spectrum sensing data falsification attack (SSDF). In this chapter we compare the performance of the centralized cooperative cognitive radio network in the presence of SSDF and PUEA attacks by applying three hard decision fusion rules AND, OR and K-out-of-N rules. Simulation results show that the K-out-of-N rule performs well in the presence of PUEA attack and AND rule performs well under SSDF attack. A packet delivery ratio of unity is achieved as the simulation time progresses.

Keywords Energy detection · SSDF attack · PUEA attack

1 Introduction

To serve the demands of present day wireless communication systems, cognitive radio was proposed as a promising solution. Studies have shown that about 15–85% [1] of the time, the spectrum is remaining idle because of the usage by only licensed users/primary users. To make efficient utilization of the electromagnetic spectrum, cognitive radio technology facilitates the unlicensed users/secondary users/cognitive users to utilize the spectrum in the absence of primary users. Secondary users have to

D.L. Chaitanya (✉)
GRIET, Hyderabad, India
e-mail: chaitanya.mekapati@gmail.com

K. Manjunatha Chari
GITAM University, Visakhapatnam, India
e-mail: manjunath4005@gitam.edu

S.C. Satapathy et al. (eds.), *Computer Communication, Networking and Internet Security*, Lecture Notes in Networks and Systems 5,
DOI 10.1007/978-981-10-3226-4_21

transmit their signals only in the absence of primary users without causing any interference to the primary users. To achieve this, the secondary users have to sense the spectrum perfectly to recognize the presence or absence of the primary users. Cognitive radio technology has four main functions to perform [2]:

- Spectrum sensing observes the RF spectrum for white spaces and allocates them to CRs. Sensing not only involves the measurement of spectral content but also gets the characteristics of spectrum in dimensions like time, frequency, space etc.
- Spectrum decision chooses the most suitable band from the list of available vacant bands.
- Spectrum mobility performs a proper hand-off when a licensed user is detected.
- Spectrum sharing allocates vacant bands to CRs reasonably thereby preventing collisions.

Among the four tasks of cognitive radio, the most vital task is spectrum sensing. Secondary users have to perform spectrum sensing to recognize whether the spectrum is occupied by the primary user or not, which can be done in various ways by making use of different spectrum sensing techniques such as energy detection, cyclostationary detection, matched filter detection etc. [3]. Any one of the spectrum sensing techniques can be utilized to detect the presence of the primary user. Energy detection is the simplest method of all [4]. Here, primary signal presence is detected by comparing the energy of the signal sensed in the band of interest with a predefined threshold value. The problem with the energy detection is its incapability to make a distinction between primary user's signal and noise signal when the sample size is large. Cyclostationary detection technique is a non-blind detection technique as it requires prior knowledge of cyclostationary features of primary user's signal such as mean, variance and autocorrelation. Cyclostationary features are also known as periodicity features of the signal that repeat at specific time intervals. This technique clearly discriminates noise signal from primary user's signal even in low signal to noise ratio regions. The disadvantage of this technique is its complexity in implementation and also it requires some knowledge of primary user's signals which is difficult to obtain. Another spectrum sensing detection technique called matched filtering is also a non-blind technique that requires exact knowledge of primary user's signal characteristics such as modulation type and order, operating frequency, frame format etc.

The sensing information obtained from a single cognitive user may not be reliable due to the effects of shadowing and multipath fading. To determine the presence or absence of the primary user in the band of interest accurately, sensing information from more than one cognitive user is to be considered which is known as cooperative spectrum sensing technique [5]. Here each cognitive user employs its own spectrum sensing technique to decide the presence or absence of the primary user and send this information to the fusion center. In the fusion center, hard or soft decision rules are applied to the sensing reports to obtain a global decision. This phenomenon is called as centralized cooperative spectrum sensing. Major security threats to cooperative spectrum sensing are primary user emulation attack (PUEA) [6] and spectrum sensing data falsification attack (SSDF) [7]. In primary user emulation attack, malicious secondary users mimic the characteristics of the primary signal and transmit them, thereby misguiding the genuine secondary users to report to the fusion center that the spectrum is already occupied when it is actually vacant. In spectrum sensing data falsification attack, malicious users intentionally send incorrect sensing reports to the fusion center and makes it to take wrong final decision. In this chapter, cooperative sensing network is considered and its performance is evaluated in the presence of PUEA and SSDF attacks by employing three hard decision fusion rules: AND, OR and K-out-of-N [8]. Section 2 describes the system model and results are presented in Sect. 3 and finally, Sect. 4 concludes the chapter.

2 System Model

Local sensing at each cognitive user is performed using energy detection technique; the detection problem is formulated using binary hypothesis testing as,

$$H_0 : z[n] = w[n] \tag{1}$$

$$H_1 : z[n] = x[n] + w[n] \tag{2}$$

where $z[n]$ is the received signal at the cognitive user, $x[n]$ is the transmitted primary signal and $w[n]$ is the additive white Gaussian noise with zero mean and n being the sample index. Hypothesis H_0 indicates the presence of noise signal whereas H_1 denotes the presence of primary user's signal along with noise. The parameters used for the performance evaluation of the proposed cooperative cognitive radio network are the probability of error (P_e) and packet delivery ratio (PDR).

In primary user emulation attack, some corrupted secondary users try to imitate the characteristics of primary users deliberately, which makes fair secondary users assume that primary signal is present in the spectrum. In PUEA attack, the attackers also change their radio transmission frequency sometimes to mimic the primary signal [9]. PUEA attack occurs in the physical layer [10]. In spectrum sensing data falsification attack, some infected secondary users that are known as malicious users intentionally transmit false sensing information to the fusion center that leads to the incorrect final decision at the fusion center. The SSDF attack occurs in the link layer when data transfer takes place from one node to another [10]. To reduce the effect of PUEA attack on the cognitive radio network, hard decision fusion rules AND, OR and K-out-of-N rules are applied on the local sensing reports of cognitive users at the fusion center to make a global decision. To mitigate the effect of SSDF attack, a new clustering based algorithm is proposed; divides the nodes present in the network into clusters. A reputation value is assigned to every node in the cluster based on its sensing history. The reputation value of each node is adjusted after its participation in the decision at the fusion center. The new reputation value of the node becomes a deciding factor for the elected cluster head to either keep the node in the cluster or to expel it. Here at the fusion center we applied three hard decision fusion rules AND, OR, K-out-of-N rules on the local sensing reports of the cognitive users to make a global decision. The cognitive users in our proposed system employ energy detection technique to perform local spectrum sensing. The energy detection technique is a blind detection technique which doesnot require any prior knowledge of primary signal and it performs sensing by comparing the energy of the signal sensed in the required frequency band with a predefined threshold value [11]. The performance evaluation of the proposed system is exercised both in the absence of attacks termed as 'no attack' and in the presence of attacks termed as 'conventional attack' based on the parameters probability of error (P_e) and packet delivery ratio (PDR). The probability of error, (P_e) gives the probability of making a wrong decision that the primary signal is present when the spectrum is actually vacant and the primary signal is absent when the spectrum is being used by the primary user. Packet delivery ratio is the ratio of the number of packets successfully delivered to the receiver to the total number of packets transmitted.

3 Results and Discussions

In this chapter, centralized cooperative spectrum sensing cognitive radio network with a total of 30 nodes is considered. The nodes are placed with uniform distribution randomly in the network. The effect of PUEA and SSDF attacks on the cognitive radio network is studied and fusion rules are employed at the fusion center to reduce the influence of these attacks on the global decision. The simulations are performed using NS2 simulator [12].

Figure 1 shows the performance of the network in terms of probability of error with respect to signal-to-noise ratio in the presence of SSDF attack. It clearly shows that the AND rule gives less error probability of 0.34 when compared to the OR and K-out-of-N rules with error probabilities of 0.49 and 0.38 respectively whereas in the presence of PUEA attack K-out-of-N rule outperforms AND and OR rules with a probability of error 0.3 than that of 0.48 and 0.52 as shown in Fig. 2.

Each of our simulations is run for 25 s. The performance estimation of the network when carried out using packet delivery ratio as shown in Figs. 3 and 4 in the presence of SSDF and PUEA attacks respectively, PDR approaches unity at 7 s of simulation time for SSDF attack and at 20 s for PUEA attack. Table 1 shows the

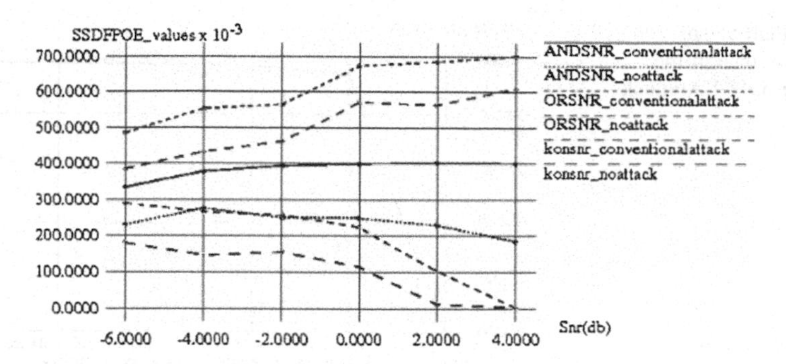

Fig. 1 Probability of error versus $SNR(dB)$ in the presence of SSDF attack

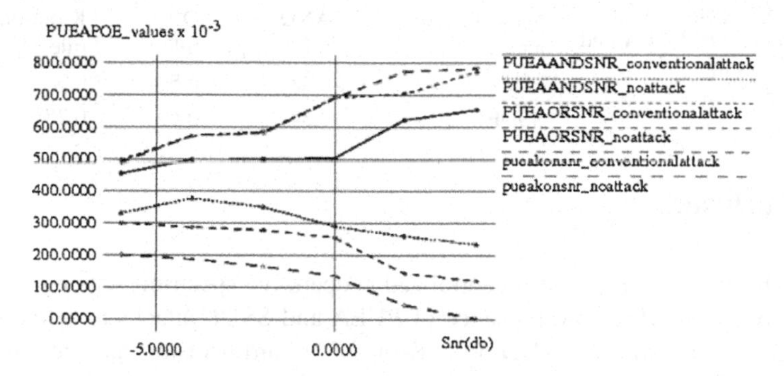

Fig. 2 Probability of error versus $SNR(dB)$ in the presence of PUEA attack

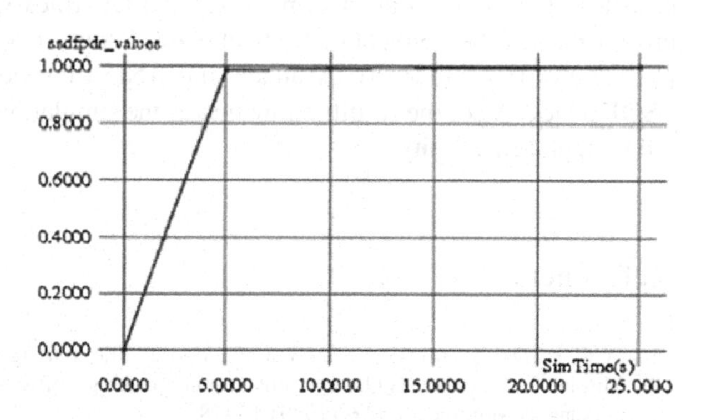

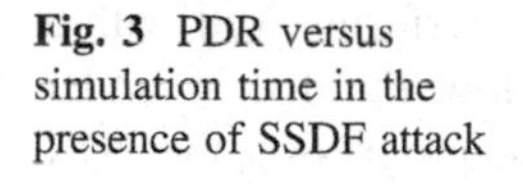

Fig. 3 PDR versus simulation time in the presence of SSDF attack

comparison of error probability values of AND, OR and K-out-of-N rules for both the attacks. It is evident from the table that in the presence of PUEA attack K-out-of-N rule yields less error probability and AND rule surpasses the OR and K-out-of-N rules in the presence of SSDF attack.

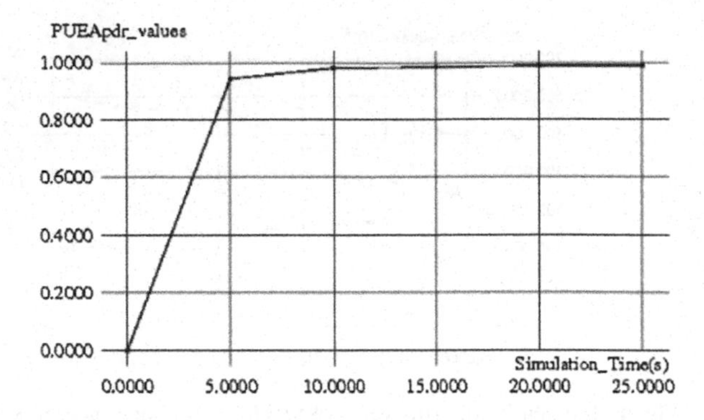

Fig. 4 PDR versus simulation time in the presence of PUEA attack

Table 1 Comparison of P_e in the presence of PUEA and SSDF attacks for SNR = −5 dB

Name of the attack	AND rule	OR rule	K-out-of-N rule
PUEA	0.48	0.52	0.3
SSDF	0.34	0.49	0.38

4 Conclusion

In this chapter, we proposed a centralized cooperative spectrum sensing cognitive radio network in which the presence of PUEA and SSDF attacks is detected. The proposed system employs AND, OR, K-out-of-N hard decision rules at the fusion center to make a global decision. Especially to mitigate SSDF attack, a reputation based clustering algorithm is applied. Simulation results show that in the no attack case K-out-of-N rule outperforms AND and OR rules yielding less probability of error value. In the conventional attack case, K-out-of-N rule performs well with a P_e value of 0.3 under PUEA attack and AND rule yields low P_e of 0.34 under SSDF attack. Also, the results show that as the simulation time progresses, the PDR value approaches unity.

References

1. FCC, E.: Docket no 03–222 notice of proposed rule making and order (2003)
2. Mitola, J., Maguire, G.Q.: Cognitive radio: making software radios more personal. IEEE personal communications **6** (1999) 13–18
3. Yucek, T., Arslan, H.: A survey of spectrum sensing algorithms for cognitive radio applications. IEEE communications surveys & tutorials **11** (2009) 116–130
4. Umar, R., Sheikh, A.U.: A comparative study of spectrum awareness techniques for cognitive radio oriented wireless networks. Physical Communication **9** (2013) 148–170
5. Akyildiz, I.F., Lo, B.F., Balakrishnan, R.: Cooperative spectrum sensing in cognitive radio networks: A survey. Physical communication **4** (2011) 40–62

6. Haghighat, M., Sadough, S.M.S.: Cooperative spectrum sensing for cognitive radio networks in the presence of smart malicious users. AEU-International Journal of Electronics and Communications **68** (2014) 520–527
7. Hyder, C.S., Grebur, B., Xiao, L.: Defense against spectrum sensing data falsification attacks in cognitive radio networks. In: International Conference on Security and Privacy in Communication Systems, Springer (2011) 154–171
8. Teguig, D., Scheers, B., Le Nir, V.: Data fusion schemes for cooperative spectrum sensing in cognitive radio networks. In: Communications and Information Systems Conference (MCC), 2012 Military, IEEE (2012) 1–7
9. Yu, R., Zhang, Y., Liu, Y., Gjessing, S., Guizani, M.: Securing cognitive radio networks against primary user emulation attacks. IEEE Network **29** (2015) 68–74
10. Sumathi, A., VIDHYAPRIYA, D.R.: Intense explore algorithm–a proactive way to eliminate pue attacks in cognitive radio networks. In: WSEAS Conference. (2015)
11. Singh, J.S.P., Singh, R., Rai, M.K., Singh, J., Kang, A.: Cooperative sensing for cognitive radio: A powerful access method for shadowing environment. Wireless Personal Communications **80** (2015) 1363–1379
12. Issariyakul, T., Hossain, E.: Introduction to network simulator NS2. Springer Science & Business Media (2011)

Design of a 3.4 GHz Wide-Tuning-Range VCO in 0.18 μm CMOS

Ningampalli Ramanjaneyulu, Donti Satyanarayana and Kodati Satya Prasad

Abstract This brief paper presents the design of a 3.4 GHz wide-tuning-range ring Voltage Controlled Oscillator (VCO). VCO is one of the most critical building blocks in Phase Locked Loop (PLL), modern high speed communication applications such as microprocessor clock generation, communications, system synchronization, and frequency synthesis. The proposed three stage ring oscillator has been implemented with 1.8 V power supply in 0.18 μm CMOS process using Cadence Virtuoso tool. The proposed VCO is generating a frequency of 3.4 GHz and the linearity is achieved over the range of frequency from 1.76 to 3.4 GHz with 48.2% tuning range.

Keywords Delay cell · Ring oscillator · VCO · PLL · Communication systems

1 Introduction

Oscillators are the fundamental component in lots of analog and digital systems. Applications make use of oscillator's variety from clock technology in microprocessors to frequency translation in mobile phones, point to point communication etc. Different application also requires specific set of oscillator performance parameters. As these day's integrated circuits are converging closer to CMOS, the design of sturdy and high-overall performance CMOS oscillators, particularly, voltage-controlled oscillators (VCOs), has end up extremely essential.

PLLs are common applications for VCOs. The general characteristic for VCOs utilized in PLLs is wide tuning range so that the entire frequency range is covered.

N. Ramanjaneyulu (✉) · D. Satyanarayana
RGMCET, Nandyal, A.P, India
e-mail: rams_ganguly@yahoo.co.in

K. Satya Prasad
JNTUK, Kakinada, A.P, India

S.C. Satapathy et al. (eds.), *Computer Communication, Networking and Internet Security*, Lecture Notes in Networks and Systems 5,
DOI 10.1007/978-981-10-3226-4_22

Also the phase noise requirement of the VCO can be loosened due to that when the loop is locked the noise generated by the VCO at the center of oscillation frequency will be filtered out by the loop bandwidth. As a result PLLs generally use wide tuning range ring topology VCO.

This brief is organized as follows. Section 2 describes the VCO types. Section 3 describes the proposed ring VCO design. The simulation results are given in Sect. 4 and the conclusion is given in Sect. 5.

2 VCO Types

2.1 Current-Starved Inverter

The simple manner to manipulate the charge and discharge time of an inverter is to manipulate the current through the inverter via a voltage controlled current source Fig. 1. This current source is pushed by way of the control voltage Vctrl, and the current will decide the charge up and discharge time of the inverter. This topology is referred to as current-starved inverter with correct sizing and current levels an odd number of stages of these current starved inverters can make a decent VCO.

This design is simple and the oscillation frequency can achieve reasonably fast and tuning range is great due to the square law change in current levels in the footer device. We can size the footer device wider so that it doesn't affect the output swing much.

2.2 Differential Pair with PFET Loads

Figure 2 shows the differential amplifier with DC biased PFET devices. The goal of this design is to increase the tuning range. Consider the PFET device as voltage controlled resistors and the bias circuit VB as an inverted version of Vctrl. As the

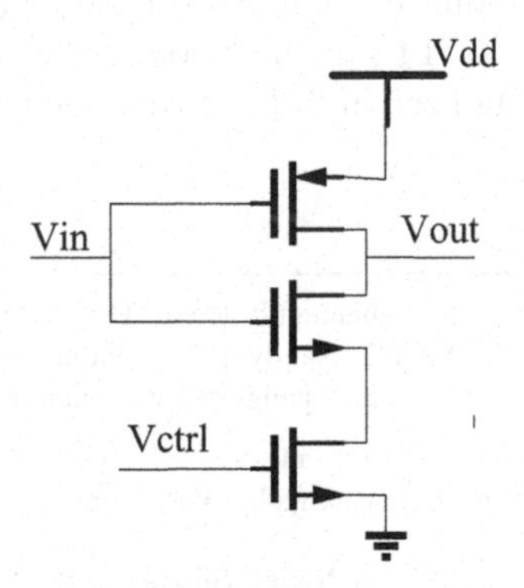

Fig. 1 Current starved inverter

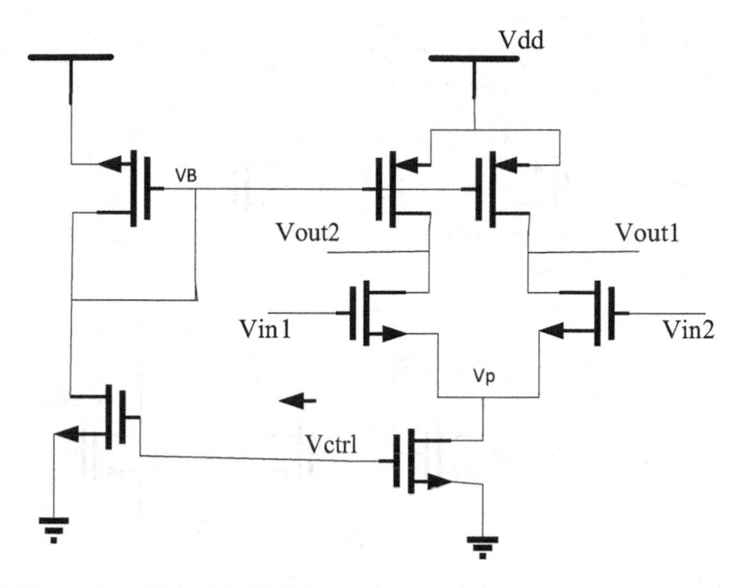

Fig. 2 Differential amplifier with PFET loads

VCO control voltage changes the resistance of the PFET device also changes, thus varying the gain of the differential pair. This method can increase the tuning range dramatically and close to the tuning range of current starved inverters. The drawbacks of this design are the small output swing and varying DC offset at the output, as current mode logic tends to produce much smaller swings. In summary differential VCOs tends to be faster than their singled-ended parts due to their current mode logic nature. However differential VCOs have smaller output swing and suffer the problems of not being able to operate at low bias voltage and varying DC offset in the singled-ended versions.

In the next proposed VCO topology we try to maintain the maximum oscillation frequency advantage of the differential pair, high output swing of the singled-ended version and try to solve the problem of operating in low bias voltage and varying DC offset.

3 Proposed Differential Ring VCO

The differential delay stage advantage is that preferably noise on the supply appears as common-mode on both outputs and is rejected through next stage in a sequence. The VCO is designed using delay cells; the schematic of the proposed delay cell is shown in Fig. 3. The delay cell consists of 4 NMOS transistors and 2 PMOS transistors. In this M1 & M2 NMOS input transistors designed for required gain and bandwidth. M5 & M6 NMOS cross coupled transistors to provide positive feedback. M3 & M4 PMOS transistors serving as current source loads and provide

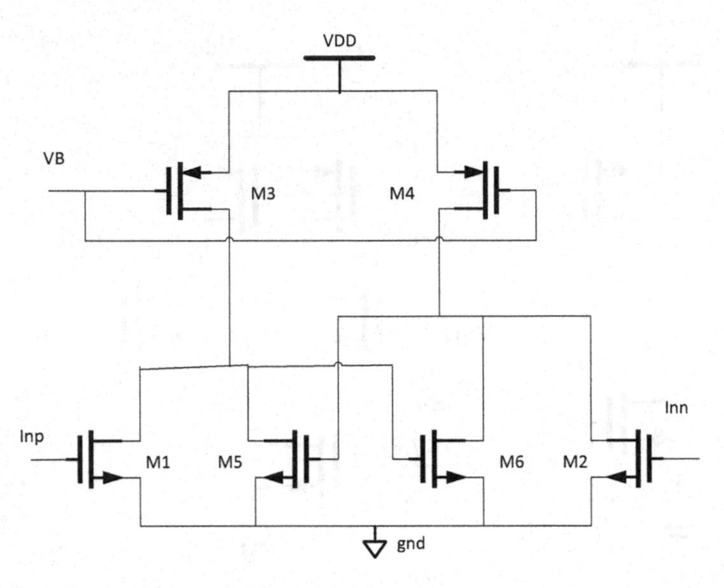

Fig. 3 Schematic of the proposed differential delay cell

Table 1 Delay cell dimension and descriptions

Device name	Dimensions (W/L)	Description
M1 (NMOS)	2.25/0.18 μm	Input transistors designed for required gain and bandwidth
M2 (NMOS)	2.25/0.18 μm	Input transistors designed for required gain and bandwidth
M5 (NMOS)	2.25/0.18 μm	Cross coupled transistor to provide positive feedback
M6 (NMOS)	2.25/0.18 μm	Cross coupled transistor to provide positive feedback
M3 (PMOS)	8/0.2 μm	Load transistor provides designed current for operating frequency
M4 (PMOS)	8/0.2 μm	Load transistor provides designed current for operating frequency

designed current for operating frequency. The current is used by the delay cell to produce output frequencies. First based on the control voltage, The VB will get corresponding voltage switch on the PMOS transistors in delay cells. Based on the VB the current driving into circuit varies and the frequency of the VCO changes. By changing the channel conductance of PMOS devices, the output frequency can be tuned between maximum and minimum frequency values. Hence the frequency generation is completely dependent on the current driving through the PMOS network. Cross-coupled transistors are adopted to guarantee oscillation with differential outputs. The design is much like the Lee–Kim delay cell. In contrast to the Lee–Kim cell, NMOS devices are used in the cross-coupled pairs instead of PMOS devices for a wide tuning frequency [1, 2]. The dimensions used in delay cell circuit are shown in Table 1.

The complete schematic of the 3 stage voltage controlled oscillator is illustrated in Fig. 4. In this figure M1, M2 and R form a Voltage to Current converter, to convert Vctrl to the current. This current acts as a bias current for all 3 delay elements in the VCO circuit.

The control voltage is changed from 0.4 to 1.6 V. The proposed architecture supports wide-tuning-range also reduces the supply sensitivity.

The complete schematic of the 3 stage voltage controlled oscillator is illustrated in Fig. 5. In the test set up V-I converter is used to convert Vctrl to the current; the current in each delay cell is 2.5 mA.

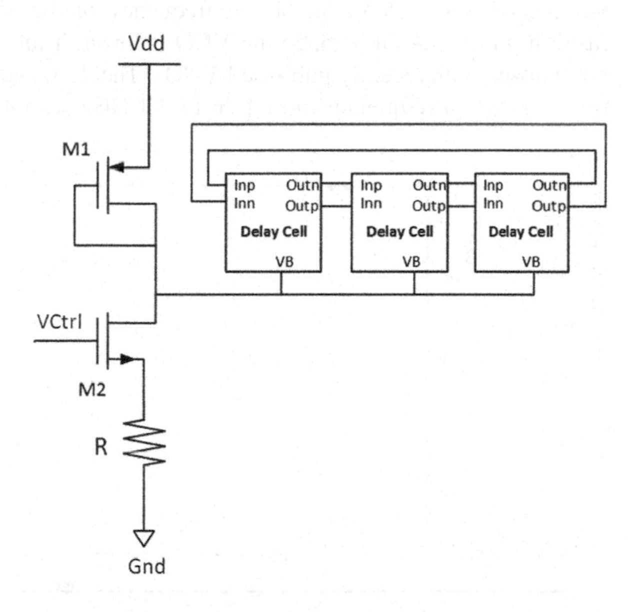

Fig. 4 Schematic of the proposed differential ring voltage controlled oscillator

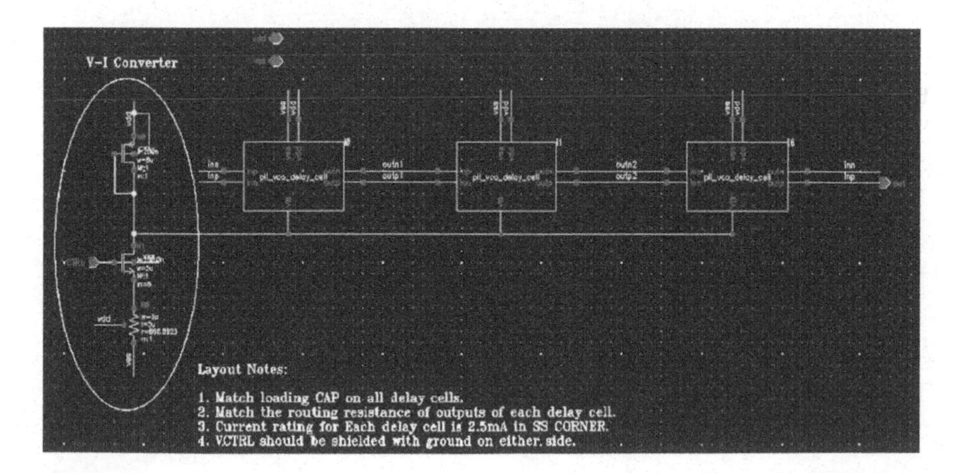

Fig. 5 Three stage ring VCO test set up

4 Simulation Results

The test case setup is done by sweeping the control voltage from 0.7 to 1.2 V and performed the transient analysis to capture the linearity of the VCO. The gain of the VCO is 3.4 GHz/V i.e., (1.7 GHz/0.5 V).

Figure 6 results illustrate VCO output frequency of 4 GHz. The test case is done by giving 1.1 V control voltage and transient analysis is performed.

Figure 7 shows that control voltage versus frequency of VCO. The transistor goes from saturation region of operation to triode region of operation if the control voltage crosses 1.15 V. So output frequency of the VCO varies in a nonlinear fashion. In the 3.4 GHz range the VCO is linear. Table 2 shows that performance comparison with recently published VCOs. The designed VCO linearity is achieved over a range of frequency from 1.76 to 3.4 GHz with 48.2% tuning range.

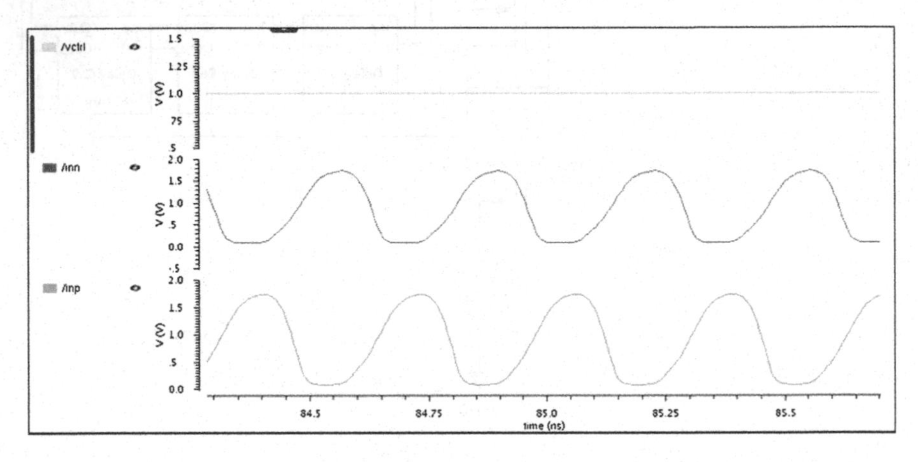

Fig. 6 VCO differential output frequency of 4 GHz

Fig. 7 Control voltage versus frequency

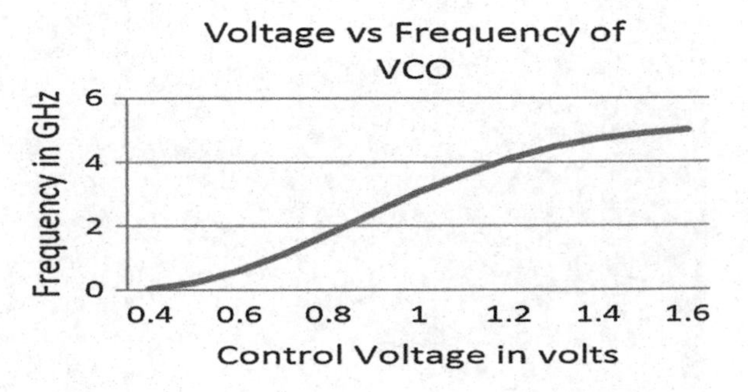

Table 2 Performance comparison with recently published VCO (n.a = not applicable)

CMOS process	Frequency (GHz)	Tuning range (%)	Supply voltage (V)	Power (mW)	Reference
0.18 μm	40	20	1.5	27	[3] JSSC'2007
0.18 μm	5.2	18	1.8	6.3	[4] MWCL'2009
0.18 μm	5.46	10.6	10.8	6.4	[5] MWCL'2009
0.18 μm	4.94	14.7	1.4	2.5	[6] TMTT'2010
0.18 μm	24	22.4	0.9	11.1	[7] MWCL'2011
0.18 μm	1.92	8.13	n. a	13	[8] TCASI'2011
0.18 μm	23.08	16.5	1.8	9	[9] TCASII'2012
0.13 μm	35	26	1.2	11	[10] RFIC'2013
65 nm	4.21	41.1	n. a	8.7	[11] TCASII'2014
0.18 μm	3.4	48.2	1.8	15.4	Proposed

5 Conclusion

In this paper three stage Wide-Tuning-Range ring VCO is designed using differential delay cell in 0.18 μm CMOS process under the voltage of 1.8 V using Cadence Virtuoso tool. Carefully choosing the component W/L values, the VCO oscillate around center frequency of 3.4 GHz with 48.2% tuning range. All resistances and capacitances were extracted from layout such that we can simulate the circuits more accurately with post layout simulations. All transistors in the present design had been sized correctly to obtain the targeted design.

References

1. Razavi, B, Lee, K, Yan, Y,: Design of high speed, low power frequency dividers and phase locked loops in deep submicron CMOS: IEEE J. Solid-State Circuits, vol. 30, no. 2. pp. 101–109 (1995).
2. Ping-Hsuan Hsieh, jay Maxey, and Chih-Kong ken yang,: Minimizing the supply sensitivity of a CMOS ring oscillator through jointly biasing the supply and control voltages: IEEE J. Solid-State Circuits, vol. 44, no. 9, pp-2488–2495 (2009).
3. Jun-Chau Chien, and Liang-Hung Lu,: Design of wide-tuning-range millimeter-wave CMOS VCO with a standing-wave architecture: IEEE J. Solid-State Circuits, vol. 42, no. 9, pp. 1942–1952 (2007).
4. Young-Jin Moon, Yong-Seong Roh, Chan-Young Jeong, and Changsik Yoo,: A 4.39–5.26 GHz LC-tank CMOS voltage-controlled oscillator with small VCO-gain variation: IEEE Microw. Wireless Compon. Lett., vol. 19, no. 8, pp. 524–526 (2009).
5. Jian-An Hou and Yeong-Her Wang,: A 5 GHz differential colpitts CMOS VCO using the bottom PMOS cross-coupled current source: IEEE Microw. Wireless Compon. Lett., vol. 19, no. 6, pp. 401–403 (2009).
6. José Luis González, Franck Badets, Baudouin Martineau, and Didier Belot,: A 56-GHz LC-tank VCO with 17% tuning range in 65-nm bulk CMOS for wireless HDMI: IEEE Trans. Microw. Theory Techn., vol. 58, no. 5, pp. 1359–1366 (2010).

7. Meng-Ting Hsu and Po-Hung Chen,: 5G Hz low power CMOS LC VCO for IEEE 802.11a application: Iin Proc. Asia-Pacific Microw. Conf., pp. 263–266 (2011).
8. Zuow-Zun Chen and Tai-Cheng Lee,: The design and analysis of dual-delay-path ring oscillators: IEEE Trans Circuits Syat. I, vol. 58, no. 3, pp. 470–478 (2011).
9. Pei-Kang Tsai and Tzuen-His Huang,: Integration of current-reused VCO and frequency tripler for 24-GHz low-power phase-locked loop applications: IEEE Trans. Circuits Syst. II, Exp. Briefs, vol. 59, no. 4, pp. 199–203 (2012).
10. Qiyang Wu, Salma Elabd, Tony K. Quach, Aji Mattamana, Steve R. Dooley, Jamin McCue, Pompei L. Orlando, Gregory L. Creech and Waleed Khalil.,: A–189 dBc/Hz FOMT wide tuning range Ka-band VCO using tunable negative capacitance and inductance redistribution: In Proc. IEEE Radio Freq. Integr. Circuits (RFIC) Symp. pp. 199–202 (2013).
11. Heein Yoon, Yongsun Lee, Jae Joon Kim, and Jaehyouk Choi,: A wideband dual-mode LC-VCO with a switchable gate-biased active core: IEEE Trans. Circuits Syst. II, Exp. Briefs, vol. 61, no. 5, pp. 289–293 (2014).

A New Iterative Hybrid Edge Technique Using Image Mosaic

D. Surya Kumari and S.A. Bhavani

Abstract This paper considers the problem of Images which is the most important medium of conveying information, Image mosaic is one of the process of partitioning a digital image into two or more segments. The goal of mosaic is to change and represent the image into more meaningful and easier to analyse and we perform secrecy of transferring data, generation of secret key and also performing the image mosaic process. By implementing Diffie Hellman key exchange protocol we can generate shared key between the users. An encryption technique Reverse Binary XOR Algorithm used which converts plain format data into unknown format. Least Significant Bit Technique for storing the cipher data then performs mosaic technique for segmenting image. In this paper we are using edge based technique for performing image mosaic process from these concepts we can provide secure transfer of data and also it improves the efficiency of network.

Keywords Image mosaic · Diffie Hellman key exchange algorithm · Edge based technique · Reverse binary XOR encryption

1 Introduction

In image processing, division is frequently the first venture to pre-process pictures to concentrate objects of enthusiasm for further investigation. Division systems can be classified into two structures, edge-based and locale based approaches it is the most common approach An edge is the boundary line between the regions with relative distinct gray-level properties [1].

D. Surya Kumari (✉) · S.A. Bhavani
Department of CSE, ANIL Neerukonda Institute of Technology and Sciences (ANITS), Visakhapatnam, India
e-mail: dsuryakumari.14.cstmtech@anits.edu.in

S.A. Bhavani
e-mail: sabhavani.cse@anits.edu.in

S.C. Satapathy et al. (eds.), *Computer Communication, Networking and Internet Security*, Lecture Notes in Networks and Systems 5,
DOI 10.1007/978-981-10-3226-4_23

Fig. 1 An example of image mosaic

Image mosaic is one of the process for partitioning an digital image into two or more segments to avoid falsely object selection and denoising is done before the mosaic for mosaic to segment the image into multiple parts without loss of information. The main aim of mosaic is to change the represent an image in more meaningful and easier way to analyse. Image mosaic is used to locate the boundaries and objects (lines, curves, etc.) in images. Mosaic technique can be done on different techniques namely, Region Based, Edge Based, Threshold Based etc. we perform secret key generation, transferring data in secure manner and also performing the image mosaic n image along with hided data [2]. By using Diffie Hellman key exchange protocol we can generate a shared key between the users (Fig. 1).

In this paper we are using a Reverse Binary XOR encryption technique for converting plain text format data into unknown format data. After converting data into cipher format we can put into image by using least significant bit technique. After completion of data to be hidden into image and perform mosaic technique for segment image.

2 Related Work

Performing of mosaic image is very essential technique in image processing field, computer vision and which generates panoramic images. Image mosaicing is widely used in editing images ex: photoshop and the algorithms which used are being improved over one another and these algorithms works with image mosaicing based on edge based method [1]. This algorithm follows the concept edges between two images. which gives the selected position of two images by finding matching points of two images. When the relevant portion is shown by comparing the matched portion of the image and that matched portion is overlapped by one image with other image and then it forms a composite image. Now, Image registration is the concept of matching or recognizing two or more images. As we see in medical imaging there are large panoramic images which can help to conduct visual observations and focus on surrounding, neighbouring parts, to the doctors and Carefully it is very accurate and high resolutions of pixels of camera may be used to eliminate the need for an auto registration of an image. When User interaction is done then also is very reliable source for manually registering images. The other concept of feature based mosaicing algorithm has done a very significant contribution in auto image mosaic [3]. This algorithm works by matching the edges of points in two more images through Harris edge detection which is used to detect edges in the images. After detecting the edges in the both the images. The Panoramic visibility can been seen that is constructed in high resolution images that cover large field in generating 3-d, 2-d view by image mosaic, the Video content over-view, Medical image processing, Virtual reality technology (VRT) are techniques which are generally used in the process of image mosaic and in image mosaicing the Image registration, Image stitching, Finding line of cut in new image, Image blending, which are needed in image mosaic to produce the outcome clearly to the vision.

3 Proposed System

Image mosaic is generally enhancing the granular information in images for viewers and offering improved input for different automated image processing techniques. The primary aim of segmenting an image is to enhance quality and suitability for presenting the image for a specific given task in front of an observer. Mosaic is a process of partitioning of color or grey scale image into various set of segments. The major benefit of image mosaic is to provide a convenient way of image representation and analysis. In this process, whole image is distributed and categorized into different group of image sectors. These sectors consist of similar image level on a pixel basis. Thus, displaying same level pixels prominent and making the image outlines brighter which can be used for further analysis. Application of image mosaic is vast and could be used in many fields. It enhances clarity in the algorithms and innovating new methods of analysis is interested edges and supports better object recognition (Fig. 2).

Architecture

There are number of various image mosaic algorithms which are currently used and applied for different purposes. In this paper I also proposed other concept for transferring data into image. Before performing image mosaic process the sender will perform encryption process for data and that cipher formatted data put into image. After that the sender will use the edge based image mosaic technique we can segment image into number of parts. The sender will send those parts to specified receiver and receiver will receive parts will generate single image. After generating single image the receiver will get data from the image and decrypt cipher format

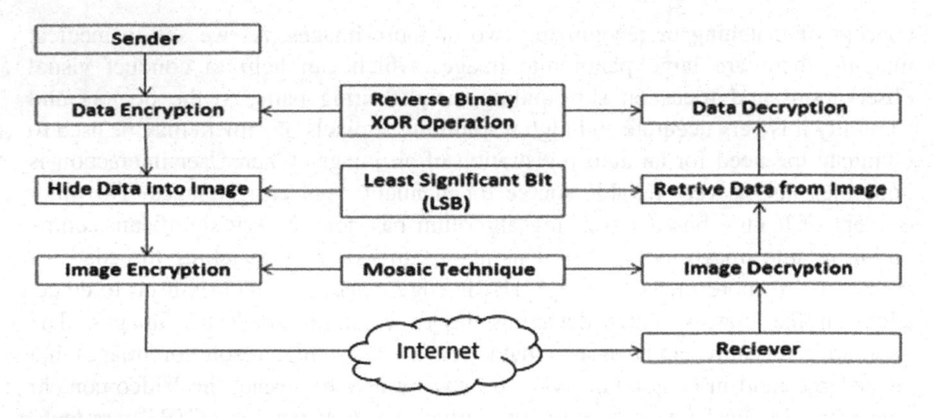

Fig. 2 Architecture

data. So that the receiver completes decryption process it will get original data and original image. By implementing those concepts we can propose four concepts in this papers i.e. shared key generation, data encryption and decryption process, least significant bit and edge based image mosaic technique. In this paper the first concept is generation of shared key by using Differ Hellman key exchange protocol. The implementation procedure of Diffie Hellman is as follows.

3.1 Diffe Hellman Key Exchange Protocol

In this module the sender and receiver will generate same shared key for encryption and decryption of transferring message. The generation of shared key is as follows.

1. The sender and receiver will agree to use modulus P and base G.
2. The sender will choose private key a and calculate public key by using following formula.

$$\text{Public key} = G^a \bmod P$$

3. After generating public the sender will send that public key to receiver.
4. The receiver will retrieve public key and choose private key.
5. Using that private the receiver will generate public by using same formula and send that public to sender.

6. The sender will retrieve receiver public key and generate shared key by using following formula.

$$\text{Shared key} = \text{receiver public}^{a} \bmod P$$

7. The receiver also generate shared key by using following formula.

$$\text{Shared key} = \text{sender publickey}^{a} \bmod P$$

After generating shared key those keys are same for both users. By using that shared key the sender will encrypt transferring message. The encryption process Reverse Binary XOR Encryption Algorithm is as follows.

3.2 Reverse Binary XOR Algorithm

Encryption process

We will be presenting the steps of the encryption algorithm of the reverse binary XOR Algorithm.

1. Input secret key and transferring message to encryption process.
2. Get each character from the message and convert into ASCii values.
3. After converting ascii values each value XOR with key until the length of message is completed.
4. The completion of XOR operation each ascii value can converted into binary format.
5. Reverse previous binary data until completion length of message.
6. After reversing binary data that data can be perform once complement.

Decryption process

1. The previous binary data can be perform the once complement.
2. After performing once complement that binary data will be reverse until the completion message length.
3. After reversing that binary data will convert into ascii format.
4. The previous ascii values will be XOR with secret key until completion of message length.
5. After performing XOR operation that ascii values can be converted characters get the original message.

The receiver will get original plain formatted data and also get original image without loss of color.

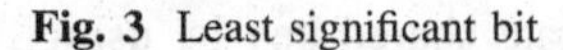

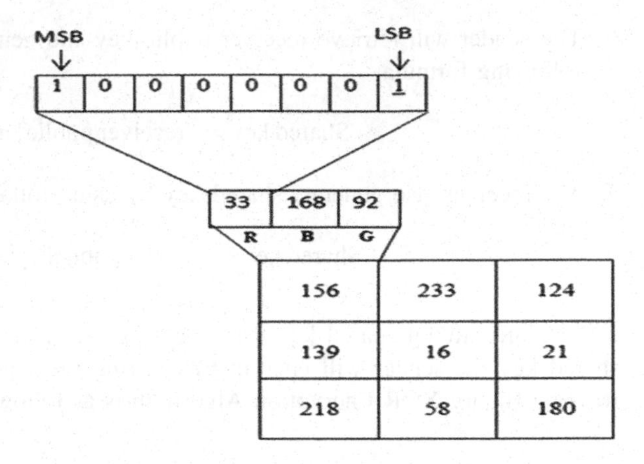

Fig. 3 Least significant bit

3.3 Least Significant Bit Technique

In this module the sender will take binary formatted data of cipher data and image pixel values. The sender will take transferring image and convert into binary format. After conversion of binary the sender will take cipher format binary data and put into binary pixel value of least significant bit. The sender will take that stored binary pixel values and again generate data hide image (Fig. 3).

3.4 Edge Based Image Mosaic Technique

In this module the sender will segment data hide image into number of parts by using edge based image mosaic technique. In this technique we are segment image using edge based. In this paper we are taking some amount of pixel will be consider in a edges and split that edges into one segment. After that take another part from previous edge of some pixel values and next edge of original image. Likewise we can segment image into specified parts and those segment will be send to receiver.

The receiver will receive parts from the sender and generate single image by applying reverse of process of edge based image mosaic technique. The completion of generating data hide image the receiver will convert image into binary format. After converting image into binary format the receiver will get all binary formatted cipher data from the image. The receiver will get all binary formatted cipher data and convert into plain format by using decryption process using genetic operation. The implementation of decryption process is as follows.

4 Experimental Results

Experiments are conducted to observe the detection applied on the edges of an image by edge technique in image mosaic The results obtained from a standard image. Experiments were done in java programming by Java SDK with eclipse/Net Beans IDE system setup is required for programming it can be seen in the FIG: result output (Figs. 4 and 5).

Fig. 4 Original image

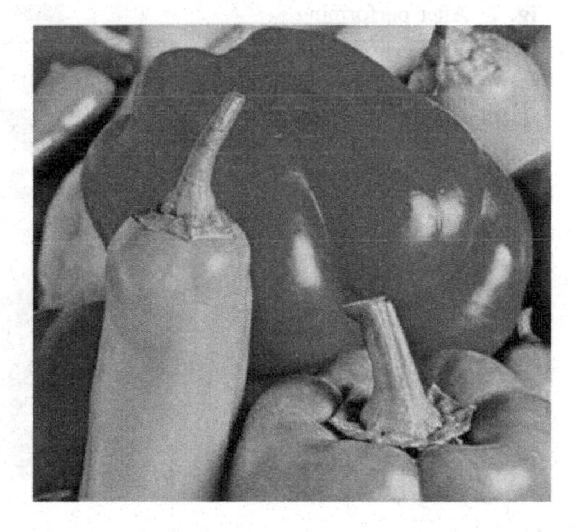

Fig. 5 Data hide image

After performing image mosaic
See Figs. 6 and 7.

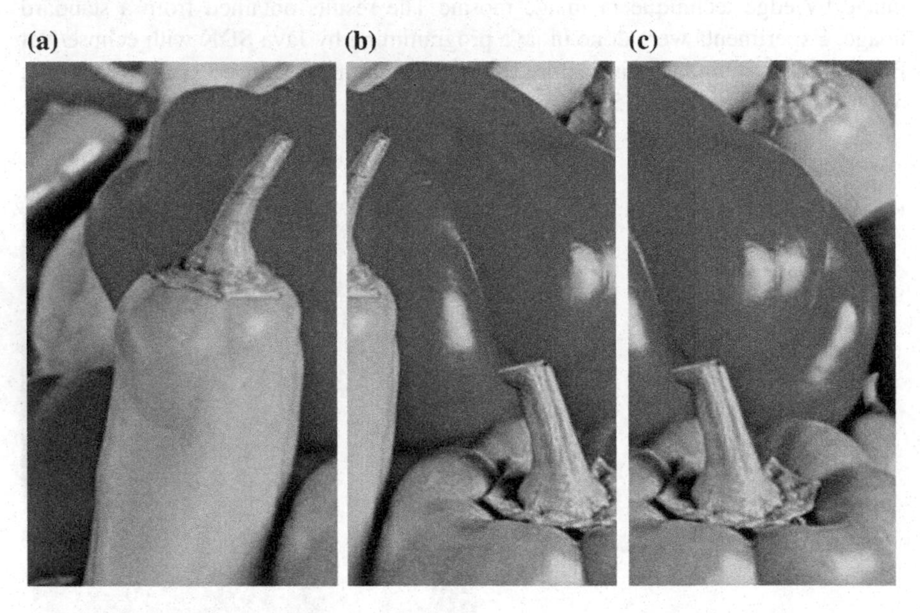

Fig. 6 After performing mosaic(a, b, c)

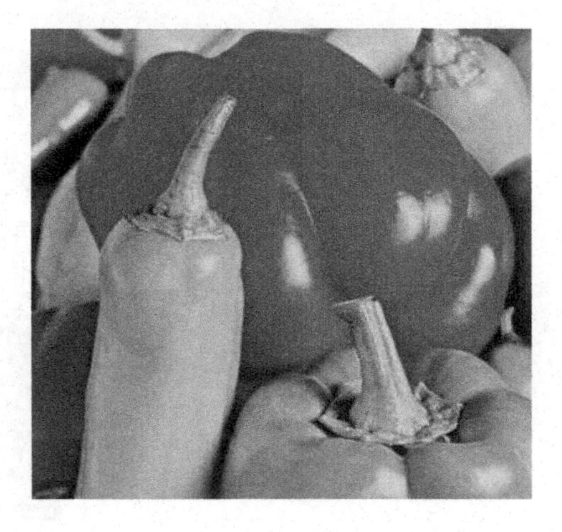

Fig. 7 After performing mosaic and encryption the original image

5 Conclusion

In this proposed work we present an efficient technique for performing image mosaic and also provides privacy of transferring data into image. Before performing image mosaic process the sender will enter transferring data and convert into unknown format by using encryption and decryption of data using Reverse binary XOR to convert data into unknown format we used least significant bit technique for storing the cipher data and perform image Mosaic technique using edge based. The receiver will receive those parts and combine parts will generate single image After getting original data the receiver will get data and also get original image. By implementing those concepts we can improve more efficiency and provide privacy of transferring message.

References

1. Pandey, et al. "Volterra Filter Design for Edge Enhancement of Mammogram Lesions," Proc. of (IEEE) 3rd International Advance Computing Conference, p. 1219–1222, February 2013. "Non-Linear Polynomial Filters for Edge Enhancement of Mammogram Lesions," Computer Methods and Programs in Bio-medicine, vol. 129C, pp. 125–134, January 2016.
2. D. I. Barnea; and H. F. Silverman,"A class of algorithms for fast digital registration," IEEE Trans. Comput, vol. C-21, pp. 179–186, 1972.
3. C. D. Kuglin and D. C. Hines," The phase correlation image alignment method", in Proc. IEEE Int. Conf. Cybernet. Society, New York, NY, pp 163–165, 1975.
4. Lisa G. Brown. A survey of image registration techniques. ACM Computing Surveys, 24(4); pp 325–376, December 1992.
5. J. B. A. Maintz and M. A. Viergever, "A survey of medical image registration," Med. Image Anal., vol. 2, no. 1, 1998, pp. 1–36.
6. A Can, C. V. Stewart, B. Roysam, H. L. Tanenbaum, "A Feature-Based Technique for Joint, Linear Estimation of High-Order Image-to-Mosaic Transformations: Application to Mosaicing the Curved Human Retina", *IEEE Conference on Computer Vision and Pattern Recognition,* vol. 2, pp: 585–591, 2000.
7. Inampudi, R. B.," Image mosaicing" in International Conference on Geoscience and Remote Sensing Symposium Proceedings, vol.5, pp 2363–2365,1998.
8. Battiato, G. Di Blasi, G. M. Farinella, and G. Gallo, "Digital mosaic framework: An overview," Eurograph.—Comput. Graph. Forum, vol.26, no. 4, pp. 794–812, Dec. 2007.
9. C. H. Bindu and K. S. Prasad, "An efficient medical image segmentation using conventional OTSU method," *Int. J. Adv. Sci. Technol.*, vol. 38, pp. 67–74, Jan. 2012.

Time Series Analysis of Oceanographic Data Using Clustering Algorithms

D.J. Santosh Kumar, S.P. Vighneshwar, Tusar Kanti Mishra and Satya V. Jampana

Abstract With the availability of huge data sets in device fields like finances to weather, it becomes very important to quality analysis and interprets the results. In such scenario K-Means and DBSCAN clustering algorithms are used for effective data grouping to get insight into the hidden structure in the data. In this paper focus on the application of clustering to ocean data observations. An attempt is made to correlate the resulting clusters to the variability focused during cyclones.

Keywords Data mining · Clustering · Temperature · Salinity · Buoy · Time series

1 Introduction

It is now well accepted that climate change is for real. Ignoring this change and not taking corrective measures can lead to disastrous consequences. No wonder, there is a focused research in simulating the climate for the short and long term predictions. India is not insulated from the changing weather and climate scenario. There are sustained efforts to have early warning systems and observational platforms in the oceans surrounding India. INCOIS is one such center which collects and decimates ocean observations. A large number of ocean observation buoys gather data and it is uplinked through satellites to the data centers at INCOIS. The climate models and

D.J. Santosh Kumar (✉) · T.K. Mishra
Computer Science and Technology, Department of CSE, ANIL Neerukonda Institute of Technology and Sciences (ANITS), Visakhapatnam, India
e-mail: dj.santosh666@gmail.com

T.K. Mishra
e-mail: tusar.k.mishra@gmail.com

S.P. Vighneshwar · S.V. Jampana
Indian National Centre for Ocean Information Services (INCOIS), Hyderabad, India
e-mail: vighneshwar@incois.gov.in

S.V. Jampana
e-mail: raju.jvs@incois.gov.in

S.C. Satapathy et al. (eds.), *Computer Communication, Networking and Internet Security*, Lecture Notes in Networks and Systems 5,
DOI 10.1007/978-981-10-3226-4_24

remotely sensed data from satellites and host of other observation platforms provide a huge amount of data. These datasets need to be reduced to smaller dimensions to infer meaningful statistics and conclusions. The goal is to detect underlying patterns and correlations among them show how these patterns contribute to the variability in the data. For reduction of data dimensionality, there are a number of tools and techniques which can bring the data to the manageable form. One such popular technique is the K-means clustering algorithm which along with other density-based clustering algorithm [1] help in finding noise and correlations in the data in a robust manner. Clustering is not a fool proof technique. The interpretation of the results need background domain knowledge and some ambiguity is also present. This technique should be used in conjunction with other results to make analysis more robust [1]. Section 1.1 describes the K-means and density-based clustering algorithms, Sect. 2 is proposed work and Sect. 3 is for data, and Sect. 4 is on results and conclusions and future work follow in the last section.

1.1 Related Work

Different tools are used to predict the climate. The statistical models are used for the initial efforts. Mostly SST is main parameter for prediction [2]. Another statistical model is Canonical Correlation analysis [3, 4] are used for ocean forecast. For climate effects variation in vegetation the suitable data mining techniques are added recently [5]. Independent component analysis [6], regression tree techniques are used for relations and predict the climate efforts [6]. Nonlinear canonical correlation analysis are used for sea level presser and temperature at different levels based on neural networks [7].

2 Proposed Work with Block Diagram

The proposed work shows how time series data forming clusters, based on clusters we correlate with other clusters and predict the seasonal cyclones. By using clustering algorithms, predicting the seasonal rains and cyclones. The clusters show how the data points are forming in different quad rates at different depth levels. Data mining teachings are using for analyzing in ocean domain also. Basically, the data set contains different parameters like sea surface temperature, salinity, current speed, current direction and pressure etc. Actually sea surface temperature, wind speed is the main parameter for ocean field so the water temperature at different depth levels like wtf at 1 m depth, Ltd at 5 m depth etc. Air temperature, wind speed is other parameters. The data set given as input and data cleaning techniques are used by using distance based noise removal technique and remove null values in ocean data sets in the second stage. And apply the clustering algorithms to current data and form clusters. İf the data set not properly fitted to the algorithm then repeat

Fig. 1 Block diagram

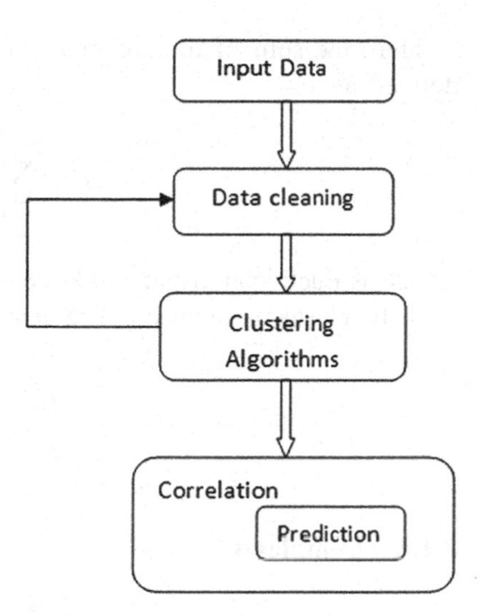

the data cleaning. After applying the algorithms, these clusters are correlate with the other observations. Based on the data analysis for the clusters we go for prediction. The block diagram shows Fig. 1 how the experiment works.

2.1 Clustering Algorithms

Data Clustering [8, 9] is classification procedure wherein objects are tagged with class labels. The labels are derived from the data. In other words, it is a data-driven procedure and comes under the class of unsupervised classification. The classification itself is of different types like e.g., well separated, prototype based, graph based or based on density. To generalize, clusters are a set of objects that change a common property. In K-means clustering [1, 8] which is prototype based classification, where a user specified number of clusters are found, the clusters are represented by their centroids.

2.1.1 K-Means Algorithm

The general procedure of the K-Means algorithm is

1. Select K points as initial centroids.
2. Repeat.
3. Assign all points to the closest centroid.
4. Recompute the centroid of each cluster.
5. Untill the centroied doesn't change.

Here the sum of the squared error (ϙ) in the data is minimized. The SSE is defined as.

$$E = \sum_{i=0}^{k} \sum_{x \in C_i} D(C_i, x)^2 \tag{1}$$

'D' is Euclidean distance between two objects. The centroid that minimized the ϙ of the cluster is its mean. The centroied of the ith cluster is

$$C_i = \frac{1}{m_i} \sum_{x \in C_i} P(x) \tag{2}$$

2.1.2 Complexity

The complexity is rather modest. Only the data and centroids needs to be stored. The storage is O((m + K)n) where m is number of data points and n is number of attributes. The time requirement is O(I × K×m × n), where I is the iteration number for convergence. The algorithm is efficient when the number of clusters are small compared to m.

2.1.3 DBSCAN Algorithm

As we know data mining is extracting hidden and interesting patterns from large data sets from the scientific data. To extracting, the useful patterns from spatial data. The density based algorithms are very useful in some point areas. If the data set is in any orbiter shape and if it contains any noise and it is in large size data. The clusters are formed according to density based connectivity analysis [10]. Two issues are addressed in DBSCAN make effective with a large volume of spatial data objects [11]. In 1996 DBSCAN is proposed by Ester et al. [10]. The cluster neighborhood is a given with radius (Eps) and at least maintain minimum points (Mints) that mean cardinality of the neighborhood has to exceed some threshold ϵ neighborhood of an arbitrary point 'p'.

$$N_{Eps} = \left\{ \frac{q\varepsilon D}{dist(p, q)} < E_{ps} \right\} \tag{3}$$

Є-neighborhoods of P, D = data set. Core object refers to such point that its neighborhood of a given radius (Eps) has to contain at least a minimum number (MinPts) of other points.

$$N_{Eps}(p) > Minpt \tag{4}$$

Eps and Min points are user specified. DBSCAN is used to remove redundant test cases so it reduce test cases [12].

3 Data

The data is from a series of deployments of the OMNI (Ocean Moored Buoy Network for Northern Indian Ocean) buoys of NIOT (National Institute of Ocean Technology). The OMNI buoys are equipped with a suite of sensors to record near surface and subsurface ocean parameters made at various depths. Ocean currents are obtained from a 150 Hz Teledyne Acoustic Doppler Current profiler. The white noise level in the data is identified by examining the noise floor of the spectra of the time series. Various filtering techniques are used to minimize the noise levels arising from the sensors. Figure 2 shows the time-depth representation of the ocean currents data over a month period. Similar time depth representations can be constructed for other variables from high-resolution observations and model output of climate simulations. In the next section, we show the clusters that arise from such complex time series data.

4 Experimantal Results

Figure 3 shows an example of small subset of data that has been clustered using K-means clustering algorithm (Fig. 4).

Clusters: Table 1 shows the cluster analysis of buoy data over a year. The near surface SST shows four clusters of similar class labels. The table shows how the clusters are correlated with another cluster. Accordingly, September–January forms a class label, the post monsoon, and winter of the year. February forms one class label, followed by class labels March–April and May–June–July. The next column shows the air temperature which shows a distinct class label for the period June–

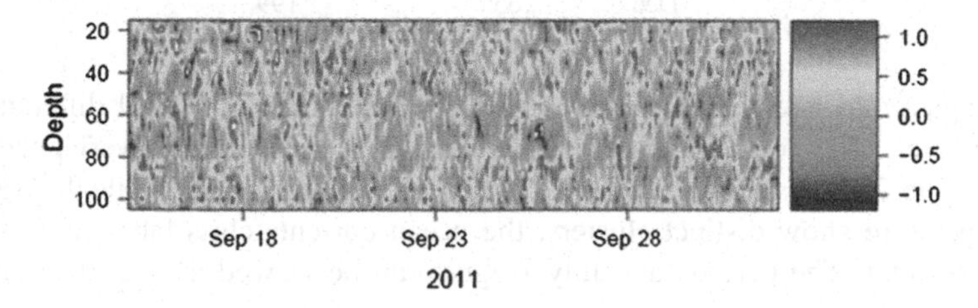

Fig. 2 Complex structure of currents in ocean

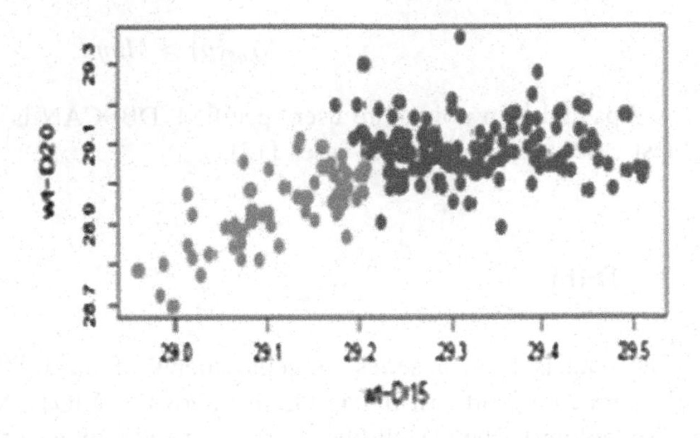

Fig. 3 Clustering of subsurface data

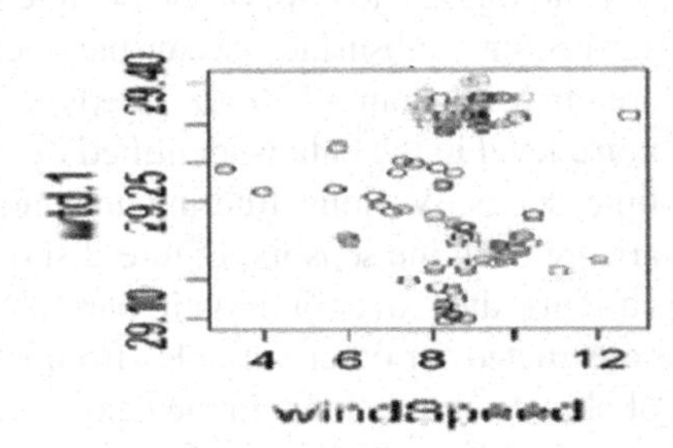

Fig. 4 Clustering of DBSCAN for wind speed and surface data

Table 1 K-means partitioning of near surface climatic variable sea surface temperature, air temperature and solar irradiance from the Buoy data across different months of a year. The monsoon period show distinctly small clusters of temperature and air temperature data

Months	Water temperature	Air temperature	Solar irradiance
Jan	246	245	997
Feb	171	199	683
Mar	147	136	461
Apr	147	131	442
May	116	118	457
Jun	128	49	543
Jul	129	82	533
Aug	117	94	714
Sept	230	209	1001
Oct	257	241	990
Nov	262	212	966
Dec	265	199	1024

July–August and nearly follows the class labels of SST with small differences in their number. The solar irradiance, however, show much more overlapped class labels. Table 2 shows the class labels over a seasonal time scale. While the SST and air temperature show distinct clusters, the ocean currents class label show similar size class label. The period June-July-August can be viewed as a distinct climatic pattern.

Table 2 The clusters of SST, salinity currents and surface winds for nearly yearlong data

Months	Water temperature	Salinity	Currents	Wind
Dec, Jan, Feb	97	289	643	647
Mar, Apr, May	0	0	0	0
Jun, Jul, Aug	358	728	796	259
Sep, Oct, Nov	35	214	761	1174

5 Conclusions

A year long oceanic data which has the significant amount of variability across multiple time scales has been analyzed using K-means clustering algorithm. The analysis suggests that distinct events like the June-July-August cluster can be identified. This method can be complemented by other data mining methods like the Principal Component Analysis and decision trees for more robust analysis.

5.1 Future Work

The ocean devices measure every point of time with different parameters. When the data is in large amount it takes more time for clustering and prediction. The featured work uses Hadoop technology used for handling a large amount of data.

Acknowledgements Authors from ANITS would like to thank HOD ANITS for support. Authors from INCOIS would like to thank the Director INCOIS for providing all necessary facilities to carry out this work.

References

1. Data Mining: Concepts and Techniques (The Morgan Kaufmann Series in Data Management Systems) Hardcover – Import, 25 Jul 2011 by Jiawei Han (Author), Micheline Kamber (Author), Jian Pei Professor (Author).
2. Landman, W.A.: A canonical correlation analysis model to predict South African summer rainfall. NOAA Experimental Long-Lead Forecast Bulletin 4(4), 23–24 (1995).
3. Landman, W.A., Mason, S.J.: Forecasts of Near-Global Sea Surface Temperatures Using Canonical Correlation Analysis. Journal of Climate 14(18), 3819–3833 (2001).
4. Rogel, P., Maisonnave, E.: Using Jason-1 and Topex/Poseidon data for seasonal climate prediction studies. AVISO Altimetry Newsletter 8, 115–116 (2002).
5. White, A.B., Kumar, P., Tcheng, D.: A data mining approach for understanding control on climate induced inter-annual vegetation variability over the United State. Remote sensing of Environments 98, 1–20 (2005).
6. Basak, J., Sudarshan, A., Trivedi, D., Santhanam, M.S.: Weather Data Mining using Component Analysis. Journal of Machine Learning Research 5, 239–253 (2004).

7. Hsieh, W.W.: Nonlinear Canonical Correlation Analysis of the Tropical Pacific Climate Variability Using a Neural Network Approach. Journal of Climate 14(12), 2528–2539 (2001).
8. Hartigan, J.: Clustering Algorithms Wiley, New York (1975).
9. Jain, A.K, M.N. Murty, P.J. Flynn.: Data Clustering : A Review, ACM Computing Surveys, 31(3):264:323, September (1999).
10. M. Ester, H.P. Krigel, J. Sander, and X. Xu, "A Density-Based Algorithm for Discovering Clusters in Large Spatial Databases with Noise, "Proc.of the 2nd International Conference on Knowledge Discovery and Data Mining, Portland, WA, 1996, pp. 226–231.
11. B. Borah and D. K. Bhattacharyya, "An Improved Sampling- Based *DBSCAN* for Large Spatial Databases," presented in the international Conference on Intelligent Sensing and Information Processing, Chennai, India, January 2004.
12. B. Borah and D. K. Bhaftacharyya, "A Clustering Technique using Density Difference," IEEE - ICSCN 2007, MIT Campus, Anna University, Chennai, India. Feb. 22–24, 2007. pp. 585–588.
13. Jain, A. K, Dubes. R. C.: Algorithms for clustering Data Prentice Hall Advanced references Series, Prentice Hall, (1988).
14. Shai Shalev-Shwartz, and Shai Ben David: Understanding Machine Learning. From theory to Algorithms (2014).
15. Karsten Steinhaeuser, Nitesh V Chawla and Auroop R Ganguly: Comparing Predictive power in Climate Data: Clustering Matters (2014).

Cluster Based Prediction of Keyword Query Over Databases

K. Geethanjali, K. Suresh and R. Kanaka Raju

Abstract In this paper, We using the cluster-based prediction, to predict the keywords over database in a efficient way. By using the cluster-based prediction, efficiency of searching query is improved and time complexity is reduced. In this paper, we proposed Text preprocessing, MVS matrix, k-means clustering and character shuffle preprocessing searching algorithm in order to improve efficiency. In text preprocessing, we eliminates all the tags and find the relative frequencies of each document then weights is calculated. By using MVS matrix, similarities of each document is calculated then formed into matrix. Based on similarities, Clusters are formed by using K-means clustering, then the keyword is searched in clustered instead of several documents. Then the searching is performed in efficient way.

Keywords Text preprocessing · MVS matrix · Keyword

1 Introduction

Data mining is to extracting knowledge from large amounts of data. For example, bank have very huge amount of data. In this case data mining is used to extracting the data from the huge amount of data in the banking sectors. Another one is in

K. Geethanjali (✉) · K. Suresh
Department of CSE, ANIL Neerukonda Institute of Technology and Sciences (ANITS), Visakhapatnam, India
e-mail: kgeethanjali.14.cstmtech@anits.edu.in

K. Suresh
e-mail: kurumallasuresh@gmail.com

R. Kanaka Raju
GVP college for degree & Pg courses, Visakhapatnam, India
e-mail: rkinraju@gmail.com

S.C. Satapathy et al. (eds.), *Computer Communication, Networking and Internet Security*, Lecture Notes in Networks and Systems 5,
DOI 10.1007/978-981-10-3226-4_25

educational systems, there are many number of students and their details, marks, how many students take college buses, and the fees details and etc. All that information is stored is not possible. So, we are using data mining to extract the data from the large amount of data.

Data mining is used in marketing, educational institutes, banking sectors, MNC companies and etc. By using data mining, we reduce space and size of the stored data. So that, we use this technique to efficient search query. By implementing this we reduce the time complexity and get efficient search query when the user search a query [1–10].

Clustering is the grouping of set of objects into groups is called as clusters so that objects in one cluster are high similar than the objects in another cluster. Clustering is the one of the method to reduce the time in order to search a particular query. If the user want to search a query then this method gives directly what we want without time taken. Based on clustering algorithm, reduce time complexity and get the efficient search query.

2 Existing System

A user searches a keyword, from the set of documents in the database. If database have 1000s or lakhs of documents, if the searched keyword is present in 1000 document or in the last document then we have to search that keyword in all documents from first to last document. So, based on this case time complexity increases. The Main problem in this paper is time complexity. we proposed an one method i.e. character shuffle preprocessing searching process. Based on this method we can get directly the searched query with in a less time when compare to previous methods that are used.

2.1 Problem Identification

Time complexity is the main problem, When a user search a keyword, it searches all the documents in the database. If database have 1000s or lakhs of documents, if searched keyword is present in last document then we have to search that keyword in all documents from first document to last document. So, it takes time to search particular query.

3 Proposed System

In this system we can implement better query related search technique and avoid time complexity by performing search operation over the document. We are implementing character shuffle preprocessing algorithm for searching query similar document in the database. First, we can take one document and to perform the search operation for searching similar documents. The character shuffle preprocessing algorithm mainly contain following process.

3.1 Preprocessing Document Text

Before performing searching operation over document each document can be preprocessed over the text. By performing text preprocessing we can reduce all document related tags and only get text related documents. After getting all the text related documents we can find out relative frequency r_{freq} of each document. Before identifying relative frequency we also find out local and global frequency of each word in document. The local frequency (l_{freq}) of each document can be calculated by counting number same words in the document. The global frequency is total number of word in particular document. After that we can find out relative frequency of each document by using following formula.

$$r_{freq} = l_{freq} + g_{freq}/2.0 \quad (1)$$

Using that relative frequency we calculate document weight by using following formula.

N = Document size
l_{freq} = Each word in the document can be referred as local frequency
g_{freq} = Total number of word in document can reffered as global frequency

$$Weight(W) = l_{freq} * \text{Math.Log}\left(\frac{\text{N}}{g_{freq}}\right) + 0.01 \quad (2)$$

After calculating each document weight using above formula we can build MVS Matrix of each document related to other documents.

3.2 Construction of MVS Matrix

After calculating document weight we can find out the cosine similarity of each document related to other document. After calculating cosine similarity we can construct MVS (Multi-View Similarity) Matrix and using that matrix we perform

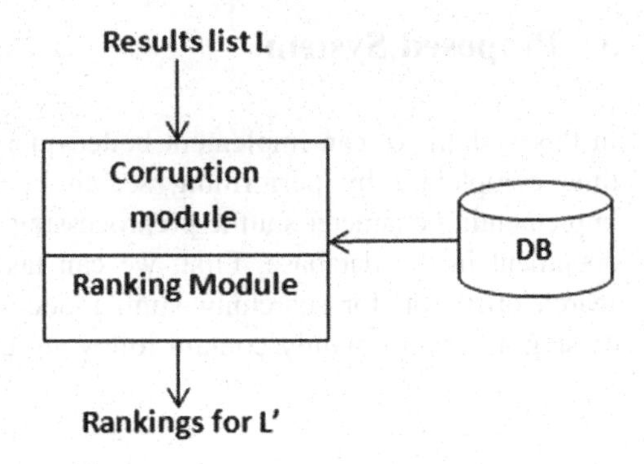

Fig. 1 Ranking approximation

the clustering process of document. the similarity of two documents can be calculated by using below formulas (Fig. 1).

$D1$ = Total number of words in first document
$D2$ = Total number of words in second document

$$D1_{sqr} = D1 * D1 \tag{3}$$

$$D2_{sqr} = D2 * D2 \tag{4}$$

$$D_{sqrprd} = D1_{sqr} * D2_{sqr} \tag{5}$$

$$similarity = D_{prd}/D_{sqrprd} \tag{6}$$

By using those formulas we calculate cosine similarity of each document and also construct MVS matrix formatted data. By calculating cosine similarity related document will be arranged in matrix format.

3.3 Grouping Related Documents Using K Means Clustering Algorithm

After generation of MVS Matrix using cosine similarities we can perform the clusterization of document. by implementing clustering process we can grouping related documents into one group. In the clusterization process we consider cosine similarity of each document. the implementation procedure of K means clustering Algorithm is as follows.

(i) Before performing clusterization user enter the number of clusters can be performed.
(ii) After entering number of clusters we can Randomly choose the number of centroids required for performing clusters.
(iii) After finding centroid document we can take cosine similarity of centroid document and also take remaining document of cosine similarity.
(iv) By using both cosine similarities we can Calculate distance of each centroid document to other document by using following formula.

By performing above operation we can group the all related documents into single group. After completion of clustering process we can perform the searching process for query related documents. The implementation procedure Character shuffle preprocessing searching Algorithm is as follows.

3.4 *Character Shuffle Preprocessing Searching Process*

Character shuffle preprocessing algorithm is used to processes the pattern P and also build alphabet Σ related last-occurrence function, L mapping Σ to integers, where L is defined as the largest index i such that P[i] = c or −1 if no such index exists The last-occurrence function can be represented by an array indexed by the numeric codes of the characters. The last-occurrence function can be computed in time O (m + s), where m is the size of P and s is the size of Σ (Fig. 2).

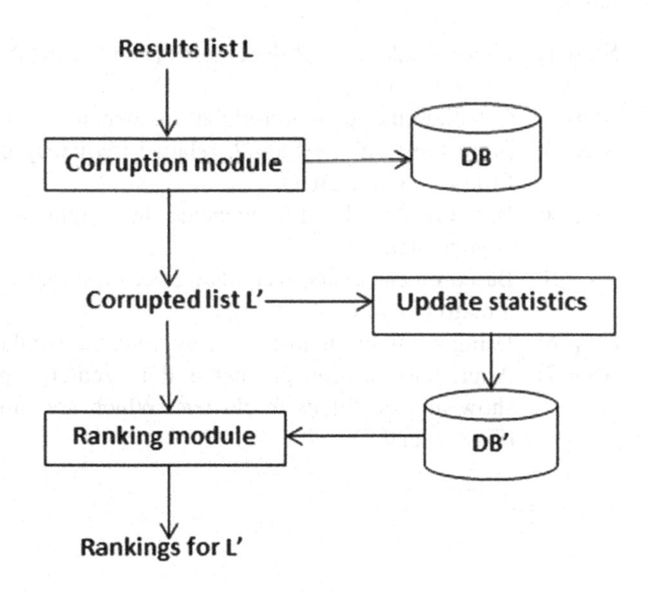

Fig. 2 Character shuffle preprocessing searching process

Algorithm:

CharactershuffleMatch (T, P, Σ)
L ← lastOccurenceFunction (P, Σ)
$k_i \leftarrow m - 1$ $n_j \leftarrow m - 1$
repeat
if $T[k_i] = P[n_j]$
if $n_j = 0$
return
k {match at k }
else $k_i \leftarrow k_i - 1$ $n_j \leftarrow n_j - 1$
else
{ character-jump }
$l \leftarrow L[T[k_i]]$
$k_i \leftarrow k_i + m - \min(n_j, 1 + l)$
$n_j \leftarrow m - 1$
until $k_i > n - 1$
return −1 { no match }

4 Implementation Results

For experimental results, We search a keyword from different sets of documents from database. It shows efficient search query. Those steps are shown following below:

Step 1: First we take sets of documents from the database. In that all documents we reduce tags.

Step 2: Calculate the local and global frequencies from all the documents.

Step 3: From the result of Step 2, relative frequency of each word is generated from every document.

Step 4: Based on calculated frequencies, the weights of each and every document is generated.

Step 5: Based on calculated weights we get similarities. Based on those similarity a matrix is formed.

Step 6: Using similarity matrix, we can group the similar documents into clusters.

Step 7: After clusterization process is done search a particular keyword it will show all documents in clusters. Which are similar to searched keyword (Figs. 3 and 4).

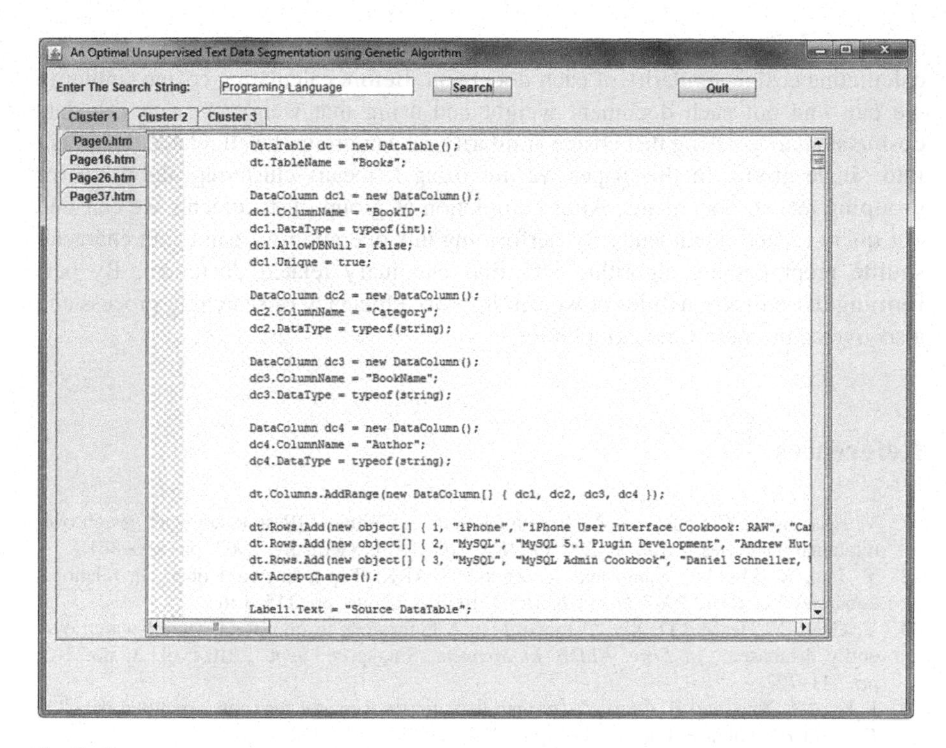

Fig. 3 Output

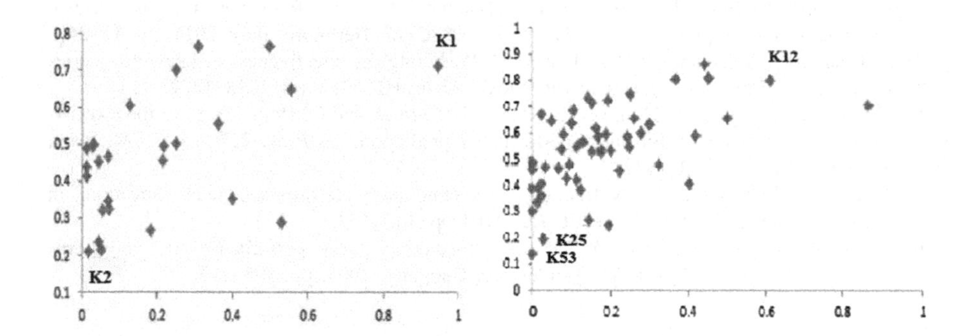

Fig. 4 Custering

5 Conclusion

In the proposed system we are design an effective searching operation over query related documents. By performing that process we can implement character shuffle preprocessing algorithm for finding query related document. Before performing that process we can preprocessing text for reduce tags related content over the documents. After completion text preprocessing we can build MVS Matrix by calculating cosine similarity of each document. Before calculating cosine similarity we can find out each document weight and using that weight we can calculate cosine similarity. Using that cosine similarity we can group the all related document into single group. İn this paper we are using k means clustering algorithm for grouping related documents. After completion of group of documents we can find out query related documents. By performing this operation we can using character shuffle proprocessing algorithm and find out query related document. By performing those functionalities pt we can improve efficiency of searching process and also overcome more time complexity.

References

1. V. Hristidis, L. Gravano, and Y. Papakonstantinou, "Efficient IR style keyword search over relational databases," in *Proc. 29th VLDB Conf.*, Berlin, Germany, 2003, pp. 850–861.
2. Y. Luo, X. Lin, W. Wang, and X. Zhou, "SPARK: Top-k keyword query in relational databases," in *Proc. 2007 ACM SIGMOD*, Beijing, China, pp. 115–126.
3. V. Ganti, Y. He, and D. Xin, "Keyword++: A framework to improve keyword search over entity databases," in *Proc. VLDB Endowment*, Singapore, Sept. 2010, vol. 3, no. 1–2, pp. 711–722.
4. J. Kim, X. Xue, and B. Croft, "A probabilistic retrieval model for semi structured data," in *Proc. ECIR*, Tolouse, France, 2009, pp. 228–239.
5. N. Sarkas, S. Paparizos, and P. Tsaparas, "Structured annotations of web queries," in *Proc. 2010 ACM SIGMOD Int. Conf. Manage. Data*, Indianapolis, IN, USA, pp. 771–782.
6. O. Kurland, A. Shtok, D. Carmel, and S. Hummel, "A Unified framework for post-retrieval query-performance prediction," in *Proc. 3rd Int. ICTIR*, Bertinoro, Italy, 2011, pp. 15–26.
7. S. Cheng, A. Termehchy, and V. Hristidis, "Predicting the effectiveness of keyword queries on databases," in *Proc. 21st ACM Int. CIKM*, Maui, HI, 2012, pp. 1213–1222.
8. O. Kurland, A. Shtok, S. Hummel, F. Raiber, D. Carmel, and O. Rom, "Back to the roots: A probabilistic framework for query performance prediction," in *Proc. 21st Int. CIKM*, Maui, HI, USA, 2012, pp. 823–832.
9. K. Collins-Thompson and P. N. Bennett, "Predicting query performance via classification," in *Proc. 32nd ECIR*, Milton Keynes, U.K., 2010, pp. 140–152.
10. A. Shtok, O. Kurland, and D. Carmel, "Predicting query performance by query-drift estimation," in *Proc. 2nd ICTIR*, Heidelberg, Germany, 2009, pp. 305–312.

An Efficient Data Encryption Through Image via Prime Order Symmetric Key and Bit Shuffle Technique

R.V.S.S. Lalitha and P. Naga Srinivasu

Abstract In this paper, we had proposed a technique that enriches the data privacy and integrity of the data to transfer over a unsecured channel. In this technique we had done encryption and then compression, Data is encrypted to generate the cypher text and then the resultant cypher text is been embedded in image and then compressed for effective utilization of the bandwidth. To encrypt the data, prime order symmetric key algorithm is used and then the encrypted data is embedded within the image using LSB (Least Significant Bit) technique. Then the image is been encrypted at the next stage to reinforce the security at multi-level by using bit shuffle technique. And in the final stage the resultant encrypted image is been compressed using arithmetic coding technique. By using this proposed technique there would be significant improvement in the security of the data.

Keywords Data encryption · Image encryption · Image compression · Cryptography

1 Introduction

Now-a-days, usage of network is increasing exponentially. Innumerable users are transferring data over the network. So, the Security of the information becomes more important. Transferring the plain data is vulnerable and leads to security issues. For reducing the malicious attacks, various cryptographic techniques are used.

R.V.S.S. Lalitha (✉) · P.N. Srinivasu
Computer Science and Technology, Department of CSE,
ANIL Neerukonda Institute of Technology and Sciences (ANITS),
Visakhapatnam, India
e-mail: rlalitha.14.cstmtech@anits.edu.in

P.N. Srinivasu
e-mail: parvathanenins.cse@anits.edu.in

S.C. Satapathy et al. (eds.), *Computer Communication, Networking and Internet Security*, Lecture Notes in Networks and Systems 5,
DOI 10.1007/978-981-10-3226-4_26

Cryptography is most often used to convert the data and image from plain text format to cipher text format using any technique. Cryptography should consider: Confidentiality, Integrity, Non-repudiation and Authentication. Stenography is a process of hiding information by embedding the secret message within other message. Kadam et al. proposed that, data encryption is performed firstly by using Advanced encryption standard algorithm then embedded that data into the ordinal text, digital images, audio or video via stenographic algorithm [1]. Compression is the process of reducing the size of data to save space and time required to transfer the data. Zhang et al. states that, the compression performed on the encrypted data for effective utilization of bandwidth [2].

In this scenario, Alice wants to send information to Bob in efficient and secure way. For that, we are using concepts of data hiding, image encryption and decryption and image compression and decompression, by these concepts, more security and efficiency of system is provided. Alice first encrypt the data by using encryption algorithm with some secret key then hide encrypted data into the image through stenographic algorithm then data embedded image is encrypted with same key which is used for data encryption then the encrypted image is compressed by using the compression algorithm then send it to the receiver Bob. After receiving, Bob perform the decompression followed by image decryption and data extraction and finally data decryption then the message is obtained.

The rest of this paper is organized as follows. Section 2 represents the details of the existing work. Section 3 represents the proposed work and techniques used in that work. Experimental results are provided in Sect. 4. We conclude in Sect. 5.

2 One-Time Padding, AES, RSA

In conventional approach data was transfer as a plain text over the network, which is vulnerable to the integrity of the data. To create a chaos for the attacker, a plenty of encryption techniques were available. One-time password is one such cryptographic technique. Liu et al. [3] proves that one-time pad encryption algorithm as unbreakable by considering two assumptions: the random key is used and can't be used more than once. But one-time padding suffer with key size which is almost equal to the key size and the same key must be maintained by both the parties, where transferring the key itself is a prolonged task, which is impractical at times. Zin zhou [4], proposed RSA which relay on *public key cryptosystem* that shares the key with every one which is a public key and for decryption process it uses a secret key known as private key. RSA is mostly used in mobile nodes to overcome the vulnerable attacks. Due to high time complexity, it is not suitable for wireless sensor networks apart from that, évery individual must maintain a private key which may not be feasible in all the instances. Kadam et al. [5], proposed a AES (*Advanced Encryption Standard*) which is assumed to be the best symmetry key cryptographic technique which is used in encryption of text, audio and image files using the same key [128/192/256 bits] at both the ends, which perform encryption on multiple [10/12/14] rounds depends on the key size.

In the traditional approach, to transfer the image through the internet has suffered vulnerable attacks. To protect from such attacks, the image which we want to send is encrypted using the Prediction error clustering and random permutation. The size of the image is more when compare to data, so the image has to be compressed. Then the compressed image is being transferred which could be easily transferred and storage space is also reduced by which the efficiency of the system is enhanced. The compression can performed by arithmetic coding technique after the encryption process i.e., compression is performed on encrypted domain [6, 7].

3 Proposed System

In this proposed system, Sender is able to send the sensitive data to receiver in efficient and secure approach by encryption the data and then compressing. For that, we are employing practices like data hiding, image encryption/decryption and image compression/decompression. At the earlier stage Sender encrypt the data by using encryption algorithm with some secret key then hide encrypted data into the image through stenographic algorithm then data embedded image is encrypted with the same key which is used for data encryption then the encrypted image is compressed by using the compression algorithm then the resultant image will be sent to the receiver. After receiving, receiver performs the decompression followed by image decryption and data extraction and finally data decryption then the original message is obtained (Fig. 1).

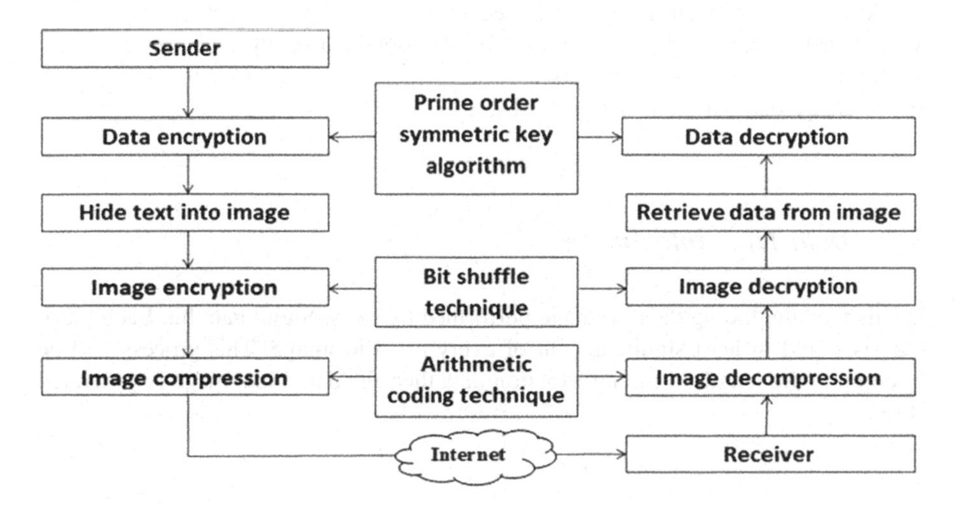

Fig. 1 System design

3.1 Prime Order Symmetric Key Algorithm

Encryption
In this module, the information will encrypt via prime order symmetric key algorithm by the sender.

- Add randomized characters for every 3 characters of plain data.
- Get ASCII codes of those characters.
- Convert the obtained ASCII codes into Binary formatted data.
- Perform complement for obtained binary values.
- Consider any prime numbers series and transform into Binary values.
- Perform first level XOR operation for the characters of plain data and of prime numbers series.
- Select last value in the prime number series and consider it as 2^8 bit key.
- Perform Second level XOR operation for result of first level XOR operation and key.
- Convert the obtained result into decimal, then the cipher text is obtained.

Decryption
In this module, the information will decrypt via prime order symmetric key algorithm by receiver.

- Get the last value as a key from prime number series table which is used in encryption process and convert it into binary format.
- Perform first level XOR operation for cipher data and key.
- Consider any prime numbers series and transform into Binary values. (Same process as done on encryption process).
- Perform second level XOR operation for obtained result and considered prime numbers series.
- Perform complement for the obtained result.
- Convert to decimal format from previously obtained result.
- Remove randomized characters.
- Now, we can get plain data as a result.

3.2 Data Hide into Image

In this module, hiding data in image performed by Least Significant bit. Each bit of data is stored in least significant bit of every pixel in image. This process will be done up to the length of the information/data then the data hide image is generated (Fig. 2).

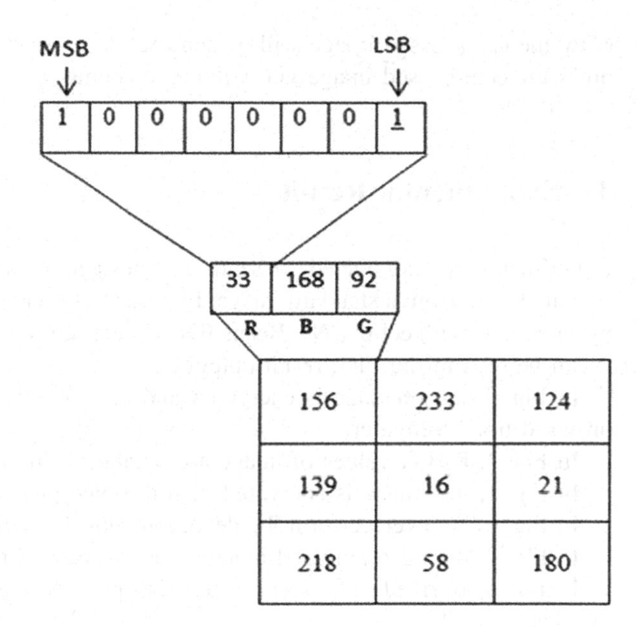

Fig. 2 Least significant bit

3.3 *Bit Shuffle Encryption Technique*

Encryption

- Input will be data hidden image.
- Generate the R, B, G values of each pixel of image.
- Now the binary values will be obtained from R, B, G values of each pixel.
- The obtained each bit of binary results will be shuffle bits from left to right.
- Now apply addition operation for the shuffled data and key.
- From the result of the previous step, encrypted image is generated.Decryption

- Take R, B, G values of each pixel from encrypted image.
- Now apply subtraction operation for binary values of R, B, G and key.
- The obtained results from previous step will be shuffle bits from right to left.
- The shuffled data is converted into decimal values.
- The data hide image is generated.

3.4 *Compression and Decompression Image*

In this module, the compression can be performed after image encryption. The Encrypted mage is compression is performed by the Arithmetic coding technique.

Now the compressed image will be send to the receiver then the receiver decompress the compressed image via Arithmetic coding technique.

4 Experimental Result

Experiments are conducted to send the message in a secure way. The results obtained were from a standard image. Experiments were done in java programming by Java SDK with eclipse/Net Beans IDE system setup is required for programming it can be seen in the FIG: result output.

In Fig. 3, the message is encrypted and convert in to binary format. Image is browsed from computer.

In Fig. 4, R,B,G values of image are obtained, hide message into image.

In Fig. 5, the image is encrypted then the encrypted image is compressed.

In Fig. 6, reciever perform the decompression for received message.

In Fig. 7, the decompressed image is decrypted and retrive data from image.

In Fig. 8, retrived data is converted into plain format, the output is displayed (Tables 1 and 2).

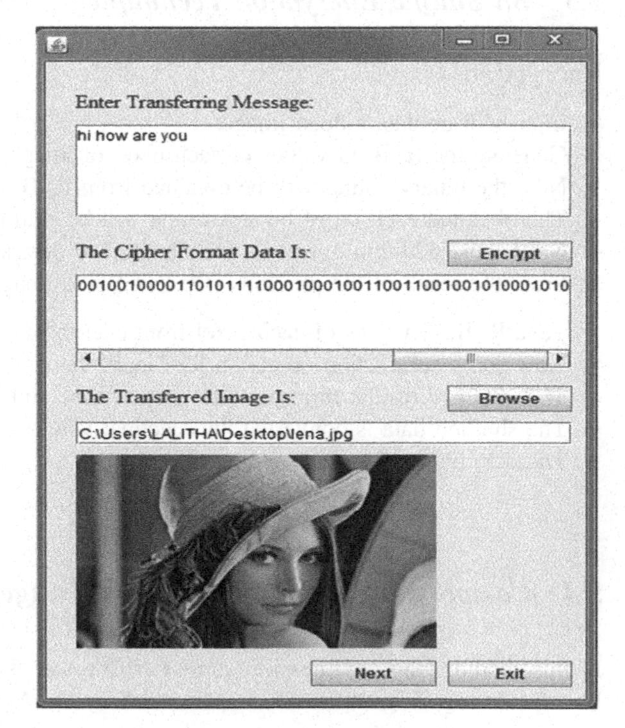

Fig. 3 Data encryption

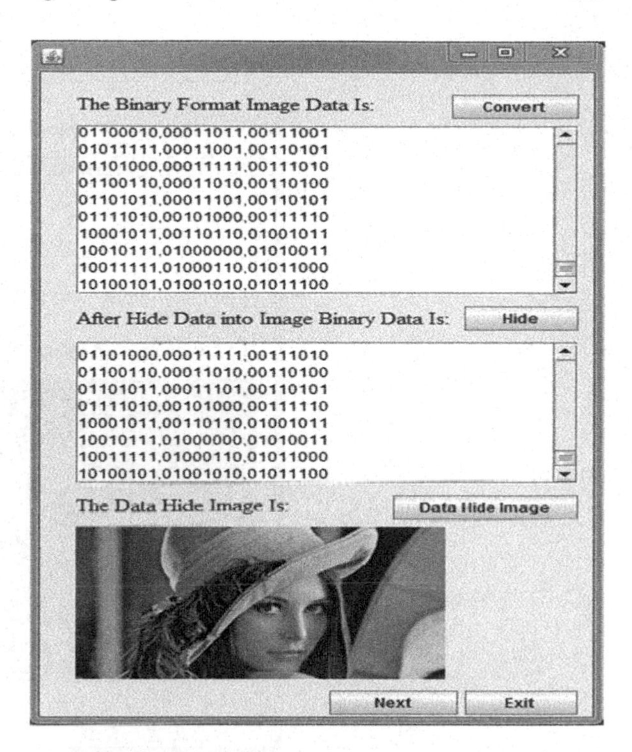

Fig. 4 Data hide into image

Fig. 5 Image encryption and compression

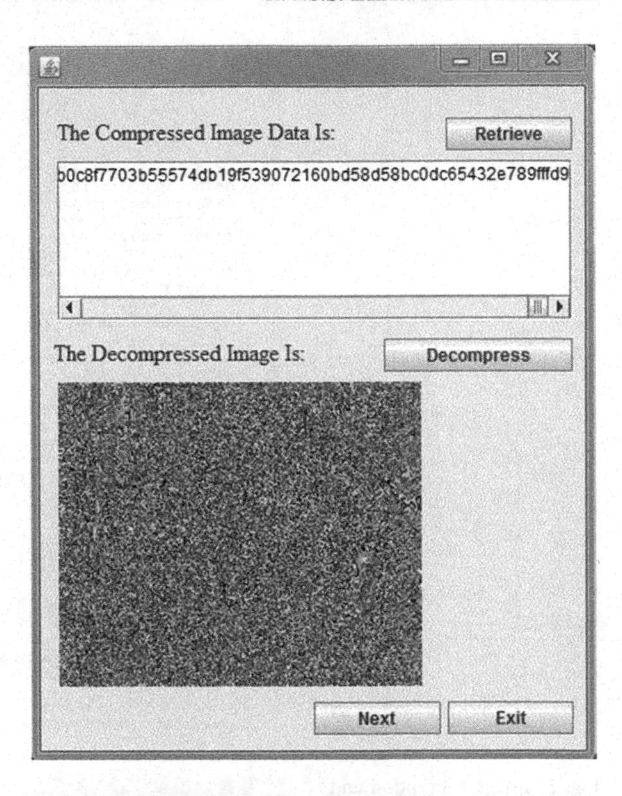

Fig. 6 Decompressed image

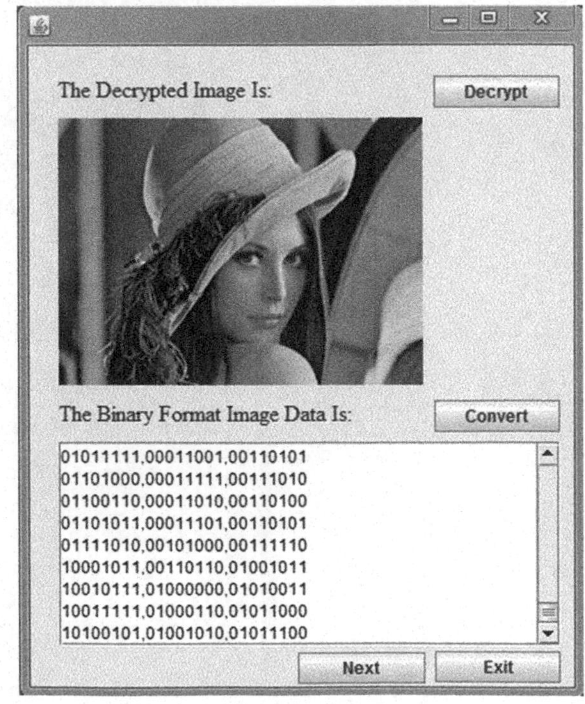

Fig. 7 Decrypted image

Fig. 8 Decrypted message

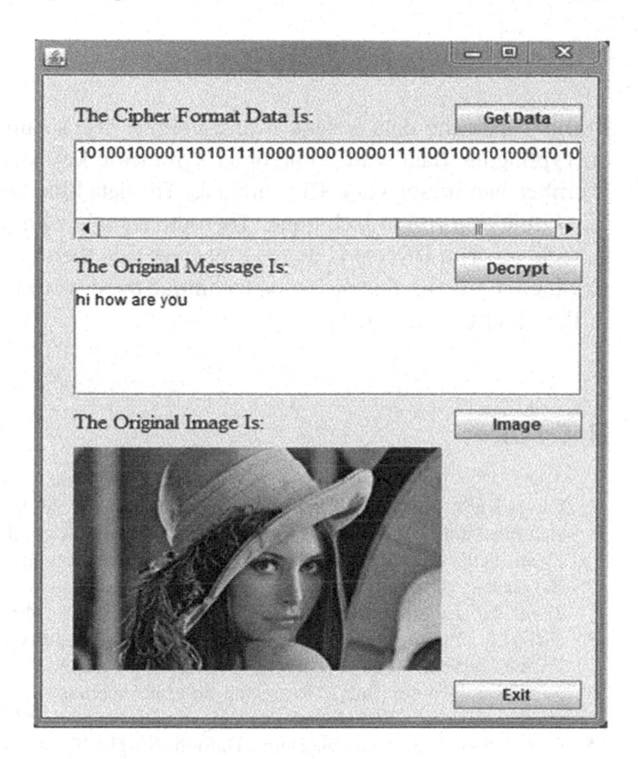

Table 1 Time complexity analysis, the time complexity of our proposed methods is compared by AES which is having complexity of $O(n^2)$, one-time padding and the entropy of each method is calculated for the 128*128 image Lena

Image (Lena)	Entropy	Encryption (s)	Decryption (s)
One-time padding	7.7620	2	10
AES	7.8683	10	70
Prime order symmetric key	7.5672	4	10
Bit shuffle technique	7.1579	2	6

Table 2 Compression performance, The compression efficiency of our proposed method applied to the encrypted images is compared with the lossless rates given by Zhou et al. [6]

Image	Zhou et al.	Proposed
Lena	134267 B (4.096 bpp)	134283 B (4.096 bpp)
Barbara	150369 B (4.589 bpp)	150434 B (4.589 bpp)
Boat	134732 B (4.112 bpp)	134861 B (4.112 bpp)
Harbor	160567 B (4.900 bpp)	160614 B (4.900 bpp)
Airplane	121377 B (3.704 bpp)	121493 B (3.704 bpp)

5 Conclusion

In this paper, the data is send over a internet in a secure and efficient way, we are encrypting the data with prime order symmetric key algorithm then encrypted data is embed into image via LSB technique. The data hided image is encrypted by using bit shuffle encryption technique. Then the resultant image is compression and sent to receiver. The reverse is done at receiver side then the original data and image is obtained. By using this proposed technique there would be significant improvement in the security of the data.

References

1. Parag Kadam, Akash Kandhare, Mangesh Nawale, Mukesh Patil "Separable reversible encrypted data hiding in encrypted image using AES Algorithm and Lossy technique," *IEEE Trans.* Pattern Recognition, Informatics and Mobile Engineering (PRIME), Feb 2013.
2. X. Zhang, "Lossy compression and iterative reconstruction for encrypted image," *IEEE Trans. Inf. Forensics Security*, vol. 6, no. 1, pp. 53–58, Marh 2011.
3. Huiyi Liu, Yuegong Zhang, "An improved one-time password authentication scheme," *IEEE Trans. Communication Technology (ICCT).*, Nov 2013.
4. Xin Zhou, Xiaofei Tang, "Reasearch and implementation of RSA algorithm for encryption and decryption," *IEEE Trans. Strategic Technology (IFOST),* vol. 2, Aug. 2011.
5. O P Verma, Ritu Agarwal, Dhiraj Dafouti, Shobha Tyagi, "Performance of data encryption algorithms," *IEEE Trans. Elwctronics Computer Technology (ICECT),* vol. 5, Aug. 2011.
6. Jiantao Zhou, Xianming Liu, Oscar C. Au and Yuan Yan Tang, "Designing an Efficient Image Encryption-Then-Compression System via Prediction Error Clustering and Random Permutation," *IEEE Trans. Signal Process.*, vol. 9, no. 1, Nov 2014.
7. J. Zhou, X. Liu, and O. C. Au, "On the design of an efficient encryption-then-compression system," in *Proc. ICASSP*, 2013, pp. 2872–2876
8. Xinpeng Zhang, Yanli Ren, Liquan Shen, Zhenxing Qian, "Compressing Encrypted Images with Auxiliary Information," *IEEE Trans. Multimedia,* vol. 16, issue: 15, April 2014
9. A Nadeem, M Y Javed, "A Performance comparision of Encryption Algorithms," *IEEE Trans. Information and communication Technology,* vol. 2, Aug. 2011.
10. Xu Shu-Jiang, Wang Ying-Long, Wang Ji-Zhi, Tian Min "A novel image encryption scheme based on chaotic maps," *IEEE Trans. Signal processing,* vol. 2, Dec 2008.
11. Ling Wang, Quen Ye, Yaoqiang Xiao, Yongxing Zou, Bo Zhang, "*IEEE Trans. Image an signal processing,*" vol. 3, May 2008.

Peak Detection and Correlation Analysis in Noisy Time Series Data

L. Trinadh, R. Venkat Shesu, M. Kranthi Kiran and Satya V. Jampana

Abstract This paper focus on, the study of correlation (dependency) between the extreme trends (peaks) in multi-variant noise time series data, In some sense, the extreme events disrupt the underlying structure distribution in the data. The peaks are identified using a data-driven algorithm. It is observed that all majority of peak locations are identified using this method. We also evaluate its robustness by giving the different size of data records.

Keywords Peaks · Correlation · Buoy · Multi-variant time series

1 Introduction

Peaks in some measure indicate the event that can affect or lead to the significant change in the data distribution. The detection of peaks in time series data [1] has been a long-standing problem in many areas of research. In order to identify the extreme trends in time series data, we study two approaches. The first approach is geometric approach and the second approach is of statistical definitions [2]. In this paper, we use the statistical approach in order to find extreme trends.

L. Trinadh (✉) · M.K. Kiran
Computer Science and Technology, Department of CSE,
ANIL Neerukonda Institute of Technology and Sciences (ANITS),
Visakhapatnam, India
e-mail: ltrinadh.14.cstmtech@anits.edu.in

M.K. Kiran
e-mail: mkranthikiran.cse@anits.edu.in

R.V. Shesu · S.V. Jampana
Data and Information Management Group,
Indian National Centre for Ocean Information Services (INCOIS), Hyderabad, India
e-mail: venkat@incois.gov.in

S.V. Jampana
e-mail: raju.jvs@incois.gov.in

S.C. Satapathy et al. (eds.), *Computer Communication, Networking and Internet Security*, Lecture Notes in Networks and Systems 5,
DOI 10.1007/978-981-10-3226-4_27

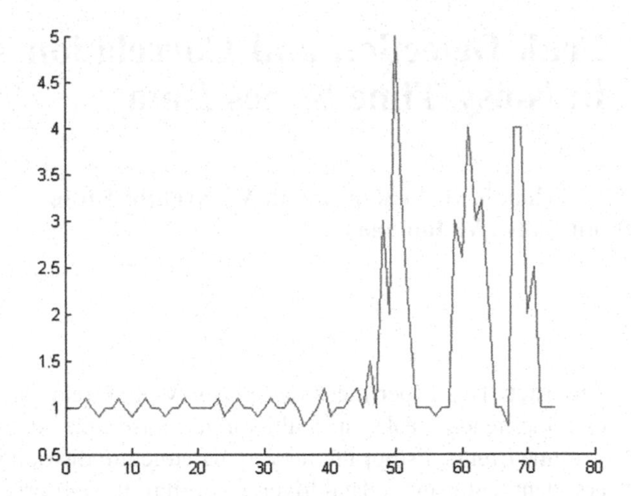

Fig. 1 An example of peak [4]

Time series data arise in a variety of domains, such as environmental, medical, and financial etc [2]. For example, in the field of oceanography, specially designed instruments (Argo, Buoy) are used to measure sea state parameters like sea surface temperature, salinity, and oxygen levels at various depths in the ocean. These data have a lot of fluctuations. These significant fluctuations in this context are called peaks and more generally events.

These peaks are used in time series analysis techniques like dependence (correlation), missing values prediction [3] etc. In this work, we focus on a dependency between peaks of two variables of the data set (Fig. 1).

Analyzers' easy to visualize and identified peaks in the dataset having less number of data points by observation but the problem is they did not analyze the exact peak locations efficiently, and consuming lot of time while an analyze peaks if the data is having huge no of data points.

In order to overcome this problem, we use an existing method to detect the peaks in time series automatically and accurately.

After finding peaks, we analyze the dependence (correlation) between detected peaks of both time series data.

1.1 Flow Chart of Proposed Work

In Fig. 2 The flow chart provides the information about how the algorithm works to find out the peaks in input data generated from the buoy.

In this process first, we took the raw data from the buoy. Raw data is highly susceptible to null values, and missing values, so in order to remove null values and missing values, we perform smoothening. After smoothening, we pass validated data as input to the algorithm. The algorithm processes the input data and generates

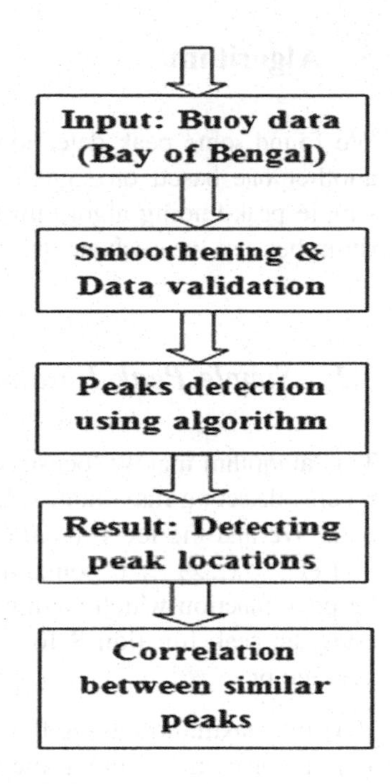

Fig. 2 Flow chart of proposed work

possible peaks. After that, apply correlation between detected peak locations in time series data.

2 Smoothening and Data Validation

We considered the Nov 2013–Dec 2014 buoy data [5] from Bay of Bengal mooring at 18°N and 90°E as an input. This dataset consists of several fields like observation date, the temperature at various depth levels, and salinity at various depth levels. We had chosen the water-temperature; salinity values; at 4 m depth and also at 7 m depth.

After that, we perform smoothening on the noisy data i.e. removing null values and fill up missing values if it is there in data using simple moving average [6].

The yearlong data is partitioned based on patterns driven by climatic variability. In this region, the variability is dictated by monsoons. Accordingly, there are four distinct seasonal patterns; namely Dec–Jan–Feb, March–April–May, June–July–Aug, Sep–Oct–Nov. The analysis is carried out based on this partitioning of the data. This helps to isolate events driven by seasonality. So, the bias is based on seasons is removed. Further, the data is analyzed at smaller months by scales. The objective in both the case is the detection of peaks and analyzes the correlation between these two ecological variables.

3 Algorithm

We found some peak detection algorithms in literature survey, like APD [7] and another one based on signed distance in relation to the neighboring points i.e. simple peak-finding algorithm [3]. In this work we use simple peak-finding algorithm because we evaluate its robustness by giving the different size of data records.

3.1 Simple Peak-Finding Algorithm

This algorithm mainly focuses on to find out the local maximum. Local maximum means, detecting maximum value when we compare the two neighbor values in the data. We find the local maximum through peak function S.

Let $T = x_1, x_2 \ldots x_n$ containing N values, and $\mathbf{x_i}$ is the ith data point in T. Let S_1 be peak function which having the parameters like T, $\mathbf{x_i}$, k. where $k > 0$. First, we compute peak function S for a given point $\mathbf{x_i}$ in T. the function S computes the average of

(i) the maximum among the signed distances of x_i from its *k* left neighbors
(ii) the maximum among the signed distances of x_i from its *k* right neighbors

In mathematical representation, the peak function S_1 is

$$S_1(k, i, T) = \frac{max\{x_i - x_{i-1}, x_i - x_{i-2}, \ldots, x_i - x_{i-l}\} + max\{x_i - x_{i+1}, x_i - x_{i+2}, \ldots, x_i - x_{i+l}\}}{2} \quad (1)$$

Pseudo code:
Input: $T = x_1, x_2, \ldots, x_n$. T is input time-series of *N* points, k (window size), H, typically $1 \leq h \leq 3$.
Output: Output (O) consists of detected peaks in T.

1. Compute the peak function for each and every point in time series.

$$\text{i.e. } \mathbf{for}\ (i = 1;\ i < n;\ i++)\ \text{do} \quad a[i] = S1(k, i, T); \quad (2)$$

2. Compute the mean (m) and standard deviation (s) of all positive values in array a.
3. Remove local peaks which are small in the array.

$$\begin{aligned} &\text{i.e. } \mathbf{for}\,(\mathrm{i}=1;\ \mathrm{i}<\mathrm{n};\ \ \mathrm{i}++)\mathbf{do} \\ &\mathbf{if}(\mathrm{a[i]}>0\ \&\&\ (\mathrm{a[\,i\,]}-\mathrm{m})>(\mathrm{h}*\mathrm{s})) \\ &\qquad \mathbf{then} \\ &\qquad \mathrm{O}=\mathrm{O}\ \cup\{\mathrm{x}_i\} \end{aligned} \tag{3}$$

4. Order the peaks in output O in terms of increasing order, and for every adjacent pair of peaks x_i, x_j in O remove the minimum value.

$$\text{i.e. if } |j-i| \leq k \,\text{then remove the smaller value of}\, \{x_i,\ x_j\} \tag{4}$$

4 Correlation Analysis

The peak detection algorithm [3] is use full in many ways and has novel applications. Here we briefly describe its usefulness in evaluating dependencies among important oceanographic variables. More specifically we evaluate the correlation (dependencies) between salinity and temperature with reference to the peaks detected by the implemented algorithm.

Correlation is expressed in terms of the correlation coefficient. The magnitude of coefficient lies in −1 to +1. If the coefficient is positive, then both time series have a strong dependence. And if coefficient is negative, then both time series have strong inverse relationship, and if the value of the coefficient is zero then there is no correlation at all.

Here, we find the correlation [8] (dependency), in terms of total similar peak locations, are identified in temperature and salinity out of total detected peak locations in both the variables at different depth levels and in various time periods.

$$\text{Dependency} = \lfloor \text{no of similar peaks in detected peaks/total detected peaks(round of)} \rfloor * 100. \tag{5}$$

In the next section, we show how the correlation varies across seasons as well as across different months.

5 Experimental Results

5.1 Detection of Peaks

We run experiments to evaluate the results of this algorithm with different size of data records, different parameters (temperature, salinity) against different levels of depths (4, 7 m) in the process of generating the results.

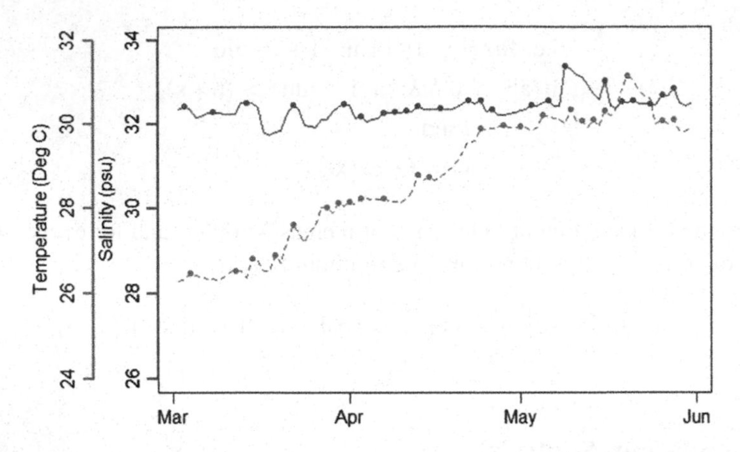

Fig. 3 Peaks detected in sample data

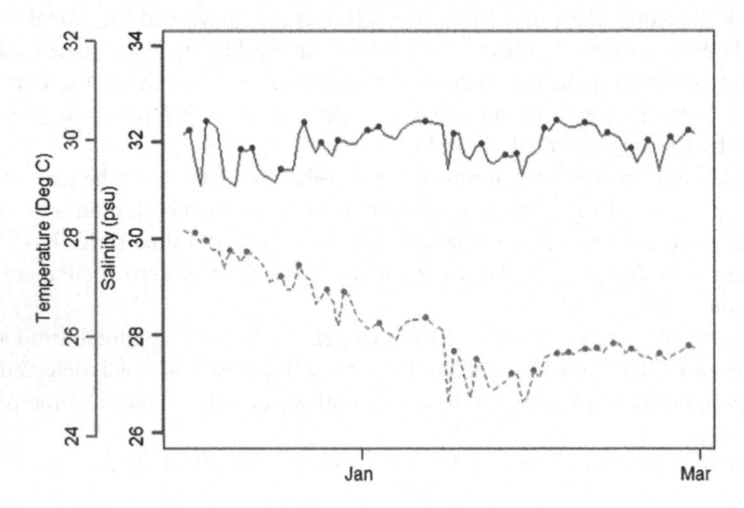

Fig. 4 Dec–Jan–Feb plot at 4 m

In Fig. 3 we observe the dots that indicate peaks. From Fig. 3 we conclude that the algorithm detects all possible variations in the data.

5.2 *Results of Correlation Analysis*

5.2.1 Long-Term Variation

Now we carry out the analysis of the similarity in detected peak locations of temperature and salinity in every season at 4 m depth and 7 m depth.

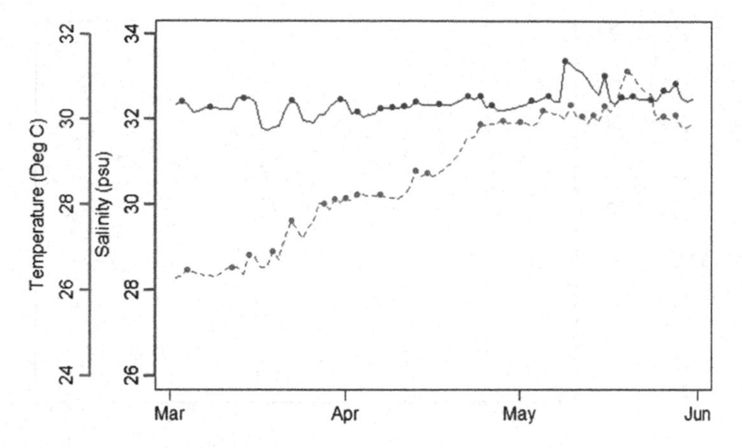

Fig. 5 March–April–May plot at 4 m

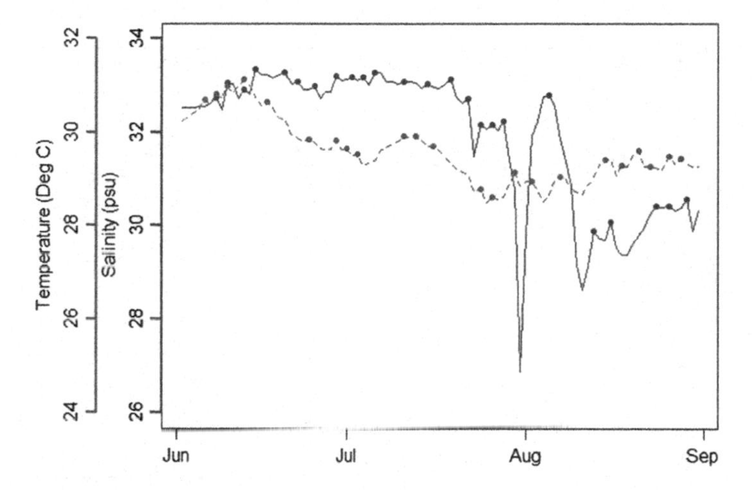

Fig. 6 June–July–August plot at 4 m

Plots at 4 m depth:
The above results represent the Plots at 4 m level depth. In Figs. 4, 5, 6 and 7 observe the dots that similarly occurred in temperature and salinity and find out the correlation using dependency.
Plots at 7 m depth:
The above results represent the Plots at 7 m level depth. In Figs. 8 and 9, analyze the dots that similarly occurred in temperature and salinity and find out correlation using dependency.

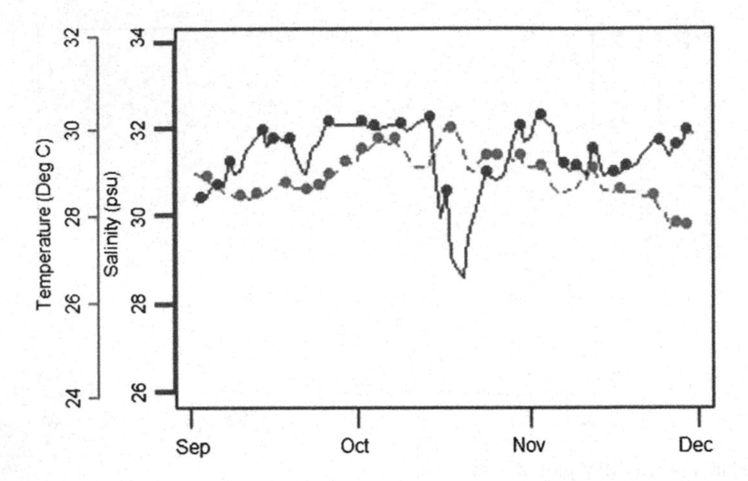

Fig. 7 Sep–Oct–Nov plot at 4 m

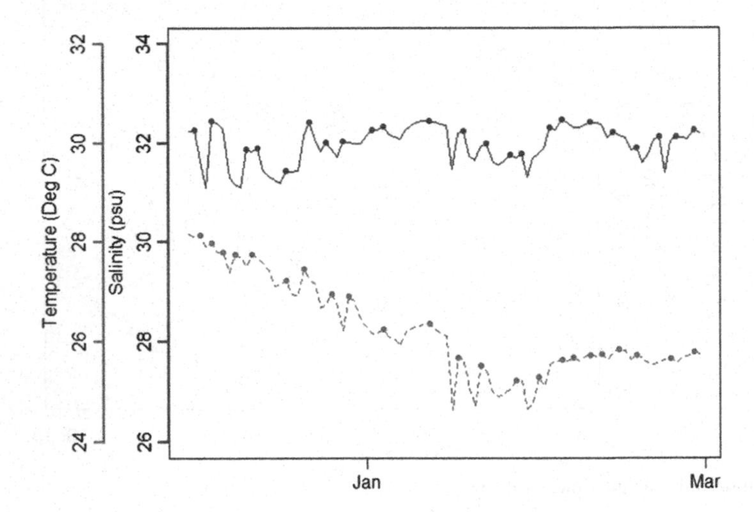

Fig. 8 Dec–Jan–Feb plot at 7 m

The above results represent the Plots at 7 m level depth. In Figs. 10 and 11, analyze the dots that similarly occurred in temperature and salinity and find out correlation using dependency.

Theoretical:

We provide the details of dependency between peaks of temperature and salinity in Table 1. So, we concluded that the dependency between temperature and salinity is strong in Dec–Jan–Feb season at both 4 and 7 m depth. In a similar way, we also

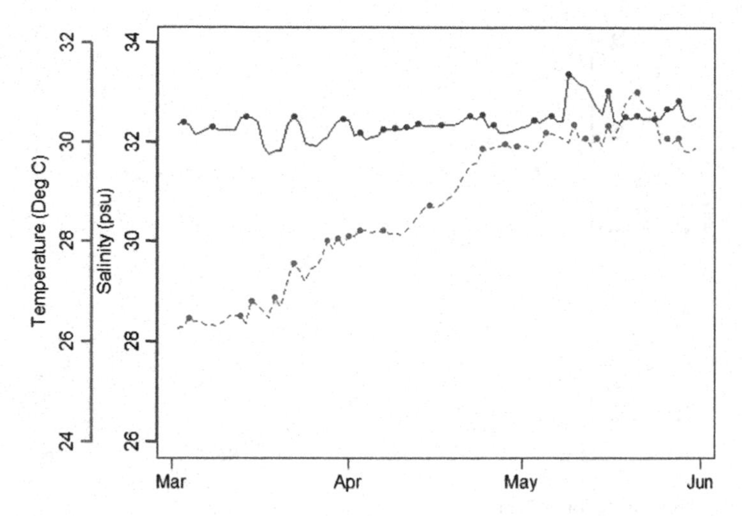

Fig. 9 March–April–May plot at 7 m

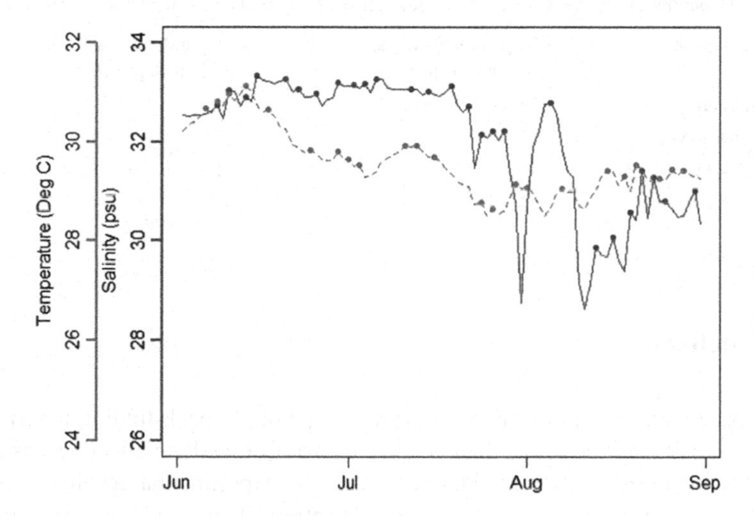

Fig. 10 June–July–August plot at 7 m

compute the dependency for short term variation i.e. months. In that January month shows the strong dependence (correlation) between peaks of temperature and salinity at both 4 and 7 m depth.

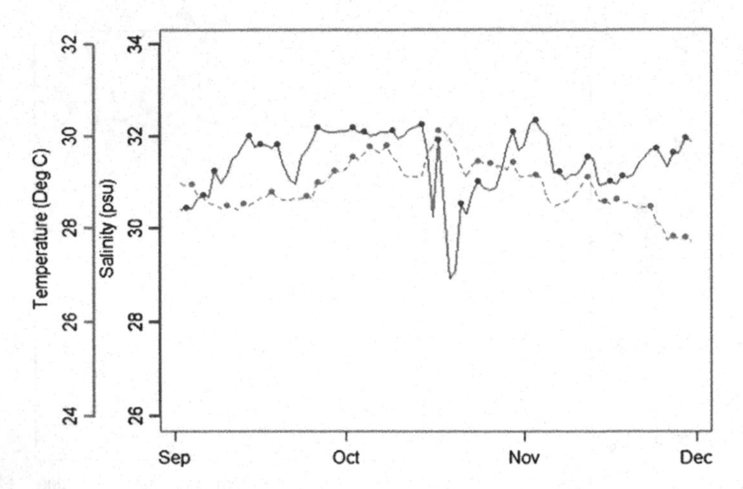

Fig. 11 Sep–Oct–Nov plot at 7 m

Table 1 Dependency between peaks of temperature and salinity at 4 m depth and 7 m depth

Name of season	Dependency of peaks at 4 m depth (%)	Dependency of peaks at 7 m depth (%)
Dec–Jan–Feb	73.91	70
March–April–May	52.17	52
June–July–August	45.17	50
Sep–Oct–Nov	54.16	48

6 Conclusion

In this paper, we use an existing algorithm i.e., Simple peak-finding algorithm to detect the peaks in time series data. We have also observed the effectiveness of this algorithm by giving different data records. The experimental results verify the promising performance of peak finding algorithm. This work is a part of data mining and we improve the functionality of the detecting sudden changes/trends in oceanic data using the algorithm. We have concluded that correlation in between Dec–Jan–Feb season is high compared to other seasons.

Acknowledgements This work was completed in INCOIS Hyderabad. Authors wish to thank Director INCOIS, Hyderabad for the encouragement and facilities provided. Authors also acknowledge the support and guidance of other INCOIS scientists throughout working on this project and preparing this manuscript. We would also like to express our gratitude to Prof. S.C. Satapathy (Head of Department), ANITS, Visakhapatnam for his continuous support and encouragement.

References

1. L.M Bhar and V.K Sharma, "Time Series Analysis", Indian agriculture statistics research institute, New Delhi.
2. Roger Schneider, "Survey of Peaks/Valleys identification in Time Series", August 23, 2011.
3. Girish Palshikar, "Simple Algorithms for Peak Detection in Time-Series", article, Jan 2009.
4. Sayanti Chattopadhyay, Susmita Das, "Design and Simulation Approach Introduced to ECG Peak Detection with study on different cardiovascular Diseases", IJSRP, Vol. 2, December 2012.
5. www.incois.gov.in/portal/datainfo/mb.jsp/.
6. www.inside-r.org/packages/cran/TTR/docs/GD.
7. Felix Scholkmann, Jens Boss and Martin Wolf, "An Efficient Algorithm for Automatic Peak Detection in Noisy Periodic and Quasi-Periodic Signals", article, 2012.
8. Julian D. Olden, Bryan D. Neff, "Cross correlation bias in lag analysis of aquatic time series", Springer Marine Biology pp. 1063–1070, 2001.

Novel Hash Based Key Generation for Stream Cipher in Cloud

K. DeviPriya and L. Sumalatha

Abstract Cloud Computing is an advanced technology which provides services to the users on rental basis. Cloud minimizes the installation cost of hardware, software, applications setup at client side and these services are available at cloud server, accessed by any one, any time, any place through the internet. Apart from the benefits one big challenge faced by the cloud is security problem as the data and resources are not under the control of data owner. Security techniques are required to protect data from the unauthorized access. In this paper, we proposed simple efficient stream cipher to protect information which is stored in the cloud. Also a hash based key is generated for encryption. The size of key is 128 bit and the first 32 bit of the key is used for encrypting each character. This provides strong security to the each character. We implemented our algorithm in azure cloud and the results are evaluated.

Keywords Cloud · Stream cipher · Symmetric encryption · Decryption · MD5 hashing · Key · Azure cloud · Virtual machine

1 Introduction

Cloud storage refers to the storing of data in a remote place which the user is not aware of the stored location. This means instead of storing data into user's computers or other storage devices, client save it to a remote database or remote virtual machine where internet provides the communication between user desktop to remote database or virtual machine. Cloud protects the user's data by configuring

K. DeviPriya (✉)
Department of Computer Science & Engineering, Aditya Engineering College, Surampalem, AP, India
e-mail: k.devipriya20@gmail.com

L. Sumalatha
Department of Computer Science & Engineering, UCEK, JNTUK, Kakinada, AP, India
e-mail: lsumalatha@jntucek.ac.in

S.C. Satapathy et al. (eds.), *Computer Communication, Networking and Internet Security*, Lecture Notes in Networks and Systems 5,
DOI 10.1007/978-981-10-3226-4_28

security rules in the firewalls, establishing virtual private networks and by using encryption algorithms. The use of encryption algorithm enables to protect data from the outside attacker and inside attackers as well [1]. Cloud providers, cloud employees themselves may not be trusted entities. Security is therefore a major concern in utilizing the services of the cloud. Today cloud services are provided by several companies, some of the popular company names are: Google (Google App Engine), Microsoft (Azure), Amazon (Amazon Web Service) etc.

Our proposed stream cipher is executed on one of the public cloud called windows azure. Windows Azure is Microsoft's application platform for the public cloud. Cloud users can use this platform in many different ways. Based on the type of the subscription the user will able to create their own virtual machines, web applications, storage applications etc., easily in the azure cloud. Azure cloud provides a flexible user interface to access the cloud services. Some of the available azure cloud services are web service, storage service, traffic manager, HDInsight etc. [2]. Azure provides several security mechanisms for providing confidentiality, integrity and availability for customer data and to build applications in a secure way. We proposed a stream cipher encryption algorithm as a security service in azure.

A stream cipher is a symmetric cipher where the plain text stream is combined with key stream to generate cipher stream. Stream ciphers are divided into two types. (1) Classical stream ciphers and (2) Modern stream ciphers. In classical stream cipher, each bit is encrypted at a time and in modern stream cipher, encryption and decryption are done on 'r' bits at a time. Modern stream ciphers are mainly categorized as synchronous and asynchronous. Synchronous stream cipher uses key stream which is independent of plain text or cipher text stream, in asynchronous stream cipher, key stream is based on either plaintext or cipher text. Examples of stream cipher are RC4, A5/1 etc. RC4 is a byte oriented stream cipher in which the plain text byte is exclusive-ored with a key stream cipher. A5 is another type of stream cipher which is used in GSM cellular communication to provide confidential voice communication.

Contribution. The proposed encryption algorithm is the design of a Novel Hash based Key Generation for Stream Cipher (NHKS). The proposed stream cipher is based on the concept of modern stream cipher where each character is encrypted at a time and the size of the each character is 32 bits. In proposed, the key is generated recursively from the user's secret key and salt value. The new key size is 128 bit which is obtained by applying MD5 hash function. MD5 Algorithm is chained block digest algorithm developed by Ron Rivest which generates 128 bit hash value. Initially the message is padded with additional bits for congruent to 448 modulo512 [3–5]. A 64-bit representation of block is appended to the message. Initialize MD5 buffer registers with initial seed values. The first block is processed with an initial seed value and generates digest value. The digest value of this block is considered as seed for the next block. Similarly all the blocks are processed and final block digest is the final hash value for entire data stream. The message is processed as 16 word blocks and generates 128 bit hash value. Among the 128 bits the first 32 bits are used for encrypting the character in the proposed method. The strength of the algorithm is based on the generated hash based key. Here different cipher texts are generated for the same type of key and plain text due to concatenation of key with salt. This feature makes the cipher more complex and difficult to analyze by the Cryptanalyst.

This paper is organized as follows: Sect. 2 discusses the security concerns in Cloud computing, Sect. 3 related work in this area, Sect. 4 details about the Proposed Method, Experimental Results are shown in Sects. 5, and 6 concludes the proposed method.

2 Cloud Computing Security Concerns

Security is an important requirement to the cloud computing. Whenever the cloud providers designing and establishing new cloud, they need to follow the proper security measurements and the users who are availing the services of cloud must ensure that the security policies and mechanisms provided by the providers are according to standards or not. The following security issues [6] mostly identified by the Gartner from the perspective cloud designing and usage.

2.1 Security Issues in Cloud

Privileged access : Privileged access indicates who will access the data and who decides about the hiring and management of such administrators.

Regulatory compliance : Is the cloud provider performs willing to go audits and certifications related to security?

Data location : Whether the cloud provider allow for handle over the location of data?

Data segregation : Whether the encryption mechanisms are available at all stages. Were these mechanisms tested by experts?

Recovery : If the cloud servers and backup is disaster. What are the recoveries measurements handle by the cloud provider?

Investigative support : Whether the providers have the capability to investigate any wrong or illegal activity?

Long-term viability : What happen to data if the cloud providers leave the business, is the users data safely returned and which format?

Data availability : Whenever the cloud provider moves the cloud data from the existing environment to new environment. Whether the existing environment is compromised or unavailable?

By identifying the above specified considerations executives can gain better understanding of properties of cloud computing and to develop better solutions to their cloud strategy.

2.2 Information Security Requirements

According to International Standard Organization (ISO) 7498-2, information security covers different concepts. The Cloud computing security should also suggest in this way to become a secure and effective technology. The following are security mechanisms highlighted in the context of cloud computing:

Identification & authentication : In cloud computing, identifying and authenticating the cloud user is very important requirement. The cloud user proves their identity to the cloud provider through the username and password initially. Some providers also implemented multi factor authentications for strong security. Current research area from the perspective of security in cloud is authentication.

Authorisation : Authorisation decides what types of tasks and actions performed by the users. Based on the type of subscription the administrator assigns the authorisation permissions to the users for accessing cloud resources.

Confidentiality In Cloud computing, confidentiality plays a major role to protect data from the unauthorised attacks. En-cipherment can provide confidentiality of cloud data. The proper en-cipherment mechanisms' provides security to the data which is stored in the cloud.

Integrity: : The integrity requirement ensures whether the data which is stored in the cloud is modified. Traditional integrity mechanisms are not suitable for the cloud domain due to more computation. Special types of integrity mechanisms are required in cloud computing to preserve integrity.

Non-repudiation : Non-repudiation in cloud means the applications and services are provided by cloud will not changed by other cloud providers or users. Cloud computing uses the concepts like digital signatures, time stamps etc. for supporting non-repudiation.

Availability : Availability is one of the important requirements in the cloud. The data which is stored in the cloud must be available at all the time. Attacks on availability service are denial of service, distributed denial of service attack etc. The cloud provider needs to design proper security algorithms for preventing these attacks.

3 Related Work

Several researchers worked on stream ciphers. Ronald Rivest proposed RC4 stream cipher which is used in data communication and networking protocols, including SSL/TLS etc. RC4 is a byte oriented stream cipher in which the plain text is exclusive-ored with a key stream and produced a cipher text stream. The secret key is one-byte key which is selected randomly from the 256 bytes. RC4 based on the concept of a state. A state which contains 256 bytes at each moment from that one of the byte is randomly selected as a key for the encryption. The security of Rc4 cipher based on the key size, if the key size is at least 128 bits then the cipher is secure.

Srikantaswamy et al. [7] proposed Recursive Key Generation Technique. In this each character of plain text is converted into ASCII code. Initially the key value K [0] is set by the user for encrypting first character. For the next character the key value is doubled. i.e., K[1] = K[0] + K[0]. For nth character the key value is K [n] = K[n−1] + K[n−1]. In this, the encryption algorithm works as follows: Adding the ASCII value of plain text P_i to K_i where $0 < i < n$ value and result is considered as C1. Applying the right shift operation on c1 and the result is c2. Perform 1's complement on c2 and the result is final cipher text which is transmitted to the receiver. The decryption procedure is performing 1's complement on cipher text and the result is p3. Performing left shift operation on p3 and the result is p2. Subtracting key from the p2 and the result is p1. Converting p1 into ASCII symbols and the result is final plain text. The drawback of this scheme is if the initial key is compromised then all the sub keys are easily revealed due to statistical relation between keys i.e., doubling the key value which does not satisfy confusion property of the cipher. The cipher text has a statistical relationship with plain text which doesn't satisfy diffusion property of security cipher.

4 Proposed Method

The proposed method describes the process of key generation, encryption and decryption algorithms.

4.1 Key Generation

The process of key generation for creating n keys is as follows:

1. Initially cloud user sets the secret key (*Ks*) and *Salt* value. *Salt* is an additional value which is set by the user.
2. The first character of plain text is encrypted by the new key value instead of the original key. The new key is generated by concatenating salt to original key and applying hash function on this value. The result of hash function is 32 digit (128-bit) hash value.
   $new\_K[1] \leftarrow MD5(Ks||Salt)$ where $new\_K[1]$ is a new key for first character.
3. From the second to n characters, the key value is doubled value of $K[i-1]$ where $2 \le i \le n$ and new key is calculated for encrypting each character

$$new\_K[i] \leftarrow MD5(K[i-1] + K[i-1])\ where\ i = 1\ to\ n$$

In this, the procedure is simple, easy to implement and provides strong security due to large key. The objective of using MD5 hash function is to make the key value complex.

4.2 Encryption

The proposed NHKS encryption algorithm converts each plaintext character into cipher text. The stream of plaintext is represented as $P = P_1, P_2, P_3 \ldots P_n$ and key stream $K = K_1, K_2, K_3 \ldots Kn$. The key is unique for each character and the first 32 bits of key stream is used for masking each character. The generated cipher text stream is $C = C_1, C_2, C_3 \ldots C_n$.

The procedure of encryption is described in the following steps.

1. NHKS is a modern stream cipher, where the encryption is performed on each character.
2. The plaintext stream is represented as $P = P_1, P_2, P_3, \ldots, P_n$, key stream $K_i = K_1, K_2, K_3, \ldots K_n$ and cipher stream is $C = C_1, C_2, C_3, \ldots, C_n$.
3. The characters of plaintext are encrypted by first 32 bits of 128 bit key which is unique for each charcter. The generation of key is based on MD5 hash function.
4. The following mathematical formula is used for encrypting each character.

$$C_i = P_i * new\_key_i \text{ where } 1 \leq i \leq n$$

4.3 Decryption

The procedure of decryption is described in the following steps.

1. In decryption, the cipher text stream is converted into plaintext stream by using the same key which is used in the encryption algorithm.
2. The input of NHKS decryption algorithm is cipher text stream $C = C_1, C_{2.} \ldots C_n$ key stream $K_i = K_1, K_2, K_3, \ldots K_n$ and generated output is plaintext stream is $P = P_1, P_2, P_3, \ldots, P_n$
3. The following mathematical formula is used for decrypting each character.

$$P_i = C_i / new\_key_i \text{ where } 1 \leq i \leq n$$

5 Experimental Results

We have implemented the proposed algorithm in Azure cloud virtual machine (VM) with Net beans IDE and java API. Azure provides a set of VM images where the user chooses the VM according to their choice. The pricing of VM is based on the type of user subscription and type of virtual machine. In computing, VM is an emulation of computer. Azure VM is created at Microsoft remote centers and accessed from anywhere any place through the internet. We have chosen Windows R2 virtual machine with D1core, 3.5 GB memory. The virtual machine is successfully created and connected to the VM through the remote desktop application by specifying essential credentials. We executed the encryption algorithm in cloud VM and evaluate the results. The evaluated encryption and decryption results are shown in the tables (Tables 1 and 2).

Table 1 Encryption algorithm results

Plain text	ASCII	Key	New key	Cipher text
I	49	69	3d9f8ee1db299aa712a029a0e3a2d6f4	11927FBE29
N	4E	138	013d407166ec4fa56eb1e1f8cbe183b9	0060A9A26E
F	46	276	db8e1af0cb3aca1ae2d0018624204529	3C08DB5DA0
O	4F	552	94c7bb58efc3b337800875b5d382a072	2DE9A2D028
R	52	1104	4da04049a062f5adfe81b67dd755cecc	18DD549762
M	4D	2208	cd3afef9b8b89558cd56638c3631868a	3DBABEB0E5
A	41	4416	1da546f25222c1ee710cf7e2f7a3ff0c	0786F70372
T	54	8832	060fd70a06ead2e1079d27612b84aff4	01FD328F48
I	49	17664	8d98ea39261415654200fc3faa058283	28609ACA41
O	4F	35328	a01cfb486435838979bef7c4a7899539	3168F18B38
N	4E	70656	fe3160ea7e3092ca593f36717280ea2f	4D730B874C

Table 2 Decryption algorithm results

Cipher text	Key	New key	ASCII	Plain text
11927FBE29	69	3d9f8ee1db299aa712a029a0e3a2d6f4	49	I
0060A9A26E	138	013d407166ec4fa56eb1e1f8cbe183b9	4E	N
3C08DB5DA0	276	db8e1af0cb3aca1ae2d0018624204529	46	F
2DE9A2D028	552	94c7bb58efc3b337800875b5d382a072	4F	O
18DD549762	1104	4da04049a062f5adfe81b67dd755cecc	52	R
3DBABEB0E5	2208	cd3afef9b8b89558cd56638c3631868a	4D	M
0786F70372	4416	1da546f25222c1ee710cf7e2f7a3ff0c	41	A
01FD328F48	8832	060fd70a06ead2e1079d27612b84aff4	54	T
28609ACA41	17664	8d98ea39261415654200fc3faa058283	49	I
3168F18B38	35328	a01cfb486435838979bef7c4a7899539	4F	O
4D730B874C	70656	fe3160ea7e3092ca593f36717280ea2f	4E	N

6 Conclusion

In this paper, proposed a NHKS stream cipher for data security in the cloud computing by taking the advantages of modern stream cipher and MD5 hash function. We also discussed security issues and requirements in the cloud computing. The proposed key generation algorithm generates 128 bit unique key for each character. Among the 128 bits, the first 32 bits are used for encrypting each character. Even if the two characters are same different cipher text is generated because the key value is doubled for the each character and then MD5 hash function is applied on these key. The scheme allows the user to encrypt a message before storing in a cloud.

Our proposed scheme provides unique 128 bit key generation with simple encryption and decryption algorithms. The security of encryption algorithm is based on the key size. In this algorithm, it is very difficult to identify key from the cipher text and plain text due to hash based key and which provides strong confusion and diffusion properties. We implemented proposed scheme in Microsoft Azure cloud virtual machine and evaluated the results. A possible way to extension this work would be to provide an encryption and decryption scheme which would have less computation cost and strong security proof under standard security assumptions.

References

1. Sandeep k sood.: A combined approach to ensure data security in cloud computing: Journal of Network and Computer Applications (2012).
2. http://managewindowsazure.com.
3. Joseph D. Touch: Performance Analysis of MD5: ACM SIGCOMM Computer Communication Review, 1995 - dl.acm.org.
4. J. Deepakumara., H. Hey: FPGA Implementation of Md5 hash algorithm: Electrical and Computer, 2001:ieeexplore.ieee.org.
5. William Stallings: Cryptography and Network Security, 3rd Edition, Prentice Hall, 2003.
6. Ramgovind S., Eloff MM., Smith E.: The Management of Security in Cloud Computing: International Symposium on Computer Architecture (ISCA), in Portland.
7. S. G. Srikantaswamy, H. D. Phaneendra.: A Cryptosystem Design with Recursive Key Generation Techniques: Procedia Engineering, International Conference on Communication Technology and System Design 2011.

PSNM: An Algorithm for Detecting Duplicates in Oceanographic Data

L. Srinivasa Reddy, S.P. Vighneshwar and B. Ravikiran

Abstract This work discusses a new method of identifying duplicates in surface meteorology data using PSNM (Progressive Sorted Neighborhood Method) Algorithm. Duplicate detection is the process of identifying the same representations of the real world entities in the data. This method needs to process a large amount of ocean data sets in shorter time. PSNM algorithm increases the efficiency of finding duplicates with lesser execution time and get the efficient results much earlier than traditional approaches. It is observed that all possible duplicates associated with the data can be identified using this method, and also this work proposes a new way to access the resulted (Duplicate eliminated) data using authorization restrictions based on the type of user and their need with different file conversion formats.

Keywords Duplicate detection · PSNM · CTD · Data cleaning

1 Introduction

In real time, Oceanographic data (Scientific) generated from different devices like Argo's, buoys and satellites etc. This type of data we called as raw data. Data scientists main task is to preprocess this raw data received from different oceanographic devices which are being operated in ocean surface and inside the ocean.

L.S. Reddy (✉) · B. Ravikiran
Computer Science and Technology, Department of CSE, ANIL
Neerukonda Institute of Technology and Sciences (ANITS), Visakhapatnam, India
e-mail: lsrinivasa.14.cstmtech@anits.edu.in

B. Ravikiran
e-mail: bravikiran.cse@anits.edu.in

S.P. Vighneshwar
Computational Facilities and Web Based Services Group (CWG), Indian National Centre for Ocean Information Services (INCOIS), Hyderabad, India
e-mail: vighneshwar@incois.gov.in

S.C. Satapathy et al. (eds.), *Computer Communication, Networking and Internet Security*, Lecture Notes in Networks and Systems 5,
DOI 10.1007/978-981-10-3226-4_29

Data pre-processing means to process the data, that is directly received from devices called as raw data. Raw data is highly suspectable to null values, missing values, and duplicates. So, Data analyzers are analyzing this type of raw data. This may lead to wrong results. Because of this, we are trying to eliminate duplicates in the raw oceanographic data.

The Data analyzers are analyzing the data, it must be in good quality. If it is in good quality it leads good results. For achieving the good quality data we are performing the data pre-processing step on the scientific data. Duplicates in scientific data can be of various causes and various reasons, those can be receiving the same data from different sources, device malfunctioning, manual entries in the data, Data Maintenance issues…etc. To Handle large volumes of oceanographic data that are receiving from different oceanographic devices hourly, monthly, yearly. It is very difficult task to maintain and quality controlling large volumes of data in oceanographic organizations like ESSO-INCOIS, NIOT…etc. [1]. When a device (specifically designed device to capture scientific data) fails to capture data from different locations in the ocean, it repeatedly gives previously captured data. In this way, duplicates may occur in scientific data. In this work, we are considered CTD as scientific data [2].

The problem of identifying duplicate data from the huge volumes of CTD (Conductivity, Temperature, and Depth) data sets comprising of ocean parameter collections such as Conductivity, Temperature, Depth, Air temperature, humidity, and wind speed is a difficult task because of its size and large amount [3, 4]. This paper presenting a new method of identifying duplicate observations in surface meteorology data using PSNM (Progressive Sorted Neighbourhood Method) Algorithm. PSNM is such an efficient algorithm for finding duplicates in the ocean data compared to traditional approaches like manual observation, manual comparison…etc. [5].

2 PSNM Algorithm

PSNM algorithm is a sorted neighborhood method. it is working based on predefined sorted key value. The key value is treated as the input value in the data and the records are compared with the window of records in sorted order. When the records arranged into the sorted order based on the magpie sort (Selection Sort). If the records are similar or close to actual records that can be treated as the duplicates based on sorted key value. The PSNM calculates the distances of two records and assigns rankings based on sorted order (rank distances). The records are sorted by using state approach [5, 6]. In this we take key as argo id and buoy id. Using these key we search the duplicates whereever it occur. This algorithm takes five input parameters that are data, key (combination of two attributes in the dataset), window size which corresponds to the traditional sorted Neighborhood method. This algorithm cannot work on entire dataset so that we are partition the data based on maximum number of records that fit into the memory.

3 Data Validation

We considered the CTD data for a period of Dec 2013–Nov 2014 from world ocean database (NODC). This data sets comprising of ocean parameter collections such as device-id, Gts-id, observation time, Conductivity, Temperature, Depth, Air temperature, humidity, and wind speed etc. To handle this large volume of data in terms of null values detection, duplicate detection is the difficult task because of its size and large volume. Here we can detect the duplicates in CTD data based on Device Id and its Id in monthly data and year wise data. Device Id represents the Id of special design devices to read all oceanographic parameters as mentioned above from the particular ocean region [1, 2]. After that, we integrate this quality data into web platform and giving access to users who needed this data for specific requirements like student research, Industry Need, Researcher Need…etc.

4 Flow Chart of Duplicate Detection

Figure 1 represents the flow chart of the duplicate detection, in this first, the input raw CTD data passed as input to the preprocessing step. After preprocessing the raw data the resulted data passed as input to the PSNM Algorithm. PSNM Algorithm classifies the preprocessed data based on the Magpie (Selection Sort) sort method. It classifies the data as duplicate and non-duplicate data. The result from the PSNM algorithm is Quality Controlled (Duplicate Eliminated) Data. This quality controlled data we are integrating to the web- based platform. In this work, we are providing access restriction to the users for accessing the data, based on the type of user and their need. we are providing the quality controlled data to registered users [5].

There are different oceanographic devices, that are capturing data in real time with various time periods. The devices Argos, buoys, and satellites observe ocean and capture the parameters. So, there is a possibility to get noise value if there is any device failures, repeated measurements and algorithm failures to deliver dirty data. Here we are using CTD data, that can an acronym as conductivity, temperature, and depth. İt is an electronic instrument measures the above parameters. But we are not concentrating on how it works and its structure. But it is an electronic device so there is a chance to measure repeated values at different depth levels. And also the possibility of circuit failures. The instrument collect the data in different locations at different depths. İt generates the raw data. Here we are considering CTD data.In this data, we are identifying the duplicates [4]. CTD is a homogeneous data that is collected from different locations of the Indian ocean, that data we are used for experimental work. İnsecond phase how PSNM algorithm identifying the duplicates shows in Sect. 5. The PSNM algorithm (progressive sorted neighborhood method) is used to identify the duplicate values in data. Ocean forecasting mainly depends

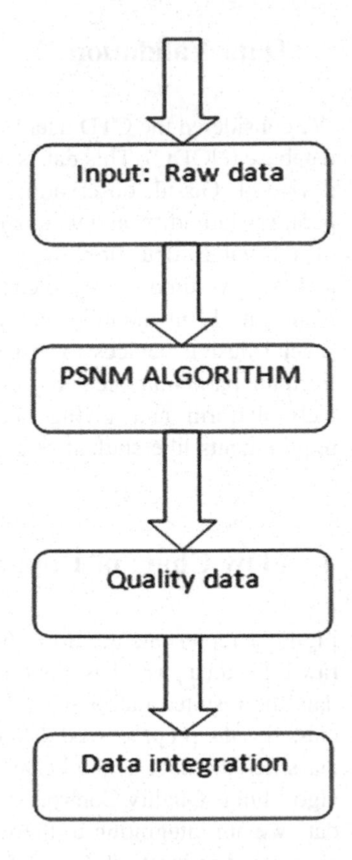

Fig. 1 Flow chart representation of working

on the good quality of data. So most of the time Data analysts are concentrate on this preprocessing area [7]. This paper mainly concentrates on identifying the duplicates and null values by using PSNM algorithm. İt gives the best results for a large amount of data.

Data integration is combining the data from several heterogeneous (or homogenous) data from different sources, which can be sorted using various technologies and visualize into a uniform manner. İn this work we collected homogeneous data from different sources and integrated all the data into one platform. The experimental results show in Sect. 6. There are several techniques to integrate data into one platform [8]. Those are manual integration, application-based integration, Middleware data integration…etc. and finally, the quality data provided to users with different extensions based on user specifications.

Algorithm (Progressive Sorted Neighbourhood Method)
Required Inputs: Dataset DA, sorting key KE, window size WS, enlargement interval size IS, number of records N.

1 PSNM(DA,KE,WS,IS,N)
2 pSize ← calcPartitionSize(DA)
3 pNum ← [N/(pSize - WS + 1)]

4 array order size N as Integer
5 array recs size pSize as Record
6 order sort (DA, KE, IS, pSize, pNum) [5]

5 Experimental Results

Figure 2 shows multiple duplicates in the ocean data. In the above figure the yellow rows indicates the duplicate values which are generated from the special designed devices.

Figure 3 shows the data with out duplicate values. After performing PSNM algorithm on the data with duplicates (Fig. 2) it occurs. So the data (Fig. 3) we termed as cleaned data.

Figure 4. shows the detection of duplicates in scientific data using PSNM algorithm and we represent it in barchart. The above bar chart shows the duplicate detection in one month by date wise. In the above barchart we do experiments in only tow days of may month. It represents the how many duplicates are detected in these two days. This the way we detect the duplicates in scientific data using PSNM and representation of that data in barchart.

1	Parameter / Buoy ID	GTS ID	Date & Time	Batterycharge [V]	Batterydischarge [V]	Batteryvoltage [V]	Latitude [deg]	Longitude [deg]	Humidity [%]	Airpressure [hPa]	Airtemperature [degC]	Winddirection [deg]	Windspeed [m/s]	Windgust [m/s]	Currentspeed [cm/s]	Currentdir [deg
2	AD02	23097	23-May-2016,23:46:14	-	-	12.678	14.878	68.927	88.78	1008.856	29.35	229	4.10	6.00	**29.88**	177
3	AD04	23494	23-May-2016,23:53:18	-	-	12.649	8.482	73.104	82.65	1010.015	29.14	247	5.07	8.45	**0.00**	-640
4	CALVAL	23497	23-May-2016,23:54:59	-	-	12.678	10.598	72.240	73.67	1009.635	30.26	319	6.93	8.93	**35.36**	137
5	CB01	23491	23-May-2016,23:48:46	-	2109.757	14.435	11.589	92.596	80.47	1007.218	30.33	249	9.85	12.59	-	-
6	CB02	23492	23-May-2016,23:48:45	-	-	12.649	10.874	72.209	77.00	1009.598	30.22	304	3.82	6.59	**32.30**	144
7	CB04	23170	23-05-16,23:53:00	-	-	-	15.404	73.769	76.32	1008.867	29.88	315	3.83	4.70	**38.66**	155
8	CB06	23099	23-May-2016,23:48:44	-	-	12.619	13.106	80.318	79.65	1004.952	29.10	227	5.83	7.24	**39.04**	180
9	AD02	23097	23-May-2016,23:46:14	-	-	12.678	14.878	68.927	88.78	1008.856	29.35	229	4.10	6.00	**29.88**	177
10	AD04	23494	23-May-2016,23:53:18	-	-	12.649	8.482	73.104	82.65	1010.015	29.14	247	5.07	8.45	**0.00**	-640
11	CALVAL	23497	23-May-2016,23:54:59	-	-	12.678	10.598	72.240	73.67	1009.635	30.26	319	6.93	8.93	**35.36**	137
12	CB01	23491	23-May-2016,23:48:46	-	2109.757	14.435	11.589	92.596	80.47	1007.218	30.33	249	9.85	12.59	-	-
13	CB02	23492	23-May-2016,23:48:45	-	-	12.649	10.874	72.209	77.00	1009.598	30.22	304	3.82	6.59	**32.30**	144
14	CB04	23170	23-05-16,23:53:00	-	-	-	15.404	73.769	76.32	1008.867	29.88	315	3.83	4.70	**38.66**	155
15	CB06	23099	23-May-2016,23:48:44	-	-	12.619	13.106	80.318	79.65	1004.952	29.10	227	5.83	7.24	**39.04**	180
16	AD02	23097	23-May-2016,23:46:14	-	-	12.678	14.878	68.927	88.78	1008.856	29.35	229	4.10	6.00	**29.88**	177
17	AD04	23494	23-May-2016,23:53:18	-	-	12.649	8.482	73.104	82.65	1010.015	29.14	247	5.07	8.45	**0.00**	-640
18	CALVAL	23497	23-May-2016,23:54:59	-	-	12.678	10.598	72.240	73.67	1009.635	30.26	319	6.93	8.93	**35.35**	137
19	CB01	23491	23-May-2016,23:48:46	-	2109.757	14.435	11.589	92.596	80.47	1007.218	30.33	249	9.85	12.59	-	-
20	CB02	23492	23-May-2016,23:48:45	-	-	12.649	10.874	72.209	77.00	1009.598	30.22	304	3.82	6.59	**32.30**	144
21	CB04	23170	23-05-16,23:53:00	-	-	-	15.404	73.769	76.32	1008.867	29.88	315	3.83	4.70	**38.66**	155
22	CB06	23099	23-May-2016,23:48:44	-	-	12.619	13.106	80.318	79.65	1004.952	29.10	227	5.83	7.24	**39.04**	180
23	AD02	23097	23-May-2016,23:46:14	-	-	12.678	14.878	68.927	88.78	1008.856	29.35	229	4.10	6.00	**29.88**	177
24	AD04	23494	23-May-2016,23:53:18	-	-	12.649	8.482	73.104	82.65	1010.015	29.14	247	5.07	8.45	**0.00**	-640
25	CALVAL	23497	23-May-2016,23:54:59	-	-	12.678	10.598	72.240	73.67	1009.635	30.26	319	6.93	8.93	**35.35**	137
26	CB01	23491	23-May-2016,23:48:46	-	2109.757	14.435	11.589	92.596	80.47	1007.218	30.33	249	9.85	12.59	-	-
27	CB02	23492	23-May-2016,23:48:45	-	-	12.649	10.874	72.209	77.00	1009.598	30.22	304	3.82	6.59	**32.30**	144
28	CB04	23170	23-05-16,23:53:00	-	-	-	15.404	73.769	76.32	1008.867	29.88	315	3.83	4.70	**38.66**	155

Fig. 2 Scientific data with duplicates

pid	BuoyId	GTS IDHAI	DATE	time	BATTERYC	BATTERYD	BATTERYV	LATITUDE	LONGTITU	HUMIDITY	AIRPRESSI	AIRTEMPE	WINDDIRE	WINDSPEI	WINDGUS	CURRENTS	CURRENTI	SSTSKIN	SST3M	CONDI
2	AD04	23494	23-May-2(	-	-	12.649	8.482	73.104	82.65	1010.015	29.14	247	5.07	8.45	0	-640	-1000.3	30.53	21.35	-
4	CB06	23099	23-May-2(	-	-	12.619	13.106	80.318	79.65	1004.952	29.1	227	5.83	7.24	39.04	180	29.02	29.69	56.36	-
1	AD02	23097	23-May-2(	-	-	12.678	14.878	68.927	88.78	1008.856	29.35	229	4.1	6	29.88	177	29.86	30.46	60.6	-
10	CB04	23170	23-05-16,2	-	-	-	15.404	73.769	76.32	1008.867	29.88	315	3.83	4.7	38.66	155	30.65	31.15	60.97	-
3	CB02	23492	23-May-2(	-	-	12.649	10.874	72.209	77	1009.598	30.22	304	3.82	6.59	32.3	144	30.44	30.88	54.29	-
7	CALVAL	23497	23-May-2(	-	-	12.678	10.598	72.24	73.67	1009.635	30.26	319	6.93	8.93	35.35	137	30.39	-1000	-1000	-
8	CB01	23491	23-May-2(	-	2109.757	14.435	11.589	92.596	80.47	1007.218	30.33	249	9.85	12.59	-	-	-	30.35	54.68	-

Fig. 3 Data with out duplicates

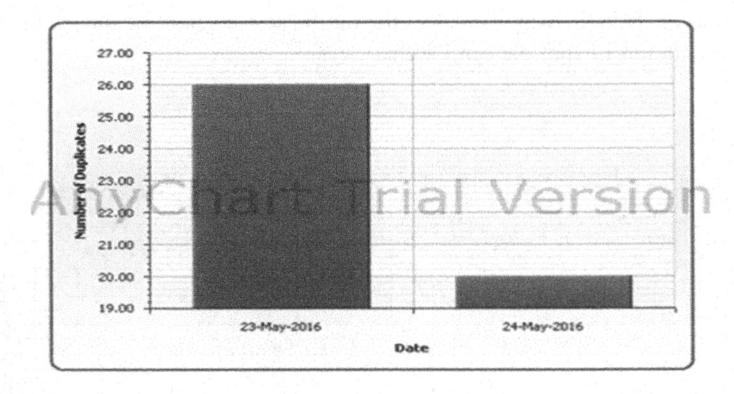

Fig. 4 Bar chart representation of detecting duplicates in 1 month data by date wise using PSNM

6 Conclusion

In this paper, we presented a novel algorithm i.e. PSNM algorithm to detect the duplicates and null values in the oceanographic CTD data from world ocean database. We have also observed the effectiveness of this algorithm by giving a different size of inputs. The experimental results verify the promising performance of PSNM algorithm. we have also illustrated how effectively these algorithm works on different input sizes in Sect. 6. We have initiated the data mining concepts to improve the functionality of the detecting duplicates in oceanographic data using this algorithm in ESSO-INCOIS and also we integrate this quality data into the web platform and giving access permissions to users for their specific requirements.

Acknowledgements This work was completed in INCOIS Hyderabad. The Authors wish to thank Director ESSO-INCOIS, Hyderabad for the encouragement and facilities provided and also Authors wish to thank scientists for their support and guidance throughout working on this project and preparing this manuscript. We would also like to express our gratitude to our Professors in the college and Prof. S.C. Satapathy (Head of Dept.), ANITS, Visakhapatnam for his continuous support and encouragement.

References

1. http://www.incois.gov.in/argo/argo.jsp.
2. https://www.nodc.noaa.gov/access/index.html.
3. Richard E. Thomson, William j. Emery, "Data Analysis Methods In Physical Oceanography", Elsevier.
4. L. Boehme, p Lovell, M. Biuw, F Roqucet, J Nicholson, S.E. Thorpe, M.p. Meredith, and M. Fedak, "Technical Note: Animal-bornce CTD-Satellite Relay Data Loggers for real-time Oceanographic data collection.
5. Thorsten Papenbrock, Arvid Heise, and Felix Naumann, "Progressive Duplicate Detection", IEEE Transactions on Knowledge and Data Engineering, Vol. 27, pp. 1316–1329, 2015.
6. Ashwini. V. Lakote, Lithin k, "A Study And Survey on Various Progressive Duplicate Detection Mechanisms", IJRET International Journal of Research in Engineering and Technology, vol. 05, pp. 454–456, 2016.
7. Su Yan, Dongwon Lee, Min-Yen Kany, C. Lee Giles, "Adaptive Sorted Neighbourhood Methods for Effcient Record Linkage", Proceedings of the ACM/IEEE–CS joint conference, pp. 185–194, 2007.
8. Erhard Rahm, Hong Hai Do, "Data Cleaning: Problems and Current Approaches", IEEE Data Engineering Bulletin, vol. 23, 2000.
9. Arfa Skandar, Mariam Rehman, Maria Anjum, "An Efficient Duplication Record Detection Algorithm for Data Cleansing", International Journal of Computer Applications, vol. 127, pp. 27–38, 2015.
10. Mauricio A. Hernandez, Salvatore J. Stolfo, "Real-world Data is Dirty: Data Cleansing and The Merge/Purge Problem", Data Mining and Knowledge Discovery, vol. 2, pp. 9–37, 1998.

Detecting and Correcting the Degradations of Sensors on Argo Floats Using Artificial Neural Networks

T. Satyanarayana Raju, T.V.S. Udaya Bhaskar, J. Pavan Kumar and K.S. Deepthi

Abstract Argo floats are autonomous floats designed to measure temperature and salinity of the world oceans. Once deployed these floats goes to as deep as 2000 m and while coming up measure temperature and salinity of the underlying ocean automatically. These floats act as a substitute to the ship-based data sets and currently as many as ~3800 are active in the global oceans. These instruments being autonomous in nature, measure and transmit data seamlessly irrespective of the weather, season, and region. However, the salinity sensors on these floats are sensitive to bio-fouling and can cause degradation to the data. As these are one time deployed and data is continuously obtained they are not available for calibration unlike the instruments on the ship. In this work ANN is used to check the degradation of the sensors and correct the same so that the data can be use in scientific analysis.

Keywords Artificial neural network · Back propagation · Feed forward · Argo floats · Salinity · CTD

T. Satyanarayana Raju · K.S. Deepthi
Computer Science and Technology, Department of CSE,
ANIL Neerukonda Institute of Technology and Sciences (ANITS),
Visakhapatnam, India
e-mail: raju.tirumani43@gmail.com

K.S. Deepthi
e-mail: selvanideepthi.cse@anits.edu.in

T.V.S. Udaya Bhaskar (✉) · J. Pavan Kumar
Data and Information Management Group,
Indian National Centre for Ocean Information Services (INCOIS), Hyderabad, India
e-mail: uday@incois.gov.in

J. Pavan Kumar
e-mail: pavankumar.j@incois.gov.in

S.C. Satapathy et al. (eds.), *Computer Communication, Networking and Internet Security*, Lecture Notes in Networks and Systems 5,
DOI 10.1007/978-981-10-3226-4_30

1 Introduction

Argo is an global program aimed at seeding the ocean with 3000 profiling floats [1]. The problem of identifying the degradation of Argo float parameters such as salinity, temperature at various depths (e.g.: range 1800–2000 m depth) from a huge amount of CTD data sets is a tricky task because of its size. CTD data sets are collected from the latest available World Ocean Database 2013 (WOD13) and used for correcting the sensor degradations. CTD stands for Conductivity, Temperature, and Depth and is the primary high quality subsurface data for cross comparing and analysing the physical properties of sea water. This paper presents a new method of identifying degraded salinity values obtained from data collected by Argo floats using Artificial Neural Network (ANN). Here ANN is used to determine the degraded values of the sensors fitted on Argo floats (e.g.: salinity) and correct the same so that the data can be further used in scientific analysis of various projects and applications.

Argo program is an global collaboration program that collects high-quality temperature and salinity profiles from the upper 2000 m depth of the ice-free global ocean and currents from intermediate depths, and these floats act as a substitute to the ship- based data sets. Currently as many as ~ 3800 are actively working in the global oceans. These Argo floats primarily measure Conductivity-Temperature-Depth (CTD) data. In this CTD data, salinity parameter is found to be degraded over time due to bio-fouling caused by algae and oil sleeks in the ocean. Here in this work, salinity values are corrected by using Neural Networks with the use of back propagation algorithm [2]. An Artificial Neural Network (ANN), often called a "Neural Network" (NN), is an arithmetical model or computational model, based on biological neural networks. It consists of a unified group of artificial neurons and processes the information using a connectionist approach to computation. Each neuron has multiple inputs as well as multiple outputs. The system that receives signals from the input produces a resultant signal and transmits that signals to all outputs.

Figure 1 represents the Artificial Neural Network, consisting of three layers viz., input layer, hidden layer, and an output layer. In the input layer, each neuron is connected to multiple inputs, hidden layer is connected to multiple input layer outputs and the output layer is connected to multiple hidden layer outputs. The final system processes all these layers and produces an output.

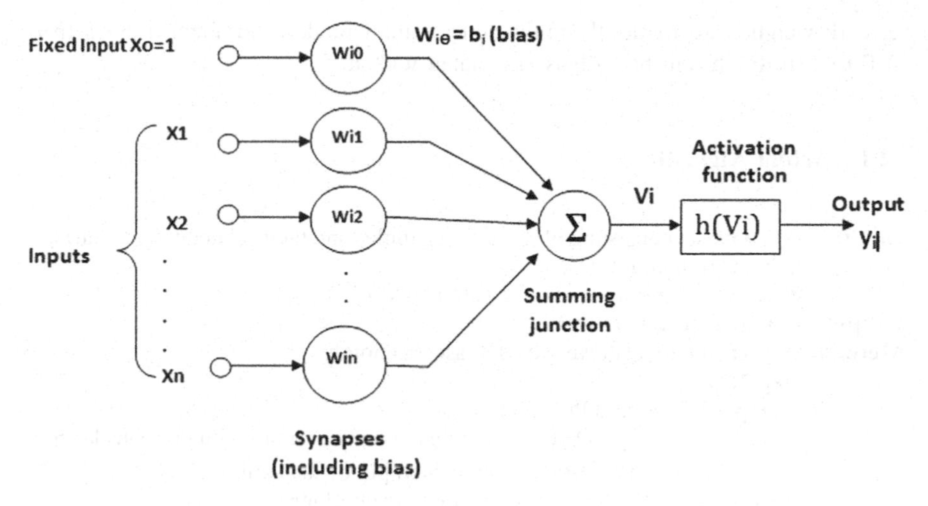

Fig. 1 Artificial neural network

2 Artificial Neural Network

2.1 Feed Forward Neural Network

The word "feed forward" neural network has links that connect in only one direction. Except during training, there are no backward links in a feed forward network. All links proceed from input nodes to the hidden nodes and to the output nodes [3]. The output of each node is called its "activation". In this, every connection has weights which are associated with each vector and node in the network and these weight values are used to assess how input data is related to output data. The weights associated with individual nodes are also known as biases. Weight values are determined by the iterative flow of training data through the network. Once trained, the neural network can be applied towards the classification of new data. After training, the neural network is to be tested by using the feed-forward pass with updated weights.

2.2 Back Propagation Neural Network

Back propagation algorithm is a frequent method for teaching artificial neural networks to perform a given task [4]. The neural network is a set of linked input/output units in which each link has a weight associated with it. Throughout the learning phase, the network learns by adjusting the weights. So, we can be able to predict the correct class label of the input tuples. In this back propagation, neural network weights are randomly initialized to small random numbers (ranges from −1.0 to 1.0). Each unit has a bias associated with it.

2.2.1 Actual Algorithm

Input: D, a data set consisting of the training tuples and their related target values;
l, the learning rate;
network, a multilayer feed-forward network.
Output: A trained neural network.
Method: **Step 1:** Initialize all weights & biases randomly
Step 2: Do
Step 2.1: Forward Propagation:
$I_j = \sum_i w_{ij} O_i + \Theta_j$ //compute the input of unit j with previous layer, i
$O_j = 1/1+e\text{-}Ij$ //compute the input of each unit j
Step 2.2: Back propagation with Errors checking:
$Err_j = O_j(1\text{-}O_j)(T_j\text{-}O_j)$ //compute the error
$Err_j = O_j(1\text{-}O_j)\sum_k Err_k w_{jk}$ //compute the error with next layer, k
Step 2.3: Increment and Update the Weights
$\Delta w_{ij} = (l)Err_j O_i$ // weight increment
$W_{ij} = wij + \Delta w_{ij}$ // weight update
Step 2.4: Increment and Update the Bias
$\Delta\Theta_j = (l)Err_j$ // bias increment
$\Theta_j = \Theta_j + \Delta\Theta_j$ // bias update
Step 3: Continue until network is trained [7].

2.3 Training Neural Network

There are three types of learning techniques by feeding it teaching patterns and change its weights according to some learning tasks. These are supervised learning, unsupervised learning, and reinforcement learning. Here in this work, we used supervised learning for teaching the neural network.

2.3.1 Supervised Learning

Supervised learning is a machine learning task of inferring a function from labeled training data, that sets parameters of an artificial neural network from training data. The task of the learning artificial neural network is to set the input corresponding to the output patterns. These input-output pairs can be provided by an exterior teaching or by a system which contains the neural network [4].

3 Data

Data used in this study is obtained from latest version of World Ocean Database 2013 (WOD13). From this database we collected high quality CTD data sets which have undergone many quality control procedures [5]. Temperature and salinity data for the depths between 1800–2000 m are collected from the WOD13. CTD data which is in encoded format is decoded using programming techniques.

Figure 2 represent the CTD data that we considered for the years starting from 1980 to 2013 and profile (PFL) data for the years starting from 1999 to 2013 for the Indian Ocean region only. It consists of 3,30,000 temperature and salinity data points. Out of this data 70% is used for training 15% is for testing and remaining 15% is for validation.

Neural network understands values only in the range between −1 to 1 or 0 to 1 [8]. The data need to be converted to this form to make it meaningful to a neural network or it understands continuous values like monthly data 1–12. Table 1 represents normalizing the values for better understanding by the neural network.

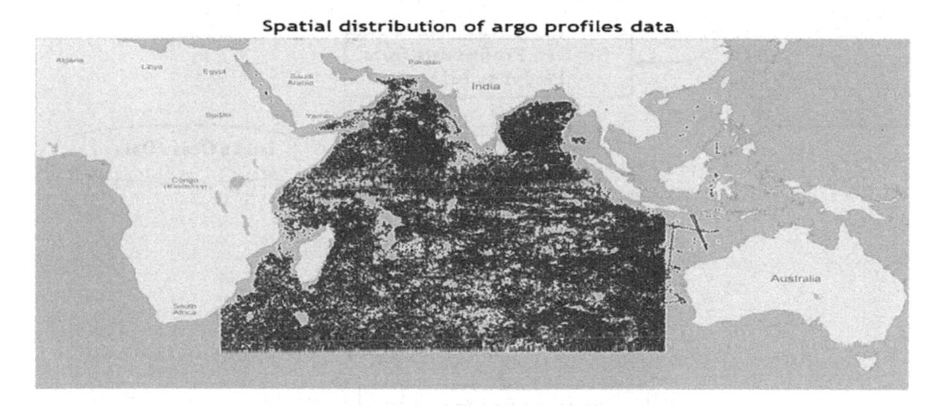

Fig. 2 Spatial distribution of argo profile data

Table 1 Inputs and outputs of emulating NNs [6]

Variable	Units	Input	Size
Year		Yr	1
Day of the year		Sin((2.day.(3.14))/366) Cos((2.day.(3.14))/366)	1
Longitude		Sin(long), Cos(long)	1, 1
Latitude		Sin(lat)	1
Sea surface height	M	SSH	1
Sea surface salinity	g/kg	SSS	1
Sea surface temp	C	SST	1

4 Flow Chart of Algorithm

Figures 3 and 4 provides the information about how the algorithm works in the form of flow charts. The process of correcting the degraded salinity values by using back propagation algorithm and feed forward algorithm, works as follows. First, we collect huge amount of raw CTD data from WOD13. Then we given this raw data as input to the proposed back propagation algorithm for training the neural network. In this process, we obtained a stable set of weights. These weights are then given as input to the feed forward neural network for correcting the degraded salinity values.

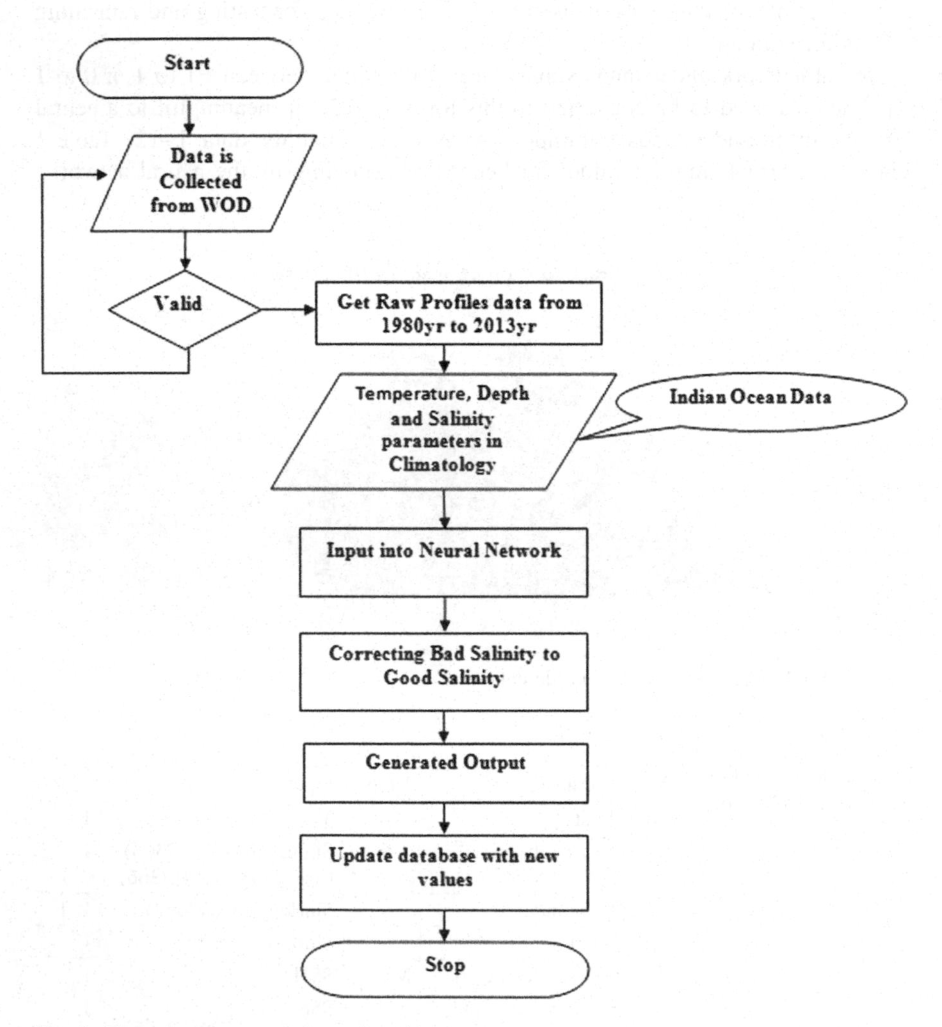

Fig. 3 Flow chart representation of working

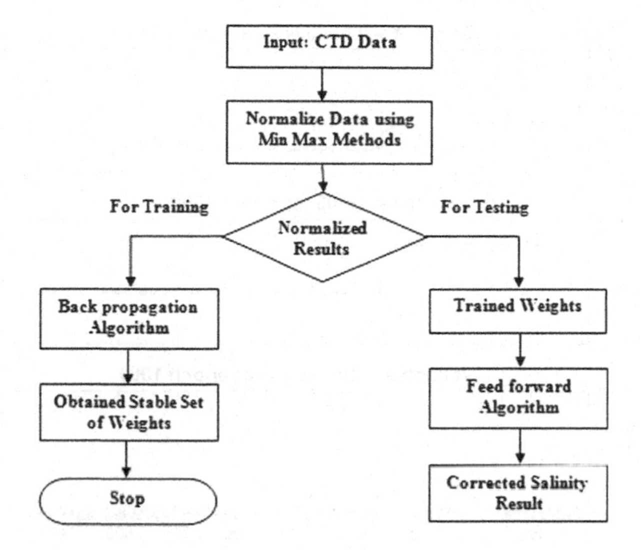

Fig. 4 Flow chart representation of NN

5 Results and Discussion

The neural network is trained with different types of input, such as longitude, latitude, day, month, year and time in the course of generating results. The physical parameters includes salinity, temperature and depth data. As mentioned before for this purpose CTD data pertaining to Indian Ocean obtained from WOD13 sets are used, starting from the year 1989 to 2013. Out of this data, 70% was used for training, 15% for testing, and 15% for validating.

Figure 5 represent the training of the neural network using gradient 0.00069452 at epoch 1000, to search for a set of weights that fits the training data, so as to minimize the error in that algorithm [8]. The learning rate Mu = 1e-06 at epoch 1000 is helpful to avoid local minimum and encourages the global minimum.

Figure 6 represent the results of the Argo salinity values with the use of back propagation algorithm. Here the neural network is tested for a good Argo float data. From the figure we observe that the results of the data sets are perfect and matching well in case of good floats which did not have any drift during life time. From this, we can say that the back propagation algorithm is well trained and the low error between the original and forecasted salinity values is quite encouraging.

Figure 7 represent the typical case where a degradation is observed in salinity after 45th time stamp. The degraded salinity values are corrected using back-propagation algorithm using latitude, longitude, date, time, temperature and ocean depth parameters. The NN is observed to do a good job by matching the observed data up to 45th time stamp which is clearly reflected in the error close to zero. Argo salinity sensor has a accuracy of 0.005 psu and this method is found to clearly detect the degradation and correct the same to great degree of accuracy.

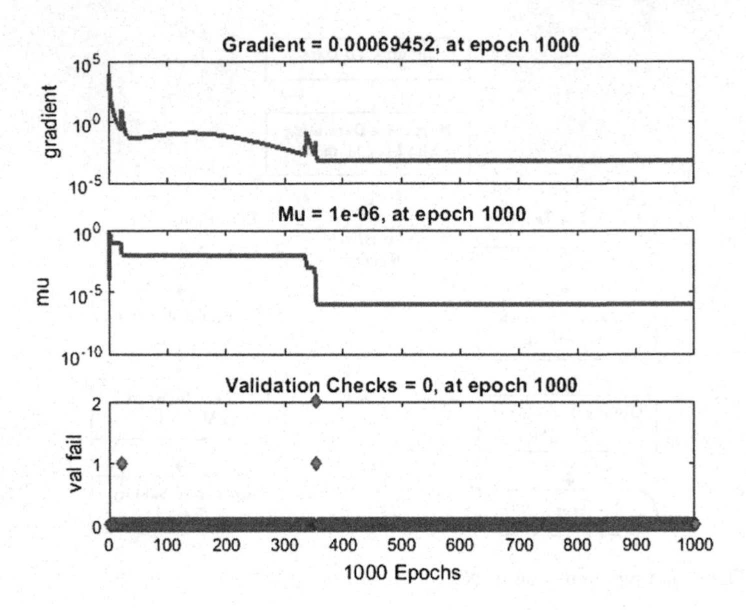

Fig. 5 Training neural network

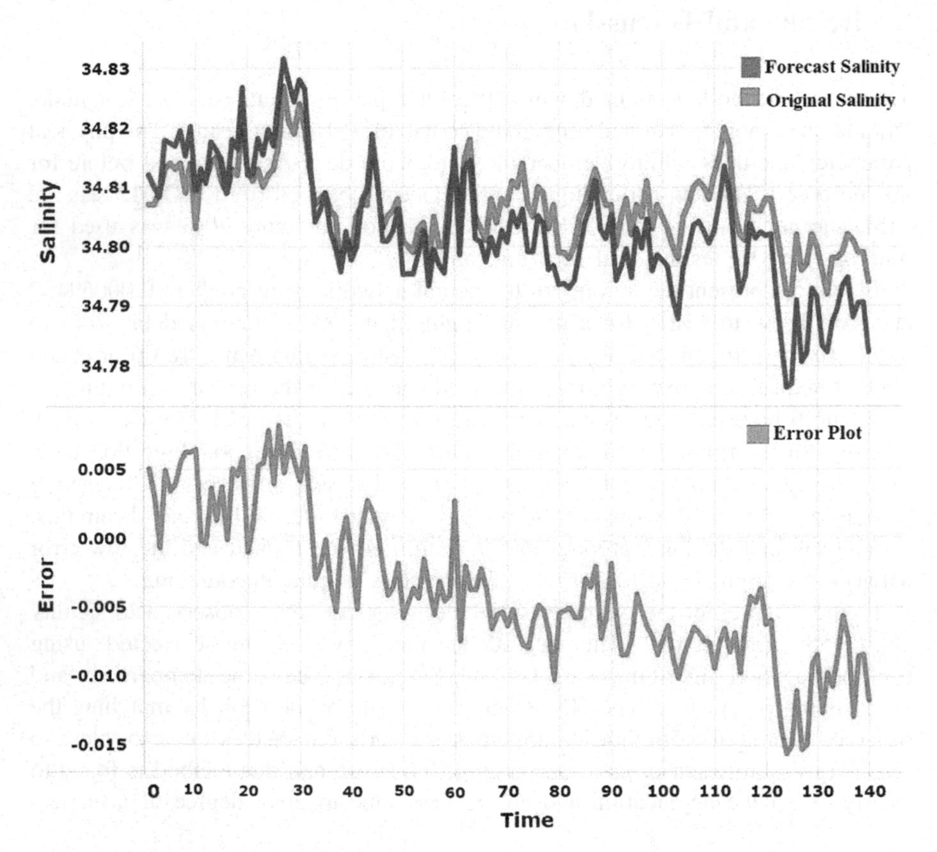

Fig. 6 Argo good data validation

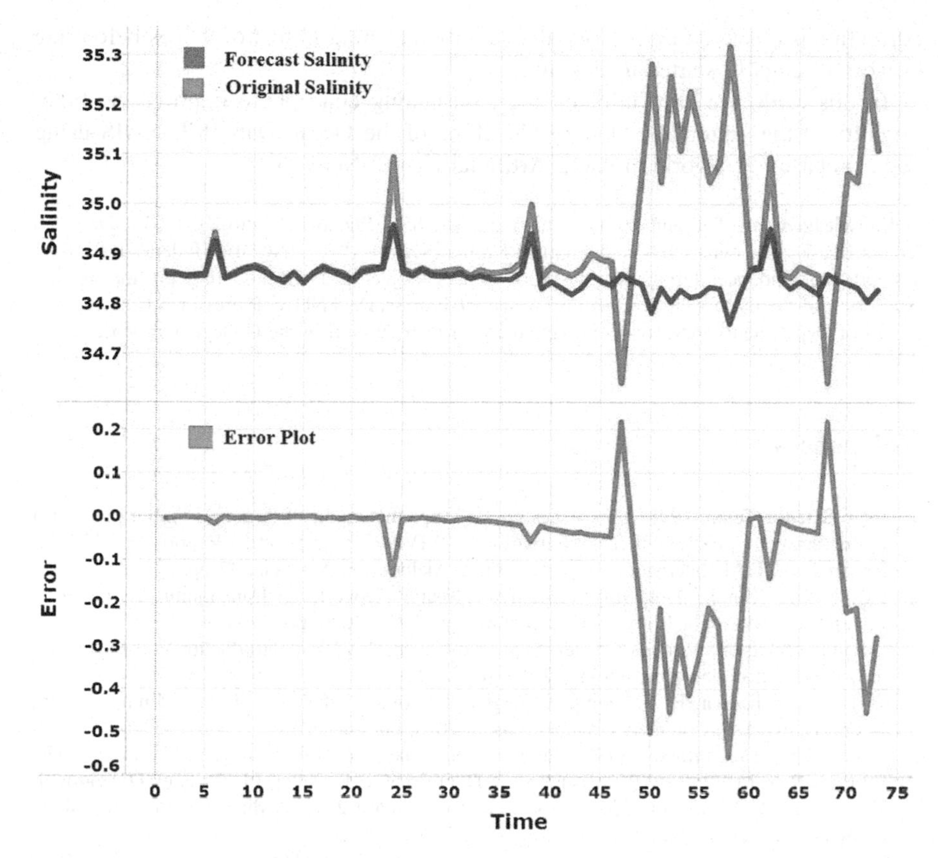

Fig. 7 Argo bad data validation

After the correction the trend in the data is observed to match well with that of the previous cycles. Hence the degradation values can be corrected within acceptable ranges by employing neural networks method.

6 Conclusion

In this paper, we proposed a novel method for correcting degraded parameters of Argo float sensors using an artificial neural network method by implementing it with using back propagation algorithm. In this method we first, train the neural network with supervised learning, after training we have tested and validated the method. Further to check the validity, it is applied to a good float with no sensor degradation and found to match the observed values closely. Once the confidence is obtained the method is applied to correct the data from a degraded float. The correction is observed to improve the salinity matching to that of values before the sensors is actually started to degrade.

In this work, we have initiated the data mining concepts to improve the functionality of the Automatic Quality Checking of the Ocean Data in INCOIS using back propagation algorithm using Artificial Neural Networks.

Acknowledgement The authors wish to thank Dr. SSC Shenoi, Director, INCOIS, Dr. TVS Udaya Bhaskar (Scientist-"E"), and J. Pavan Kumar (Scientist-"B"), INCOIS, Hyderabad for their support and guidance throughout the work in this project and for preparing the manuscript. Authors also thank Mrs. K.S. Deepthi (Assistant Professor), ANITS, Professor S.C. Satapathy (Head of Department), ANITS, Visakhapatnam for their support in the College and work.

References

1. Argo Science Team, 1998. On the design and implementation of Argo: An initial plan for a global array of profiling floats. International CLIVAR Project Office Report 21, GODAE Report 5. GODAE International Project Office, Melbourne, Australia, 32 pp.
2. DR. Yashpal Singh, Alok Singh Chauhan, "Neural Networks in Data mining", Journal of Theoretical and Applied Information Technology, 2005–2009, pp. 37–42.
3. David Leverington, Associate Professor of Geosciences, "A Basic introduction to feed forward back propagation neural networks", 2016, pp. 1–25.
4. D. Prashanth Kumar, B. Yakhoob, N. Raghu "Improving Efficiency of Data Mining through Neural Networks", Conference paper October 2013, pp. 14–18.
5. Boyer, T.P., J.I. Antonov, O.K. Baranova, C. Coleman, H.E. Garcia, A. Grodsky, D.R. Johnson, R.A. Locarnini, A.V. Mishonov, T.D. O'Brien, C.R. Paver, J.R. Reagan, D. Seidov, I. V. Smolyar, M.M. Zweng, 2013, World Ocean Database 2013. Sydney Levitus, Ed.; Alxey Mishonov, Technical Ed.; NOAA Atlas NESDIS 72, 209 pp.
6. Haykin, S., *Neural Networks*, Prentice Hall International Inc., 1999.
7. Vladimir Krasnopolsky, Sudhir Nadiga, Avichal Mehra, Eric Bayler, and David Behringer., "Neural Networks Technique for Filling Gaps in Satellite Measurements: Application to Ocean Color Observations", Volume 2016, pp. 1–9.
8. Jiawei Han, Micheline Kamber, Jian Pei, "Data mining Concepts and techniques" Third edition, Elsevier 2012, pp. 398–408.

Bit Resultant Matrix for Mining Quantitative Association Rules of Bipolar Item Sets

Dileep Kumar Koda and P. Vinod Babu

Abstract Association rule learning is a familiar technique for finding interesting relations that exist between variables in immense databases. Existing mining techniques which are available at present cannot pay attention to negative dependencies and considering only one evaluation criteria for measuring quantity of obtained rules. But for better results, data sets demanding for populating negative associations. To overcome the problem, proposing a technique called Bit Vector (BV) generation for mining quantitative association rules of bipolar (Positive and Negative) item sets together. Proposed system can reduce time complexity of finding recurrent item sets of bipolar association rules and provide more flexibility for finding the frequent item sets.

Keywords Association rule mining · Frequent item sets · Infrequent item sets · Negative association rule mining · Bit vectors

1 Introduction

Data mining is the activity of exploring and examining data from different sources such as data warehouses, repositories, large databases and summarizing it into utilitarian information, the obtained information used to improve the revenue. Data mining software is a logical tool for analyzing the valuable hidden information from databases and used for decision making. Data Mining is a task of exploring, analyzing, discovering meaningful patterns among the existed fields in large databases. Disk storage, processing power, statistical software are the continuous revolutions in the field of computer science and increasing the accuracy of analysis while taking down the cost. Association rule mining is the major job in knowledge discovery

D.K. Koda (✉) · P. Vinod Babu
Anil Neerukonda Institute of Technology and Sciences (ANITS), Visakhapatnam, India
e-mail: dileepkoda@gmail.com

P. Vinod Babu
e-mail: vinodbabusir@gmail.com

S.C. Satapathy et al. (eds.), *Computer Communication, Networking and Internet Security*, Lecture Notes in Networks and Systems 5,
DOI 10.1007/978-981-10-3226-4_31

process. Association rule mining is mainly focused on Market Basket Analysis such as finding coincide associations of collected items. Association rules are evolved by examining data and uses the benchmark support and confidence to recognize the relationship between the variables in database.

Apriori algorithm is earliest algorithm in association rule mining concept then after so many number of algorithms were evolved from apriori algorithm. Apriori algorithm introduced for mining frequent items of item sets and boolean association rule learning across the transactional databases. All the extracted association rules contains the transactional database should satisfy towards support and confidence. İf the obtained association rules unexpected by user, may diminish the performance of support is known as low support value. If the support approach is increase, additional number of algorithms and extra number of discovering rules are needed and cause low interestingness among the variables. The support approach must be minimal while separating information from the databases. Some times the rules are unbearable to use because of exceeding the rules limit.

To overcome the drawback, various techniques and algorithms were proposed in the relevant work. Many algorithms were introduced to optimize the number of item sets by initiating optimal item sets or maximal item sets. Most of the existing post processing techniques to minimize the number of rules works on mechanism of statistical data available in the database. Post processing will improve the choosing of rules which is obtained from the database, different post processing methods are: Pruning, Visualization, Grouping, Summerizing. Pruning removes the unnecessary rules and uninteresting rules. Visualization going improve the readability of rules and represents in graphical notation. And in the grouping process, group of rules are evolved. Set of rules are initiate in summerizing module.

Existing post processing techniques for reduce the number of rules, are works according to statistical information available in the databases. Interestingness rules mainly focus on user's knowledge, and can not say that all interesting rules are be extracted. For instance, the user require to pay attention on specific strategy of rules, only specific rules only selected from the existed rules. The post processing rule technique should be imperatively concentrate on strong interactive. The important thing is to identify strong rules obtained in databases using different measuring techniques of interestingness.

2 Related Work

Number of EAs (Evolutionary Algorithms) are proposed in the related work to accomplish set of QARs (Quantitative of Association Rules) from the transactional database. Genetic Algorithms (GAs) are the most victorious and familiar search technique for solving the complex problems and have demonstrated to be key technique for knowledge extraction. The algorithms existed at present considering only one evaluation for calculating the quality of obtained rules. In order to obtain negative association rules requires maximum number of infrequent item sets. So that the extracted association rules need to treated as multi objective instead of single objective in the process of association rules extraction. The existed bit vector method performing the Bitwise AND operation hence the time complexity is more. Proposed Bit Vector (Bit Resultant) approach minimizes the restrictions and imperfection of the single objective algorithms, hence allows user to optimizing and measuring together for mining set of interesting rules and able to implementation simply with appropriate intersect with database.

3 Proposed Work

Association rules are taken as binary values or discrete values as per interestingness measure criteria. Patterns are defined by the user which is considered as neither single objective nor multi objective. Existing techniques for mining bipolar association rules maintains frequent as well as infrequent item sets this leads to scalability problem. To overcome scalability problem and for obtaining effective results, it is imperative to develop effective technique for mining bipolar association rules. More efforts have been commited to develop algorithm for extracting association rules efficiently, whose transactions (Transactional Database), user defined Minimum Support (ms), and Minimum confidence (mc). Proposing a technique called BV process, for finding frequent and infrequent item sets in a given transactional database.

Proposed Bit Vector process consists of two steps, and they are as follows:

step 1:
Bit Vectors initiation for all item sets existing in the transactional database.

1. Read all item sets from the transactional database.
2. Create a Bit Vector matrix for each item in the transactional database as follows:
   - 2.1. If item is in the transactional database then put one in the Bit Vector matrix.
   - 2.2. If item is not exists in transactional database then put zero the Bit Vector matrix.
3. Repeat the process until end of transactional database (Fig. 1).

step 2:
Find out frequent and negative frequent item sets using BV matrix. The process of finding the frequent and negative frequent items set is as follows.

1. Calculate number of ones in each column and store it in Column Bit Vector.
2. Sort the values in column bit vector.
3. Calculate number of columns with same number of value in Column Bit Vector.
4. For each value in Column Bit Vector.

Item Sets	Transaction Records'ID										
	0	**1**	**2**	**3**	**4**	**5**	**6**	**7**	**8**	**9**	**10**
a	1	1	1	1	1	1	1	1	0	1	1
b	1	1	1	1	0	1	0	1	0	0	0
c	1	0	1	1	1	0	1	1	0	0	0
d	0	0	1	1	0	1	0	0	0	0	1
e	0	0	0	0	0	0	0	0	1	1	1

Fig. 1 Example of bit vector initiation

4.1. If the value is greater than minimum support value, Then Calculate maximum length of frequent length item sets.

4.1.1. For each item sets, calculate support count of each item.

4.1.2. If the support count of an item is exceeding user defined minimum support (ms) value, then item set is known as frequent and the item is added to frequent item set.
Else the item is known as infrequent item and the item set is added to negative itemset.

5. After completion of total transaction we can stop the process and display frequent and negative frequent items.

By performing with this process, we can acquire both frequent as well as infrequent item sets from the given transactional database. So proposing this process we can also provide more flexiblity and more efficiency.

4 Result and Analysis

Proposed system can reduce time complexity of finding frequent item sets of positive and negative association rules and provide more flexibility for the generation of frequent item sets. Where as in apriori algorithm database is scans in iteration manner, if the number of iterations are more obviously the time complexity is more and requires more I/O Operations. Proposed Bit Vector technique search the database only once for extracting association rules, hence minimize the working process and of I/O operations (Figs. 2 and 3).

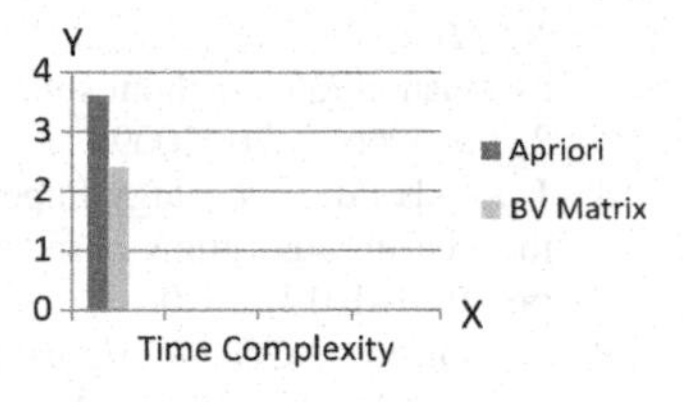

Fig. 2 Time complexity of Apriori algorithm and BV matrix

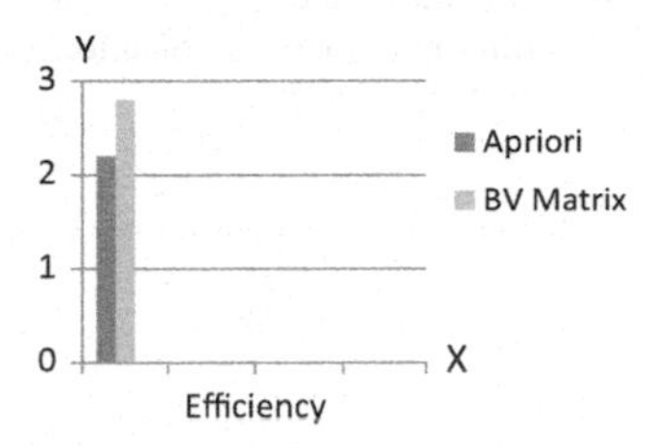

Fig. 3 Efficiency of Apriori algorithm and BV matrix

5 Conclusion

Infrequent item sets are play vital role for achieving typical results. A new BV (Bit Vector) structure stores both the frequent as well as infrequent item sets for mining bipolar association rules has been proposed. Proposed technique inspected the database only once for mining bipolar association rules, hence time complexity and the number of I/O operations are optimized. Additional malleability of the BV process is, if new frequent item sets are appear it will be mined by reducing the user approach of minimum support (ms) value, and allows us to append new item sets and reperforming for finding polarity association rule. Minimum support is defined by the user and support count is to be calculated for each pair of association rules between vectors.

Bibliography

1. R. Renesse, K. Birman, and W. Vogels, "Astrolabe: A robust and scalable technology for distributed systems monitoring, management, and data mining," *ACM Transactions on Computational System*, volume. 21, No. 2, pp. 164–206, 2003.
2. J. Han and M. Kamber, "*Data Mining Concepts and Techniques*", 2nd ed. Burlington, MA, USA: Morgan Kaufmann, 2006.
3. C. Zhang and S. Zhang, "Association rule mining: Models and algorithms," in *Lecture Notes Computer Science* (LNAI 2307). Berlin/Heidelberg, Germany Springer-Verlag, 2002.
4. R. Agrawal, T. Imielinski, and A. Swami, "Mining association rules between sets of items in large databases," in *Proc. SIGMOD*, 1993, pp. 207–216.
5. R. Agrawal and R. Srikant, "Fast algorithms for mining association rules," in *Proc. Int. Conf. Large Data Bases*, 1994, pp. 487–499.
6. R. Srikant and R. Agrawal, "Mining quantitative association rules in large relational tables," in *Proc. ACM SIGMOD*, 1996, pp. 1–12.

7. X. Yan, C. Zhang, and S. Zhang, "Genetic algorithm-based strategy for identifying association rules without specifying actual minimum support," *Expert Syst. Appl.*, vol. 36, no. 2, pp. 3066–3076, 2009.
8. J. Alcala-Fdez, N. Flugy-Pape, A. Bonarini, and F. Herrera, "Analysis of the effectiveness of the genetic algorithms based on extraction of association rules," *Fund. Inf.*, vol. 98, no. 1, pp. 1001–1014, 2010.
9. D. Goldberg,"*Genetic Algorithms in Search, Optimization and Machine Learning*". Reading, MA, USA/White Plains, NY, USA: Addison- Wesley/Longman, 1989.
10. B. Alatas and E. Akin, "MODENAR: Multi-objective differential evolution algorithm for mining numeric association rules," *Appl. Soft Comput.*, vol. 8, no. 1, p. 646, 2008.

Application of Ant Colony Optimization Techniques to Predict Software Cost Estimation

V. Venkataiah, Ramakanta Mohanty, J.S. Pahariya and M. Nagaratna

Abstract In modern society, machine learning techniques employed to predict Software Cost Estimation viz. Decision Tree, K-Nearest Neighbor, Support Vector Machine, Neural Networks, and Fuzzy Logic and so on. Every technique has contributed good work in the significant field of software cost estimation. The Computational Intelligence techniques also contributed a great extent in standard-alone. Still there is an immense scope to apply optimization techniques. In this paper, we propose Ant colony optimization techniques to predict software cost estimation based on three datasets collected from literature. For each datasets, we performed tenfold cross validation on International Software Benchmarking Standards Group (ISBSG) dataset and threefold cross validation performed on IBM Data Processing Service (IBMDPS) and COCOMO 81 datasets. The method is validated with real datasets using Root Mean Square Error (RMSE).

Keywords Software Cost Estimation (SEC) · Ant Colony Optimization Technique (ACOT) · Travelling Sales Person (TSO) · Root Mean Square Error (RMSE)

V. Venkataiah
Computer Science and Engineering, CMR College of Engineering and Technology, Medchal, Hyderabad, India
e-mail: venkat.vaadaala@gmail.com

R. Mohanty (✉)
Computer Science and Engineering, Keshav Memorial Institute of Technology, Narayanaguda, Hyderabad, India
e-mail: ramakanta5a@gmail.com

J.S. Pahariya
Computer Science and Engineering, Rustamji Institute of Technology, Tekanpur, Gwalior, India

M. Nagaratna
Computer Science and Engineering, JNTUH College of Engineering, Kukatpally, Hyderabad, India
e-mail: mratnajntu@gmail.com

S.C. Satapathy et al. (eds.), *Computer Communication, Networking and Internet Security*, Lecture Notes in Networks and Systems 5,
DOI 10.1007/978-981-10-3226-4_32

1 Introduction

Software cost estimation is an important tool which can impact the planning and budgeting of a project. Effective monitoring and controlling of a software project is required to estimate the cost, accuracy and quality. Accordingly, in modern society many machine learning techniques employed to find out the software cost estimation i.e. Decision Tree, K-Nearest Neighbor, Support Vector Machine, Neural Networks, and Fuzzy Logic and so on. Neural network [1–4] contributed good work over a decade in the significant field of software to predict cost, effort, and size estimation. These were trained and tested by Back propagation, gradient descent algorithm employed on different prominent datasets viz. ISBSG, IBMPDS, COCOMO 81, DESHARNAIS, CF, and so on. Followed by Fuzz Logic [2, 5–7] and rest of the techniques has also contributed. Further, the Computational Intelligence Standard-alone techniques [8] are Multiple Linear Regression (MLR), Polynomial Regression, Classification and Regression Tree (CART), Multivariate Adaptive Regression Splines (MARS), Radial Basis Function Neural Network (RBF), Counter Propagation Neural Network (CPNN), Dynamic Evolving Neuro–Fuzzy Inference System (DENFIS), Tree Net, Group Method of Data Handling (GMDH) and Genetic Programming (GP) carried out to a great extent and these techniques were tested on different data sources. Still there is immense scope of research, practices in software cost estimation using evolutionary computing techniques such as Genetic programming, particle swarm optimization techniques and ant colony optimization techniques. The predicted method used into a Travelling Sales Person Problem with effort estimation datasets to predict the cost, effort and the methodology results are carried out. Thus the main focus of explanation to be presented here is ACOT for discrete optimization that employed to predict the software cost using TSP where size, product delivery rate given as input and the result determined through RMSE to predict the SCE.

In this paper, we propose Ant Colony Optimization Techniques to predict the software cost estimation. The rest of the paper is organized in the following manner. A brief discussion about literature survey is presented in Sect. 2. In Sect. 3 describes about datasets. Methodology of the paper is presented in Sect. 4. In Sect. 5, presents a detailed discussion of the results and discussions. Finally Sect. 6 concludes the paper.

2 Literature Review

Machine learning techniques have been dominating in the last three decades. The state-of-the-art review published by Mohanty et al. [5] justifies this issue. In the Software Development study of Literature is the majority of significant step. In

Software Engineering significant of SCE is an incredible essential aspect to predict its methods. The recent work carried by Venkataiah et al. [9] application of particle swarm optimization to predict SCE. The author has employed k-means algorithms to clustering of given datasets are as the input. Further PSO applied to clustered data for predicting SCE. Patil et al. [10] has proposed hybrid model consists of two parts are training and classification. The training of Feed Forwarded Artificial Neural Networks using delta rule is based on the Principal Component Analysis (PCA). The PCA is a type of classification method which can filter multiple input values into few certain values. Idri et al. [1] has described about validation and comparison of Fuzzy analog, and Classical analog. The Fuzzy analog was validated on ISBSG, COCOMO 81 datasets and which, given better results compared with Classical analog. Manikavelan et al. [11] has proposed enhancement of Expert Judgment using Deferential Evaluation for accurate Software cost. Azzeh [12] has highlighted and illustrated the possible importance role of UCP in SCE method at multi-site development. Attarzadeh et al. [2] described to handle uncertain and imprecise software attributes using the incorporation adaptive artificial neural networks in the COCOMOII (ANN-COCOMO II). Bardsiri et al. [13] Proposed C-means clustering technique employing the principles of fuzzy logic, to make a suitable method to deal with the uncertainty and complexity of software project features. The performance of proposed method evaluated based on MMRE and PRED (0.25) using threefold cross-validation. Its performance was compared to that of viz. ANN, ABE, CART, SWR, MLR, and so on. They described that the proposed method outperformed of all standard-alone and hybrid techniques. Dejaeger et al. [3] highlighted that data mining techniques can make a better contribution to software effort estimation techniques but it should not replace the Experts Judgment. Attarzadeh et al. [4, 6, 14] presented neuro-fuzzy techniques to handling imprecision and uncertainty. Hari et al. [7] proposed a hybrid method that to construct clustered data using CAK-means algorithm. The PSOIW algorithm applied to clustered data and their parameters also required determining estimate effort. In order to implement the process of classification of a dataset is clustered using NN toolbox. Jianget et al. [8] presented that in the software cost evaluation model, the driver factors were initiate by the grey association degree analysis and it's calculated the relationship between attributes and effort. Hari et al. [15] handling uncertainty and imprecision employed to proposed model structure is Interval Type-2 Fuzzy in a better way. The inputs are fuzzified by using Takagi-Sugeno fuzzy controller of the totality of known with statistical measurements are mean and standard deviation values influence the control performance. Zhang et al. [16] highlighted ample fuzzy-grey theory evaluation method which merges the fuzzy evaluation and grey theory. Pahariya et al. [17] suggested that the recurrent architecture is a hybrid model for genetic programming (RGP), in which output of GP as the input to GP. We also implemented liner ensemble system. Throughout study ten-fold cross validation was performed and the proposed method outperformed compared to all other hybrid techniques were tested on ISBSG dataset.

Malathi et al. [18] the proposed hybrid approach is analogy based on reasoning integrated with fuzzy logic, and logistic variables to handle uncertainty and impression. Mittas et al. [19] proposed a framework for visualization and statistical comparing various cost estimation methods employing an automated tool like StartREC. It is provide strategies for an intelligent decision making. Zhang et al. [20] proposed Bayesian Regression and Expectation Maximization method for software effort prediction and both Missing Data Toleration approach ignores missing data and missing data Imputation approach employs observed data to impute missing data. Miandoab et al. [21] proposed hybrid model of Cuckoo Optimization and KNN with six different datasets i.e. KEMERER, MIYAZAKI1, NASA93, and NASA60. They used different evaluation criteria to predict software cost estimation.

3 Data Description and Data Preparation

In this paper, we have used 3 datasets. Firstly, IBM data processing Service (DPS) consists of 24 projects developed by 3rd generation languages and each project entails of five numerical features is Input Count (IC), Output Count (OC), Enquiry Count (EC), Master File Count (FC), and Adjustment Factor (ADF) employed to estimated project effort. Secondly, in order to encourage reusable, and improvable predictive models of software engineering, the COCOMO 81 dataset [22] collected from the PROMISE Software Engineering data warehouse made available publicly. This data set encompasses of 63 instances and 15 attributes for the effort multipliers, one for KLOC and Actual Effort from 17 attributes. Third dataset is the International Software Benchmarking Standards Group (ISBSG) has been build based on metrics and refined over a decade data collection. The ISBSG-10 dataset contains knowledge about 4106 projects and each project has total numbers of 105 attributes are divided into 18 sub-attributes. In this paper, we predict the work effort summary, i.e. total number of manpower is required to complete the work and then we can easily calculate time and software cost. However, during data preparation and data preprocessing there are a number of issues to be considered into account.

Data cleaning was first step to delete the projects having null values for the attribute work Effort summary. The second step relevant to summary of work effort given 1531 project values for the 5 attributes are Input count, Output Count, Enquiry Count, Mastered File and Interface File. We get only a few projects, if considered more attributes, and then which are not enough for machine learning techniques. Hence we considered five attributes of ISBSG dataset that contains 1531 project values are used for train and test swarm intelligent methods. Finally, the datasets are normalized for simulation.

4 Methodology

4.1 Ant Colony Optimization

Meta-heuristic algorithm is an Ant colony optimization introduced by Dirogo [23], which take inspiration from real ants foraging behavior. The real ants can drop pheromone trail on the ground in order to identify the path by other members of colony while searching for food source and return back on the same way to the nest. It exploits identical mechanism but can dropped pheromone trail on the ground while return trip. When exploring for the food, ants initially look into the surrounding area their nest in the random manner. As soon as ants discover food source, it assess the quality of the food and takes some back to the nest. Ants coordinate their activities via indirect communication and an interaction in the form of a hidden mediated by changes in the environment. For instance the witnessing of social insects to observe colony level behaviors by many biologists has been described in comparatively simple models that use only indirect communication. The idea behind ant colony algorithms is to employ a form of artificial indirect communication to synchronize societies of artificial agents. The ants come after with high probability pheromone streams their sense on the ground.

Each Ant evaluates the next move to another vertex based on Gambardella et al. [24, 25]

$$p_{ij}^{k} = \begin{cases} \frac{[\tau_{ij}]^{\alpha}[\eta_{ij}]^{\beta}}{\sum[\tau_{ij}]^{\alpha}[\eta_{ij}]^{\beta}} & \mathit{if\, j \in allowed\, k} \\ 0 & \mathit{otherwise} \end{cases} \tag{1}$$

p_{ij}^{k} is the probability of worker K to move to vertex "i to j"

τ^{ij} is the amount of pheromone deposited on the edge to "i and j"

η_{ij} is the inverted distance

The cost of tour for each ant is given by d_{ij} the tour cost from the city i to city j is calculated and hence the shortest path is found. This is applied to the Travelling Sales Person Problem and optimized solutions are obtained using

$$d_{ij} = \sqrt{(x_i - x_j)^2 + (y_i - y_j)^2} \tag{2}$$

The amount pheromone deposited is updated by

$$\tau_{ij}^{k}(t+1) = \rho\tau_{ij}^{k}(t) + \sum_{k=1}^{m}\Delta\tau_{ij}^{k} \tag{3}$$

where ρ is a coefficient such that $(1-\rho)$ represents the evaporation of trail rate in period t to t + n.

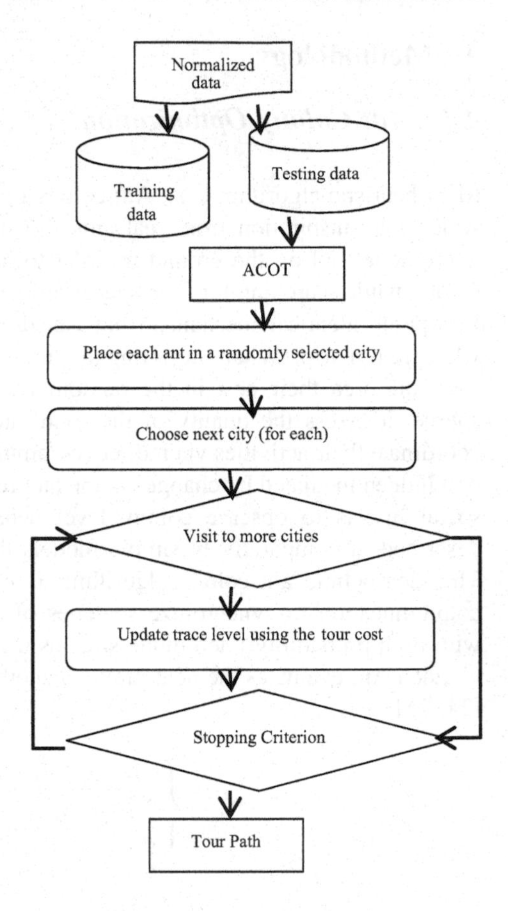

Fig. 1 Flow chart for testing phase of ACOT

$\Delta\tau_{ij}^{k}$ The amount of each pheromone deposited by each trail of ant

$$\Delta\tau_{ij}^{k} = \begin{cases} \frac{Q}{L_k} & if(i,j) \in bestTour \\ 0 & otherwise \end{cases} \quad (4)$$

where, Q is a constant parameter. L_k is the total path length which the kth ant finds, that is, it is the index value of the solution. The work flow of ACOT is depicted in Fig. 1.

5 Results and Discussions

In this paper, we used ISBSG dataset, which contains 1531 projects having five independent variables and one dependent variable. Decimal Scaling [26] is one of the data normalization concepts used to normalize by moving an attribute B of the decimal point values. The number of decimal points moved depending on the maximum absolute value of B. A value s of B is normalized to s' is computing by

$$s' = \frac{s}{10^k} \tag{5}$$

where k is the least integer such that Max $(|s'|) < 1$

In this study, we measure the project effort estimation [27] as follows:

$$\text{Effort} = \text{C} * \text{Size} \tag{6}$$

where C is a constant (Product delivery rate), effort is measured in terms of person-hours and size also measured in terms of KLOC (Kilo Line of Code).

Function Point Analysis (FPA) [28] is evaluated and applied to calculate size.

$$\text{FP} = \text{UFP} * \text{VAF} \tag{7}$$

where VAF is value adjusted point for evaluating the environment and complexity of processing the project and Unadjusted Function Point (UFP) is expressed as:

$$\text{UFP} = (\text{External Input} * 4 + \text{External Output} * 5 + \text{External Enquiry} * 4 + \text{Log File} * 10 + \text{Interface File} * 7) \tag{8}$$

where EI, EO, EQ, LF, and IF are functional type multitier by weight is resented by fictional complexity.

We simulated our experiment by writing code in java and executed in eclipse integrated development tool. The Simulation of ACOT using Traveling Sales Problem. The experiments are performed on tenfold cross validation by dividing data into training and testing set in the ratio of 90:10. The performance of estimated methods carried out based on Root Mean Square Error (RMSE), which defined as follows:

$$\text{RMSE} = \sqrt{\frac{1}{\text{n}}\sum_{\text{i}=1}^{\text{n}} (\text{Effort}_{\text{actual}_\text{i}} - \text{Effort}_{\text{estimated}_\text{i}})^2} \tag{9}$$

where n is the number of projects Effort $_{\text{actual}}$ is actual effort is given in datasets and Effort $_{\text{estimated}}$ predicted effort from proposed methodology.

While simulating, we chose the following parameters for our experimental analysis by using ACOT are presented in Table 1.

We presented the results of the RMSE values of ACOT in Table 2. The average accuracies of RMSE values of 11 techniques viz., GP, CPNN, CART, TREENET, MLP, MLR, DENFIS, MARS, SVR, RBF, and POLYNOMIAL REGRESSION are 0.03794, 0.04499, 0.04561, 0.04565, 0.04817, 0.04833, 0.04837, 0.04871, 0.04922, 0.05167, and 0.05327. We found that our results by employing ACOT is outperformed other techniques in case of ISBSG dataset is concerned.

Our simulation results of COCOMO-81 and IBMDPS datasets are presented in Tables 3 and 4 respectively. In Table 3 the RMSE value of Fold1, Fold2 and Fold 3 are 0.023184, 0.012899, and 0.007884, respectively and Average RMSE value of ACOTT is 0.014656, which is compared with Average RMSE value of PSO [9] value is 0.157119. We found that ACOT is performed better than other technique [9].

Further from Table 4. The RMSE values of Fold1, Fold2 and Fold 3 is 0.007036, 0.008485, and 0.007036, respectively and Average RMSE values of ACOT is 0.007519, compared with Average RMSE values of PSO value is 0.139438 [9] (Table 5).

Table 1 Ant colony parameters

S.L No	Name of the parameter	Value
1	Alpha	−0.2d
2	Beta	9.6d
3	Pheromone_ Persistence	0.3d
4	Initial_ Pheromone	0.8d
5	Q	0.0001d
6	No. of agents	2048*20

Table 2 Average RMSE of tenfold cross validation on ISBSG

S.L. No	Method	RMSE (test)
1	ACOT	**0.00817**
2	GP	0.03794
3	CPNN	0.04499
4	CART	0.04561
5	TREENET	0.04565
6	MLP	0.04817
7	MLR	0.04833
8	DENFIS	0.04837
9	MARS	0.04871
10	SVR	0.04922
11	RBF	0.05167
12	POLYNOMIAL REGRSSION	0.05327

Table 3 Fold wise and Average RMSE values of threefold cross validation on COCOMO 81

S.L. No	Method	RMSE (test)
1	Fold1	0.023184
2	Fold2	0.012899
3	Fold2	0.007884
4	Average ACOT	**0.014656**
5	Average PSO	0.157119

Table 4 Fold wise and Average RMSE values of threefold cross validation on IBMDPS

S.L No	Method	RMSE (test)
1	Fold1	0.007036
2	Fold2	0.008485
3	Fold2	0.007036
4	Average ACOT	**0.007519**
5	Average PSO	0.139438

Table 5 Comparison of RMSE values

Dataset	Sheta [29]	Proposed model
COCOMO 81	8.1023	0.014656
IBMDSP	7.0123	0.007519

We compare our simulation results of ACOT employed on COCOMO 81 and IBMDPS datasets with Sheta et al. [29]. We found that the RMSE values of our proposed model outperformed the results obtained by Sheta [29].

6 Conclusion

In this paper, we employed ACOT to predict Software Cost Estimation using different effort estimation datasets. Throughout the study, we performed tenfold cross validation on ISBSG dataset and threefold cross validation on IBMDP and COCOMO 81 datasets. We compared performance of ACOT that of PSO, GP, TREENET, MLP, DENFIS, MARS and SVR. Our experimental results in terms of RMSE value were better to that of other stand-alone techniques. Hence, we conclude that after extensive experimentation that the ACOT model is relatively best predictor among all the other techniques.

References

1. Ali, I., Azeddine, Z.: Software Cost Estimation by Classical and Fuzzy Analogy for Web Hypermedia Applications: A replicated study. IEEE Symposium on Computational Intelligence and Data Mining (CIDM), pp. 117–121, (2013)
2. Attarzadeh, I., Merhanzadeh, A., Ali, B.: Proposing an Enhanced Artificial Neural Network Prediction Model Improve the Accuracy in Software Effort Estimation. IEEE Fourth International Conference on Computational Intelligence, Communication Systems and Networks, pp. 167–172, (2012)
3. Dejaeger, K., Verbeke, W., David, M., Bart B.: Data Mining Techniques for Software Effort Estimation: A Comparative Study. IEEE Transactions on Software Engineering, Vol. 38, No. 2, March/April (2012)

4. Attarzadeh, I., Hock, O. S.: Proposing a New Software Cost Estimation Model Based for Software Cost Estimation. IEEE 2nd International Conference on Computer and Electrical Engineering, pp. 112–116, (2009)
5. Mohanty, R. K., Ravi, V., Patra, M. R.: The Application of Intelligent and Soft-computing Technique to Software Engineering Problems: A state of the art Report. International Journal of Information and Decision Sciences, Vol. 2, Number 3, pp. 232–272 (2009)
6. Attarzadeh, I., Merhanzadeh, A., Ali, B.: Proposing an Enhanced Artificial Neural Network Prediction Model to Improve the Accuracy in Software Effort Estimation. IEEE Fourth International Conference on Computational Intelligence, Communication Systems and Networks, pp. 167–172, (2012)
7. Hari, CH. V. M. K., Tegjyot S. S., Kaushal B.S. S., Abhishek S.: CPN-a hybrid model for software cost estimation. IEEE International Conference on Recent Advances in Intelligent Computational Systems (RAICS), pp. 902–906, Sep 22, (2011)
8. Jiang, G. Wang,Y., Haitao.: Research on Software Evolution Model on Case Based Reasoning. IEEE 2nd International Conference on WRI World Congress on Software Engineering, pp. 338–341, (2010)
9. Venkataiah, V., Mohanty, R.K., Nagaratna, M.: Application of Practical Swarm Optimization to predict Software Cost Estimation. 6th IEEE International Conference on Communication Systems and Network Technologies, 05–07, March (2016)
10. Lalit Patil, V., Nitin Shivale, M., JoshiJ, D., Khanna, V.: Improving the Accuracy of CBSD Effort Estimation using Fuzzy Logic. IEEE International Advance Computing Conference, pp. 1395–1391, (2014)
11. Manikavelan, D., Ponnusamy, R.: To Find the Accuracy Software Cost Estimation Using Differential Evaluation Algorithm. IEEE International Conference on Computational Intelligence and Computing Research, (2013)
12. Azzeh, Mod.: Software Cost Estimation Based on Use Case Points for Global Software Development. IEEE 5th International Conference on Computer Science and Information Technology (CSIT), pp. 214–218, ISBN: 978-1-4673-5825-5, (2013)
13. Khatib Bardsiri, V., Jawawi, D.N.A., Hashim, S.Z.M., Khatibi, E.: Increasing the accuracy of software development effort estimation using project clustering. The Institution of Engineering and Technology Journal, Vol.6, Iss.6, pp. 461–473, (2012)
14. Attarzadeh, I., Hock, O. S.: Proposing a New Software Cost Estimation Model Based on Artificial Neural Networks. IEEE 2nd International Conference on Computer Engineering and Technology, Vol. 3, pp. 287–291, (2010)
15. Hari, CH. V. M. K., Prasad Reddy, P. V. G. D., Jagadeesh, M., SriRam Ganesh, G.: IntervalType-2 Fuzzy Logic for Software Cost Estimation Using TSFC with Mean and Standard Deviation. IEEE International Conference on Advances in Recent Technologies in Communication and computing, pp. 40–44, (2010)
16. Zhang, B., Zhang, R.: Evolution Model of Software cost estimation methods based on Fuzzy-Grey Theory. IEEE Fourth International Conference on Internet Computing for Science and Engineering, pp. 52–55, (2009)
17. Pahariya, J.S., Ravi, V., Carr, M.: Software Cost Estimation using Computational Intelligence Techniques. IEEE Conference on World Congress on Nature & Biologically Inspired Computing (NaBIC 2009), pp. 849–854, (2009)
18. Malathi, S., Lijin, B.S.: An Efficient Method for the Estimation of Effort in Software Cost. International Journal of Advance Research in Computer Science and Management Studies Volume 2, pp. 330–335, February (2014)
19. Nikolaos, M., Mamalikidis, I., Angelis, L.: A framework for comparing multiple cost estimation methods using an automated visualization toolkit. Information and Software Technology Vol. 57, pp. 310–328, (2015)
20. Zhang, W., Yang, Y., Wang, Q.: Using Bayesian Regression and EM algorithm with missing handling for software effort prediction. Information and Software Technology, pp. 58–70, February (2015)

21. Miandoab, E., Gharehchopogh, F. G.: A Novel Hybrid Algorithm for Software Cost Estimation Based on Cuckoo Optimization and K- Nearest Neighbors Algorithms. International Journal of Engineering, Technology & applied Science Research. Vol. 2, No. 3, pp. 1018–1022, (2016)
22. Boehm, B.: Software Engineering Economics. Prentice Hall, (1981)
23. Coloni, A., Dorigo, M., Maniezzo, V.: Ant system: Optimization by a colony of cooperating agent. IEEE Trans. Systems Man and Cybemetics-Part B: Cybemetics, vol. 26, No. 1, pp. 29–41, (1996)
24. Dorigo, M., Dicaro, G.: The Ant Colony Optimization Meta-Heuristic. In Corne, D., Dorigo, M., Glover, F. editors, New Ideas in Optimization, McGraw-Hill, pp. 11–32, (1999)
25. Dorigo, M., Gambardella, L. M.: Ant Colony System: A cooperative learning approach to the Traveling Salesman problem. IEEE Transactions on Evolutionary Computation, vol. 1, No.1, pp. 53–66, (1997)
26. Han, J., Kamber, M., Pei, J.: Data Mining: Concepts and Techniques. Morgan Kaufmann Series in Data Management Systems, (2006)
27. Bhardwaj, M., Ajay, R.: Estimation of Testing and Rework Efforts for Software Development Projects. Asian Journal of Computer Science and Information Technology, ISSN.2249–5126, pp. 33–37 (2015)
28. Pressman, R. S.: Software Engineering: A Practitioner's Approach. McGraw-Hill series in Computer Science, New York, (2001)
29. Sheta, A.F., David, R., Ayesh, A.: Development of software Effort and Schedule Estimation models using Soft Computing Techniques. IEEE Conference on Evolutionary Computation, pp. 1283–1288, (2008)

Cloud-Based e-Healthcare Service System Design for On-Demand Affordable Remote Patient Care

Parvathy Dharmarajan and B. Rajathilagam

Abstract In this paper, a low-cost system design for e-healthcare service including software and hardware components is presented. Vital signs of the human body are measured from the patient location and shared with a registered medical professional for consultation. Temperature and heart rate are the major signals obtained from a patient for the initial build of the system. Data is sent to a cloud server where processing and analysis is provided for the medical professional to analyze. Secure transmission and dissemination of data through the cloud server is provided and an authentication system, a secure storage server for the cloud is included for control by the patient from a smart phone. A prototype of the system is built with all the components for testing and the challenges of implementing the system in real time have been discussed.

Keywords Arduino UNO · Healthcare · Heart rate sensor · Temperature sensor · FFT

1 Introduction

Medical treatment has become a very costly, complicated and time consuming nowadays. With growing population hospitals are always crowded and patients are made to wait a long time before meeting a doctor. Though emergency medical care is given to deserving patients, others are always in a queue. Existing systems work

P. Dharmarajan (✉)
Department of Computer Science and Applicaion
Amrita School of Engineering Amritapuri, Amrita Vishwa Vidyapeetham
Amrita University, Amirtapuri, India
e-mail: parvathydvlp@gmail.com

B. Rajathilagam
Department of Computer Science and Engineering, Amrita School of Engineering
Coimbatore, Amrita Vishwa Vidyapeetham, Amrita University, Coimbatore, India
e-mail: b_rajathilagam@cb.amrita.edu

S.C. Satapathy et al. (eds.), *Computer Communication, Networking and Internet Security*, Lecture Notes in Networks and Systems 5,
DOI 10.1007/978-981-10-3226-4_33

in such a way that patients have to go through a rigorous procedure even if they do not require serious attention. Telemedicine systems are a boon to people living in remote areas where they are connected to hospitals in cities for consultation with expert doctors. If a person or a community is equipped with simple cost effective devices where a measure of the vital signs is collected and sent to a registered doctor for consultation online, then every case need not rush to a hospital. Doctors can judge the severity of the patient's ailment and recommend a direct consultation and further tests if necessary. In this work, an end-to-end system with a patient portal and doctor portal hosted on a cloud server is proposed to be connected to a simple healthcare device that can post the measured data on the secure cloud. The system scalable to include more components which can be added to the device for additional measurements and data can be pushed to the cloud server. Being low-cost system is capable of working with low bandwidth internet connectivity while connecting a registered patient with a registered doctor 24 $\times$ 7. The system may be used for post medical care and follow up services that may be extended to their home instead of frequent hospital visits. To build a system of this kind needs solutions to many challenges. It requires the support of doctors, a secure cloud service provider and a standard to protect the privacy of patient data. In spite of the difficulties involved many service providers have started providing software support for extending hospital services online.

2 Background and Motivation

Work on remote medical aids has been consistently happening in the past few decades. Healthcare units for monitoring the elderly patients were carried out in [1, 8], using bio-signals measured with a Smartphone. Most of the Arduino projects for medical aids deal with body temperature, body positioning, heart rate etc. [2]. This is basically used for measuring the ECG of the patient. The focus of the projects has been carried out in a most effective manner using a well-equipped tool [3, 7]. Analyzing the data, doctors can identify and advise exact medicines for the patients. Using latest technologies, they are able to reduce the workload of the both patients and doctors [4, 6]. Researchers have proposed for creating a remote medical health-care for people in rural areas where availability of hospitals is profound [5].

3 Proposed System Design

The proposed system contains an arduino board and sensors as hardware components and a cloud service which connects to a patient portal and doctor portal running a smart phone as software components. The design of the model is shown in the Fig. 1. The patient's data is measured using sensors and arduino board takes care of sending the data to cloud service. The data is viewable by a healthcare

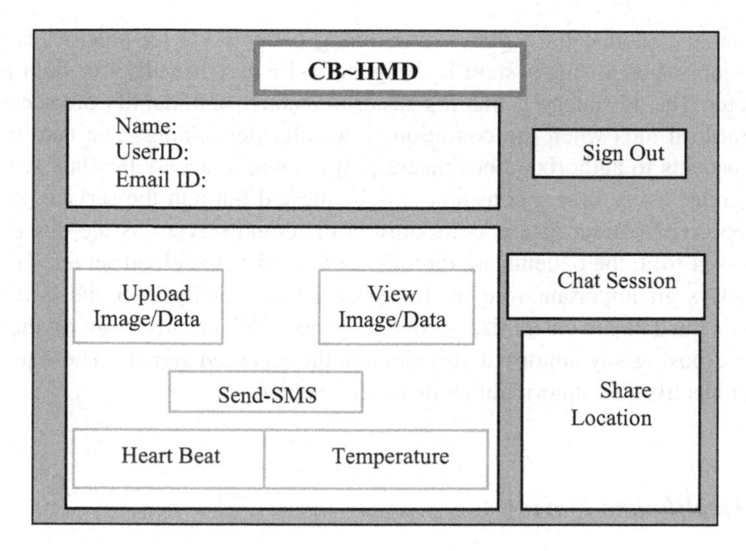

Fig. 1 Patient interface

professional via smart phone. Analysis is done on the cloud service. Initially both patient and doctor should register to the cloud service. The patient is required to explicitly permit the healthcare professional to view his profile and data for any consultation. Only after the patient is registered into the system, data could be sent to the cloud service. A healthcare personal can analyze the data and take a decision on the severity of the patient's health condition. Its cost effective, as the components in this system are open sourced, and have been used for varies experiments. Therefore it's easily accessible, affordable, and at the same time easier to be assembled. They can reduce the time and cost consumption for both patients and doctors. As the doctor is able to analysis more than one patient at a time and the patient don't have to spend time for the consultation procedures. Moreover a patient or a family member can handle the system themselves and reduces the cost of getting assistance from a medical assistant at home or nearby hospital. The patient need not be educated for any special medical equipment usage, they can use the regular smart phones and webpage for the application as well as simply insert their finger for the heart beat in the specified slot. The system usage is not very complicated.

3.1 Interactive Cloud Storage

The signal that is received from the Arduino board is forwarded to the cloud service and authorized patient is requested to upload their data to the server for further processing. Continuous uploading of data ensures that the data is appended to the earlier data, to store all the observations over a period of time is stored. The data is processed and plotted on a graph. Examining patients via the Internet is a challenging job, whereas this system is designed to be user-friendly, for both patients and doctor. The advantage is that it's easier to manage and identify the necessity of direct medical aids when the condition is actually demanding. The user interface allows patients to authorize a healthcare professional to access the data sent to the cloud service every time a consultation is requested through the service. Analyses of the observed patient data is done only in the cloud server. As the raw data that are received from the patients are directly forwarded to the cloud server. The cloud server plays an important role in this system as an analyst for the doctor. The reports on the data are customizable by the doctor. FFT is carried out on the patient data for exposing any abnormal variations in the recorded signals. These technique helps to identify the abnormalities in the frequency.

3.2 *Healthcare Device*

The hardware components are capable to measure the heartbeat and temperature of patients. The signals are sent to the server after encryption for security of patient data during transfer. Data is encrypted during patient registration, access control, device authentication, and data access. The patient can use a smart-phone for all the services. It's simple and user-friendly and allows patients to keep a close association with a healthcare professional for adequate medical assistance. Encryption–decryption of data must not take much time to work in a real-time environment. More professional authentication feature must be identified for the reliability of the data that have been sending for registering into the system. Validation of the data that are entered have be carried out efficiently.

4 Software Components of the System

System basically constitutes of two main portals namely patient portal and doctors portal. In patient portal, each patient can create profile and user account in the cloud service. Patients can also view their uploaded data or profile anytime. The system acts as a preliminary healthcare assistant before actually going to a hospital for a serious ailment. The system shares the location of the patients on demand. The system also enables a chat session for simple communication between the two parties.

The interface of the doctor includes the ability to view any assigned patients and their data for analysis. The basic analysis reports on the measured data will be displayed to the doctor. If there is any abnormality in the data, graphical analysis helps in easy identification. The plotting of data is done without much latency for the doctor to check the fluctuations in data real-time. Doctors can comment data, and these comments are sent to the patients along with the analysis report. The system supports plotting multiple graphs together as well as individually for the convenience of doctor's diagnosis.

4.1 Data Security

Patient signals are forwarded to the cloud service from a smart device using Wi-Fi. Once the patient is registered and authenticated by the system, they are able using their login credentials to upload their data to the server. Once they have done the process, an SMS notification is automatically sent to the doctor. Every data sent to the server and received from the server is encrypted on both the patient portal and doctor portal for secure data transmission. Location of the patient can be viewed by the doctor. As if the vital status of the patient shows any abnormality, then their location can be sketched for any critical situations. It's also customized in such a way that the patient takes the initiative to share their current location. Chat session for both doctor's and patient's interface in order to communicate with each other if required. It's an easy mode for both parties to share the information within a particular session. The patient's data is stored in the cloud server in an encrypted form. This authentication ensures that nobody accesses the patient data without permission of the patient. Every doctor and patient has to be registered as a user for the services (Fig. 2).

5 Experimental Setup

Heartbeat and temperature sensors are used for sensing with the fingertip of a person. A microcontroller such as Arduino and various other tools have been used for developing the device. Figure 4, shows the hardware setup of the healthcare device.

The experimental setup of this system constitute of a microcontroller, sensors and a device that's connected to the internet. The microcontroller used here is

Fig. 2 Physical view of the device

Arduino UNO which has a number of services for communicating with the computer. ATmega328 attains a serial communication through the USB and has a virtual com port to the computer. The heart beat sensor works by giving 660 nm digital signals when the finger is being placed. The detector makes the LED light blink for each heartbeat. The signals attained from the sensor are connected to the microcontroller which helps to measure the rate of the heartbeat every minute (BPM). Arduino uses two consecutive operational amplifiers to establish a baseline of signals, identifying the peaks and filter out the noise. The temperature sensor used is LM35 series which has a linear temperature sensor. The sensor characteristics depend on the environment (physical/electrical), the range of temperature, accuracy, response speed and as well as the thermal coupling.

6 Result and Analysis

In Fig. 3, the temperature variation of a particular patient in given period of time. Using this data doctor will be able to analyze the frequency of variation of the body temperature of the patient. The peculiarity of this graphs that the graph can be zoom and the values can be viewed for even between 2 s. From the Fig. 4, the heart rate variation of the patient is shown over a specified set of the time period. Resultant data measured from the sensors are uploaded to the server where they are processed. The graph is plotted such that they can be zoomed into monitor smaller variations of the signal. From the observed data, it has been possible to identify the abnormality in the patients. The standard value of both temperature and heart rate is known. The doctors can identify any minute fluctuation is the data. In this system, temperature is measured in terms of Fahrenheit and heart rate in beats per minute. The temperature of a human body in terms of Fahrenheit ranges from 97.3–99 F and heartbeat ranges from 60–110 BPM for a healthy person.

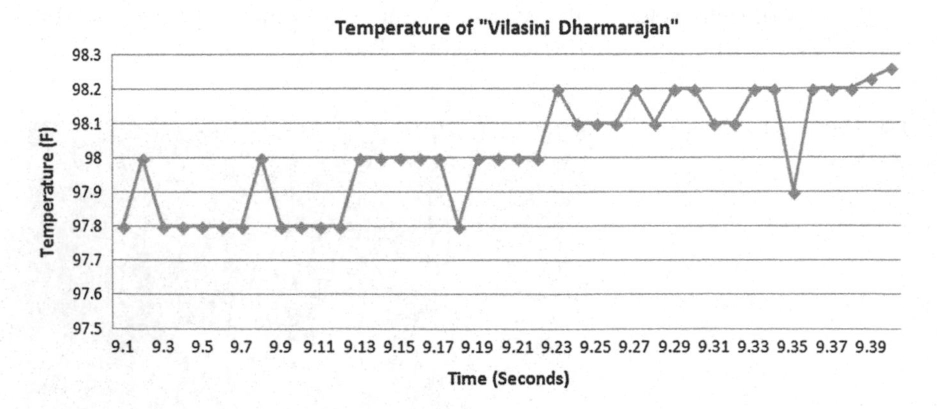

Fig. 3 Temperature sensor data

These value can differ in terms of gender, age, physical ability etc. therefore here we have identified the basic conditions of the patients and for the future works tries to incorporate modification on the data that are attained and tries to check the condition of the patients and analyzed accordingly. Fast Fourier Transform (FFT) methods have been used for analysis of signal data. As this technique is used for analyzing even very small fluctuation in the attained data which help the medical experts to analyses. The generated FFT graphs also prove to be helpful for the any future analysis for the patient. Figure 5, shown the depiction of FFT graph in a particular time period for temperature data similar way have worked on heart rate data. ASE algorithm has been used for securely storing data to the cloud. ASE algorithm has got various benefits over this system, as the execution time has been much faster compared to any other. They take very low memory space and provide best authentication facility. Moreover, this algorithm is capable of handling a large data for encryption and at the same time can secure both user and provider.

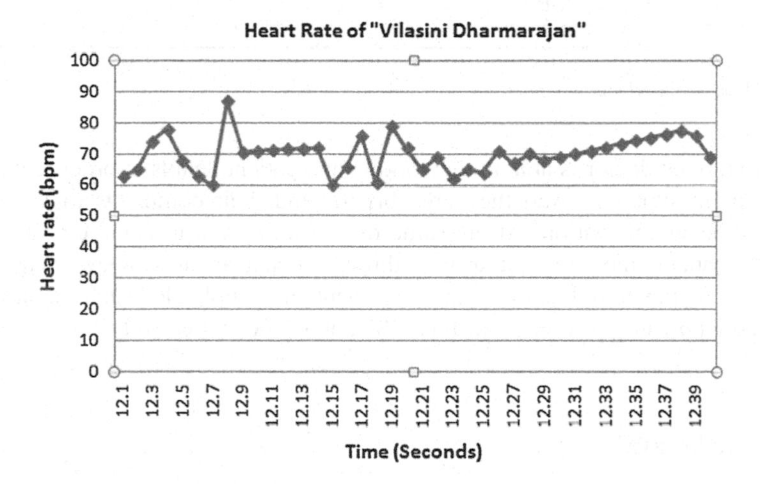

Fig. 4 Heart rate sensor data

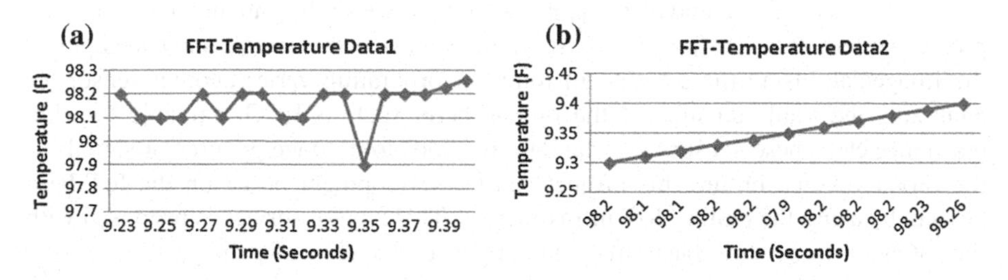

Fig. 5 Analysis using FFT in the temperature graph **a** peaks during the time interval **b** variation during the time intervals

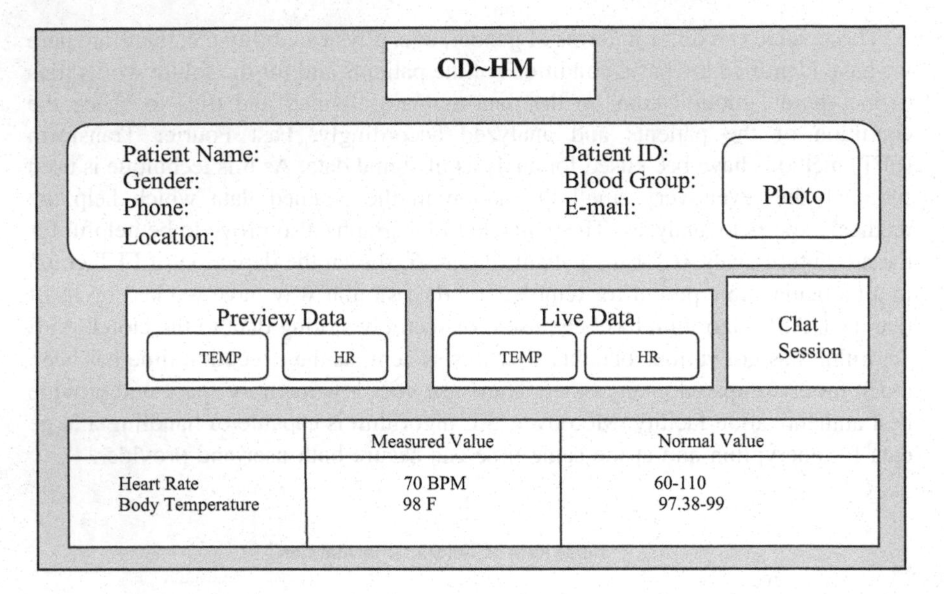

Fig. 6 Doctors interface

In Fig. 6 the doctor is able to view individual patient details in order to analyses their current situation. And they are able to send their comments to the patient through the submit button. At the same time if they want to communicate longer with the patients they are able to chat through a chat window. View image tab is used to view the complete data that the patients have uploaded such pictures, file, data and report in jpg, document file, CSV, and PDF respectively.

7 Conclusion

The system is developed for cost effective patient care using sensor data routing. Patient's data is sent to and stored in a cloud server for processing. The analysis of patient data is available and also enables keeping track of the patient data. Using the present system, the user (patient/assistant) has access to the device, and can captures the image/data using the sensors. In future incorporating varies another sensor for analyzing the vital statistics of the human body such as ECG, galvanic for skin resistance etc. These sensors use standard methods to produce pictures or signals of the blood vessels. Future this project can be development based on the IoT base device, by using the exiting IoT infrastructure. This type of service is proposed with the intention to provide basic medical care at the location of patient.

References

1. Agham, N. D., Thool, V. R., & Thool, R. C. (2014). Mobile and Web Based Monitoring of Patient's Physiological Parameters using LabVIEW. 1, p. 6. doi:978-1-4799-5364-6/14.
2. I. Orha., & S. Oniga. (2014, October 23). Study regarding the optimal sensors placement on the body for human activity recognition. doi: 10.109/SIITME.2014.6967028.
3. Mallick, B., & Patro, A. K. (2016, January). Heart Rate Monitoring System using Finger Tip Through Arduino and Processing Software. International Journal of Science, Engineering and Technology Research, 5(1). Retrieved June 15, 2016.
4. Julio, M. T. (2015). Development of a Prototype Arudino Mobile in Area of Telemedicine for Remote Monitoring Diabetics People. 1, pp. 36–40. doi: 10.1109/APCASE.2015.14.
5. Kumari, P., & Vinita, Y. (2015). Heart Rate Monitoring and Data Transmission via Bluetooth. International Journal of Innovative and Emerging Research in Engineering, 2(2).
6. Hamed, S. A., & A. Anand. (2013). Data acquisition and control using Arduino-Android Platform. IEEE, 241–244.
7. S.U, U., C.O, O., & M.E, U. (2015, August). Heart Monitoring And Alert System Using GSM Technology. International Journal of Engineering Research and General Science, 3(4).
8. Zainee, N. M., & Chellappan, K. (2014, December). Emergency Clinic Multi-Sensor Continuous Monitoring Prototype Using e-Health Platform. pp. 32–37. doi: 978-1-4799-4084-4/14.

References

1. [illegible] Thool, V.R., & Thool, R.C. (2016). [illegible] Web Based Monitoring [illegible] Physiological Parameters using LabVIEW [illegible] p. 6, doi:978-1-[illegible]
2. [illegible] (2015). [illegible] the optimal sensor placement on the body for human activity recognition [illegible] doi:10.1109/[illegible]
3. [illegible] A. K. (2015, January). Heart Rate Monitoring System using [illegible] and Processing Software. International Journal of Science, Engineering and [illegible] June 15, 201[illegible]
4. [illegible] (2015). Development of [illegible] Android Mobile Application [illegible] 2015.
5. [illegible] (2017). [illegible] Health Monitoring and Data Transmission in [illegible] Journal of [illegible] Emerging Research and Engineering, 2(1).
6. Hamed, S. A., [illegible] (2016). [illegible] Arduino [illegible] Platform, 11(3), [illegible]
7. [illegible] (2015, August). Heart Monitoring And Alert System Using GSM Technology. International Journal of Engineering Research and General Science, 3(4).
8. [illegible] Challenges [illegible] Multi-Sensor [illegible] Platform [illegible] 978-1-[illegible]

A Streamlined Approach of Cloud Certificate Less Authentication Scheme Using ASCII Code Data Encryption Technique

AyodhyaRam Mohanthy and Jagadish Gurrala

Abstract Cloud storage public services changed the way of the user access in the internet i.e. (1) Click (2) Download (3) Install (4) Use (time taking 4 steps) to click and use (time saving 2 steps). This easier accessibility for wider users lead to a threat of data confidentiality. Encrypting the data before uploading into public clouds is not possible all the times. So to provide the confidentiality of stored public cloud data, the encryption mechanism inside the cloud should be able to support the access of encrypted data. For balancing the security and efficiency we are proposing a public key encryption scheme for generating a secret key for both encryption and decryption using ASCII codes in the encrypting process.

Keywords Secret key auditing protocol · ASCII based encryption · Group key generation · Certificate-less authentication · Cloud computing

1 Introduction

Cloud data encryption is a kind of data transformation strategy followed by cloud providers by some encrypting algorithms [1] for transforming plain data into cipher text. Many providers provide basic encrypting only that to on a few data fields, i.e. passwords and secret account numbers. As encryption consumes much processor overhead. Encrypting a user's whole DB can be of much cost that it might make more sense to save the whole data in house or encrypt the whole data before sending it to the public cloud. For lower costs, some of the cloud providers are offering alternative encryptions [3–5, 10, 12] that doesn't require much processing power i.e. redacting/obfuscating [8] and use of the provider's own encryption

A. Mohanthy · J. Gurrala (✉)
Computer Science and Technology, Department of Computer Science and Engineering, ANIL Neerukonda Institute of Technology and Sciences (ANITS), Visakhapatnam, India
e-mail: gjagadish.cse@anits.edu.in

A. Mohanthy
e-mail: ayodhyarammohanthy@gmail.com

S.C. Satapathy et al. (eds.), *Computer Communication, Networking and Internet Security*, Lecture Notes in Networks and Systems 5,
DOI 10.1007/978-981-10-3226-4_34

algorithms created by the provider himself. Encryption is all about protecting enterprise cloud data from unauthorized access and cyber criminals.

1.1 Cloud Data Encryption Issues

For various reasons, enterprises rely on their cloud service providers to maintain ownership and management of the keys with mathematical transformation of data which is undecipherable without the "key" and must be able to get data back to its initial form.

1.2 Importance of the Subject, Who Holds Encryption Keys

Law enforcement authorities may take control. Ceding control [6] of the encryption keys might also make the firm more susceptible to cyber/rogue employees.

1.3 Using Well Vetted Algorithms with Very Strong Proofs of Security

Firms IT Security teams should ensure the strength of the encryption algorithm peer reviewed security proofs and implications on the end users of the applications in the cloud.

2 Related Work

2.1 Deployment of Infrastructure

In Public Key Infrastructure [12], a form of a certificate delivered the assurance i.e. A signature by a (CA) Certification Authority on a public key. There are many issues associated with CA and management including revocation, storage, distribution, computational cost of verifying the certificate. Identity-Based Public Key Cryptography (ID-PKC) tackles the problem of authenticity in different ways from traditional PKI.

In ID-PKC, certain aspects of its identity of an entity's public key is derived directly. Private key is generated by a trusted/authorized 3rd party i.e. Private Key Generator. Direct derivation of public keys in ID-PKC, eliminates the need of certificates [3, 15, 16]. On the other hand, PKG uses a system-wide master key to generate private key. ID-PKC can't offer true non-repudiation as traditional PKI can. Key escrow problem can be solved by introducing multiple PKGs and threshold techniques. But this involves extra communication and infrastructure.

PKG's master key compromising could be very disastrous in ID-PKC, and usually more severe than the compromise of a CA's signing key. Public cloud shouldn't learn any data and must rely on cloud users instead of cloud providers. Improvement Policy of Access control either demands decryption/re-encryption of the data, which exposes plain data inside the cloud/the data owner again have to encrypt the whole data and upload it again to the cloud bringing bandwidth, computation cost to the data owner.

2.2 *Secret Key Management*

Another tough task for access control based encryption is Key management. Data owner may manage the key himself, e.g. by sending encryption/cipher key to each user assuming that each client has a public-private (key pair) and data is encrypted with the client's public key introducing lot more computational load at the data owner side. In addition, as there will be membership changes frequently, it makes the group key management complex for general users. Cloud should deliver the encryption keys respecting pre-defined policies of access control from data owner.

3 Existing System

With rapid increase of cloud storage offerings, critical issue of data confidentiality raised. In order to assure security and confidentiality a common approach is encrypting the data before uploading it. As cloud storage is not aware of the keys used for decrypting, the confidentiality is expected. Many providers force access control of the data to be fine grained, as encryption mechanism must support it. Encrypting dissimilar data item sets to which the same AC Policy applies with dissimilar symmetric keys is the approach followed by users. Key derivation based approach lower the number of keys to be managed. Traditional PKC [7] requires a fully trusted CA to issue a Digital Certificate by (binding users and public keys) and later generate an own signature on each client's public key. Whole Certificate Management (CM) is very expensively complex for managing each client's certificate. For addressing such shortcoming Identity Based PKC came into the picture but has the same key escrow problem e.g. KGS (Key Generation Server) reads the secret keys of all clients. Attribute Based Encrypting finally allows one, for encrypting each data item based on the AC policy applicable. However, the revocation problem still exists-secret keys issued must be updated by the existing cloud users whenever a cloud user is revoked.

4 Proposed System

In the proposed system, we are implementing certificate less public key cryptography [2] by using the secret key auditing protocol for encrypting [9] and decrypting the shared data of the public cloud. Cloud mechanism authenticates the user data of each client before the data is shared. The cloud server generates the secret key and finally the cloud server will cast that secret key to all the clients with the authentication code (Fig. 1).

4.1 Secret Key Auditing Protocol

The security mechanism [4], generates the group keys and sends it to all the members.

1. The cloud server will randomly choose the secret points (xi, yi) and send it to the individual users.
2. Users will retrieve the secret points and choose the random challenge.
3. After choosing the random challenges (ri) [15], the value is sent to the server (cloud).
4. All the random challenges are retrieved by the server and generates the verification share (xi, yi ® ri) and send it to the individual clients.
5. The client retrieves the share and again generates the verification share (VS).
6. As VS is generated, now the verification of the cloud user can be done as authenticated or not.
7. The cloud server generates Langrange polynomial F(x) = secret key + bx + ax^2 Secret keys a, b are constants and generated randomly.
8. The generated six points will satisfy that Lagrange's polynomial equation.
9. The cloud server sends any three points to individual cloud users and then user will retrieve all those three points and generates Lagrange's polynomial again.

After getting the secret key each user will use that key for encryption and decryption [13–16] of the shared data. Before sending the subset of points to the

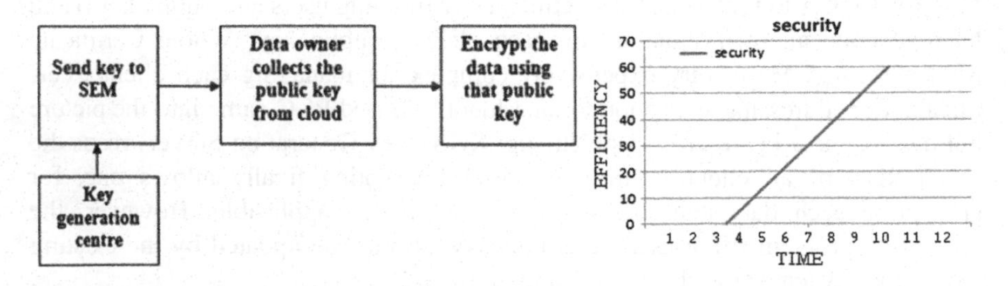

Fig. 1 Key Generation and efficiency graph

individual users, the cloud server will send the secret key to the data owner. The data owner retrieves the secret key and perform encryption [5]. The data is encrypted by the owner before storing the data into the cloud server. A user can access that data after decrypting it.

4.2 Encryption Process

The encryption process initially based on the characters of plain format [6–10], helps in calculating ASCII codes of different characters [11, 12]. In this encryption process, the plain format data is composed of lower letters, capital letters, Arabic numbers and special characters. We can transform all plain format data into cipher format by using the process.

The transformation of plain text into cipher text follows.

1. The plain format data if contains character A–M, it will be transformed into plain text ASCII code value + 45.
2. If the character of plain text contains N–Z, it will be transposed into plain text ASCII code value + 19.
3. The plain text range between a–m, will be transposed into plain text ASCII code value − 19.
4. The plain text range between n–z, will be transposed into plain text ASCII code value − 45.
5. In the 2nd process of classification, the plain character range between 0–4 of cipher (text) multiplies with 2 + 1.
6. The plain text character range 5–9, will be transposed to the cipher text equal to plain text multiplied by 2–10.
7. In the third classification process, if the plain text character contains special character, the cipher text value will also be same as the plain text character.
8. After completion of step 7, result will be taken and converted those characters' values into ASCII code.
9. Take the individual ASCII code [14] of each character and convert into binary format.
10. Take that binary format data and generate 32 * 32 matrix format.
11. After performing, take that matrix and perform the shifting operation on that matrix.
12. The outer circle of matrix will be rotated in clock wise direction and also performs rotated process alternatively as circle in the matrix for completion of all circles.
13. Take the inner circle of matrix, as outer circle will be rotated in anti-clock wise, this process is repeated with total inner circles of matrix.
14. After completion of this process, take that binary data and convert into ASCII format.

The data owner will take that ASCII format and store into storage server. The storage server will contain all cipher format related files and also maintain list of available files in the server. If any user wants that data, he must retrieve and perform decryption process.

4.3 Decryption Process

Decryption process follows as the encryption process, but the generation of sub keys is actually inverse of the encryption process i.e. sub keys.

5 Experimental Results

After registration in the cloud with the details of username, password etc. The sequence being…(Fig. 2, 3, 4, 5, 6, 7, 8, 9, 10).

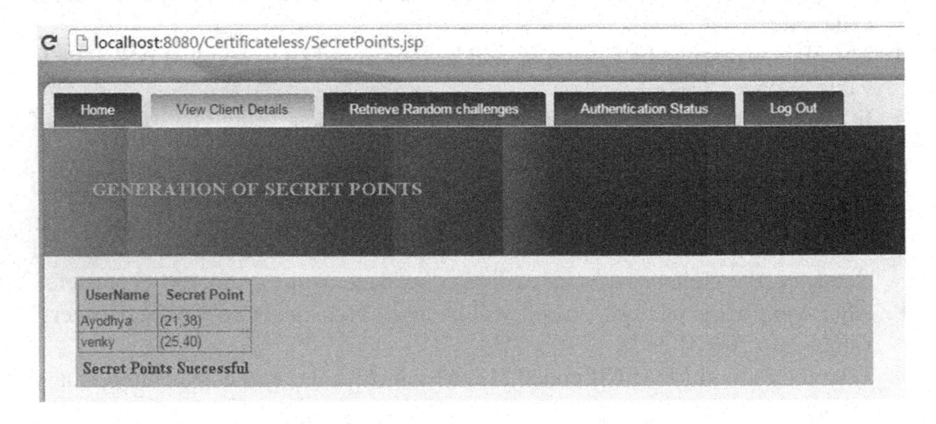

Fig. 2 Secret key points are generated for every registered user in the cloud

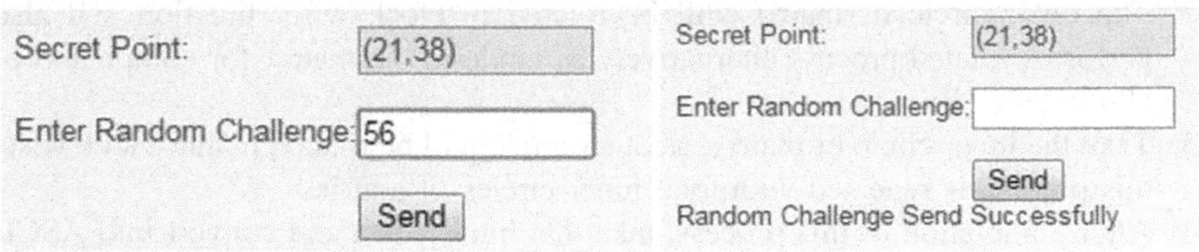

Fig. 3 Random challenge is requested to enter in the input with respect to the secret points

User Name	Random Challenge
Ayodhya	56
venky	87

Generate Verification Shares And Send

User Name	Random Challenge	Verification Shares
Ayodhya	56	(21,30)
venky	87	(25,127)

Verification shares are generated and sent successfully

Fig. 4 Verification share is generated

Verification Share: (21,30)

Generate Authentication Code And Send

User Name	Verification Share	Authentication Code
Ayodhya	(21,30)	80cd90167823bc7060d3983380376a9c

Authentication Code Generated And Send Successfully

Fig. 5 Linking the verification code authentication code is generated and sent accordingly

User Name	Verification Shares	Authentication code	Status
Ayodhya	(21,30)	80cd90167823bc7060d3983380376a9c	Authenticated User
venky	(25,127)	9afdeb8d16334d175d588eda919eb5	Authenticated User

Generate Points And Send

Fig. 6 Where points are generated for every registered authenticated user for file access

USERS	POINTS DETAILS
Ayodhya	(3,11740)(4,14770)(0,7606)
venky	(4,14770)(2,9536)(1,8158)

Points are sent to individual users successfully

USER NAME: DataOwner

PASSWORD: ••••••••

SIGN IN

New Client

Fig. 7 After generating points. Now the data can be accessed when the owner uploads

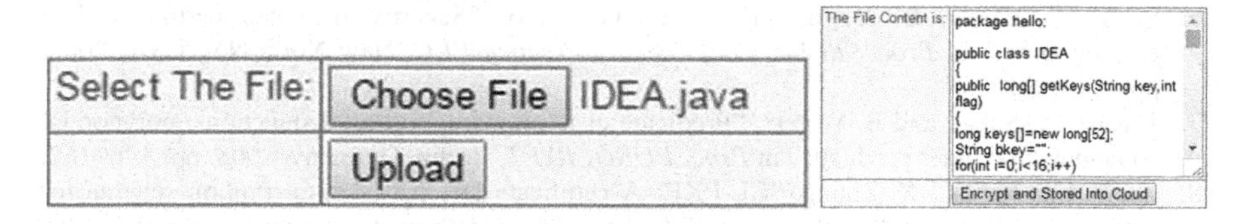
Select The File:
Choose File
IDEA.java
Upload
The File Content is:
package hello;
public class IDEA
{
public long[] getKeys(String key,int flag)
{
long keys[]=new long[52];
String bkey="";
for(int i=0;i<16;i++)
Encrypt and Stored Into Cloud

Fig. 8 Data owner uploads the file which has to be securely shared

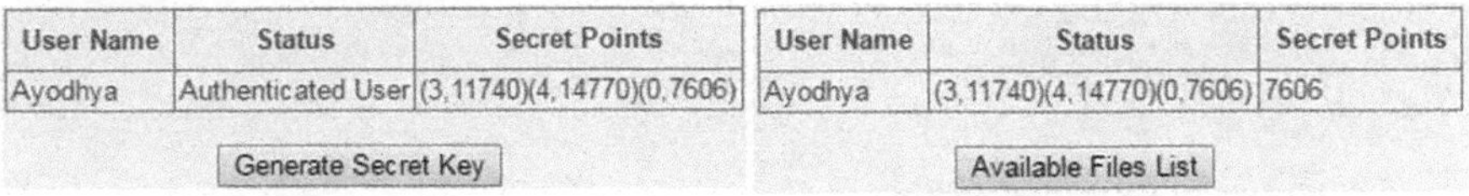

User Name	Status	Secret Points
Ayodhya	Authenticated User	(3,11740)(4,14770)(0,7606)

Generate Secret Key

User Name	Status	Secret Points
Ayodhya	(3,11740)(4,14770)(0,7606)	7606

Available Files List

Fig. 9 Each authenticated user avails the files shared with their generated secret points

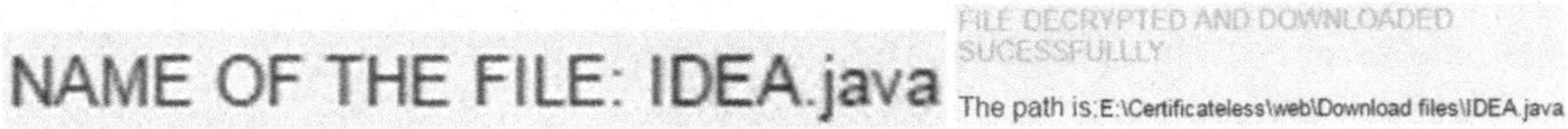

NAME OF THE FILE: IDEA.java

FILE DECRYPTED AND DOWNLOADED SUCESSFULLLY

The path is:E:\Certificateless\web\Download files\IDEA.java

Fig. 10 The file shared can be decrypted finally with individual secret points respectively

6 Conclusion

We have proposed an improved Certificate-Less Public Key Cryptography scheme key building block for securely sharing and converting plain characters into ASCII encrypted ones. We need not generate any certificate for further authentication purpose.

References

1. S. H. Seo, M. Nabeel, X. Ding and E. Bertino, "An Efficient Certificateless Encryption for Secure Data Sharing in Public Clouds," in *IEEE Transactions on Knowledge and Data Engineering*, vol. 26, no. 9, pp. 2107–2119, Sept. 2014.
2. S. Al-Riyami and K. Paterson, "Certificateless public key cryptography," in *Proc. ASIACRYPT 2003*, C.-S. Laih, Ed. Berlin, Germany: Springer, LNCS 2894, pp. 452–473.
3. J. Bethencourt, A. Sahai, and B. Waters, "Cipher text-policy attribute-based encryption," in *Proc. 2007 IEEE Symp. SP*, Taormina, Italy, pp. 321–334.
4. D. Boneh, X. Ding, and G. Tsudik, "Fine-grained control of security capabilities," *ACM Trans. Internet Technol.*, vol. 4, no. 1, pp. 60–82, Feb. 2004.
5. J. Camenisch, M. Dubovitskaya, and G. Neven, "Oblivious transfer with access control," in *Proc. 16th ACM Conf. CCS*, New York, NY, USA, 2009, pp. 131–140.
6. S. S. M. Chow, C. Boyd, and J. M. G. Nieto, "Security mediated certificate less cryptography," in *Proc. 9th Int. Conf. Theory Practice PKC*, New York, NY, USA, 2006, pp. 508–524.
7. J. Katz, A. Sahai, and B. Waters, "Predicate encryption supporting disjunctions, polynomial equations, and inner products," in *Proc. EUROCRYPT*, Berlin, Germany, 2008. pp. 146–162.
8. X. W. Lei Xu and X. Zhang, "CL-PKE: A certificate less proxy re-encryption scheme for secure data sharing with public cloud," in *ACM Symp. Inform. Comput. Commun. Security*, 2012.
9. B. Lynn. *Pairing-based cryptography* [Online]. Available: http://crypto.stanford.edu/pbc.
10. T. Dillon, C. Wu and E. Chang, "Cloud computing: issues and challenges," 24th IEEE International Conference on Advanced Information Networking and Applications, AINA, pp. 27–33, Apr. 2010.

11. Li J, Zhao G, Chen X, Xie D, Rong C, Li W, Tang L, Tang Y, "Fine-grained data access control systems with user accountability in cloud computing," IEEE second international conference on cloud computing technology and science(CloudCom) 2010, pp. 89–96.
12. eung-Hyun Seo and Xiaoyu Ding, "An Efficient Certificate less Encryption for Secure Data Sharing in Public Clouds", IEEE Transactions On Knowledge and Data Engineering, Vol. 26, No. 9, September 2014.
13. S. Al-Riyami and K. Paterson, "Certificate less public key cryptography," in Proc. ASIACRYPT 2003, C.-S. Laih, Ed. Berlin, Germany: Springer, LNCS 2894, pp. 452–473.
14. Y. Sun, F. Zhang, and J. Baek, "Strongly secure certificate less public key encryption without pairing," in Proc. 6th Int. Conf. CANS, Singapore, 2007, pp. 194–208.
15. Smitha Sundareswaran, Anna C. Squicciarini, Member, IEEE, and Dan Lin, "Ensuring Distributed Accountability for Data Sharing in the Cloud" March 2012.
16. P. Sanyasi, Naidu, J. Gurrala "Investigation and analysis of Location Based Authentication and Security Services of Wireless LAN's and Mobile Devices", published in IJCA,Volume 146/Number 8 (ISBN: 973-93-80893-84-7, July2016, pp. 12–17.

Improved Identity Based Digital Signature Authentication Using Feistel Algorithm in Cloud Computing

Juluru Aruna and D. Ashwani

Abstract Though the widespread popularity of public clouds brought great channel of communication for sharing and retrieving data. The data is vulnerable to security threats as it is fully to the service provider side. So there must be lot many security checkpoints, a real world system should meet. So we are proposing an efficient method of authenticating the data by generating a group key through ID based digital signature approach where a group key is broadcasted to all the users in the group. After the authentication process the data members decrypt the data from the cloud and get their plain format back using Feistel algorithm.

Keywords Feistel algorithm · Identity based digital signature schema · Cloud secret points · Cloud computing

1 Introduction

An infrastructure technology using internet for day to day data storing and retrieval in remote servers for easy operations is Cloud computing. As it doesn't need any hardware, software installations on user machine, it is cheap. It also provides on-demand services to use various online resources. Data can be accessed from any system or place in the world providing scalability. Usage of Cloud is elevating at a rapid rate so the trust of confidentiality in cloud is dissolving. Having data is equal to having critical assets of clients. Enterprise Business competitors are fighting for the client's data confidentiality. So confidential, the flexibility of fine grained access control is needed mostly in the cloud models. An information system is strictly

J. Aruna (✉) · D. Ashwani
Computer Science and Technology, Department of CSE,
ANIL Neerukonda Institute of Technology and Sciences (ANITS),
Visakhapatnam, India
e-mail: arunajuluru@gmail.com

D. Ashwani
e-mail: ashwani.cse@anits.edu.in

S.C. Satapathy et al. (eds.), *Computer Communication, Networking and Internet Security*, Lecture Notes in Networks and Systems 5,
DOI 10.1007/978-981-10-3226-4_35

required to restrict access of clients protected records. A management system must allow access to directors of the enterprise only [1, 2]. There is a need of mechanism for tracking the cloud data usage in security perspective. Files should be linked to the server cryptographically, as server should maintain a full record log as we can use the previous old logs to know the action genuinely.

2 Related Work

A new service offered over a network for the current IT Service delivery model is cloud computing, which is by providing dynamically virtualized resources which are scalable. Cloud computing is a rapidly growing service providing ease of accessibility e.g. user's data is transferred to the very large centralized (DCs) data centers, where the management of the data and services may not be fully trustworthy to the core. Machines which actually process and host user data is not known to the users. Cloud data is outsourced often, leading to unfaithful data leak of (PINs). Cloud user data usage effective monitoring mechanism is needed. For e.g. for user—[3–5] SLA assurance agreement at the time of signing for the services. Databases or approaches which use a centralized server in the distributed environment traditional environments are not suitable due to the following.

2.1 Data Managing/Handling

Data can be easily outsourced by Direct CSPs to others and so on.

2.2 Entities

There are free for joining and leaving the cloud in a much more flexible way. So, the data management becomes a fully complex hierarchical chain service. A framework named Cloud Information Accountability (CIA) [6], is introduced to handle the complex problems. CIA framework can maintain powerful lightweight accountability which combines the access aspects, usage and authentication control. Data owners can track whether their SLA's are being honored or not with enforced access and usage control rules needed. For auditing, two modes are developed which are named as push and pull modes. Logs which are sent to the data owner are referred as push mode [7, 8] and the logs which are retrieved by users are referred as pull mode.

Statistics of the files are shared with the data customers after the data owners encrypt and save it in the cloud. Data owners are managed by an influencing domain. Neither owner of the data/data users will be online. They be online only when it is required only, whereas the CSPs and domain management are online always. Once after the Encrypt/Decryption service has completed, encrypted user data can be supplied it to an application like a CRM system. The cloud must erase specific entire client data after making sure that nothing is open to read.

3 Existing System

Statistics of data usage could be fake if it is forged. So, using message authentication code or digital signatures [9, 10] can be useful as user data is critical confidential data. Anonymity of users and any failures might lead to the SLA contract cancellation from the clients. Users could be more in a country/worldwide cloud. The system reduces the communication and computation cost to the finest. Otherwise it will surely lead to energy wastage, contradicting the main goals. So Cloud should meet these checklists

3.1 Encryption and Decryption (data transformation)
3.2 Key Management (as Encryption keys shouldn't be in cloud and managed)
3.3 Authentication (for authorized user access in cloud)
3.4 Authorization (step verification)

Many of the cloud providers fail to maintain this checklist and open gates for the cybercriminals to take control of their systems having user data.

4 Proposed System

The implementation of identity based digital signature is as follows:

1. Every user selects the secret key x selected by signer with α, β, p_i as public Keys.
   Where, $\beta = \alpha^x \bmod pi$. The Public keys are α, β, pi published to everybody publicly through a file, 'x' should be secret.
   $\alpha^x = \beta \bmod pi$ where the public keys should be α, β, pi.
   And X ($1< x < \phi(p)$) is the private key of the signer.
   p = large prime.
2. Generation of Signature
   A random number "k" is chosen i.e. $0 < k < p_i-1$ & gcd $(k, p_i-1) = 1$.
   $\Gamma = \alpha^k \bmod p_i$
   Again choose a random number "t" i.e. $0 < t < p_i-1$ & gcd $(t, p_i-1) = 1$.
   $\Lambda = \alpha^t \bmod p_i$
   $M = (x\gamma + k\lambda + t\delta) \bmod (pi-1)$
   signature of the user can be $(\gamma, \lambda, \delta)$.
   For every user, signature is generated and it is sent to cloud service. By the cloud services, the signature is retrieved and again another signature is generated and then it verifies the both signatures.

3. Verification of Signature

$$\alpha^m = \beta^\gamma \gamma^\lambda \lambda^\delta \bmod p_i$$

By using this equation, the authenticity of signature is verified by the receiver after computing both sides.
4. Key Generation
The Cloud service verifies all the user's authentication status and generate unique secret key for all the users. Then key is sent to all the users in a cipher manner by having a secret point to each users with secret point in return will retrieve the real secret key. Secret point generation be like:
Secret Point = (X_i, Y_i),
Where
K = Randomly selected range, $X_i = K/P_i$, $Y_i = K \% P_i$
Before casting the secret points to the clients, the server sends status message to each users with the secret key to the owner.
5. Encryption and Decryption
In this encryption process, the data owner will convert the data into cipher format and stored into the cloud. By using Feistel encryption process, the data is encrypted and by cloud service, the data owner will get the secret key before the encryption process.
By Feistel algorithm, each user will perform the decryption process and data can be retrieved from cloud. Before performing [11, 12] Feistel decryption process, each user will retrieve the secret point and the status of authentication from the cloud service. The secret points will be retrieved and secret key is generated only if the status of authentication is true.
The generation of secret key is as follows.

$$K = X_i * P_i + Y_i$$

From the cloud, cipher data is obtained by each user and decryption of data is performed, when secret key is obtained.
When the decryption process is completed, the original formatted data is obtained. By using this process, the efficiency of process of authentication and security of cloud data is improved.

5 Results

The process of Digital Signature authentication starts with the login credentials of registered members where their personal user ID and password are requested for authorizing the user for accessing the shared file by some data owner (Figs. 1, 2, 3, 4, 5, 6, 7, 8, 9, 10, 11, 12 and 13).

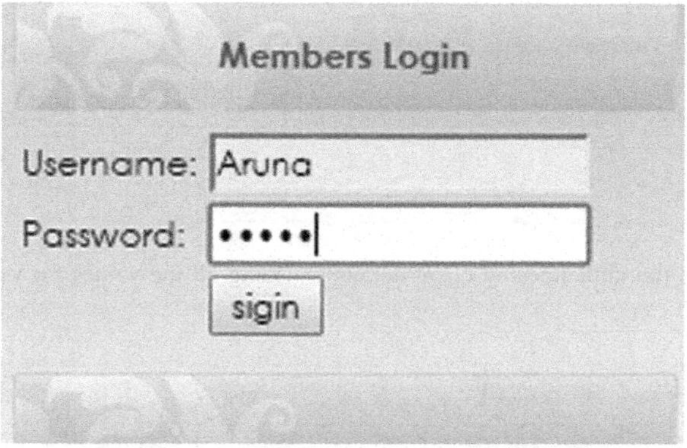

Fig. 1 Authenticated client's login with their own user registration credentials

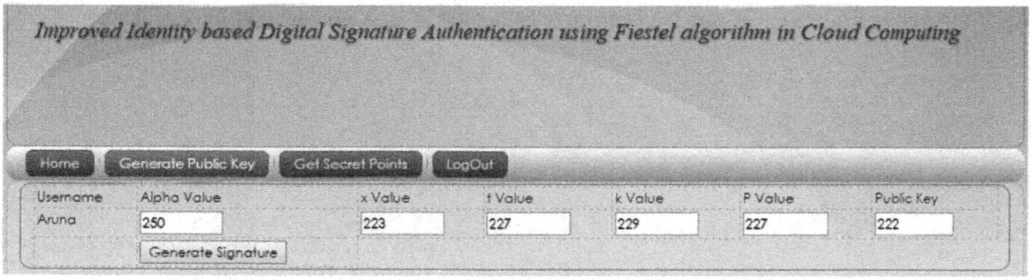

Fig. 2 Client enters user values for generating an identity based digital signature

Fig. 3 A special digital signature is generated for each user linked with public key

Fig. 4 It continues with the other users who want to access the shared files in the cloud

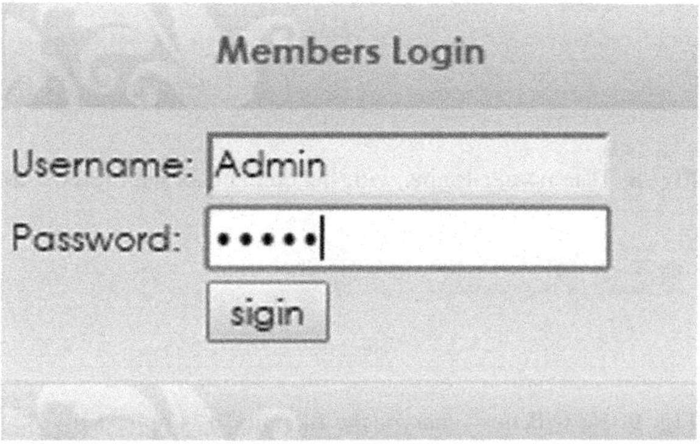

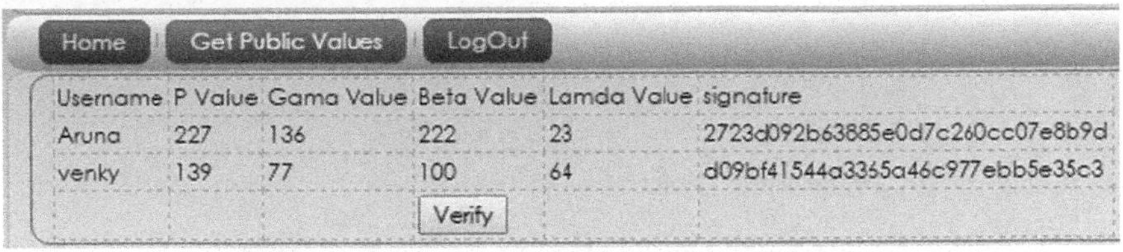

Home | Get Public Values | LogOut

Username	P Value	Gama Value	Beta Value	Lamda Value	signature
Aruna	227	136	222	23	2723d092b63885e0d7c260cc07e8b9d
venky	139	77	100	64	d09bf41544a3365a46c977ebb5e35c3

Verify

Fig. 5 Cloud has the authenticated client details including all the values for verifying

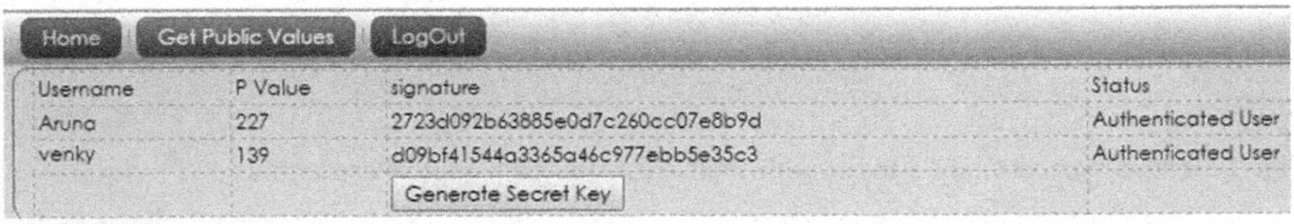

Home | Get Public Values | LogOut

Username	P Value	signature	Status
Aruna	227	2723d092b63885e0d7c260cc07e8b9d	Authenticated User
venky	139	d09bf41544a3365a46c977ebb5e35c3	Authenticated User

Generate Secret Key

Fig. 6 After verification, a user is confirmed as authentic and can generate a secret key

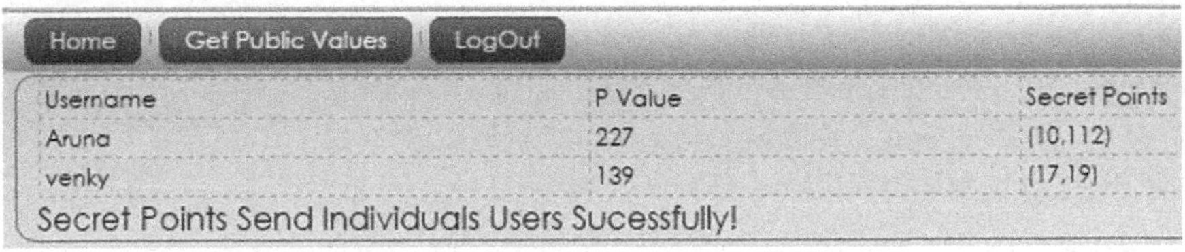

Home | Get Public Values | LogOut

Username	P Value	Secret Points
Aruna	227	(10,112)
venky	139	(17,19)

Secret Points Send Individuals Users Sucessfully!

Fig. 7 The secret points for data access are generated and sent for each authentic user

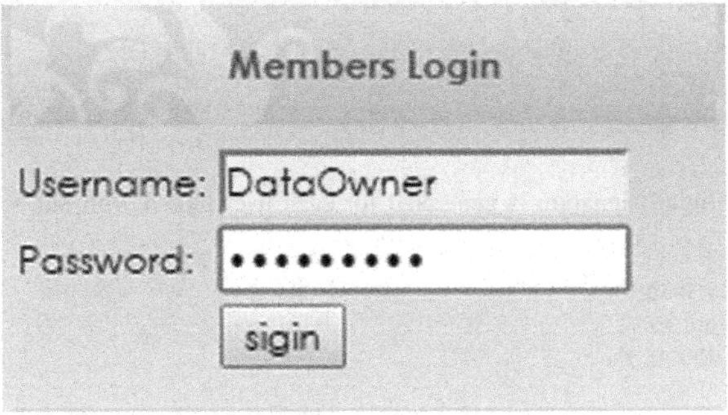

Members Login

Username: DataOwner

Password: ●●●●●●●●●

sigin

Fig. 8 Data owner logins with his credentials with respect to share a file

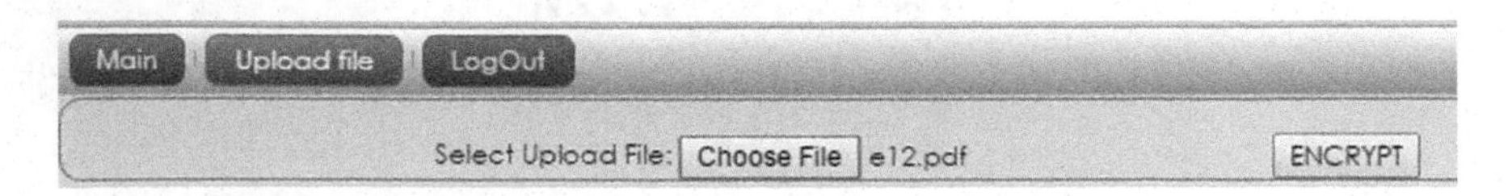

Fig. 9 He will now encrypt the file after it is uploaded

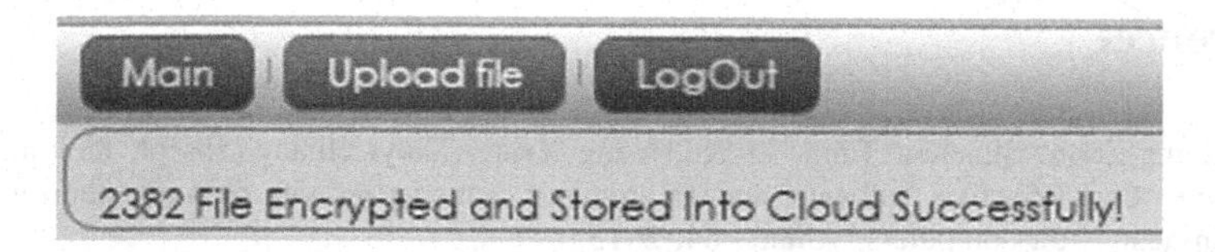

Fig. 10 As soon as the file is encrypted it is stored in the cloud successfully

Fig. 11 Now, only authenticated users with secret points can get access to the file

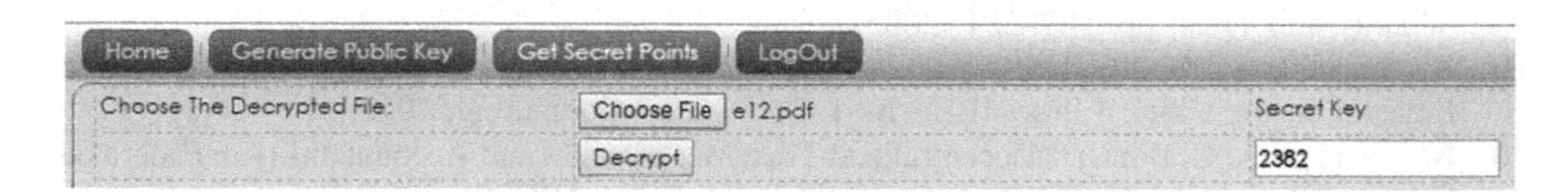

Fig. 12 Though files are available, without a secret key he can't neither decrypt/download

Fig. 13 File will be decrypted finally when entered secret key is right

6 Conclusion

We have proposed a cloud authentication approach which is efficient for sharing the data. After authentication, the status is sent to each user by the cloud service with a secret key. By Feistel Encryption, the data owner stores data into cloud in cipher format, before sharing data in the cloud. By using these approaches, privacy of shared data in cloud and efficiency of authentication process is improved.

References

1. Jianying Zhou, Shaohua Tang, Li Xu, Yang Xiang, Xinyi Huang, Joseph K. Liu, Kaitai Liang, "Cost-Effective Authentic and Anonymous data sharing with forward security", IEEE Transactions on computers, volume 64, 2015.
2. Anna C. Squicciarini, Dan Lin, and Smitha Sundareswaran, "Ensuring Distributed Accountability for Data Sharing in the Cloud," IEEE Transaction on a secure computing, volume 9, 2012.
3. S. Huang, A. Squicciarini, D. Lin, and S. Sundareswaran, "Promoting Distributed Accountability in the Cloud," IEEE Transaction on Cloud Computing, 2011.
4. Jun'e Liu, Robert H. Deng and ZhiguoWan, "HASBE: A Hierarchical Attribute-Based Solution for flexible and Scalable Access Control in Cloud Computing.
5. P.-L. Cayrel, P. Gaborit, C. A. Melchor, and F. Laguillaumie. A new efficient threshold ring signature scheme based on coding theory. IEEE Transactions on Information Theory, 57 (7):4833–4842, 2011.
6. M. Mowbray, S. Pearson, and Y. Shen, "A privacy Manager for Cloud Computing", IEEE transaction on Cloud Computing, pp. 90–106, 2009.
7. A. Charlesworth, and S. Pearson, "Accountability as a Way Forward for Privacy Protection in the Cloud," conf. Cloud Computing, 2009.
8. J.I. den Hartog, I. Staicu, G. Lenzini, S. Etalle, and R. Corin, "A Logic for Auditing Accountability in Decentralized Systems," IFIP TC1 WG1.7 Workshop Formal Aspects in Security and Trust, pp. 187–201, 2005.
9. A. Squicciarini, S. Sundareswaran and D. Lin, "Preventing Information Leakage from Indexing in the Cloud," Proc. IEEE Int'l Conf. Cloud Computing, 2010.
10. B. Chun and A. C. Bavier, "Decentralized Trust Management and Accountability in Federated System," Proc. Ann. Hawaii Int'l Conf. System Science (HICSS), 2004.
11. A. K. Awasthi and S. Lal. Id-based ring signature and proxy ring signature schemes from bilinear pairings. CoRR, abs/cs/0504097, 2005.
12. W. Lim and K. Paterson, "Multi-Key Hierarchical Identity-Based Signatures," Proc. 11th IMA Int'l Conf. Cryptography and Coding (IMA 07), S. Galbraith, ed., pp. 384–402, Dec. 2007.

Securing BIG DATA: A Comparative Study Across RSA, AES, DES, EC and ECDH

I. Bhargavi, D. Veeraiah and T. Maruthi Padmaja

Abstract Now a days cloud computing has emerged as a cost-effective platform for providing IT or business services over the Internet. However, the services provided by the cloud are of third parties, ensuring the security and privacy of the customers. BIG Data is critical at cloud storage. Several security frameworks using cryptographic methodologies have been proposed to address this issue. However, there is no wide comparative study across the cryptographic methods that ensure the security of BIG DATA at the cloud data center. This paper presents a comparative study across fundamental cryptographic methodologies used in securing BIG Data at cloud storage. This work assumes that the cloud storage is erected with HADOOP based data center. In order to carry out comparative study, we have considered ECC, RSA, AES, DES and Elliptic curve Diffie-Hellman to verify the BIG Data security at cloud based data center.

Keywords Big Data · HADOOP · RSA · AES · DES · EC · ECDH

1 Introduction

Big data is a term for datasets that are so large, complex and that cannot be analysed with traditional computing technologies. The quantity of computed data being generated is increasing exponentially from different application sources like retail, logistics and financial databases, social networks, sensors, internet of things etc.

In order to explicate the data and know its characteristics, it is very important to securely store, manage and share the huge amount of complex data. Now this sort

I. Bhargavi (✉) · D. Veeraiah · T. Maruthi Padmaja
Department of Computer Science and Engineering, VFSTR University, Guntur, India
e-mail: bhargaviinduri10@gmail.com

D. Veeraiah
e-mail: d.veeraiah@gmail.com

T. Maruthi Padmaja
e-mail: padmaja.tu2002@gmail.com

S.C. Satapathy et al. (eds.), *Computer Communication, Networking and Internet Security*, Lecture Notes in Networks and Systems 5,
DOI 10.1007/978-981-10-3226-4_36

of facility made available via the distributed platform which is popularly known as cloud computing.

The main feature of cloud computing is on-demand network access to computing resources, on pay per use basis, which are provided by cloud service providers. Common deployment models for cloud computing include Platform as a Service (PaaS), Software as a Service (SaaS), Infrastructure as a Service (IaaS). The PaaS provides platform to customers to develop, run and manage applications without owning the respective infra. The SaaS provides businesses with applications that are stored and run on virtual servers in the cloud.

In the IaaS model, client will pay on a per-use basis for the use of equipment to support computing operations that include storage, hardware, servers and networking equipment.

Security over cloud services is however in its maturing phase, the data in the cloud would be at risk for a large number of security vulnerabilities. The cloud administrators do not have any clue over the data where it is being stored and in what format. Therefore, in this scenario the users must be ensured that proper security measures have been adapted to protect their information mainly from data leakage and data tampering. Further, processing/analysing the huge data at cloud data center is a critical issue. Recently, several distributed frameworks like HADOOP [1, 2], Google File System [3] have been developed for storing and processing the BIG DATA. However, HADOOP distributed framework is quite popular among industry and research communities. HADOOP includes two sets of functionalities, (i) The HADOOP Distributed File System (HDFS) to store large and unstructured datasets, (ii) The Map Reduce framework for processing huge data. Usually, HADOOP works with applications having thousands of nodes and petabytes of data.

Security mechanisms are not incorporated at HADOOP. Several works have been reported the usage of cryptographic algorithms to encrypt the data and store the data at HDFC. Encryption is used to provide security for sensitive information. Encryption algorithm performs various substitutions and transformations on the original message or data and transforms it into cipher text which ultimately becomes a random message. Various cryptographic algorithms are available and used in information security. There are different types of algorithms: (i) Symmetric-key algorithms [4, 5] such as Data Encryption Standard (DES) [6], Triple DES [7] and Advanced Encryption Standard (AES) [7] (ii) Asymmetric-key algorithms [8] such as RSA [7] and Elliptic Curve Diffe-Hellman (ECDH).

This paper is organized as follows. Section 2 depicts the framework for HADOOP based cloud data center. Section 3 presents the work related to securing BIG DATA at HADOOP based cloud data center. Section 4 discusses the performances of encryption and decryption operations performed using different cryptographic algorithms for securing BIG DATA at HADOOP based cloud data center. Finally, Sect. 5 concludes the work.

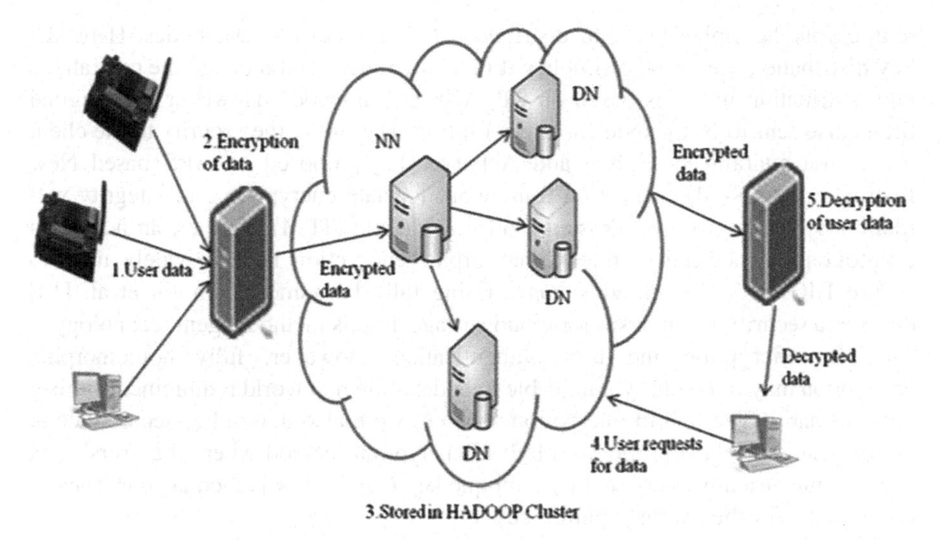

Fig. 1 A HADOOP based cloud data center architecture for big data analytics

2 HADOOP Based Cloud Data Center

As shown in Fig. 1, first, the user data are taken from different sources and encrypted on the server using either symmetric or asymmetric cryptographic mechanisms. Soon after encryption, the data is stored on cloud i.e. it will be stored in a cluster via HADOOP File System (HDFS). In HADOOP, the NameNode (NN) is responsible for the data distribution to DataNodes (DN). Whenever the user requests for data, the encrypted data is given by the server for decryption. Then user takes the encrypted data and decrypts using corresponding keys.

3 Related Work

The HADOOP architecture assumes secure network and hence no security framework is incorporated at base level. As a first step Park and Lee [2] introduced a secure HADOOP framework with AES based encryption/decryption. Research work cited in [9, 10] demonstrated the adoption of Kerberos based authentication mechanism to secure the data in HDFS storage. Zhou and Wen [11] applied 'Cipher Text Policy' and Attribute Based Encryption (CP_ABE) to provide access control credentials for valid cloud users. Here, CP_ABE uses an encrypted data access control structure rather using user's personal identity. The user can perform the decryption provided, the user identification attribute matches with access control structure. In this mechanism the cipher text and corresponding cipher key generated via CP_ABE method are transmitted to the Namenode. The Namenode further re-encrypts the cipher text and distributes the file blocks to Datanodes. Here, the key distribution seems to be simple with less user intervention due to the centralized key distribution which is based on CP_ABE at Namenode. However, the original file is also sent to Namenode for re-encryption. Therefore, the security to the client file is not guaranteed. Cohen and Acharya [12] proposed an AES based New Instruction (AES-NI) encryption framework for data encryption and integrity validation by making use of Trusted Platform Module (TPM). Further, an advanced cryptographic mechanism like homomorphic encryption is also widely used to secure BIG DATA at cloud storage. Using fully homomorphism Jin et al. [13] devised a security mechanism for cloud storage. In this method, agent technology is used for encryption and user authentication. However, fully homomorphic encryption may not be fully applicable to address the real world requirements in Big data Scenario. The hybrid encryption schemes were also devised to secure data at HDFS. Lin et al. [14] proposed a hybrid encryption method where the users' data file is symmetrically encrypted by a unique key *k* and this *k* is then asymmetrically encrypted with the owner's public key.

To encrypt users' files this mechanism uses the DES algorithm initially in order to generate the 'data key'. Later on, RSA is used to encrypt the already generated 'data key' and the user keeps the private key to decrypt the 'data key'. Here Yang et al. [15] assumed that the generated private key using RSA is still vulnerable. Consequently, they have used IDEA (International Data Encryption Algorithm) to further encrypt the secret key. Although, this hybrid encryption method seems to secure the data, it increases the computational complexity to the extent. Saini and Naveen [6] presented a steganography based hybrid scheme to make the encrypted data completely not visible to the outside users.

4 Comparative Study of Cryptographic Algorithms Over HADOOP Based Cloud Data Center

Table 1, depicts the comparison of encryption and decryption of differing file sizes using various cryptographic algorithms. Here the time unit is shown in seconds (s). From our experiments we have observed that RSA could not perform over the files

Table 1 Comparison across RSA, AES, DES, EC and ECDH with respect to encryption and decryption times

Algorithms	Key length (bits)	File size (MB)	Encryption time (SECs)	Decryption time (SECs)
AES	128	50, 100, 150	4, 6, 9	14, 14, 14
DES	56	50, 100, 150	4, 6, 9	14, 14, 14
EC	256	50, 100, 150	7, 15, 18	14, 14, 14
ECDH	256	50, 100, 150	58, 115, 286	180, 360, 555

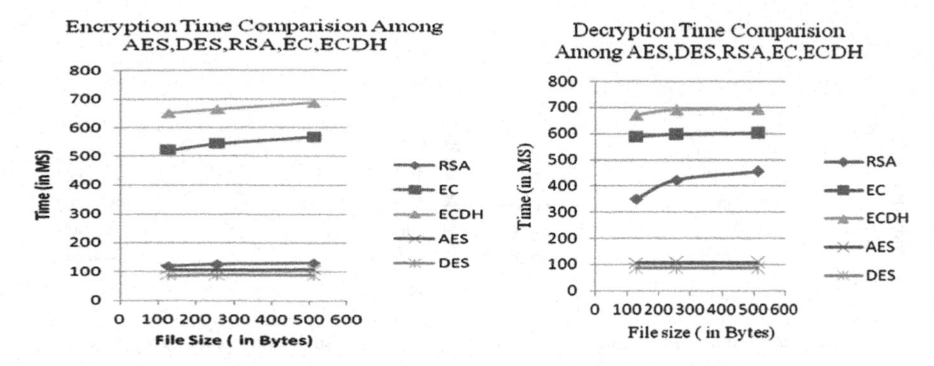

Fig. 2 Encryption and decryption times for cryptographic

with larger sizes (considered file sizes 50, 100 and 150 MB). Hence, the results reported here are in the file sizes 512, 1024, 2048, 4096 Bytes. However, the encryption and decryption using AES and DES algorithms even performed well for the files in smaller sizes (512, 1024, 2048, 4096 Bytes) which is shown in Fig. 2. Thus, AES, DES, EC and ECDH are considered for HADOOP based experiments.

In order to study the behavior of considered algorithms on HADOOP we have considered three different scenarios of HADOOP setting in securing the BIG DATA. The performance of these algorithms with respect to HDFS storage is measured in terms of writing and read speeds of the files with varying sizes in to and from HADOOP's HDFS. Usually HDFS reading takes more time than writing as data is stored in different blocks. Further, HADOOP supports replication factor that reflects how many times the original data can be replicated across HDFS nodes.

4.1 Scenario 1

Same node act as a master and slave of considering replication factor as one. From Fig. 3, it can be observed that as the plain text size increases the time required to write and read to and from HDFS is also increasing. In this scenario encrypted file writing and reading using EC and ECDH yields more time than AES and DES. This is because of larger encrypted files generation by EC and ECDH.

4.2 Scenario 2

By considering a three node cluster, among the three, one node act as Namenode and two others are data nodes. i.e., one master node and two slaves and here the replication factor are considered as one. The writing and reading speeds in this scenario are shown in Fig. 4. Here writing time is increased compared to the first

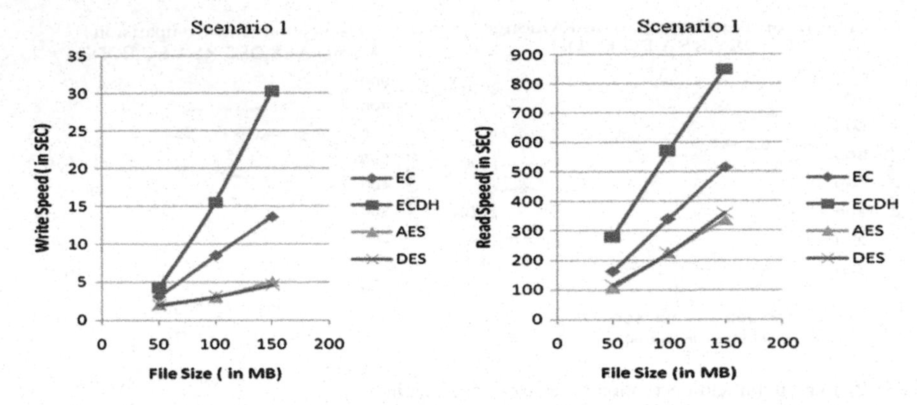

Fig. 3 Write and read speeds in Scenario 1

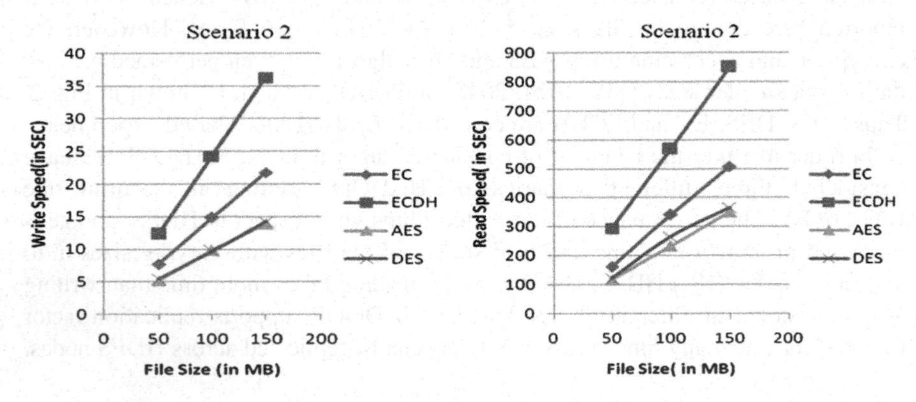

Fig. 4 The write and read speeds in Scenario 2

scenario because in previous scenario all blocks are stored in one node, but in this scenario, data blocks are stored in three different nodes. In this scenario, read speed is nearly same as the previous scenario.

4.3 Scenario 3

This scenario is similar to Scenario 2 except replication factor is set to three. The writing and reading speeds in this scenario are depicted in Fig. 5. In this scenario, the writing time is increased when compared with previous scenarios because of replication factor. In previous scenario only one copy of the data is stored, but here three copies of data are stored due to three replication blocks. Here also the read speed is similar to previous one and two scenarios.

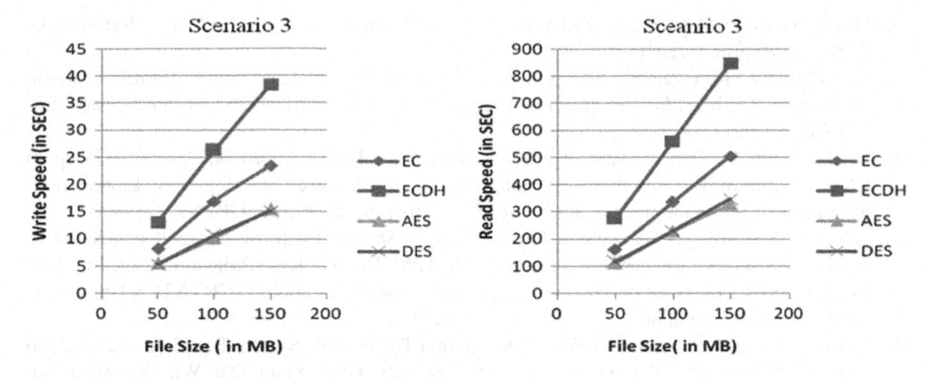

Fig. 5 Write and read speeds in scenario 3

5 Conclusion

To observe the behavior of encryption and decryption operations on BIG DATA in the HADOOP environment, we have used symmetric and asymmetric crypto algorithms over varied file sizes. We could conclude from the experiment that ECDH yielded highest reading and writing speed compared to EC, AES and DES. EC yielded a second highest speed and AES, DES yielded least speed for the same size of file blocks. Further, it is identified that the read speed is consistent even with the increase in replication factor.

References

1. Tom White, "Hadoop: The Definitive Guide", Third Edition, O'Reilly, (2012).
2. S. Park, Y. Lee, "Secure Hadoop with Encrypted HDFS," Chapter Grid and Pervasive Computing, Vol. 7861 of the series Lecture Notes in Computer Science, pp 134–141, (2013).
3. Bo Li, Mengdi Wang, Yongxin Zhao, Geguang Pu, Huibiao Zhu, Fu Song "Modeling and Verifying Google File System Modeling and Verifying Google File System", 16th International Symposium on High Assurance Systems Engineering, Pages: 207–214, IEEE, (2015).
4. Sourabh Chandra, Siddhartha B, Smita Paira. "A Study and Analysis on Symmetric Cryptography" ICSEMR, pp 1–8, IEEE (2014).
5. Rejani. R, Deepu.V. Krishnan, Study of Symmetric key Cryptography Algorithms, Volume 2 Issue 2, pp 45–50, IJCT (2015).
6. Garima Saini and Naveen Sharma, "Triple security of data in cloud computing," International Journal of Computer Science and Information Technologies, Vol. 5, No 4, pp. 5825–5827, (2014).
7. Behrouz A. Forouzan "Cryptography and Network Security", Tata McGraw-Hill Companies, (2007).
8. Sourabh Chandra, Sk Safikul Alam, Smita Paira and Goutam Sanyal. "A comparative survey of symmetric and asymmetric key cryptography", International Conference on Electronics, Communication and Computational Engineering (ICECCE), pp 83–93, IEEE (2014).

9. D. Das, O. O' Malley, Sanjay Radia, Kan Zhang, "Adding Security to Hadoop," Hortonworks Technical Report I, (2010).
10. O. O'Malley, Kan Zhang, Sanjay Radia, Ram Marti, and Christopher Harrell. Hadoop security design. https://issues.apache.org/jira/secure/attachment/12428537/securitydesign.pdf, October, (2009).
11. H. Zhou and Q. Wen, "Data Security Accessing for HDFS Based on Attribute-Group in Cloud Computing," In Proc. of International Conference on Logistics Engineering, Management and Computer Science (LEMCS 2014), pp. 525–528, (2014).
12. J. Cohen, S. Acharya, "Towards a Trusted Hadoop Storage Platform: Design Considerations of an AES Based Encryption Scheme with TPM Rooted Key Protections," IEEE 10th International Conference on and Autonomic and Trusted Computing (UIC/ATC), Ubiquitous Intelligence and Computing, pp. 444–451, (2013).
13. S. Jin, S. Yang, X. Zhu, and H. Yin, "Design of a Trusted File System Based on Hadoop," In Proc. of Trustworthy Computing and Services, ed: Yuyu Yuan, Xu Wu, Yueming Lu, pp. 673–680, (2013).
14. H. Y. Lin, S. T. Shen, W. G. Tzeng, B. S. P. Lin. "Toward Data Confidentiality via Integrating Hybrid Encryption Schemes and Hadoop Distributed File System," In Proceedings of 26th International Conference on Advanced Information Networking and Applications, IEEE Computer Society Washington, DC, USA, pp. 740–747, (2012).
15. C. Yang, W. Lin, and M. Liu, "A Novel Triple Encryption Scheme for Hadoop-Based Cloud Data Security," in Proc. of 4th Emerging Intelligent Data and Web Technologies (EIDWT), pp. 437–442, (2013).

Signal Condition and Acquisition System for a Low Cost EMG Based Prosthetic Hand

B. Koushik, J. Roopa, M. Govinda Raju, Biswajith Roy, H. Manohar, K.S. Geetha and B.S. Satyanarayana

Abstract The rehabilitation process for the physically challenged people is a real challenge and concern where technology has not yet reached all sections of the society. This paper presents a novel and a low cost multi finger Robotic prosthetic hand based on the principle of Electromyography (EMG). A three channel EMG sensor is placed across the forearm to detect the voluntary actions of the superficial muscles in the forearm. The low frequency interference and noise in the signal was removed using sensor module that eliminates the need for high end processor and bringing down the cost. The unique nature of this prototype is that, it is adaptable to different people just by adjusting a potentiometer in the sensor module. The proposed prototype of the robotic prosthetic hand is a low cost, non-invasive, painless solution for prosthesis which is capable of performing some basic actions.

Keywords Electromyography (EMG) · Robotic prosthetics · Rehabilitation · Myoelectric prosthetics

1 Introduction

The rehabilitation process for the physically challenged has become one of the biggest question to technology for development in the world in comparison to all other sectors. There were many studies and research about assistive devices for the handicapped. Especially, the prosthetic legs and hands were representative assistive

B. Koushik (✉) · J. Roopa · M. Govinda Raju · B. Roy · H. Manohar · K.S. Geetha
Department of Electronics and Communication, RV College of Engineering,
Bengaluru 560059, Karnataka, India
e-mail: koushik.kops@gmail.com

J. Roopa
e-mail: roopaj@rvce.edu.in

B.S. Satyanarayana
Munjal University, Gurugaon 122413, Haryana, India
e-mail: satyang10@gmail.com

S.C. Satapathy et al. (eds.), *Computer Communication, Networking and Internet Security*, Lecture Notes in Networks and Systems 5,
DOI 10.1007/978-981-10-3226-4_37

components or devices for the amputees. Prosthetic hand for the upper limb amputee is either the hook or the hand-shaped which are actuated by either body or external power.

Body-powered prosthetic limbs are controlled by cables which are connected to the body elsewhere. Externally powered prosthetic limbs are powered by motors and are controlled by the amputee accordingly in several ways. The switches and buttons were used in some cases that allow the patient to move his or her prosthetic device based on switch control. In recent years, the muscles in the residual limb are used for controlling the prosthetic limb. Small electrical signals are generated due the muscle contraction and relaxation can be detected by using the electrodes placed on the surface of the skin. This electrical property of muscles (myoelctric) is used to develop the prosthetic hand.

For construction of robots in healthcare systems, EMG signals [1] plays an important role for actuating the prosthetic devices, orthotic devices, wheelchairs, exoskeleton, etc., [2, 3]. The current state of research in prosthetics focus on either, developing new pattern recognition algorithm or increasing the efficiency of the mechanical design. Scientists and researchers are working to develop an EMG signal monitoring and acquisition circuit for analysing, recording the hidden information in the original signals. The major problem in recording the surface EMG signals are due to noise and interference associated with the low voltage and low frequency physiological signal with power line noise and offset DC voltage [4].

1.1 Muscle of the Forearm

Muscles in different positions of the forearm muscle are responsible for the movement of wrist and all the fingers. The contraction and relaxation of muscles in the remaining part of the limb of an amputee is almost similar to the muscles in a healthy limb, this is underlying principle of all EMG based prosthesis. Therefore, it is possible to control the hand and finger movements using the EMG signals generated in the residual limb of the amputee. The literatures [1, 5], also suggest that the placement of electrodes are of paramount importance for it to be used for prosthetics. It is important to consider only superficial muscles near the skin since we are using surface electrodes. The key region of interest, where the electrodes are to be placed, is listed below in the Table 1. For the current prototype of the prosthetic hand, surface electrodes were placed on the skin directly above the

Table 1 List of muscles in the forearm of interest, along with their functionality

SI. no.	Muscle	Motion responsible
1	Flexor digitorum superficialis	Middle finger, ring finger, pinky finger
2	Extensor pollicis	Thumb
3	Extensor indicis	Index finger

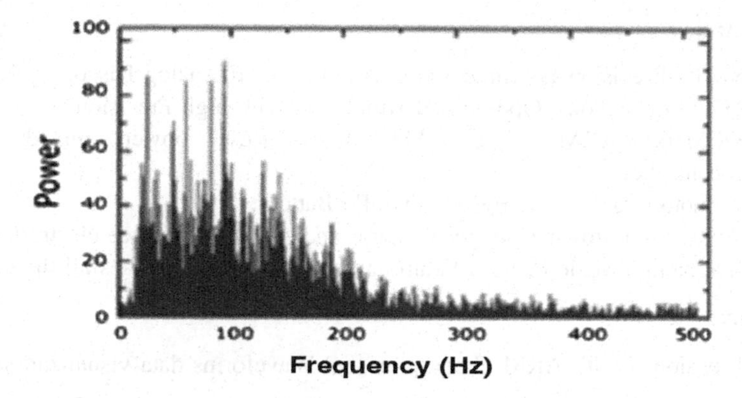

Fig. 1 Spectrum of EMG signal showing dominant power is between 50–150 Hz [5]

muscles mentioned in Table 1. The combinations of signals obtained from these points were used to decode the gestures of hand [6, 7].

1.2 Characteristic of EMG Signal

The amplitude of the EMG signal can range from 0 to 10 mV (pk-to-pk) or 0 to 1.5 mV (rms). The usable energy frequency of the signal is from 0 to 500 Hz range, with the maximum energy at 50–150 Hz frequency range [5]. Signals above electrical noise level are used for analysis and conditioning. The Fig. 1 shows a frequency spectrum of EMG signal.

2 Systems Analysis and Selection of Hardware

Based on the literature survey, a brain storming session was carried out to decide the hardware and software requirements for the prototype. Different design parameters considered for designing are listed below.

- High Performance and low cost.
- Flexibility in terms of mechanical structure.
- Software adaptability and responsiveness to optimisation.
- Inertness of equipment to changing environment conditions.
- Safety and Ease of use.

The different components used in signal acquisition system with specific details are listed below.

Hardware:

- AD8221—Precision Instrumentation Amplifier and Analog Discovery Kit.
- OP2177—Low Noise Operational Amplifiers with high Precision.
- ADUCM360—ARM cortex M3 based Low power mixed signal microcontroller.
- Servo motors and 7.4 V 1000 mah LiPo Battery.
- LM2596—Step down Voltage regulator and Ag/AgCl Surface electrodes.
- Crafted hand—Made of hard Foam and conduits which houses all the circuits.

Software:

- Keil uvision 5 with ARM C compiler and Waveforms data visualization tool.

3 EMG Signal Conditioning System

The Fig. 2 explains in pictorial form the process of capturing the signal and processing the low frequency, low amplitude EMG signal.

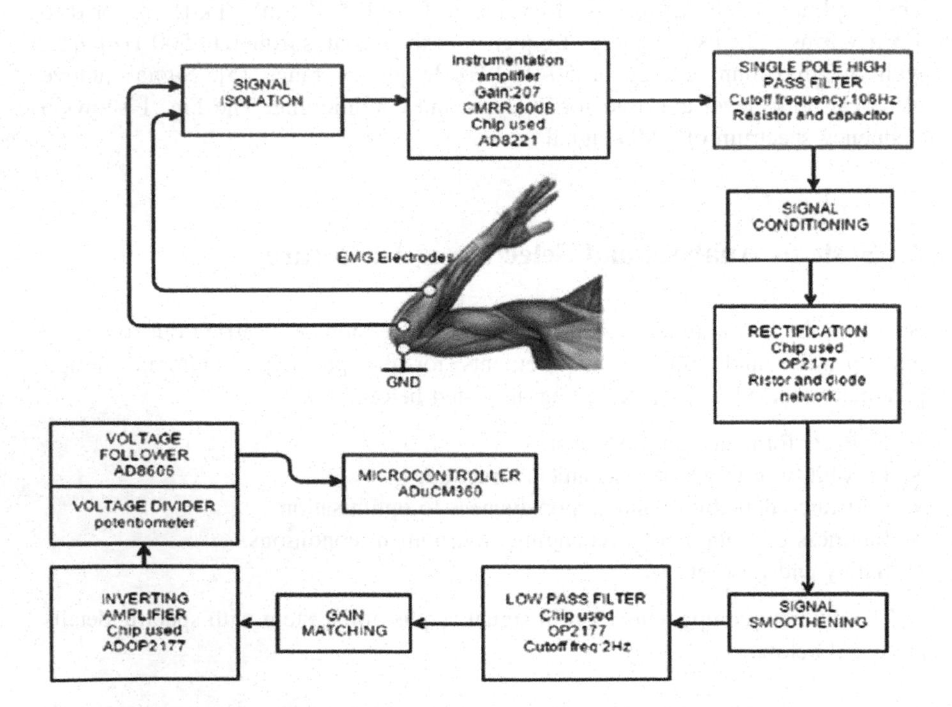

Fig. 2 Block diagram of EMG signal conditioning circuit

3.1 Trapping of EMG Signals

The EMG [8, 9] signals that can be trapped on the surface of the skin using surface electrodes are in the order of few micro-volts. Two electrodes are placed between two motor points or along longitudinal centre line of the muscle and between tendon insertion and motor point [5]. A reference electrode necessary for providing a common reference to the differential input of the preamplifier in the electrode is placed electrically neutral tissue near the elbow where it is devoid of muscles. An instrumentation amplifier with a gain of 207 is used to amplify the EMG signal. The signal is captured at two sites are subtracted and differences are amplified. This is called Bipolar Detection. As a result, the differential signals are amplified and common signals common to detection sites are removed. Relatively the distant power lines noise signals will be eliminated and relatively EMG signals are amplified.

3.2 Signal Conditioning

A capacitor (0.01 uF) is introduced to couple the AC signal for removing the DC offset error in the signal. And hence the high pass filter of cut-off frequency 106 Hz is used to get rid of low frequency noise and DC offset. Next, an active full wave rectifier is used to rectify the obtained EMG signal. The rectified signal contains lot of ripples. Therefore, the next stage is an envelope detector which produces envelope of the signal removing the ripples. This reduces the computational intensity required to decode the EMG signal in the microcontroller making it faster to decode the EMG signal. The signal is inverted in the envelop detector, therefore, to invert the signal an inverting amplifier with adjustable gain is configured using a potentiometer. The potentiometer here serves two purposes: to boost the signal to required voltage level for the microcontroller and for calibration purpose when the EMG sensor module is used on different individual. Finally, in the last stage a voltage follower is used so that there is no degradation of signal when it is used with any interfaces due to impedance matching. The output of this stage is put into one of the channels of ADC of Microcontroller ADUCM360.

3.3 Interfaces with the Microcontroller

Three EMG sensor circuits are used to decode the motion of the finger by the signal received from the surface electrodes placed on the skin of the forearm. The analog output from the EMG sensor module was connected to the ADC ports of the Microcontroller. The three ADC channels were time multiplexed in order to get the readings. UART was configured for debug purpose, in order to see the ADC values of the EMG sensor. PWM settings were initialized to control the actuation of servos in the Prosthetic hand.

Initially only one value from the ADC of the microcontroller was used to generate the control signal for the servos. If the voltage level was above the predefined threshold the servos were actuated otherwise would remain the natural state. The predefined threshold value would vary from person to person and has to be calibrated based on the user preference. This thresholding algorithm was effective for only few hand gestures, such as hand open and hand close. As the number of hand gestures increased this algorithm became quite ineffective. For certain hand gestures the value of the ADC would oscillate between the predefined threshold values, making it difficult to determine the state of the hand. This also made it difficult to set the threshold value.

In order to overcome this problem, mean of 20 continuous value of the ADCs from the microcontroller, corresponding to each EMG sensor module was calculated. The values of the mean were much more stable and helped to fix the threshold value. Based on comparing the value of the mean of 20 ADC values and threshold value control signals were sent to the servos to perform the required action of finger. This was far more accurate than the previous setup but was there was a time delay in the response. After many hit and trail, it was found that 8 readings were enough to provide accurate result with good response time. According to ADC values of three sensors, the PWM signals are generated to actuate the hand.

4 Results

Three EMG sensors were used to detect the motion of hand and fingers. One pair of electrodes was placed on flexor digitorum superficialis, another on extensor pollicis and last pair on extensor indicis to detect the motion of the fingers [8, 10]. An algorithm was used to decode the position of the fingers. The predefined threshold in the algorithm was determined based on the user preference upon calibration. The design prototype around five motion of the hand was detected-open hand, closed hand, show thumbs up, pinch and index finger pointing direction. This was achieved by observing the pattern generated by the ADC of the microcontroller through the UART monitor. The total cost in developing this project was less than 20,000 Indian rupees.

5 Observations

The design of the sensor circuit, interfacing with the microcontroller and the algorithm for signal detection and control has been explained so far. The ARM based microcontroller was coded using Keil uvision 5 software. Also, the design of

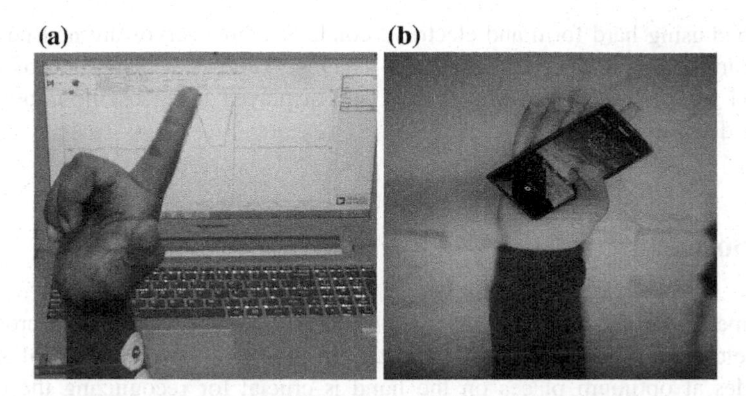

Fig. 3 **a** Movement of index finger captured using analog discovery kit and plotted on the oscilloscope of waveforms software. **b** Prototype of prosthetic arm holding a mobile phone

each EMG sensor module was verified using Analog discovery Kit, which is a USB device which has an oscilloscope, function generator, spectrum analyser etc.

The Fig. 3a shows signal captured for fore finger movement shown in waveforms tool. The Fig. 3b shows the prototype of EMG based prosthetic hand

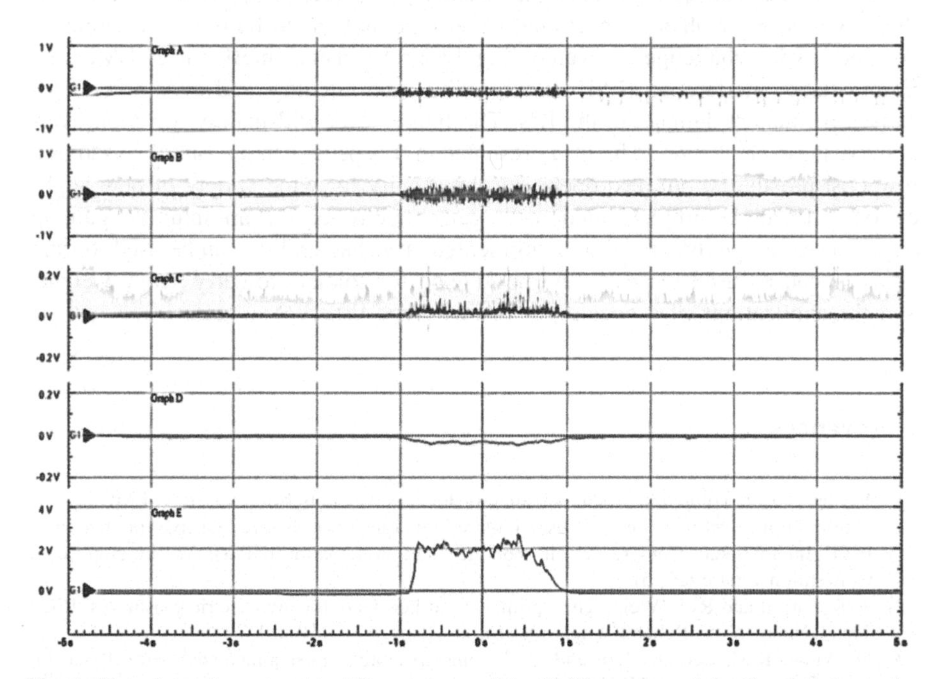

Fig. 4 Plot of signals captured in different stages of the EMG sensor module. *Graph A* shows the signal after the instrumentation amplifier. *Graph B* shows the signal after the high pass filter removing the DC offset. *Graph C* shows the signal after the rectification. *Graph D* shows the output of envelop detector. *Graph E* shows the final output which is provided to the microcontroller

developed using hard form and electrical conduits. Three servos are housed inside the forearm region of the prosthetic hand. The Fig. 4 shows the snapshots of signals captured from the EMG sensor module and displayed using oscilloscope of the Analog discovery kit.

6 Conclusion

This paper shows the how to create a platform for a low cost EMG based prosthesis for a below elbow amputee. It is important to note that the placement of surface electrodes at optimum places on the hand is crucial for recognizing the correct EMG signals each muscle mass in the forearm corresponds to different movements of the finger and wrist. The overall structure is composed by first detecting the signals, processing it, and the microcontroller decides the action and sends control signals to servos to actuate the robotic arm. Each step of this structure plays an important role in the whole system and their functions have been studied to create efficient low cost myoelectric controlled multi fingered hand prosthesis. Although the prototype worked as expected it is still faced with many challenges. The EMG sensor module is affected by radio waves and high frequency noise. Also the glue in the electrode wear off in a short amount of time and has to be replaced regularly. Another major challenge is overcoming Electromagnetic interference (EMI) and Electrostatic interference (ESI) as this affects the circuitry and even sometimes causes permanent damage to the ICs. The future work will mainly focus on overcoming these challenges. The idea presented in this paper can not only be used as a low cost prosthesis, but also opens an opportunity for wide variety of new applications. It can be implemented as a game controller for an intuitive gaming experience, can be integrated into augmented or virtual reality, can be used to train athletes, can be used to create an exoskeleton for soldiers to carry heavy weights and the applications of it goes on to make a huge list.

References

1. Jun-Furukawa, Tomoyuki Noda.: Estimating Joint Movements from Observed EMG signals with multiple electrodes under Sensor Failure Situations towards safe assistive robot control. IEEE International Conference on Robotics and Automation(ICRA), Washington State Convention Centre (2015).
2. T. R. Farrell and R. F. Weir.: The optimal controller delay for myoelectric prostheses. IEEE Trans. Neural Syst. Rehabil. Eng., vol. 15, no. 1, pp. 111–118 (2007).
3. D. Wang, K.M. Lee, J. Guo, and C. J. Jang.: Adaptive knee joint exoskeleton based on biological geometries. IEEE/ASME Trans. Mechatronics, vol. 19, no. 4, pp. 1268–1278 (2014).
4. P. Geethanjali and K.K. Ray.: A Low-Cost Real-Time Research Platform for EMG Pattern Recognition-Based Prosthetic Hand. IEEE/ASME Trans. Mechatronics, vol. 20, no. 4, pp. 1948–1955 (2015).

5. Carlo J. De Luca.: Surface Electromyography: Detection and Recording. DelSys Incorporated, (2002).
6. Pradeep Shenoy, Kai J. Miller, Beau Crawford, and Rajesh P. N. Rao.: Online Electromyographic Control of a Robotic Prosthesis. IEEE TRANSACTIONS ON BIOMEDICAL ENGINEERING, VOL. 55, NO. 3, (2008).
7. L. Eriksson, F. Sebelius, and C. Balkenius: Neural control of a virtual prosthesis. presented at the ICANN, Skoevde, Sweden, (1998).
8. Andrea Merlo and Isabella Campanini.: Technical Aspects of Surface Electromyography for Clinicians. The Open Rehabilitation Journal, 3, 98–109 (2010).
9. C.Pylatiuk, M. Müller-Riederer, A. Kargov, S. Schulz, O. Schill, M. Reischl and G. Bretthauer.: Comparison of Surface EMG Monitoring Electrodes for Long-term Use in Rehabilitation Device Control. IEEE 11th International Conference on Rehabilitation Robotics, Kyoto International Conference Center, Japan (2009).
10. Ruchika, Shalini Dhingra.: An Explanatory Study of the Parameters to Be Measured from EMG Signal. International Journal of Engineering and Computer Science ISSN: 2319–7242 Volume 2 Issue 1, Page No. 207–213 (2014).

Wireless Monitoring of NH_3 (Ammonia) Using WO_3 Thin Film Sensor

J. Roopa, S.G. Divakara, S. Lakshmi Prasad, A.M. Lakshmikanth, Rajath. B. Das, K.S. Geetha and B.S. Satyanarayana

Abstract Food and especially fresh vegetables and fruits have limited shelf life are also easily subject to damage during transport. Natural or damage induced aging lead to foul smell on account of emitted organic or inorganic gases like Hydrogen Sulfide, Ammonia, etc., Reported in this chapter is the study of Tungsten Oxide (WO_3) nanoparticles based thin film sensor for sensing Ammonia. Tungsten Oxide Nano particles are deposited as thin film on glass substrate using a sol gel/dip coating process. The chemical properties and electrical properties of these film are characterised using suitably the spectroscopic method like FTIR, XRD and digital multi-meter/circuit. At room temperature the sensors were tested for different concentration of gases. When the concentration of Ammonia gas exceeds a pre-set threshold, the circuit was designed using a microcontroller based circuit to send a message to a recorded user, through a GSM module.

Keywords Tungsten oxide · XRD (X-ray diffraction) · Synthesis · Nanoparticles · Thin film

---

J. Roopa (✉) · S. Lakshmi Prasad · A.M. Lakshmikanth · Rajath.B. Das · K.S. Geetha
Department of Electronics and Communication, RV College of Engineering,
Bengaluru 560059, Karnataka, India
e-mail: roopaj@rvce.edu.in

S.G. Divakara
Department of Chemistry, RV College of Engineering,
Bengaluru 560059, Karnataka, India
e-mail: divakarsg@rvce.edu.in

B.S. Satyanarayana
Munjal University, Gurugaon 122413, Haryana, India
e-mail: satyang10@gmail.com

S.C. Satapathy et al. (eds.), *Computer Communication, Networking and Internet Security*, Lecture Notes in Networks and Systems 5,
DOI 10.1007/978-981-10-3226-4_38

1 Introduction

India is among the top five producers of a wide variety of agricultural/food crop in the world, effectively producing close to 17–18% of the global vegetables and 14–15% of fruits. However on account of the lack of effective storage and logistic support, India also wastes an over 40% of the produce worth over $7.5 Billion annually. The rapid urbanization, has led to huge demand and supply gap, as the supply and demand originate from different geographical locations. Even with the advancement of warehousing techniques, fresh vegetables, fruits, meat and other packed food products tend to get spoilt during transport and warehousing [1]. There are several reports of unhygienic conditions at warehouses resulting in rotting of food. Such wastage could be limited and shelf life extended through timely detection and effective segregation, using developments in sensors technology and wired/wireless sensor network enabled communication of the information.

Aging food emits a foul smell comprising of a variety of gases including Ammonia, Hydrogen Sulphide, and Ethylene [1, 2]. As such, the Ethylene gas induces ripening and eventual rotting of any vegetables and fruits in the vicinity. Hence, early detection of the onset of the rotting process in food is needed to prevent further wastage of food. Detection of this odour by means of a gas sensor coupled with an embedded system can be used to prevent spoiling of food. The techniques for indication of degradation of the fruit/vegetable/food is to use gas sensors that detects the above mentioned gases or volatile organic compounds using conducting metal oxide based sensors. Metal oxide sensors uses a surface layer adsorption for sensing gases using oxides including WO_3, SnO_2, ZnO, and TiO_2 by change in the sheet resistance and thereby reflecting the conductivity and hence an experiment on sensing the ammonia gas was conducted, as NH_3 is also one of the major composition present in rotten food [2]. The Metal oxide sensors have high sensitivity and can detect adsorbed target gases even when present in very low concentrations of 100 ppb [3]. These sensors operate at room temperatures and thus do not require a micro heater. This makes them more compact, energy efficient and once securely packaged and appropriately calibrated, they provide highly secure and reliable information. The developed system from the reported study will detect the presence of the target gas produced when food is rotting and will send an appropriate warning using GSM module to the monitoring cent. This system could be used for monitoring food quality in warehouses.

In order to realize high-sensitivity detection, rapid response recovery rates and working at Room Temperature, sensing materials should exhibit very large specific surface area, high accessibility and effective diffusion channels for detecting gas molecule. Although the popular sensors based on pure tungsten oxide films or doped tungsten oxide based sensing materials has been intensively attempted for a long time, there are still some disadvantages that needs to be overcome to meet the practical requirements [4, 5]. Recently Nano material enabled sensors have attracted much interest on account of their enhanced response and ability to tailor the characteristics of the sensing materials [6]. The literature suggests that there has been a improvement in sensing properties of Nano-composites, with better selectivity and reversibility as compared to usage of pure porous silica and much lower working temperature than using pure WO_3 films [7]. However, there are limited literature highlighting the sensing properties and the sensing mechanism. Among these fabrication methods, spin coating has many advantages for the deposition of oxides, including simplicity, inexpensiveness, low temperature deposition, high deposition rate and good controllability.

The selection of WO_3 gas sensor is done based on survey which tells that among the sensor based on electro-chemical, thermal conductive, infrared absorption, the semiconductor type gas sensors have better properties.

2 Conductometric Gas Sensors

The working principle of the gas sensors are indicated by change in sheet resistance, based on amount or concentration of gas exposed in a closed chamber with only exposure of required gas [8, 9]. The change in resistance is then appropriately converted to suitable measurable signal in terms of dc voltage directly proportional to the concentration target gas. The change in sensor's resistance which is linked to surface reactions depends on many factors including specific material reactivity, microstructure, free charge concentration, and sensitive layer morphology. Most of the work on "conduction type" in this sensors was performed on n-type semiconducting metal oxides such as SnO_2, ZnO and WO_3 [10]. This is evident from the increasing number of publications addressing the sensing properties of these metal oxides.

3 Block Diagram of the Signal Conditioning System

The concentration of NH_3 is sensed by the thin film metal oxide sensor whose value is provided to the controller. The controller compares the value with the predetermined threshold value. If the concentration of NH_3 is greater than the threshold, the controller signals the GSM module to send the precautionary message to the owner. A descriptive block diagram for the above design is given in Fig. 1.

The Fig. 2 shows the testing facility developed in house using gas chamber to study the sensing property for change in resistance of the developed sensor with respective to a NH_3. The sensor was studied and sensitive to ammonia gas was tested. The gas to be sensed was allowed to flow in a controlled manner into the chamber and corresponding change in the resistance was registered using a Keystone/Agilent semiconductor analyzer and digital multimeter. The suitable controller was used for indicating the changes, the high resistance recorded was signal conditioned and monitored for different concentrations. If it is greater than threshold value, the message was sent through GSM module to registered mobiles.

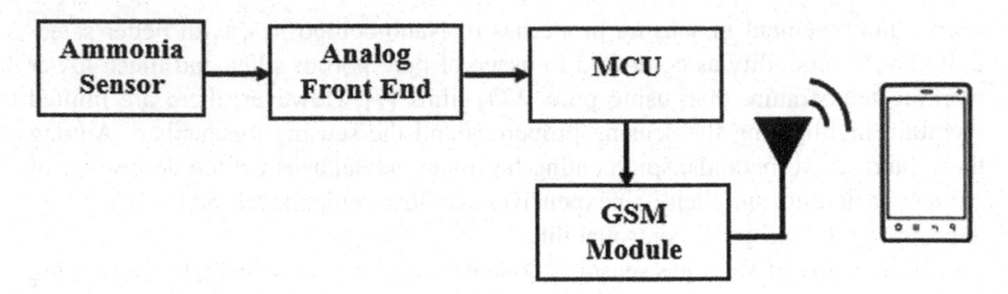

Fig. 1 Block diagram: measurement setup with GSM interface

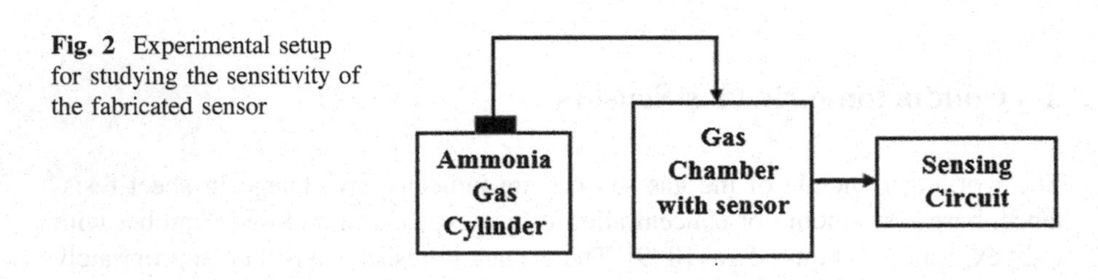

Fig. 2 Experimental setup for studying the sensitivity of the fabricated sensor

Interfacing with a controller was done and research on suitably on monitoring circuit was completed.

4 Synthesis of Tungsten Oxide (WO_3)

The sensing of Ammonia gases was carried out by using a thin film of WO_3 nanoparticles. The nanocasting route was applied for the synthesis of WO_3 nanoparticles. In the first step, ordered mesoporous silica was synthesized using ionic liquid under acidic conditions as reported by this author [11]. An imidazole type ionic liquid, 1-hexadecyl-3-methylimidazolium chloride was used as structural template for the fabrication of nanostructured inorganic materials and tetraethyl orthosilicate as silicate source to produce mesoporous silica. These ordered silica materials prepared under acidic conditions having hexagonal pores and large surface area (900 m^2/g), were used as attractive targets for the incorporation of WO_3 nanoparticles. The synthesis of WO_3 nanoparticles was carried out as follows: In a typical experiment, mesoporous silica as synthesized was dried in a vacuum oven at 80 °C for 4 h. 1 g dried silica sample was dispersed in a ethanol solution containing 1 g of phosphotungstic acid hydrate, stirred vigorously for 1 h, dried at room temperature and heated at 570 °C for 5 h. The hard template was removed by HF (2% wt.), and the product was washed with water and ethanol thrice and filtered. The WO_3 nanoparticles thus obtained was allowed to dry at room temperature and was characterized for structure and phase identification.

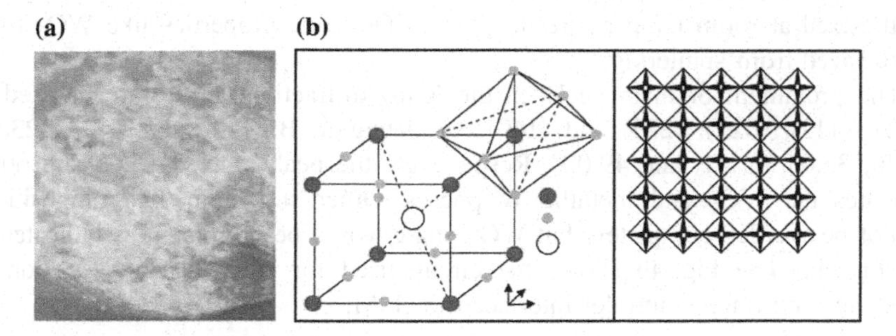

Fig. 3 **a** WO_3 thin film coated on glass and **b** Tungsten oxide crystal structure

Tungsten trioxide nanoparticles (5 mg) was dissolved in Isopropyl-Alcohol (20 mL) to get a solution. To disperse the WO_3 Nanoparticles, the solution is sonicated before dip coating and spin coat it on the glass substrate. The Fig. 3a above represents the tungsten oxide coated on glass substrate using spin coating process. The process of spin was done a steps few hundred RPM for 60 s. Annealed the coated film was done at 500 °C for 30 min.

5 Characterization

5.1 Chemical Characterization: X-Ray Diffraction (XRD)

The Fig. 4 shows XRD for spin coated WO_3 thin films. The peaks observed in the XRD patterns confirm the crystalline structures of WO_3. These peaks correspond to diffraction from the different planes of WO_3 nanoparticles. The crystallite size measurements were also calculated using the Scherrer equation, and the average particle size obtained from XRD data was found to be 10 nm. The WO_3 thin films

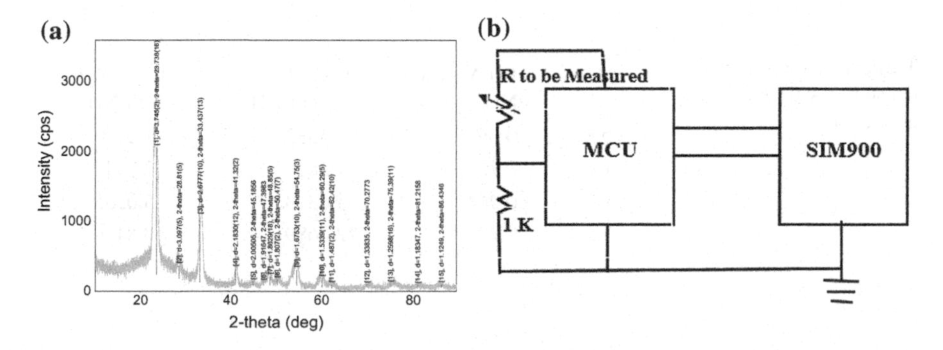

Fig. 4 **a** Shows the XRD patterns of the tungsten oxide films and **b** shows the testing circuit

synthesized at room temperature do possess the same properties like WO_3 films synthesized from sputtering.

The prominent or main peaks in the X-ray diffraction (XRD) are indexed as (020), (112), (022), (222) and (140) correlating to Bragg's angles $2\theta = 23.57$, 28.76, 33.63, 41.41, and 49.02°. Respectively the peak at $2\theta = 55.06$ and 60·4° indicates the WO_2 and metallic W phases. Other stoichiometric compositions cannot be avoided completely but WO_3 phase would be prominent as indicated in the Fig. 4a. The Fig. 4b shows the circuit used for checking the variation of resistance and microcontroller interfaced to GSM.

5.2 *Electrical Characterization: I-V Characterization*

The thin film of WO_3 coated showed the following results when it was I-V characterized in absence and presence of Ammonia gas respectively. The wireless monitoring of food quality was done using the controller-GSM module interface. The GSM module used in the interface is SIM900, which sends a message to the owner upon receiving the signal from the controller. The interface was done by connecting RX pin of the Arduino to the TX pin of SIM900, TX pin of the Arduino to the RX pin of SIM900 and GNDs of Arduino and SIM900 was made common. Separate power supply adapters (12 V/1A) were used for both. The previous chapter gave us a brief introduction to biometrics along with a review of some of the related work that has been done in the field.

The film was characterized for electrical behavior, the experiment was conducted using semiconductor analyzer and the values were tabulated as shown in Table 1. After exposing the WO_3 film for NH_3, the response of change in resistance with respect to current and voltage was found in orders of 1 or 2 G ohms. The response time was less than few seconds and recovery time after the removal of the gas was few minutes. The repeatability was observed by simple sol-gel/dip coated method without micro-heaters within a sensor as required by many metal oxide sensors. Hence the fabricated sensor can be cost effective.

Table 1 Variation of resistance with and without ammonia

Voltage	I (nA) without NH_3	R (G ohm) without NH_3	R (G ohm) with NH_3
0.5	0.06	8.3333	1.2953
1	0.78	1.282	2.5445
1.5	0.264	5.6818	5.0285
3.5	0.29	12.0689	10.8426

6 Results

The synthesized metal oxide WO_3 was characterized using XRD. The analysis of the XRD data shows that the prominent peaks indicates the crystallization and related phase of the fabricated tungsten film. The peaks indicate to be a mixed phase WO_3 material with crystalline nature. The deposited WO_3 films exhibit significant resistance change when exposed to Ammonia. The repeatability was tested. It was observed that the developed WO_3 sensor recovers to the initial state after removal of the gas in few minutes and the response time was few seconds when the gas was reintroduced. The variation in current with respect to voltage and voltage and

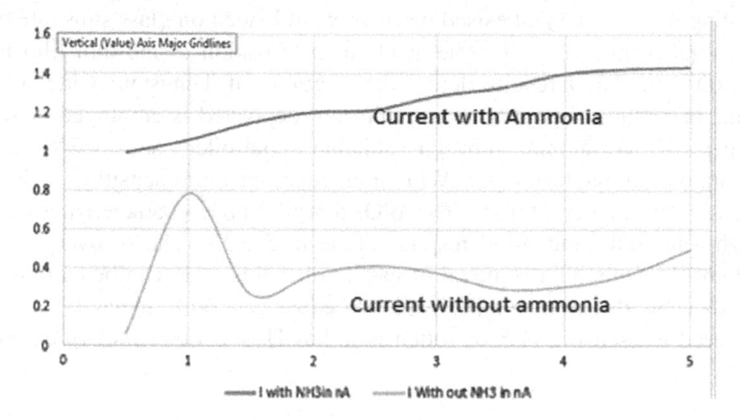

Fig. 5 I-V characteristics of the thin film response to with and without ammonia gas

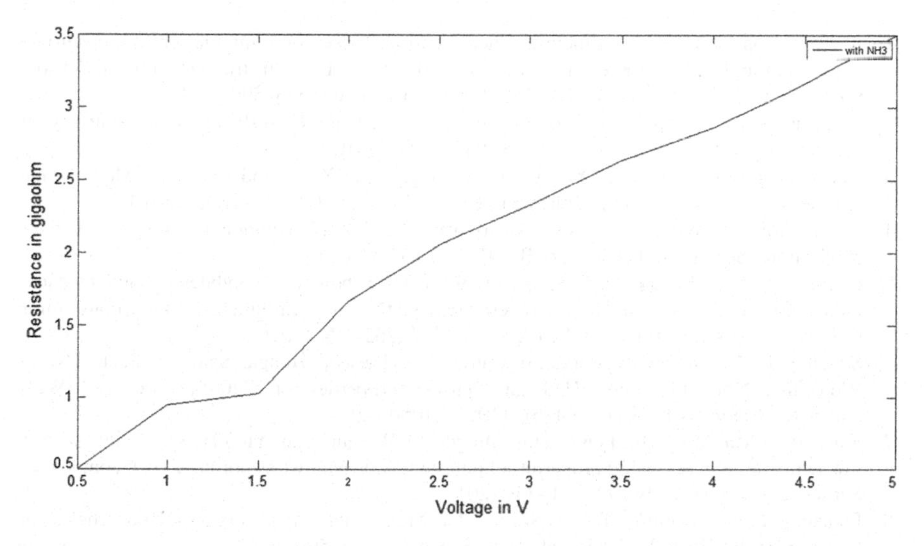

Fig. 6 Voltage and current characteristics of WO_3 films using semiconductor analyzer

resistance characteristics of the film fabricated using semiconductor analyzer B1500 when exposed to NH_3 gas as shown in Figs. 5 and 6.

Further this work could be carried out for measuring the food quality by sensing in rotten food, as the major composition in rotten food was ammonia. Sensors used for sensing target gas as ammonia could be used for food quality monitoring and sending the status of gas sensed to specific numbers for alarming or indicating.

7 Conclusion

The tungsten oxide nanoparticles average sizes of 10 nm were prepared through nano casting route. The synthesised particles were loaded on glass substrate by spin coating to get a thin film. The chemical characterization of the thin film is done using XRD. The XRD results depict the presence of Tungsten oxide of highly crystalline in nature. The metal oxide thin film deposited is employed in wireless monitoring of Ammonia through suitable controller and GSM module. A comparative study shows that WO_3 thin films are more sensitive to NH_3 and require lesser thermal conditions. The WO_3 thin film upon characterization showed very high sensitivity and good response time in 5–10 s. The recovery time was shown to be 2–5 min after removal of gas. Further this work can be carried out for sensing H_2S by different synthesizing procedure, as WO_3 metal is one of the material used for sensing H_2S as rotten food has H_2S as other major composition.

References

1. E. Kress-Rogers, C. J. B. Brimelow, Chemosensors, biosensors, immunosensors, electronic noses and tongues, in: E. Kress-Rogers, C.J.B. Brimelow (Eds.), Instrumentation and Sensors for the Food Industry, second ed., CRC Press, Boca Raton, 553–599 (2001).
2. Xiao Liu, Sitian Cheng, Hong Liu, Sha Hu, Daqiang Zhang, Huansheng Ning,: A Survey on Gas Sensing Technology Sensors 12, 5, 9635–9966 (2012).
3. Chengxiang Wang, Longwei Yin, Luyuan Zhang, Dong Xiang and Rui Gao,: Metal Oxide Gas Sensors: Sensitivity and Influencing Factors Sensors 10, 2088–2106, (2010).
4. Bjorn Timmer, Wouter Olthuis, Albert van den Berg,: Ammonia sensors and their applications Sensors and Actuators B 107, 666–677 (2005).
5. Engin Çiftyürek, Katarzyna Sabolsky, Edward M. Sabolsky.: Molybdenum and tungsten oxide based gas sensors for high temperature detection of environmentally hazardous sulfur species, Sensors and Actuators B: Chemical, 237, 262–274 (2016).
6. Suyoung Parka, Sunghoon Parka, Jihwan Junga, Taeseop Honga, Sangmin Leeb, Hyoun WooKimc, Chongmu Lee.: H_2S gas sensing properties of CuO-functionalized WO_3 nanowires Ceramics International 40, 11051–11056 (2014).
7. Shuangyun Ma, Ming Hu, Peng Zeng, Mingda Li, Wenjun Yan, Yuxiang Qi.: Synthesis and low-temperature gas sensing properties of tungsten oxide nanowires/porous silicon composite" Sensors and Actuators B 192, 341–349 (2014).
8. Franke, M. E.; Koplin, T. J.: Simon, U. Metal and Metal Oxide Nanoparticles in Chemiresistors: Does the Nanoscale Matter? Small 2, 36–50 (2006).

9. Mohammad Ahsan,: Theramally evaporated WO_3 thin films for gas sensing application, 2012.
10. Giovanni Neri,: First Fifty Years of Chemoresistive Gas Sensors Chemosensors 3, 1–20 (2015).
11. Divakara S. Gopala, Rama R. Bhattacharjee and Ryan M. Richards.: Dispersion of TiO_2 on High Surface Area Mesoporous Silica.: Functionalization with Tungstophosphoric acid and Application in Solvent-free, Aerobic Oxidation of n-Hexadecane Applied Organometallic Chemistry, 27, 1–5 (2013).

Big Data Layers and Analytics: A Survey

G. Manikandan and S. Abirami

Abstract With the advancement in the Internet technologies, the amount of data in the world has been increased dramatically. Due to this large volume of data, the traditional data analysis method fails to handle and also it is difficult to store, capture, curation, search, analyze and visualization of data. To deal with this issue we adequate tools, techniques, algorithms, architecture and the design principles. Extraction of the key information from the large data is the major issue in the big data analytical community. To discuss this issue in detail, this paper begins with the brief introduction about the big data, characteristics and structure of the big data, followed by addressing the various layers big data value chain in detail. And also this paper provides the comprehensive survey about the types and sub-types of the analytical methods of big data.

Keywords Big data analytics · Data storage · Analytical methods · Data layers · Data acquisition · Social media analytics · Audio/Video analytics

1 Introduction

With the advancement of technology and rapid growth in the Internet, the volume of the information raised explosively. It is estimated from IBM, more than 2.5 quintillion bytes of data are generated and shared in every day. So the overall will grow more than 50 times by next decade. The overall data will grow by 50 times by 2020, operated in large part by more in embedded systems such as medical device, sensors in clothing etc. And also this study predicted that the unstructured

---

G. Manikandan (✉) · S. Abirami
Department of Information Science and Technology, Anna University, Chennai 600025, India
e-mail: manitamilm@gmail.com

S. Abirami
e-mail: abirami_mr@yahoo.com

S.C. Satapathy et al. (eds.), *Computer Communication, Networking and Internet Security*, Lecture Notes in Networks and Systems 5,
DOI 10.1007/978-981-10-3226-4_39

information from social media, Internet texts, files and video will be grows more than 90% for next few years. But the available IT professionals to manage these huge data will grow by 1.5 times in today's scenario. The period of big data is not coming, it is here, the birth, growth and characteristic of the big data was defined in the year of 2000 itself. Today this might give sound to us but still we are concerning with actual doing and providing the ability for this modern era. In 1998, Google indexed around 1 million of web pages, after that it reached around 1 billion in the year of 2000, but quickly it exceeded 1 trillion in 2008, at present it increased dramatically. The reason is, the rapid expansion online social networking application such as Facebook, Weibo, Twitter etc., because that provides the users to create and share their contents freely.

Due to this rapid growth of the data the traditional database techniques becomes inefficient in storing retrieving, processing and analyzing the data. Big data management and analysis is main basis for business and management. Big data management is the new discipline which provides the data management techniques, infrastructure, tools and the platforms including the storage and security mechanisms.

In addition to this, the big data challenges include storage, capture, data curation, data transfer, visualization and security. For solving these issues we need sufficient technologies, principles. In this paper we made a study on definition of big data, characteristic of big data and brief discussion on big data architectures and the layers. The rest of the paper is organized as follows. Section 2 explains the fundamental concepts for defining the big data and characteristics of the big data; Sect. 3 discusses the big data architectures and the layers, the big data layers includes data source layer, data acquisition layer, data storage layer and data analytics layer.

2 Background

Big data definition has developed rapidly; the definition of the big data is categorized into three ways Attributive Definition, Comparative Definition and Architectural Definition [1]. Based on the study of literatures we haves summarized few best definition of the big data. Laney (2000) indicated the big data by Three V's namely Volume, Variety, and Velocity. Later, the TechAmerica Foundation (2012) defines the big data by "Big data is a term which describes large volumes of the high velocity, complex and variable data that requires the advanced techniques and the technologies to enable the capture, storage, management, distribution and analysis of the large dataset". We described the big data by three important V's namely Volume, Variety, Velocity, Veracity and Variability.

Volume refers to the size and magnitude of the data. Voluminous data is generating from the disparate sources such as sensor data, research data, social media (text, audio, video data), space image data, and so on. Definitions of the big data volumes are relatively vary by the various factors such as time and type of the data i.e. unstructured, semi-structured and structured data. **Variety** refers the various formats and structural heterogeneity of the data. Data may appear in many forms such as documents, images, audio, images, video, graphs. Storage and retrieval of these kind of data is very difficult since it does not have standard format and the type and also it is very difficult to analyze. **Velocity** concerns to the rate at which the data are generated from the source and the speed at which it has to analyze. The digital devices such as sensors and smart phones produces the unknown rate of data creation, it is very difficult to process in the real time analytics. There are two forms of the data, i.e. data-in-motion and data-in-rest. Data in motion concerns with velocity/speed of the data, here the data is to be processed and stored in the rapid manner. **Veracity** IBM originated the Veracity as the fourth V of the big data characteristics, which constitutes the unreliability and uncertainty mechanisms in some sources of the data. For example, sentiments given by the customer in social media are sometimes uncertain in nature, for this instance we cannot make the prediction over those types of data. To deal with this issue we need tools for mining the valuable information from the uncertain data. **Variability** is introduced by the SAS which refers to the variation in the data flow rate; the velocity of the data is not in the consistent manner it will vary among the periodic peaks and troughs. **Complexity** is the additional dimension of the big data which generated from the infinite number of sources. It is one of challenge to receive process, match and cleanse the data. Authors from [2] summarized the big data characteristics by "12 V's + C". It includes three traditional V's such as Volume, Velocity, Variety as we discussed in the previous section and some new V's at present are Variability, Value, Veracity, Validity, Volatility, Verbosity, Vulnerability, Verification, Visualization and Complex.

Data comes mainly in the form of four types namely structured data, Semi structured, unstructured data, and Quasi-structured data. The Structured data resides in the formal and valid data model and it can be easily accessed, stored, retrieved and analyzed. Semi-structured data is a type of the data it lacks with the strict data model, e.g. XML, RSS. Unstructured data doesn't reside in a traditional database, in opposite to the structured data. Experts estimated that 80–90% of the world data. Quasi-Structured data is slightly similar to the unstructured data. The quasi-structured data are in irregular data formats which can be formatted by doing additional effort, tools, techniques and time.

3 Architecture and Layers of the Big Data

In this section, we have described the big data value chain in detail. And also this section provides the detailed description on the four phases of big data value chain namely data generation, data acquisition, data storage and the data analytics. The diagrammatic representation of the big data its layers are shown Fig. 1.

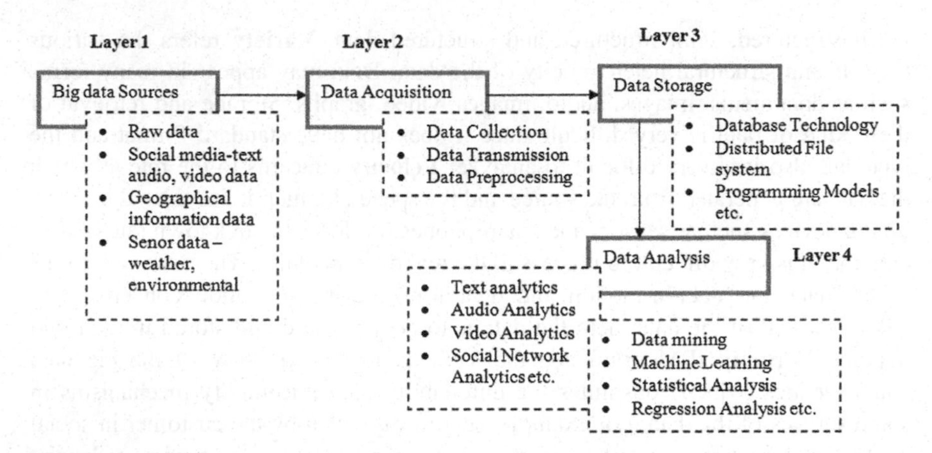

Fig. 1 Layers of big data

3.1 Big Data Sources/Data Generation

Data generation concerns where the data was generated. In general, the data may come from various sources and also it is in different format like structured, semi-structured, quasi-structured and unstructured. The data were generated from various distributed data sources [3] like Enterprise legacy systems, Data management systems, Geographical information, healthcare data, click stream data and other digital data.

3.2 Acquisition Layer

It is the second layer of the big data system, which collects and aggregates the information from the various sources and thereby it provides the mechanism for further storing and analysis of the data [1, 3]. Once the raw data is collected from the sources we need adequate technologies to process and transmit the data before storage for supporting the analytical process. The collected raw datasets may consist of noisy data, irrelevant data, redundant data which increases the storage capacity unnecessarily and affects the performance of the data storage and also analysis. For example, the log files in web server records consists of number of clicks, number of visits, click rate etc. In this records data with redundancy and inconsistent is very common, for this data deduplication, data mining and data compression techniques and operations are needed. During the big data acquisition process it undergoes three sub process namely data collection, transmission and pre processing of data.

3.2.1 Data Collection

Generally data collection is the first stage of the big data value chain; it is the process of collecting or acquiring the raw data from the specific environment. We have discussed few data collection methods [1].

Log files are the files which records everything that is happened in the particular system. Based on the recorded files, the analysis and predication can be followed. Event logs, Transaction logs, Message logs are the various kinds of log files [1, 3, 4]. Log file analysis is divided into two types, *Qualitative method:* It is conducted by either manually or computerized tools to interpret the log files. *Quantitative method:* It is automated analysis which is conducted based on the activity over time. **Sensor data** is an important component in the present Internet of Things (IoT) environment. Large amount of sensors are deployed in the various fields which may collects the different kinds of data like geographical data, agriculture sensor data, astronomical data etc. Currently, WSN have been used in various applications such as, environment monitoring [5], Video surveillance system, wildlife habit monitoring, forest fire monitoring, water monitoring etc. A **web crawler** is an automated scripting program which browses the web pages in the procedural and automated manner [3]. The process of the crawling is examined by certain policies such as politeness policy, revisit policy, selection policy and parallelization policy [5, 6]. Libpcab-Based Packet Capture Technology is one of the data collection methods the field of networking by using the library function [4, 7].

3.2.2 Data Transmission

Data transmission is the technique for transmitting the raw data into the storage infrastructure such as object storage system, data center and distributed cloud storage. After the completion of the data collection process, the data will be transmitted to the storage infra structure for further analysis. This process is classified into IP back bone transmission and data center transmission. The IP backbone transmission is carried out by high capacity pipe which helps to transmit the large data from the source to data center whereas data center transmission allows data transfer within the data center. The main importance of Data center Networks are, it provides the efficient way and connections between the thousands of servers, it supports machine to machine communication with traffic reliability and utilizes the less work load in distributed environment, it supports the virtualization methods to create virtual network, virtual function etc. The data transmission in the data center is classified into two types [1, 3]: Inter-DCN transmissions and Intra-DCN transmissions. Inter-DCN transmissions is carried out by transmitting the data from the source to the data center [3], this process is achieved by the physical network infra structure. An Intra-DCN transmission is carried out by transmitting the data within the data center. The data transmission between the data center are purely depend on the communication mechanism present in the data center [8] i.e. chips, plates, communication protocols and network architecture. Recently, optical technologies are used in interconnection of data center networks; it links the point to point with low cost and high performance. The larger part of the data center uses multi-mode fiber (MMF) [9] technique for connecting the switches over the data center networks.

3.2.3 Data Preprocessing

Data preprocessing is important part and process in data analysis phase because the collected raw contains missing values, irrelevancy, redundancy etc. Analyzing the data without preprocessing leads incorrect and invalid results, hence the quality of the data is important before preceding the analysis. Hence, the data preparation and filtering mechanisms is required before analysis phase [1, 3]. Data Preprocessing includes data cleaning, data normalization, data transformation, feature extraction and selection etc. We have explained few standard data pre processing techniques. **Data cleaning** is the process of identifying and correcting the inaccurate, inconsistent and unreasonable data thereby it improves the data quality. The main objectives of the data cleaning process are it provides the rules to detect the inconsistent, missing and incomplete data; it measures the data quality and loads cleaned data into the warehouse. The data cleansing process includes four stages [10] Auditing stage detects the anomalies presents in the data, statistical analysis tools are used to detect the anomalies. In Workflow specification process multiple operation are applied over the data to detect the anomalies and errors present in the data. Finally, Post–processing and controlling is the manual checking and data correction process based on the specific operations. **Data integration** is the technique which combines the data from the various sources and provides the coherent data store thereby it provided the unified view to the users. The data warehouse method consists of the three steps namely extraction, transformation and loading (ETL), sometimes the process may be differ based on the architecture i.e., extraction, loading and transformation (ELT) [11]. Conceptual modeling, Logical modeling, Physical modeling [11] are some of the ETL modeling techniques that has been using to deal with the big data. Also Redundancy elimination, data deduplication, dimensionality reduction [12] and data filtration methods are widely used methods for data preprocessing.

3.3 Storage Layer

Data storage system provides the platform to organize and store the collected information in the convenient format for the future value extraction and analysis. The storage infrastructure for the big data is categorized into two type's namely direct attached storage (DAS) and network storage (NS). Further Network Storage (NS) is classified into network attached storage (NAS) and storage area network (SAN) [1, 13]. In DAS, diverse hard disk drives (HDD) are directly connected with data server; each HDD receives the concerned amount in input and output resources based on the individual application. Due to the direct connection with server, it is only for suitable small scale applications. Network storage (NS) infrastructure utilizes the network to provide the common interface for accessing data and sharing mechanism to the user. Disk array and Tap library, and special storage software's are storage equipment of the Network Storage. Network Attached Storage (NAS) [4, 13] is storage device which is directly connected to the network through hub or switch via TCP/IP protocols. Finally the Storage area network (SAN) system is the independent storage system by utilizing the local area multipath data switching. The main challenge of big data system is to develop a large scale distributed storage system for efficient processing, storage and analysis. Google File System (GFS) and Hadoop distributed File system (HDFS) are widely used distributed framework in many big data corporate. Quantcast File System, Ceph Lustre, GlusterFS, Scale-out NAS file system, PVFS File systems are some of examples of Distributed File System. GFS is an extensible distributed file system which supports large-scale data in the distributed manner in many intensive applications. It provides and uses the less cost commodity servers for achieving the fault tolerance. The disadvantage of the GFS is it gives poor performance and failures when dealing the small files [1, 13].

3.4 Data Analytics Layer

Data analytics layer is the last layer of the big data value chain, the main objective of this layer is to extract the useful information from large data and to suggest conclusions, decision about the data. Some of the analysis domains are text analytics, web analytics, multimedia analytics, mobile analytics, network data analytics, social network analytics etc. Based on the literature study of [7] we have discussed four important analytics domain in detail.

3.4.1 Text Analytics

Text analytics is also known as text mining [7] is the process of extracting the useful information and from the structured/unstructured text data. Text analysis is broadly divided into four ways namely Information extraction, Text summarization [14], Sentiment analysis (opinion mining), Question answering (QA) techniques. Information extraction (IE) techniques are the potential task to extract the structured information from unstructured data [15]. Information extraction technique is further classified into three sub types namely [7] Entity Recognition (ER), Event Extraction and Relation Extraction (RE). ER extracts the names in the text data and group then into the defined categories based in date, location etc. RE technique extracts the semantic relationships between those entities text. Event Extraction (EE) techniques extract the insight information from the complex relations between entities from the textual data. Text summarization technique extract the key information of the single or multiple document thereby it produces the summary about the documents. Further it is classified into two approaches, extractive summarization approach, the summary of the document is created from the original text document and also the created document is subset of the original document and abstractive summarization approach, it extracts the semantic information and relation between the texts. Question answering (QA) techniques provides answers to questions which posed by the users. Yahoo Answers, Apple's Siri and IBM's Watson are some examples of QA systems. QA systems is further classified into three types IR-based QA systems [7], first it processes the question by examining the details of the question (question processing), question type, answer type, then it processes the document from the set of preexisted document (document processing) based on the question type and finally the answer document are ranked based on the highest rank then the output is given to the user. Second, Knowledge-based QA technique it produces the semantic description and relationship of the question then it processed by the same manner of information retrieval approach. Finally the hybrid QA technique uses both approaches i.e. the question is processed semantically and also like IR method. Sentiment analysis techniques [4] are classified into various level such as Document level [16], Word level [17], Aspect level [18], Sentence level [18], Concept level [19], Phrase level, Link based, Clause level [17], and Sense level [18].

3.4.2 Audio Analytics

Audio analytics extract and analyze the useful information from the structured and unstructured audio data. The main application of the speech analytics is customer call center and healthcare centers [7]. For example, efficient analysis of recorded calls will helps to know about the customer's requirements, complaints, customer behavior, product and service issues, privacy and security policies etc. Audio analytics system is mainly designed to analyze the live call, to know about the customers past and present interactions, feedbacks and also Interactive Voice Response (IVR) approach is used to identify frustrated callers. There are two common approaches in audio analytic namely transcript-based approach also called as large-vocabulary continuous speech recognition (LVCSR) and phonetic-based approach. LVCSR systems consist of two processes, indexing and searching. In indexing phase it uses the automatic speech recognition (ASR) algorithms to match the audio to words then the words are identified from the predefined dictionary. In searching phase the standard text based retrieval methods are used to index the file. Phonetic-based approach works the sounds or phonemes. Phonemes are the distinct units of the sound which is specified in a particular language for distinguishing the words (e.g., the phonemes/k/and/b/ for distinguishing the meanings of "cat" and "bat").

3.4.3 Video Analytics

Video analytics is the process of extracting and analyzing the useful information from the video data. It is also called as video content analysis (VCA) [7]. Due to widespread increase of closed-circuit television (CCTV) cameras and video sharing popularity in social networks websites leads the growth in development of video analysis. It is very cost effective and risk process for manually watching the thousands of hours of video, for this issue the video analytics technique is used to drawn the intelligence from the video data. Based on the architecture of the system, it divided into two categories namely server based and edge based architecture. In server based architecture, the captured video information is transferred to the centralized server, and then the server performs the video analytics. Usually the compression mechanism is done before storing the data, due to this image and video resolutions is get reduced, since this approach reduces the analysis accuracy. In Edge-based architecture the analytics is done in edge of the system where the raw data captured by the camera. So the entire content will available for doing analysis.

3.4.4 Social Media Analytics

Social media analytics is the technique for analyzing the structured and unstructured data from the social media/network channels. Based on the user generated contents, relationship and the interaction between the entities, the social media analytics is classified into two types [7] such as Content-based analytics and Structure-based analytics. Content based analytics is the analytics platform which focuses only data which is posted by the users in the social media platform, it may feedback, messages, product reviews etc. Structure based analytics concerned with the structural attributes and intelligence of the social networks. The structure and relationship of the network is modeled by nodes, edges and relationships between the edges. The network model is visualized by the graph social graph and activity graph. The relationship and links between the edges are represented by the social graph [7] (e.g., Friends, mutual friends). This graph is used to determine the direct and indirect relationship between the users. The activity graph provides the interactions (likes, comments) among the users. Community detection, Social influence analysis, Link prediction are the various social networking techniques to extract the information from the social network.

4 Conclusion

In this study we have presented the brief introduction about the big data, characteristics and structure of the big data. Further we have discussed the big data architecture and layers in detail. We have mentioned and discussed the big data architecture by the four layers namely data generation layer, data acquisition layer, data storage layer and data analysis layer. Further we have given the brief literature review with the diagrammatic representation of various analytical methods with its sub-categories such as Text analytics, audio analytics, video analytics and social media analytics.

References

1. Han Hu, Yong Gang Wentat-Seng Chua, Xuelong Li, "Toward Scalable Systems for Big Data Analytics: A Technology Tutorial", IEEE access practical innovation and smart solutions, Volume 2, pp. 652–687 (2014).
2. Manikandan G, Abirami S,2016, "A Survey: 12 V's, Challenges and Issues in Big Data" International Conference on Distributed Intelligent Computing, pp. 887–897, (2016).
3. Min Chen, Shiwen Mao and Yunhao Liu, Big Data: A Survey, Mobile Netw Appl, Springer Science + Business Media, vol. 19, pp. 171–209 (2014).
4. Kumar Ravi, Vadlamani Ravi, A survey on opinion mining and sentiment analysis: Tasks, approaches and applications, Knowledge-Based Systems vol. 89, pp. 14–46, (2015).
5. Amy Bruckman and J. Weiss, Analysis of Log File Data to Understand Behavior and Learning in an Online Community The International Handbook of Virtual Learning Environments, Springer, pp. 1449–1465 (2006).
6. Castillo. Effective web crawling, ACM SIGIR Forum, vol. 39, pp. 55–56, (2005).
7. Amir Gandomi, Murtaza Haider, Beyond the hype: Big data concepts, methods, and analytics, International Journal of Information Management vol. 35, pp. 137–144, (2015).
8. Lam, C.F, Fiber optic communication technologies: what's needed for datacenter network operations. IEEE Commun. Mag. Vol. 48, pp. 32–39 (2012).
9. Fatemeh Ahmadi-Abkenari, Ali Selamat, An architecture for a focused trend parallel Web crawler with the application of click stream analysis, Information Sciences vol. 184, pp. 266–281, (2012).
10. Aisha Siddiqa, Ibrahim Abaker, Targio Hashem, Ibrar Yaqoob, Mohsen Marjani, Shahabuddin Shamshirband, Abdullah Gani, Fariza Nasaruddin, A survey of big data management: Taxonomy and state-of-the-art. Journal of Network and Computer Applications, doi:10.1016/j.jnca.2016.04.008, (2016).
11. Sapna Dev, Dr. Arvind Kalia, Study of Data Cleaning & Comparison of Data Cleaning Tools, International Journal of Computer Science and Mobile Computing, Vol. 4, Issue. 3, pp. 360–370, (2015).
12. Winda Astriani, Rina Trisminingsih, Extraction, Transformation, and Loading (ETL) module for hotspot spatial data warehouse using geokettle, Procedia Environmental Sciences vol. 33, pp. 626–634, (2016).
13. Nawsher Khan, Ibrar Yaqoob, Ibrahim Abaker Targio Hashem, Zakira Inayat, Waleed Kamaleldin Mahmoud Ali, Muhammad Alam, Muhammad Shiraz, and Abdullah Gani, 2014, Big Data: Survey, Technologies, Opportunities, and Challenges, pp 1–18, (2014) Article ID 712826. http://dx.doi.org/10.1155/2014/712826.
14. Jiang J, Information extraction from text. In Mining text data United States: Springer, pp. 11–41, (2012).
15. Muhammad Habib ur Rehmana, Victor Chang, Aisha Batool, Teh Ying Waha, Big data reduction framework for value creation in sustainable enterprises, International Journal of Information Management vol. 36, pp. 917–928, (2016).
16. C. Bosco, V. Patti, A. Bolioli, Developing corpora for sentiment analysis: The case of irony and senti-tut, IEEE Intell. Syst. Vol. 2, pp. 55–63. (2013).

17. Y. Dang, Y. Zhang, H. Chen, A lexicon-enhanced method for sentiment classification: an experiment on online product reviews, IEEE Intell. Syst. Vol. 25. pp. 46–53, (2010).
18. S.K. Li, Z. Guan, L.Y. Tang, Exploiting consumer reviews for product feature ranking, J. Comput. Sci. Technol. Vol. 27, pp. 635–649, (2012).
19. A.C.-R. Tsai, C.-E. Wu, R.T.-H. Tsai, J.Y.-J. Hsu, Building a concept-level sentiment dictionary based on commonsense knowledge, IEEE Intelligent System, Vol. 2, pp. 22–30 (2013).

A More Efficient and Secure Untraceable Remote User Password Authentication Scheme Using Smart Card with Session Key Agreement

Ajay Kumar Sahu and Ashish Kumar

Abstract To secure the data using smart card it is very essential to authenticate the user password and provide session key agreement over the unreliable networks. To check the authenticity or validity of the remote users, smart cards based on password authentication is the best mechanism to solve this problem. There are various password authentication mechanisms proposed by different researchers have its own merits and demerits. In this proposed scheme, a more efficient and secure untraceable remote user password authentication scheme using smart card with session key agreement, we compared other password authentication schemes based on smart card and found that the proposed scheme have various security features which are not present in other schemes. The various security features include as: (1) Verifier table is not required; (2) identity of the user can not be trace out; (3) smart card having inbuilt verification mechanism; (4) facility of mutual authentication on both sides user/server is present; (5) facility of session key between user/server is present; (6) provide quick wrong password detection; (7) provide secure password select/update facility; (8) provide forward secrecy; (9) the storage cost/operational cost/authentication cost are minimum; (10) privacy of the user can be protected; (11) timestamp at every phase of authentication is required; (12) in the situation of damage/misplace of smart card, server can reissue the same without altering the identity.

Keywords Mutual authentication · Identity · Smart card · Security · Password · Session key · Un-traceability

---

A.K. Sahu (✉)
Department of Computer Science and Engineering, Raj Kumar Goel Institute of Technology and Management, Ghaziabad, Uttar Pradesh, India
e-mail: ajay4989@gmail.com

A. Kumar
Department of Computer Science and Engineering, ITS, Greater Noida, Uttar Pradesh, India
e-mail: ashishcse29@gmail.com

S.C. Satapathy et al. (eds.), *Computer Communication, Networking and Internet Security*, Lecture Notes in Networks and Systems 5,
DOI 10.1007/978-981-10-3226-4_40

1 Introduction

In any organization, the primary issue is how to access any resources securely. As the popularity of the internet is increasing day by day, network attacks, data security, virus attacks, phishing are the main concern with the unreliable networks. To provide the efficient security as a solution it is necessary to authenticate each and every individual's identity before giving the permission to access the resources as network based application systems. The password authentication is the easiest and suitable validation mechanism to protect our secret data and providing privacy over unreliable networks. The smart card based authentication mechanism provides the best solution regarding the validity/authenticity, security and privacy preservation.

When a user wants to access any network, application or resources in any organization, he must be authenticated first then he can login/access that network. For this user must have an identity (D) and a password (E). To access the services of the server the user must login the system with his unique identity and password, then system send this login request to the server with the corresponding identity and password. After getting this request by the server it will match the corresponding parameters in the stored verifier table stored on server side. After verification, if the corresponding parameters i.e. (D, E) matched with the verifier table, the server will grant the permission to user for accessing the resource. If the parameters i.e. (D, E) are not matched, then the server will refuse to grant the permission for accessing to the user. There are several issues present in the previous schemes. The first issue is the stealing the password by network/system administrator of the server, as there is no any encryption techniques are used in the verifier table so that all the passwords are in the plain text form. The other issue is to misusing the user's D and E from verifier table by an unauthorized user as poses as a legal user of the system. So, there is a big concern, how to protect our password table securely.

There are various authentication mechanisms which are based on password [1–6] have been proposed by various researchers without storing verifier table. All of these schemes having its own advantages and disadvantages. Due to the absence of verifier table in the server side the identity of each user's are static so there are chances to disclose some information regarding the authentication request generated by user on the remote system so that the intruder can use this information in some illegal operations or it can misuse it. To overcome this type of misuse, various researchers [4, 6–16] have adopted dynamic (D) based validation mechanisms based on smart card (C) which is free from password table. In these schemes, the identity (D) of the user is randomly generated each and every time when a user login to the server for accessing the resource to protect the identity (D) from the unauthorised users.

Since, there is possibilities of various different types of attacks from attackers/intruders in remote user password authentication mechanism, so our main goal is to propose such a mechanism which having the capability to protect our data and system from such types of attacks. There are various features/keys necessary for the successful implementation of any secure password authentication mechanism.

(a) There is no need to store the verifier table on server side for verification procedure.
(b) There is a freedom for the user to select/update its own password at any time.
(c) The facility of mutual authentication must be present in the scheme by which the user verifies that the server is legal and also the server verifies that the user to whom it will provide the services is not false one.
(d) For the privacy preservation there must be the facility of session key generation between the user and the server to provide secure data access.
(e) The storage/operational/authentication cost should be minimum.
(f) The scheme having the facility of time instant in login/authentication message communicating on both user and server sides must be present.
(g) When an unauthorized user enters incorrect password for login to the server, it must be detected easily.

If any illegal person found the smartcard of any registered user, by using this card he can extract any information of legal user such as identity (D). In this paper, our main goal is to present an efficient scheme which can not only provide the security against various attacks as well as handle the situation of misplaced smartcard.

Here we propose a mechanism which avoids pose attacks by imposing proper mutual authentication between user (D) and server (S_{ser}), and also provide the session key establishment facility between user (D) and server (S_{ser}) for secure data transmission. The proposed scheme's in this paper deal with many security threats/attacks identified in various previous mechanisms given by various researchers and free from those security threats/attacks. After comparison with other schemes proposed by various researchers with our scheme we can show that our scheme will adopted as most efficient, secure and will provide privacy preservation.

The remaining part of the paper as: The proposed scheme in Sect. 2; Analysis of security in Sect. 3; Goals and achievements in Sect. 4; Security requirement characteristics and performance analysis in Sect. 5. In the last conclusion is given in Sect. 6.

2 The Proposed Scheme

Here we proposed a more secure and untraceable smart card based password authentication mechanism which not only prevent from various security threats as well as provide proper mutual authentication and agreed upon a common session key for secure data transmission. The proposed scheme having several phases like: (a) registration phase; (b) login phase; (c) authentication phase and (d) password change phase (Table 1).

Table 1 Symbol used

Notations	Description
A/B	A system user/an attacker
C	Smart card
S_{ser}	Server
D, E, E_{update}	Identity of user, current password and updated password of user
F	Random identity of user (A)
t, t', t″	Different time instant acquired by A/S_{ser}
t_a	Current time instant acquired by B
Δt	Time interval
hash(.)	Hash function
XOR	XOR operator
q	A key (number) maintained by Server (S_{ser})
y_i, β_i	Unique random number assigned to user (A) by remote server (S_{ser})
$\|\|$	Concatenation operator (OR operator)
t_h	The access time of hash(.) operation
$t_\oplus$	The access time of XOR operation
r	Any number selected by A

2.1 Registration Phase

To access any resource of the network the user (A) first registers himself to the server by sending the request in the form of registration. The steps of registration are as follows:

Step 1: In the initial stage (A) select its own (D, E) and choose any random number (r). After selection it will compute ME = hash (r||E) then transfer this registration request {D, ME} to the S_{ser} via a reliable network.

Step 2: When the server (S_{ser}) found this request as in the form of {D, ME} then it will select unique random number (β_i). This number will be unique as it is different for every user.

Step 3: Then the server (S_{ser}) calculate some values as H_i = hash (D||q) XOR ME, X_i = β_i XOR hash (D||q), Y_i = hash (D||β_i||ME) and Z_i = β_i XOR hash (β||q), here q is the key maintained by S_{ser}.

Step 4: S_{ser} stores some parameters as in the form of {X_i, Y_i, Z_i, hash(.)} inside the (C) then it will delivers {C, H_i} to user (A) through reliable network.

Step 5: After getting the parameters as in the form of {C, H_i} by the S_{ser}, user computes L_i = (D||E) XOR r and M_i = H_i XOR r. User inserts L_i and M_i in Smart card then there is not required to keep this number r in future use finally the contents of Smart card = {L_i, M_i, X_i, Y_i, Z_i, hash(.)}.

2.2 Login Phase

After registration, the user must require to login himself as an authorised user, if he wants to access the resources of the network. For the same user (A) must have two parameters such as (D, E). After inputting these two parameters the C computes some values as given:

Step 1: It will calculate $r = L_i$ XOR $(D||E)$, ME = hash $(r||E)$. After that it will retrieves hash $(D||q) = M_i$ XOR ME XOR r and $\beta_i = X_i$ XOR hash $(D||q)$ and then computes $Y_i^* =$ hash $(D||\beta_i||ME)$.

Step 2: Then C will compare the values of Y_i^* and Y_i. If $Y_i^* \neq Y_i$, C deny the process. If any user attempts the login request with more than three wrong passwords then his C will be blocked and it will require (Private Unblocking Key) to again activate his C.

Step 3: When $Y_i^* = Y_i$, the inserted (D, E) is correct then C perform the following: it will compute hash $(\beta||q) = \beta_i$ XOR Z_i and $H_i = M_i$ XOR r. Notifies current time instant t then calculate F = D XOR hash $(H_i||\beta_i||t)$, $H_i' = H_i$ XOR hash $(\beta_i||t)$, $A = H_i$ XOR ME = hash $(D||q)$, V_i = hash $(H_i||\beta_i||A||t)$ and $W_i = \beta_i$ XOR (hash$(\beta||q)||t)$.

Step 4: At the last C transfer this login message = $\{F, H_i', V_i, W_i, t\}$ towards the S_{ser} through the reliable network.

2.3 Authentication Phase

After getting the request in the form of login by the server $\{F, H_i', V_i, W_i, t\}$ at time t' both A and S_{ser} validate to each other as follows:

Step 1: First of all S_{ser} verify the time interval $(t' - t) \leq \Delta t$, and check the availability of other login request having parameter $\{F, H_i', V_i, W_i, t\}$, in the time intervals between $(t - \Delta t)$ and $(t + \Delta t)$. When time interval is correct and there is no other request in same time intervals this process will continue. If any condition will false the S_{ser} stops all processes and abort this phase immediately with the same parameters $\{F, H_i', V_i, W_i, t\}$.

Step 2: S_{ser} calculate the parameter as $\beta_i = W_i$ XOR (hash $(\beta||q)$ $||t)$, $H_i = H_i'$ XOR hash $(\beta_i||t)$ and D = F XOR hash $(H_i||\beta_i||t)$. Then computes A^* = hash $(D||q)$ and V_i^* = hash $(H_i||\beta_i||A^*||t)$.

Step 3: If $V_i^* = V_i$, user (A) verified by server (S_{ser}) at time instant t″ then S_{ser} perform a = hash $(A^*||\beta_i||t'')$. If $V_i^* \neq V_i$, S_{ser} rejects A's login request $\{F, H_i', V_i, W_i, t\}$, records identity of user and the number of wrong password attempts. If there are more than three times wrong password attempt then S_{ser} denies this request in the same time interval.

Step 4: S_{ser} transmits the parameters $\{a, t''\}$ towards the C through reliable network.

Step 5: When the C receives the parameters {a, t″}, it will verify the correctness of t″. After verifying the correctness of timestamp t″ the smart card calculate a' = hash ($A||\beta_i||t''$) then after match the values of a′ and a received before. When both values are same then the user ensures for server that it is legal one.

Step 6: When S_{ser} and A authenticates to each other then after both of them individually agree upon a same session key as $S_{session}$ = hash ($A||\beta_i||t||t''||$hash $(\beta||q)$) as well as $(S_{session})^*$ = hash ($A^*||\beta_i||t||t''||$hash $(\beta||q)$) respectively.

2.4 *Password Change Phase*

When a user want to change or update his smart card password then it is required to input the previous password E one time then it will ask to input to enter the new password E_{update} two times from the A. There is no need for the server to verify this process. Only smart card is sufficient to check the validity of this request for password change.

Step 1: To make any changes regarding the password, user must enter the two parameters into the device as (D, E).

Step 2: C calculate some parameters like $r = L_i$ XOR $(D||E)$ and computes ME = hash $(r||E)$, hash $(D||q) = M_i$ XOR ME XOR r and $\beta_i = X_i$ XOR hash $(D||q)$, and computes Y_i^* = hash$(D||\beta_i||ME)$.

Step 3: Then C compare the values Y_i^* and Y_i receives. If $Y_i^* \neq Y_i$, C rejects this request. If any user inputs wrong password more than three times then there is a facility of blocking the C, it will activate by using PUK only.

Step 4: When $Y_i^* = Y_i$, the inputted D and E is verified by Smart card.

Step 5: After verifying the password change request by the C, user must enter the updated password (E_{update}) two times. When these two entries are same, C calculate the following parameters as: $(ME)_{update}$ = hash $(r||(E)_{update})$, $(L_i)_{update} = (D||(E)_{update})$ XOR r, $(M_i)_{update} = M_i$ XOR (ME) XOR $(ME)_{update}$ and $(Y_i)_{update}$ = hash $(D||\beta_i||(ME)_{update})$. Stores $(L_i)_{update}$, $(M_i)_{update}$ and $(Y_i)_{update}$ and replace instead of the following parameters L_i, M_i, and Y_i respectively.

3 Security Analysis of Proposed Scheme

When we analyse our proposed scheme we found that this scheme fulfil various security requirements and it will provides the best solution to prevent various threats over the network. There are various major points regarding the same are as given below:

3.1 Prevent (Insider) Conspirator Attack

In our scheme all the parameters are stored in the form of encryption not in plaintext mode. In this scheme the password is concatenated with random number r then submits the hashed value in updated form as ME = hash (r||E). So, the Conspirator cannot obtain a user's password. As in the form of (ME), both values of {r, E} are not known by the Conspirator, and for him it is not possible to guess both value randomly and verify it. Hence, this scheme is totally free from conspirator attack.

3.2 Prevent Destruction/Loss of Smart Card Attack

Suppose adversary extracted the parameters {L_i, M_i, X_i, Y_i, Z_i, hash (.)} which are stored in the C as [17, 18] but it cannot get any useful information from these like {D, β_i, r, E} as Y_i = hash (D||β_i||ME) because all values are unknown. To obtain hash (D||q), from M_i = H_i XOR r = hash (D||q) XOR ME XOR r, password and random number {E, r} must be known, but adversary neither guess both values simultaneously nor obtain from L_i = (D||E) XOR r, as all three values are unknown to B. The adversary also cannot obtain β_i from X_i = β_i XOR hash (D||q) as it does not know hash (D||q) as well as from Z_i = β_i XOR hash (β||q), here also hash (β||q) unknown to B. Therefore, B, cannot obtain either hash (D||q) or β_i from all the values {L_i, M_i, X_i, Y_i, Z_i, hash (.)} extracted from the C. Hence, Adversary cannot obtain any useful information like {D, β_i, r, E} from lost/stolen smart card.

3.3 Prevent User Pose and Server Mask Attack

When a unauthorised person act as a valid user of the system, B must have a valid login request with the values {F, H_i', V_i, W_i, t}. By which B (a) calculate D and H_i, the values {β_i, D, X_i} must be known to adversary. (b) To compute W_i, hash (β||q) (or q and β) of the server must be known to server, otherwise computation cannot perform. (c) To compute V_i, the values of {ME (r, E), hash (D||q)} must be required. Here, the adversary cannot obtain these values by either calculation or not directly extract these values from arbitrary smart card found by attacker. Here we adopt such a mechanism in which the values {L_i, M_i, X_i, Y_i, Z_i, and hash (.)} are stored in the encrypted form so that adversary cannot extract any information. Hence for B it is not possible to achieve successful login to the system. These parameter values {β_i, H_i, hash (β||q), hash(D||q)} must be known to B for successful login which are stored in other parameters like {X_i, M_i, Z_i, H_i} for security purpose. The acknowledgement message {a, t″} transmitted by S_{ser} to A is also useless for B as a = hash (A*||β_i||t ″) = hash (hash (D||q)||β_i||t″) because B cannot obtain any value from it and except timestamp t″ remaining values {(D||q), β_i} are unknown to B. Thus, it is not feasible for B, as to act like as user pose attack. Because B also cannot compute {a, t″} as an acknowledgement to this, hence for B this is not possible to server mask attack.

3.4 Prevent Online Password Imagination/Guessing Attack

If B found any misplaced smart card and he wants to guess the password in connected mode but due to inbuilt verification of C this is not possible for the B to guess the passwords as many as times he wants. As there is a limit of maximum three attempts for login request so there is no any chance to get the login by B because after three wrong passwords the C will be locked. So, in this way it will prevent password imagination attack.

3.5 Prevent Repeated/Replay Attack

In our mechanism, there is a provision to include the time instant in each and every message communicated over reliable network. When A send request for login as in the form {F, H_i', V_i, W_i, t}, it will receive by S_{ser} at time instant t', S_{ser} immediately verify the time interval (t' – t) $\leq$ Δt, then it will check the other request for the same parameters in the same time interval. If both conditions true, S_{ser} continue otherwise discard the login request. In the same manner, the acknowledgement message {a, t″} transmitted from S_{ser} to A. If this response message {a, t″} passes the time instant correctness test, then only session has established between A and S_{ser} otherwise connection aborted. All the components of the login request {F, H_i', V_i, W_i, t} having current time instants. If the adversary B repeated the remaining components along with current time instant t'/t″, V_i* = V_i and a' = a cannot hold. In this way, we prevent the repeated attack.

3.6 Prevent Lifted Verifier Attack

Since there is no any provision for maintaining the records of registered users in the server side so there is no any possibility to leakage of security as in the form of password of any registered users. The S_{ser} uses only the key q and random number β for extracting the random number β_i given by S_{ser} to the authenticate user. After that it will extract other values to validate the authenticity of the A.

3.7 *Prevent Denial of Service (DOS) Attack*

Since there is a provision for checking the authentication of the A in the C itself, there is no any involvement of the S_{ser}, therefore if any user inputted wrong password more than three times by mistake then C itself deny to provide the services to user. There is no any possibility that C will forward this login request to S_{ser} and then after verification the S_{ser} will deny to such a request for wrong password entry. So with this provision we can save maximum time and energy as well as resources of the server.

4 Goal/Achievement of Proposed Scheme

4.1 *Provides Forward Secrecy*

In our scheme forward secrecy plays very crucial security features for authentication mechanism, it guarantees that our data will secure during transmission once session between user and server has established. After authenticating to each other S_{ser} and A agreed upon a common session key $S_{session}$ = hash $(A||\beta_i||t||t''||$hash $(\beta|| q))$ = hash (hash $(D||q)||\beta_i||t||t''||$hash $(\beta||q))$. If anyhow key q is disclosed, the random number β still not known by B. If key q and random number β both are disclosed anyhow, $S_{session}$ still remains secure because B not know the values of D and random number β_i, hence there is no any chance to generate the $S_{session}$ even disclosing the parameters like q, β.

4.2 *User Identity is Unidentified/Untraceable*

After analysing the scheme we found that B is not capable of achieving the identity of authorised person anyhow. In the previous section we shown that B is capable of extracting the parameters $\{L_i, M_i, X_i, Y_i, Z_i,$ hash (.)} from any C. To obtain D from parameter $\{L_i, M_i, X_i, Y_i\}$ adversary having the parameters as {(E, r), (q, E, r), $(\beta_i$, q), $(\beta_i$, E, r)} which is not feasible. The other possibility for B to interfering the login parameters as {F, H_i', V_i, W_i, t}. To obtain the D of the user from {F, H_i', V_i}, it requires the knowledge of H_i, β_i to retrieve D. In our mechanism, there is no any provision to achieve random number β_i, H_i, for extracting the value of D, from F = D XOR hash $(H_i||\beta_i||t)$. Therefore it is not possible for B to obtain the secret number β_i, retrieve H_i then obtain D from F because this random number is unique for each and every user. Hence, the user identity is unidentified in our scheme.

4.3 Inherent/Inbuilt Verification Mechanism of Smart Card

Any registered user can access the services of the server after providing the correct (D, E) to the server. Then C calculate some parameters as $r = L_i$ XOR $(D||E)$, ME = hash $(r||E)$. After that calculate some parameter as hash $(D||q) = M_i$ XOR ME XOR r and $\beta_i = X_i$ XOR hash $(D||q)$ and lastly, calculate $Y_i{}^* =$ hash $(D||\beta_i||ME)$. If $Y_i{}^* = Y_i$ holds, then it confirms that inputted D and E are correct and further proceed otherwise smart card drops the session. If the password provided by user continuously wrong more than thrice then C will blocked. In this way, the smart card verifies the correctness of D or E, there is no need of server involvement in this mechanism.

4.4 Provides Proper Mutual Authentication

There is a provision for proper mutual authentication between user and server. When user provide the correct (D, E) then login request must be generated and server will check the authenticity of the time instant correctness by validating the equivalence $V_i{}^* = V_i$, if found correct then authenticate the user. Then server will generate the acknowledgement as $\{a, t''\}$ and transmit the same through reliable network towards C. If time instant t'' is correct within timeslot, then C verify the validity of a' = a, if found correct, the smart card/User can ensure that server is legal. Therefore, in this way we can successfully implement the provision of mutual authentication.

4.5 Provision for Common Session Key Generation

Our scheme provides the facility of generation of session key as $S_{session} =$ hash (hash $(D||q)$ $||\beta_i||t||t''||$hash $(\beta||q)$) between both sides so that the server and user can communicate confidentially to each other. Hence, the role of the session key plays a very important security feature for privacy and confidentiality which is not present in many other schemes.

4.6 Provides Better Password Update Facility

The password update facility in our scheme is to provide the easiest way to change/update the password of registered user without the involvement of the server. When a user wants to change his password he must enters (D, E). After which the C calculate some parameters as $r = L_i$ XOR $(D||E)$, ME = hash $(r||E)$.

Then retrieves hash (D||q) = M_i XOR ME XOR r and β_i = X_i XOR hash (D||q), and then calculate Y_i* = hash (D||β_i||ME). Then Smart card compare Y_i* with stored Y_i. If Y_i* = Y_i, the inputted D and E is verified by Smart card then after A is capable to enter the updated password twice as $(E)_{update}$. In this way, C, update/change the stored value of old password with new value of password as $(E)_{update}$. Hence, our scheme provides user friendly password change facility.

5 Performance Analysis and Security Requirement Characteristics

Here we analyze the various security features, efficiency of various other schemes proposed by different researchers and compare the features of the scheme as different parameters in terms of the storage cost, operational cost, efficiency/security characteristics, achievements/goals. There are some assumptions on various parameters are supposed to be 164-bit like {D, E}, arbitrary values {r, β_i}, time instant {t, t″}, {hash (D||q), hash (D||β_i||ME), etc.}. In this section we have compare the storage capacities required by the smart card/server with other schemes as in Table 2. The communication cost for any network represents various parameters transferring between server and user through unreliable network (Table 3).

Here in Table 4 we compare the operational cost of different schemes proposed by various researchers with our mechanism with respect to various parameters on both server and user sides such as registration, login, and authentication phases. Here operational cost can be calculated as time elapsed in various processes of different phases. Here $t_{h(.)}$ is represented as time required for hash operation and $t_{\oplus}$ denoted as time elapsed in XOR operation.

In Table 5, we compare various security features of different schemes proposed by researchers with our scheme. Here we found that our scheme achieve almost all the features of security which is necessary for efficient mechanism for privacy preservation and protect our system with various threats available in the network. In Table 6 we represent various goals and achievements of our mechanism and compare all of them with various schemes already proposed by various researchers.

Table 2 Storage capacity comparison

Storage/ Scheme	Hsieh and Leu [19]	Lee [1]	Wang et al. [8]	Wen and Li [6]	Chang et al. [7]	Our scheme (bits)
Smart card (bits)	3 * 164 = 492	3 * 164 = 492	4 * 164 = 656	3 * 164 = 492	3 * 164 = 492	6 * 164 = 984
Server (bits)	2 * 164 = 328	1 * 164 = 164	2 * 164 = 328	2 * 164 = 328	1 * 164 = 164	2 * 164 = 328

Table 3 Communication cost

Communication/scheme	Hsieh and Leu [19]	Lee [1]	Wang et al. [8]	Wen and Li [6]	Chang et al. [7]	Our scheme
Authentication (bits)	6 * 164 = 984	7 * 164 = 1148	6 * 164 = 984	9 * 164 = 1476	6 * 164 = 984	7 * 164 = 1148

Table 4 Operational/Complexity cost

Phase/Scheme	Hsieh and Leu [19]	Lee [1]	Wang et al. [8]	Wen and Li [6]	Chang et al. [7]	Our scheme
Registration phase (user side)	$2t_{h(.)} + 1t_{\oplus}$	Nil	Nil	Nil	Nil	$1t_{h(.)} + 2t_{\oplus}$
Registration phase (server side)	$6t_{h(.)} + 4t_{\oplus}$	$2t_{h(.)} + 1t_{\oplus}$	$2t_{h(.)} + 2t_{\oplus}$	$5t_{h(.)} + 4t_{\oplus}$	$3t_{h(.)} + 2t_{\oplus}$	$3t_{h(.)} + 3t_{\oplus}$
Login phase (both user and server side)	$7t_{h(.)} + 6t_{\oplus}$	$12t_{h(.)} + 11t_{\oplus}$	$7t_{h(.)} + 14t_{\oplus}$	$17t_{h(.)} + 18t_{\oplus}$	$9t_{h(.)} + 5t_{\oplus}$	$13t_{h(.)} + 12t_{\oplus}$
Mutual authentication	Nil	Nil	$2t_{h(.)} + 2t_{\oplus}$	$2t_{h(.)} + 2t_{\oplus}$	$1t_{h(.)}$	$1t_{h(.)}$
Session key	Nil	Nil	Nil	$1t_{h(.)}$	Nil	$2t_{h(.)}$
Sum of computational complexity	$\mathbf{15t_{h(.)} + 11t_{\oplus}}$	$\mathbf{14t_{h(.)} + 12t_{\oplus}}$	$\mathbf{11t_{h(.)} + 18t_{\oplus}}$	$\mathbf{24t_{h(.)} + 24t_{\oplus}}$	$\mathbf{13t_{h(.)} + 7t_{\oplus}}$	$\mathbf{20t_{h(.)} + 17t_{\oplus}}$

Table 5 Comparison of efficiency/security features

Security features/Scheme	Hsieh and Leu [19]	Lee [1]	Wang et al. [8]	Wen and Li [6]	Chang et al. [7]	Our scheme
Prevent repeated/replay attack	X	√	√	X	√	√
Prevent lifted verifier attack	√	X	√	X	√	√
Prevent user pose attack	√	√	X	√	√	√
Prevent server mask attack	√	X	√	√	√	√
Prevent connected/online password attack	√	√	X	√	X	√
Prevent SC loss/ destruction attack	X	√	X	X	X	√
Prevent insider/conspirator attack	√	X	√	X	√	√
Prevent DOS attack	X	X	X	X	√	√
Prevent the key x maintain by server	X	X	X	√	X	√

Table 6 Comparison of achievements/goals

Goals/Schemes	Hsieh and Leu [19]	Lee [1]	Wang et al. [8]	Wen and Li [6]	Chang et al. [7]	Our scheme
Provide proper mutual authentication	√	X	√	√	√	√
Provide privacy preservation features	X	X	X	X	N/A	√
Features of quick wrong password detection	X	X	X	X	X	√
Session key establishment facility	X	X	X	√	√	√
Availability of password change facility	√	√	√	X	X	√
Provides freely Pw choosing facility	√	√	√	√	X	√
Availability of Un-traceability features	X	√	X	X	X	√
Availability of verification feature in smart card	X	X	X	√	X	√
No need to maintain the verifier table by server	√	√	√	√	√	√
Storage, operational and communication cost should be low	√	√	X	X	√	√

6 Conclusion

Since the scheme having the features of establishment of common session key after the proper mutual authentication for secure communication and privacy preservation, hence the proposed scheme is more efficient, secure and avoids tracing of user password authentication. The scheme succeeds to prevent security vulnerabilities and also the requirement of server password is removed which compromise the scheme. To avoid denial of service, provision for verification procedure existing in smart card is implemented. The results obtained in Table 6 shows that scheme outperformed other related existing schemes [1, 6–8, 19]. During analysis, it is observed that scheme incurs increased computational complexity cost. However, it pays off in preventing more attacks against other schemes [1, 6–8, 20, 21, 10–16] making it more secure. The scheme implements the characteristics of user anonymity with hidden identification then achieve the requirement to maintain a database on the server is removed. The computation cost is kept to minimum by involving only hash/XOR operations. The applications requiring privacy preservation, enhanced security and having low computation power are the target users of the scheme.

References

1. Lee Y C. A new dynamic ID-based user authentication scheme to resist smart-card-theft attack. An international journal of applied mathematics & Information Sciences 6 No. 2S pp. 355S–361S (2012).
2. Zhen-Yu Wu, Dai-Lun Chiang, Tzu-Ching Lin, Yu-Fang Chung. A reliable Dynamic User-Remote password Authentication scheme over Insecure Network. 26th Int. Conference on Advanced Information Networking and Applications, 2012.
3. Manoj Kumar, Mridul Kumar Gupta, Saru Kumari. An Improved Efficient Remote Password Authentication Scheme with Smart Card over Insecure Networks. International J. of Network Security, Vol. 13, No. 3, PP. 167–177, Nov. 2011.
4. R. Madhusudhan, R. C. Mittal. Dynamic ID-based remote user password authentication schemes using smart cards: A review. J. of Network and Computer Application 35; 2012, p. 1235–1248. doi:10.1016/j.jnca.2012.01.007.
5. Kwang Cheul Shin, Won Whoi Huh. Improvements of a Remote User Password Authentication Scheme using Smart Card. International J. of Security and its application: Vol. 7, No. 4, July, 2013.
6. Wen F, Li X. An improved dynamic ID-based remote user authentication with key agreement scheme. Journal of computers and electrical engineering 38 (2012), 381–387. Doi:10.1016/j.compeleceng.2011.11.010.
7. Chang Y F, Tai W L, Chang H C. Untraceable dynamic identity based remote user authentication scheme with verifiable password update. International Journal communication system 2013. http://dx.doi.org/10.1002/dac.2552.
8. Wang Y Y, Liu J Y, Xiao F X, Dan J. A more efficient and secure dynamic ID-based remote user authentication scheme. Journal of computer communication 32 (2009), pp 583–585. http://dx.doi.org/10.1016/j.comcom.2008.11.008.

9. Khan M K, Kim S K, Alghathbar K. Cryptanalysis and security enhancement of a more efficient & secure dynamic ID-based remote user authentication scheme. Journal of computer communication 34 (2011), pp 305–309. http://dx.doi.org/10.1016/j.comcom.2010.02.011.
10. Qi Xie. Improvements of a security enhanced one-time two-factor authentication and key agreement scheme. Scientia Iranica D; 2012, 19 (6), p. 1856–1860. doi:10.1016/j.scient.2012.02.029.
11. Hung-Min Sun. An efficient remote user authentication scheme using smart cards. IEEE Transaction on consumer Electronics, Vol. 46, No. 4, November 2000.
12. Li Yang, Jian-Feng Ma, Qi Jiang. Mutual Authentication Scheme with Smart Cards and Password under Trusted Computing. International J. of Network Security, Vol. 14, No. 3, PP. 156–163, May 2012.
13. Ahirwal D, Raghuwanshi S. A noble remote authentication protocol based on smart card using hash function. International journal of emerging trends & technology in computer science (IJETICS), Volume 1, Issue 4, Nov–Dec 2012.
14. Dr. P. Kumar, Dr. B. Indrani, Dr. M. Amuthaprabakar. An efficient password based authentication scheme using time hash function and smart card. IJETAE: Volume 4, Issue 6, June 2014.
15. Xiong Li, Jianwei Niu, Muhammad Khurram Khan, Junguo Liao. An enhanced smart card based remote user password authentication scheme. J. of Network and Computer Applications 36; 2013, p. 1365–1371. http://dx.doi.org/10.1016/j.jnca.2013.02.034.
16. Da-Zhi Sun, Jin-Peng Huai, Ji-Zhou Sun, Jian-Xin Li, Jia-Wan Zhang. Improvements of Juang et al.'s Password-Authenticated Key Agreement Scheme Using Smart Cards. IEEE Transactions on Industrial Electronics, Vol. 56, No. 6. June 2009.
17. Messerges T S, Dabbish E A, Sloan R H. Examining smart-card security under the threat of power analysis attacks. IEEE Trans Comp 2002: 51 (5): 541–52.
18. Kocher P, Jaffe J, Jun B. Differential power analysis. In: Proceedings of advances in cryptology CRYPTO'99'; 1999. P. 388–97.
19. Hsieh W B, Leu J S. Exploiting hash functions to intensify the remote user authentication scheme. Computer & Security 31 (2012), pp 791–798. http://dx.doi.org/10.1016/j.cose.2012.06.001.
20. Marimuthu Karuppiah, R. Saravanan. A secure remote user mutual authentication scheme using smart cards. J of information security and applications 19; 2014, p. 282–294. http://dx.doi.org/10.1016/j.jisa.2014.09.006.
21. Min-Shiang Hwang. A Simple Remote User Authentication Scheme. Mathematical and Computer Modelling 36; 2002, p. 103–107. Password Authentication scheme over insecure networks. J. of Computer and System Sciences 72 (2006), p. 727–740. doi:10.1016/j.jcss.2005.10.001.

Avoiding Slow Running Nodes in Distributed Systems

Shyam Deshmukh, Jagannath Aghav, K. Thirupathi Rao and B. Thirumala Rao

Abstract In distributed systems like Hadoop, work is segmented into various tasks and then subsequently executed in parallel on nodes in the cluster. Stragglers, the nodes which are 6–8 times slower than median nodes, can potentially degrade the overall cluster performance by increasing the job completion time. The existing solutions mainly concentrate on reactive measures after detecting stragglers but they lead to extended job completion time and resource wastage. Currently, proactive straggler avoidance techniques have introduced the application of machine learning methods to enhance the task scheduling. In this paper, a prognostic system that proactively avoids stragglers using predictive models is proposed. It has two stages: (1) To develop the prediction model for identifying the straggler nodes before allocation of the task using distributed machine learning and (2) To guide the scheduler to efficiently assign the tasks. This results in avoiding or minimizing the number of stragglers and leads to smarter scheduling. The proposed solution is compared with default Hadoop scheduler and has shown the significant improvement.

Keywords Distributed systems · Hadoop · Stragglers · Predictive model · Machine learning

S. Deshmukh (✉) · J. Aghav
Pune, India
e-mail: dshyam100@yahoo.com

J. Aghav
e-mail: jva.comp@coep.ac.in

K.T. Rao · B.T. Rao
AP, India
e-mail: kthirupathirao@kluniversity.in

B.T. Rao
e-mail: drbtrao@kluniversity.in

S.C. Satapathy et al. (eds.), *Computer Communication, Networking and Internet Security*, Lecture Notes in Networks and Systems 5,
DOI 10.1007/978-981-10-3226-4_41

1 Introduction

The Volume of data produced in various enterprises and social media is growing tremendously at an unprecedented rate. The large scale processing platforms which processes this large volume are becoming progressively complex. Concerning to process and analyse the huge volume of data, distributed frameworks like Hadoop are used. It's based on map-reduce approach where the workload is divided into small fragments, each of which may be executed on any node in the cluster. MapReduce works by splitting the job within two type of tasks viz., map task and reduce task. Whenever a MapReduce job executes, the map task divides the input data into key value pairs and the reduce task processes the partitions in parallel.

Most of the nodes finish their task execution in small duration but some nodes run slower as compared to others which can potentially degrade the job's progress. Such slow-running nodes are termed as Stragglers. The problem of stragglers has received considerable attention already and many solutions have been proposed. The two most extensively used techniques like blacklisting and speculative execution react in an 'after-the-fact' manner which have been proved to be inefficient. In contrast, proactive techniques predict straggler nodes before the tasks are launched and are more time efficient.

2 Related Work

Existing techniques to straggler mitigation are inadequate in many ways. Blacklisting identifies nodes which are not performing properly. No jobs are scheduled on such blacklisted nodes. Up to 50% of the nodes get blacklisted in a cluster, in turn causing resource wastage [1]. The MapReduce paper [2] suggested Speculative Execution technique that observes the progress of the tasks of a job and launches duplicates of those tasks that are slower. Due to wait and speculate, there is an increased contention over the available resources resulting into higher latencies for new tasks. LATE [3] uses progress score to enhance the performance as compared to speculative execution. But it exerts pressure on other running tasks by competing for the resources and presumes that tasks make development at a roughly constant rate which is not always the case. Mantri [4] focuses more on saving computing resources of a cluster, i.e., task slots. If the backup job has an extremely large probability to finish early, Mantri will stop the initial task while cluster is active (kill-restart method). However, kill-restart method may not guarantee that the new task will be completed earlier than the original one. In all reactive techniques, the problem gets even worse when some tasks start straggling when they are well into their execution. Cloning mechanism like Dolly [5] is proactive but focuses only on interactive jobs and being replicative in nature, incur additional resources.

Instead of reacting after detection of stragglers, proactive approaches try to avoid creation of straggler nodes. These methods do not replicate tasks and support smarter scheduling. Therefore, we use Wrangler [6], a similar proactive approach that utilizes machine learning technique like Support Vector Machine (SVM) which helps to model the relation between tasks. One of the limitations of wrangler is the slow training process for straggler prediction models.

3 Proposed Solution

Using resource utilization metrics of nodes and considering the job resources required, a generalized prediction model is built using distributed SVM to predict whether a node will be a straggler or not. The scheduler uses the prediction based on this model in making decisions that avoid straggler formation proactively. In learning phase, as shown in Fig. 1a, the obtained set of feature vectors (e.g. node resource utilization metrics and job history traces) of hadoop cluster using Ganglia-based [7] node monitor on each node are used to separate the obtained set of feature vectors within two categories (e.g. non-straggler, straggler class). The continuous modeling technique, SVM with linear kernel [8–10] on distributed environment for model building is used which attains a linear function separating a likely set of feature vectors into two classes. This linear function, called separating hyperplane; separates the spaces into two different classes. In the phase of model building, this hyperplane is computed in such a manner so it divides the vectors of node's resource utilization metrics associated with one category (stragglers) from the indicated other category (non-stragglers) with maximum distance (called margin) amidst them. In this paper, the attempt to speed up the training process time, one of the limitations of Wrangler, required for straggler prediction model have been partially resolved through the use of Large-scale Linear Classification libraries on Distributed environment. As well, the high dimensionality issue is resolved by using dual solver SVM. In prediction phase as illustrated in Fig. 1b, a latest observed resource utilization vector (a test vector) as it may be classified to observe whichever partition of the separating hyperplane it lies, with a score to assess the assurance in classification depending upon the distance from hyperplane. The scheduler before execution of a task on a particular node inquires about the class of node, if it is found to be straggler with confidence exceeds the minimum required confidence then the task is delayed. This reduces the chances of creating stragglers. We have used 22 features most of them related to CPU utilization (e.g. CPU idle time, user time, system, CPU wait I/O and CPU speed, etc.), Disk utilization (e.g. amount of free space, local read/write statistics from the data nodes, maximum percent used for all partitions, etc.), Memory utilization (e.g. Amount of buffered, cached, shared, free and total amount of available memory etc.), Network utilization (e.g. Packets in and out per second etc.), and System-level features (e.g. total

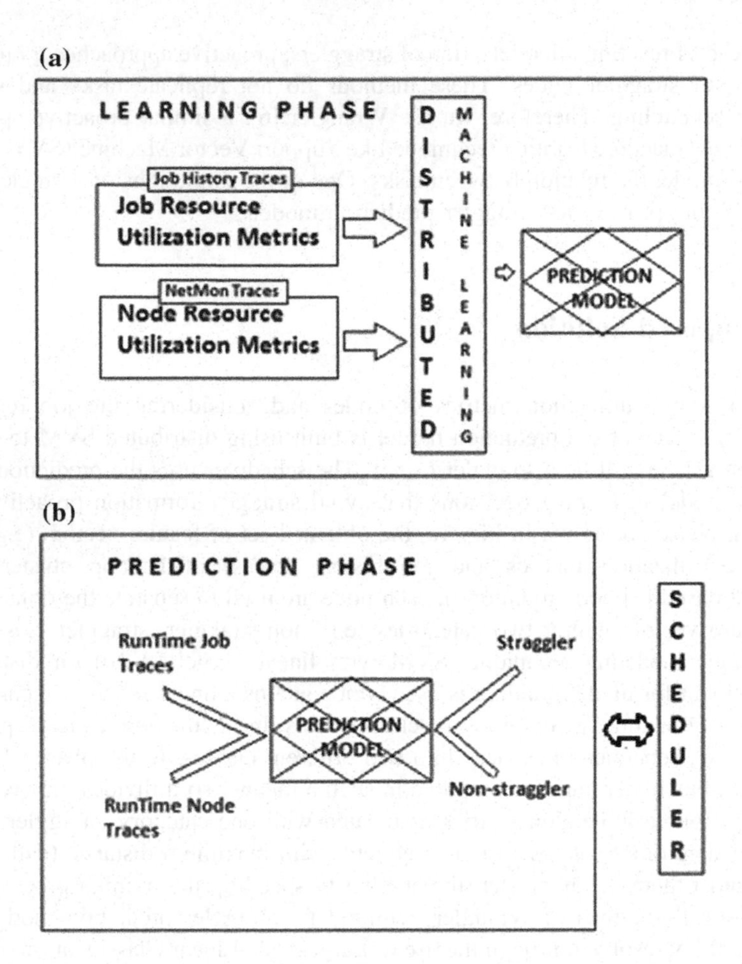

Fig. 1 **a** Learning phase. **b** Prediction phase

number of processes, total number of running processes, total amount of swap memory, amount of available swap memory, etc.). The job history server traces Job execution time, Task execution time, Read data in bytes, Write data in bytes, Start time, Finish time and Elapsed time are obtained using REST APIs. The notion of certainty measure in addition to the prediction of linear models is introduced to ensure reliable predictions. More the distance of a node from the separating hyperplane; greater the chances of it associating to the predicted class.

Once the proposed model is developed, it can be hooked to any scheduling algorithm. The generalized algorithm is as follows:

Algorithm

Let $D = (d_i : i=1, \dots, \#\text{ data nodes})$ be the set of data nodes.
Let $willStraggle_i \in (1,0)$ be the prediction using a snap of resource utilization metrics of ith data node using its model.
Let $c \in [0,1]$ be the minimum acceptable confidence of predictions.

1: /* *S-PREDICT runs every small interval in background* */
2: **algorithm** S-PREDICT
3: **for** all the data nodes in D
4: collect a snap of node's resource utilization metrics
5: $willStraggle_i$ = prediction if data node d_i will create a straggler
6: $confidence_i$ = confidence in the above prediction
7: **if** $willStraggle_i ==$ 1 with $confidence_i$ *LABEL* == fast
8: assign the *LABEL=fast*
9: **else**
10: assign the *LABEL=slow*

7: **procedure** LB-SCHEDULE
8: **for** a task chosen as per the preferred scheduling policy
9: **when** heartbeat is received from a data node indicating free slot(s)
10: **if** *LABEL* = = fast
11: assign the task to d_i
12: **else**
13: advance as per the configured scheduling policy

Researchers interest in Hadoop Job scheduling, a lot like the rest of Hadoop, is developing rapidly. The architecture employs a master node daemon called a Resource Manager consisting of two parts, a scheduler and Application Manager.

In Hadoop job scheduling, the master assign tasks to data nodes, also termed as workers, in acknowledgement to the heartbeat message sent by them after every few seconds. When the job scheduler receives a heartbeat message indicating free map or reduce slot, the configured scheduling policy like Capacity-scheduling (or fair share scheduling), selects the tasks to be designated. The proposed solution can be extended to any configurable or pluggable existing schedulers. Before assigning a task, our proposed scheduler anticipates whether a worker will complete it in a timely manner or not. The task is not assigned to the worker if it is presumed to create a straggler at that time instance. The task is appointed to the worker only after it is predicted that it is not to create a straggler. Algorithm generalizes this scheduling which includes S-PREDICT algorithm and LB-SCHEDULE policy. The S-PREDICT algorithm runs in the background after every T time-interval to guess if the workers will devise stragglers. All predictions also have a confidence measure associated with them. The procedure LB-schedule is the hook that can be embedded in the default-scheduler code. We are using the default capacity scheduling code for our proposed scheduler along with label based scheduling. An administrator can use label-based scheduling to have a control on exactly which nodes are selected to run jobs acknowledged by distinct users and groups. This is helpful when the hardware is heterogeneous (availability of different CPU, memory and disk resources) with different type of workload is running. Label-based

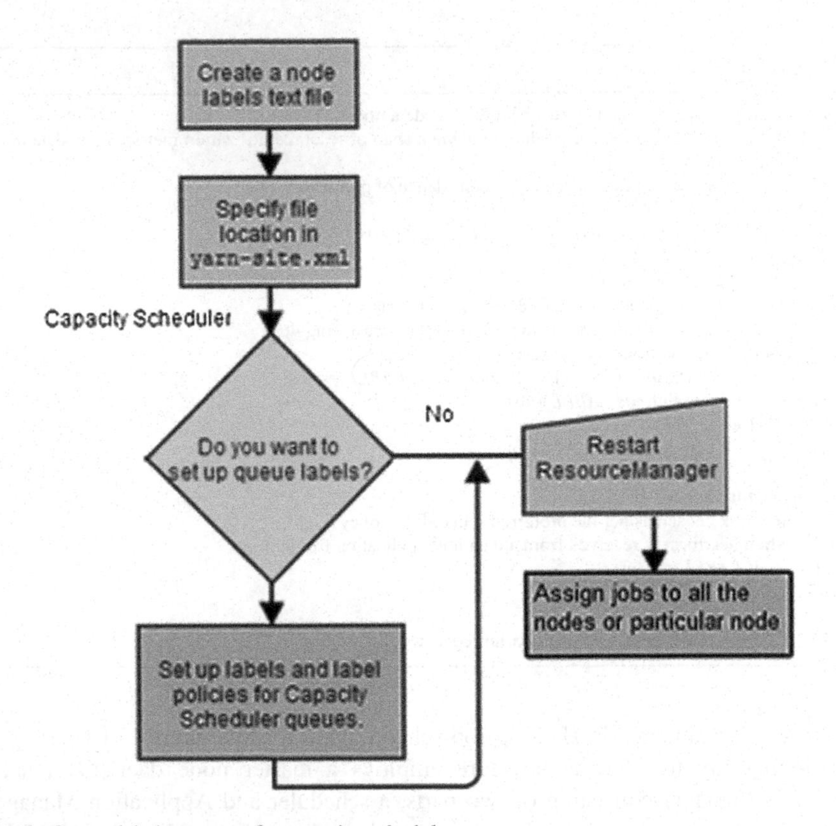

Fig. 2 Queue labels concept for capacity scheduler

scheduling [11] feature is used by assigning labels to nodes which is in a text file, then create queue labels based on the node labels. The flowchart in Fig. 2 summarizes the steps and introduces queue labels concept for the Capacity Scheduler. You can place the jobs which are running on a specified nodes at the queue level (using a queue label). To use the Capacity-Scheduler, we need to configure the ResourceManager by setting the following property in the `yarn-site.xml` file: < PropertyName:`yarn.resourcemanager. scheduler. class>`and < Value:`org.apache.hadoop.yarn.server.resourcemanager. scheduler.capacity.CapactyScheduler>.` Whenever the worker requests for a task to process through a heartbeat message, the proposed scheduler holds the assignment, if it is predicted to create a straggler with confidence higher than assumed threshold c in Algorithm else, it let the default scheduling policy to make the assignment decision. This proposed solution works as a system which provides clue to the default or configured scheduler. The decision of which task to be launched next solely depend upon the fundamental scheduler. The schematic of the same is as shown in Fig. 3.

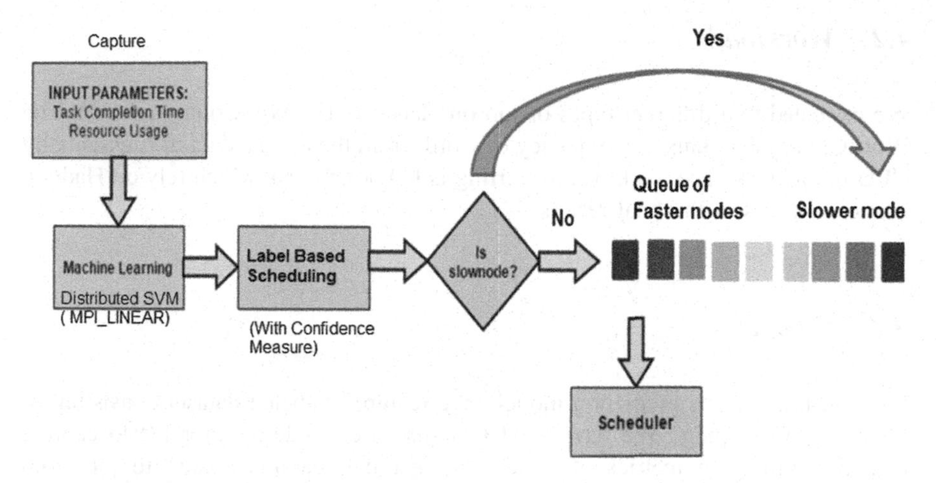

Fig. 3 Proposed solution

Table 1 Hadoop cluster hardware configuration

Node	Hardware configuration
Master node	Intel(R) Core(TM) i5-4200U CPU @ 1.60 GHz, 8 GB RAM, 1 TB Disk Space
Slave node 1	Intel(R) Core(TM) i3-3110 M CPU @ 2.40 GHz, 4 GB RAM, 500 GB Disk Space
Slave node 2	Intel(R) Core(TM) i5-3230 M CPU @ 2.60 GHz, 4 GB RAM, 500 GB Disk Space
Slave node 3	Intel(R) Core(TM) i5-3230 M CPU @ 2.60 GHz, 4 GB RAM, 500 GB Disk Space

4 Experimental Setup

We followed various steps to institute the experimental setup needed to conduct the experiments.

4.1 Cluster Setup

We have deliberated heterogeneous nodes in a Hadoop Cluster as shown in Table 1. We have built the Hadoop Cluster of 4 heterogeneous nodes to estimate the proposed solution for discovering straggler nodes. One of the nodes with very high configuration was picked as a master node and it runs the Hadoop Distributed File System (NameNode) and MapReduce runtime (Resource Manager). The remaining 3 nodes are slave nodes (DataNodes and NodeManagers). The nodes were interconnected using wireless networking. All systems in the multi-node setup use Ubuntu v14.04 operating system, JDK 1.7 and Hadoop 2.7.1 version for performance.

4.2 Workloads

We executed two different types of job on Hadoop viz. WordCount and Sort. The WordCount job counts the frequency of words from the textual data. It is essentially CPU bound task. The workload of Sorting is I/O bound task which rely on Hadoop framework to sort the final results.

4.3 Datasets

For constructing the prediction models, we require a labeled dataset consisting of {feature, label} pairs. We have used Ganglia-based node-monitor [7] to capture resource utilization metrics of nodes. We get the features related to job from Hadoop. In all 22 feature subsets are being considered as per the discussion in Sect. 3 and these features are collected into a single file. The units and metrics of all the features are different and complex. So for convenience, the features are being normalized ranging from 1 to 10. The label is assigned to the dataset by classifying straggler nodes based on the following definition.

Let *nd(t)* be normalized durations [6] i.e.

$$\left[\frac{task\ execution\ time}{amount\ of\ work\ (bytes\ read/written\ by\ task\ t)}\right] \quad (1)$$

A straggler is defined if for task t_i of a job J

$$nd(t_i) > (\beta \times median\{nd(t_i)\})$$

where,
β is threshold coefficient ($\beta \sim 1.3$) or signifies the extent to which a task is allowed to slow down before it is called a straggler.

Once we have labeled the dataset, we evaluated the accuracy of straggler prediction on all the workloads using SVM. First, we have collected features by running the workloads on the available nodes with the mentioned configuration in Table 1. To get the features related to straggler node, we have overloaded each node alternatively and then captured the features of it. This process of capturing the dataset for straggler and non-straggler vectors in the training phase required over 4–5 hours. This cleaned and normalized data is fed to the distributed SVM environment provided by MPILIBLINEAR [12]. This has reduced the model building time to the scale and speed up of number of machines available. This completes the model building phase. We have considered 2/3 of training data set for testing purpose as per the principles of model building in machine learning.

During prediction phase, while the job is executing, the scheduler delays the assignment while the node resource utilization metrics are being captured by Ganglia and sent to the model which will predict the straggler condition will occur or not. Accordingly scheduler will take a decision on assignment of a task to the respective node and thus reducing the overall job completion time. Models could be regularly updated and rebuilt once few jobs have completed implementation and new data has been collected. We have used label based scheduling feature for Capacity Scheduler which enables us to provide the labels to the nodes. The straggler node is labeled with slow and non-straggler with fast. If the node is slow then it will be sent to slow queue else fast queue. The capacity scheduler will make use of these queue further for scheduling of the job.

5 Result and Conclusion

Our proposed solution being proactive in nature, it avoid stragglers to attain faster job completion time while using slight resources. Instead of allowing tasks to execute and identifying them as stragglers while they run slow, proposed solution predicts stragglers before they are assigned to a particular node. Confidence measure strengthens our decision and help to attain a reliable task scheduling; thus cancelling the need for simulating them. The prediction model accuracy is 84% as per the evaluation. In order to assess the performance, we have unified our proposed solution with the Hadoop default scheduling algorithm to classify the straggler nodes in the heterogeneous Hadoop cluster. For the job completion time comparison with and without proposed solution, it needs to be hooked up with any default scheduler which definitely provide us with the better result as expected. The Fig. 4 in below bar chart shows the wordcount job of different sizes viz., 400, 300 and 200 MB on X-axis with job completion time on Y-axis. We have observed that the average job completion time is improved by 10–11% in comparison with the default scheduler. Also, we have demonstrated the change in the label of a

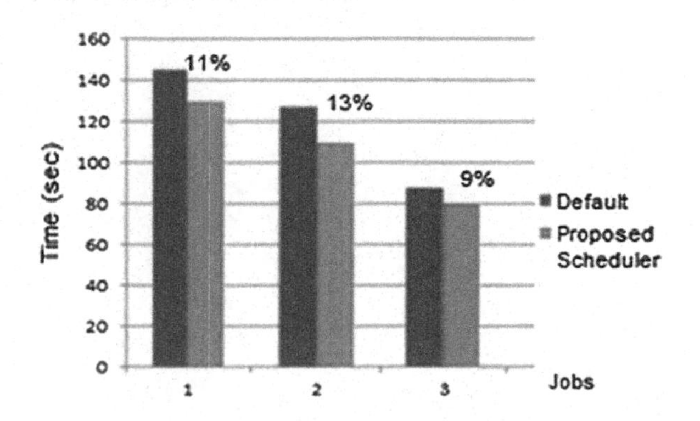

Fig. 4 Comparison of default and proposal scheduler

particular node when it switches from a straggler to non-straggler and vice versa. We also have achieved the step up of the training process by lowering the time spent for training straggler prediction models through distributed machine learning.

Acknowledgements We are indebted to Puja, Ankit, Kalpesh and Aditi for helping us in the implementation of this proposed solution.

References

1. Turkington G.: Hadoop Beginner's Guide. Packt Publishing Ltd. (2013).
2. Dean, Jeffrey, and Sanjay Ghemawat.: MapReduce: Simplified data processing on large clusters. In: Communications of the ACM 51.1 pp. 107–113 (2008).
3. Zaharia, Matei, et al.: Improving MapReduce performance in heterogeneous environments. In: Proceedings of the 8th USENIX conference on Operating systems design and implementation. USENIX Association (2008).
4. Ananthanarayanan, G., Kandula, S., Greenberg, A. G., Stoica, I., Lu, Y., Saha, B., & Harris, E. In: Reining in the Outliers in Map-Reduce Clusters using Mantri. In OSDI (2010).
5. Ananthanarayanan, Ganesh, et al.: Effective Straggler Mitigation: Attack of the Clones. NSDI. Vol. 13 (2013).
6. Neeraja J. Yadwadkar, Ganesh Ananthanarayanan, and Randy Katz. Wrangler: Predictable and faster jobs using fewer resources. In: Symposium on Cloud Computing (2014).
7. Matthew L. Massie, Brent N. Chun, and David E. Culler. In: The ganglia distributed monitoring system: Design, implementation and experience. Parallel Computing, vol. 30 (2004).
8. Christopher M. Bishop.: Pattern Recognition and Machine Learning (Information Science and Statistics). Springer-Verlag New York, Inc., Secaucus, NJ, USA (2006).
9. Christopher J. C. Burges.: A tutorial on support vector machines for pattern recognition. In: Data Min. Knowl. Discov., vol. 2(2), pp. 121–167 (1998).
10. Fan, Yuanquan, et al.: Performance Prediction Model in Heterogeneous MapReduce Environments. In: Computer and Information Technology (CIT), 2014 IEEE International Conference on. IEEE (2014).
11. Label-based Scheduling for YARN Applications. http://doc.mapr.com/display/MapR/Label-based+Scheduling+for+YARN+Applications.
12. Chih-Chung Chang and Chih-Jen Lin.: Distributed LIBLINEAR: Libraries for Large-scale Linear Classification on Distributed Environments (2016). https://www.csie.ntu.edu.tw/~cjlin/libsvmtools/distributed-liblinear/mpi/guide_virtualbox_mpi.html.

Region Based Semantic Image Retrieval Using Ontology

Morarjee Kolla and T. Venu Gopal

Abstract Extracting Semantic images from the large amount of heterogeneous image data is a quiet challenge in Content Based Image Retrieval (CBIR). Search space and Semantic gap reduction are two major issues in extracting semantic images. The proposed method of Region based semantic image retrieval considers both Search space and Semantic gap reduction. The proposed methodology first does the region based clustering as it reduces retrieval search space. Later it reduces the semantic gap with the support of ontology framework. The ontology framework shares the information among image seekers and domains. Our experimental results reveal the efficacy of the proposed method.

Keywords CBIR · Ontology · Search space · Semantic gap · Semantic image retrieval

1 Introduction

With the advancement of visual content and technologies, image plays a vital role in many of the applications and areas like medical, education, web, social media, entertainment, etc. Storage and processing of images requires efficient methods to retrieve information from image databases. With the increase in the large amount of image data, efficient retrieval mechanisms of visual information are in huge demand [1]. Efficient searching of a relevant image from different varieties of image datasets is a daunting task. Human beings are capable of interpreting image contents in high level perception, whereas a computer can interpret the image content with low level features extracted from image pixels. Hence, there is a wide gap between human

M. Kolla (✉)
CMR Institute of Technology, Hyderabad, Telangana, India
e-mail: morarjeek@gmail.com

T. Venu Gopal
JNTUH College of Engineering Sultanpur, Sultanpur, Telangana, India
e-mail: t_vgopal@rediffmail.com

S.C. Satapathy et al. (eds.), *Computer Communication, Networking and Internet Security*, Lecture Notes in Networks and Systems 5,
DOI 10.1007/978-981-10-3226-4_42

interpretation and machine interpretation. The Semantic gap is the gap between the low level representation of the machine and high level human perception. Digital processing can provide similar images, but the user is interested in semantic images as per the user query. The semantic gap drastically reduces the retrieval efficiency [2]. Most of the search engines are using a text based approach for image retrieval. This approach may return sometimes irrelevant results due to inconsistencies in image indexing. To resolve these inconsistencies manual annotations are required, which will consume more time. An automatic annotation is used to speed up the retrieval process.

Text based approaches do not satisfy the user needs in terms of extracting semantics. So an alternative approach of CBIR is a popular technique to address the problems of text based approaches. In CBIR, image retrieval is performed with either low-level or high-level features. Semantic-based image retrieval matches a user query based on high level features to retrieve query relevant results. User interaction is essential in CBIR to get relevant results. Low-level image representations are failed to represent the high-level semantics as per user query [3].

In region based image retrieval, images can be segmented into a set of regions and every region is treated as a separate image. In our proposed method of a region based semantic image retrieval, each segment of the image has its own semantics and the entire image is a bag of semantics. In general, user query is mainly focused on a particular region of the image, not on the entire image [4]. Region based retrieval approaches are more efficient than other approaches to retrieve relevant results.

Relevance feedback is used to reduce the semantic gap, which can help to model the user in different stages by correctly expressing the user needs and automatically converting them into image features and similarity measures [5]. Ontology reduces the semantic gap by providing appropriate communication between users and applications [6]. In this paper, DBpedia [7] is used as an ontology, which covers many concepts and relations. The resource description framework (RDF) allows multiple metadata schemes, and provides interpretability between applications by adding semantics to the document [6].

2 Related Work

Numerous researches have been carried on Semantic Image Retrieval. Some researches focused on integrating textual information with low level image features. Other approaches based on Relevance Feedback and Ontology.

Some researchers have addressed the challenging and current issues of CBIR. Role of semantic, use of patterns and the sensory gap of CBIR are discussed in [1]. Descriptions of various image retrieval systems with respect to their technical details are discussed in [8]. Key challenges of image retrieval are discussed in [9]. Current research issues and trends are discussed in [10, 11]. Other studies of image retrieval on relevance feedback are discussed in [5]. Similarity matching and threshold on similarity greatly affect the retrieval accuracy [12]. Retrieval accuracy can be improved by learning approaches [13].

Some systems like query by image content (QBIC) [14], VisualSeek [15], SIMPLicity [16] use low level features in order to represent semantics. Other systems like query by example such as Netra [17, 18], Mars [19], Photobook [20], VisualSeek [21], Istorama [22], WebSeek [23], Blobworld [24] and Image Rover [25] demonstrate the use of image indexing features either in a region based or in an entire image. Segmentation algorithm proposed by Blobworld [24] uses Expectation Maximization and proved that the region based indexing features are more efficient in image retrieval than global indexing under Query by Example scheme.

Semantic image retrieval in remote sensing archives [26] discussed the domain dependent ontology framework to perform semantic retrieval. An image retrieval approach presented by using domain ontology improves relevance and retrieval accuracy [27]. To ensure that CBIR systems using single ontology do not retrieve indefinite images, Core Semantic Multiple Ontology is proposed [28]. Fuzzy ontology based model [29] combines the concepts into categories and linked the images with fuzzy values in ontology.

2.1 *Our Approach*

Semantic image retrieval needs a better focus on user interested query regions and the knowledge representation required to meet high level human perception. This motivates us in combining the region based semantic image retrieval using affinity matrix and domain ontology based refinement for retrieving the results. Our approach is not only concentrating on user interested regions, but also on high level expectations of users.

2.2 *Outline of the Paper*

In Sect. 3, we explain our proposed method with the help of Block diagram. In Sect. 4, experimental results are discussed. In Sect. 5, we conclude our approach.

3 The Proposed Method

Our proposed method envisages two stage semantic image retrieval. In the first stage, initial semantic image retrieval using affinity matrix is constructed from the user's log file. These results are refined in the second stage. The block diagram as depicted in Fig. 1 of a Region based semantic image retrieval using ontology

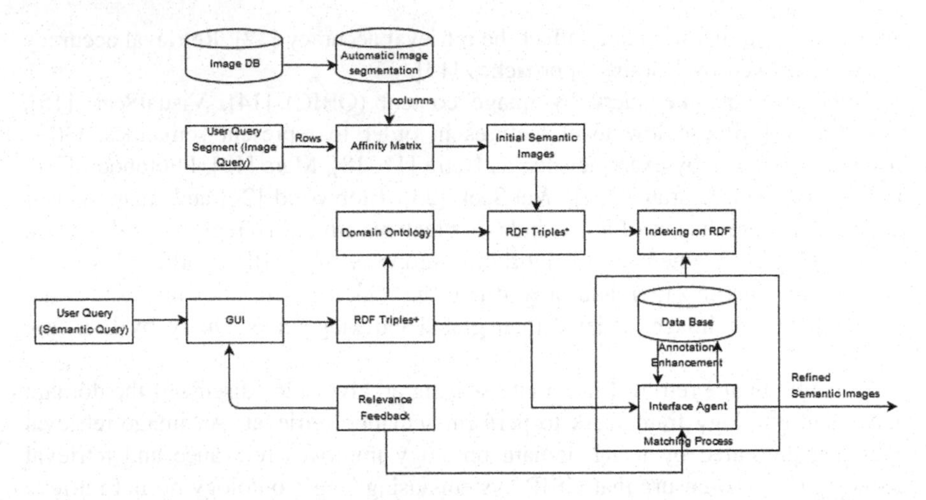

Fig. 1 Block diagram of region-based semantic image retrieval using ontology

describes our proposed method. Initial semantic images are retrieved using a region based retrieval mechanism and these results are refined using an ontology framework during the second stage. Relevance feedback is used to improve the retrieval accuracy. Images with false tags and coined term tags are removed during the second stage using ontology. Image retrieval accuracy is measured in terms of precision and recall.

3.1 Initial Semantic Image Retrieval

In the first stage, user feedback is collected over time is recorded in the form of a log file. Affinity matrix is built with query image segments as rows and database images as columns. After constructing affinity matrix [4], positive-labeled images having the shortest Euclidean distance with the query segment will be retrieved as initial retrieval results.

3.2 Refine Retrieval Results by Using Ontology

During the second stage, retrieved query relevant images obtained from the first stage are converted into RDF triples with the help of tag information and domain ontology. The translated images as RDF triples are indexed to improve retrieval efficiency. Knowledge representation is to be incorporated into initial results by using ontology to meet high level human perception. Inference agent is used in matching query ontology with domain ontology, enhancing image annotation and query reasoning. In the match process, images having strong relevance feedback are sent back to the annotation enhancement and retrieve the relevant results. Semantic keywords are derived with the help of query reasoning.

For example, a Car is a query segment, and then some of the images with a Car as a segment are retrieved from the first stage. RDF triple for a car is Car, Is-A, Vehicle (Subject, Predicate, Object). Semantic query RDF triple and retrieved image tag based RDF triple are both given to the match process. The match process, then searches for the images having a Car as the subject class, with its appropriate images which has Is-A relation in the hierarchy of the domain ontology.

4 Experimental Results

The experiment is conducted on a Flickr25 k image dataset consisting of 25,000 tagged images. Our database has 15 categories of 5297 tagged images. This comes to 44,458 segments as per Blobworld [24] automatic image segmentation method and each segment is represented by a 32-dimensional feature vector. Images are retrieved from the automatic segmented image database has the shortest Euclidean distance with the query segment. Query relevant results are retrieved from the first stage. Domain ontology is used to refine these results in the second stage. Image tags are taken as subject class and they are verified in ontology through the matching process to identify false tags. Precision and Recall of experiments are evaluated with different input queries. Table 1 provides experimental results of

Table 1 Precision and recall values of different input queries

S.no	Input query	Relevant images in the database	Total no. of images retrieved	Relevant images	Precision	Recall
1	Animal	164	158	149	0.94	0.90
2	Beach	407	394	342	0.86	0.84
3	Bird	218	199	191	0.96	0.87
4	Building	188	172	117	0.68	0.62
5	Car	212	198	189	0.95	0.89
6	Dog	372	364	348	0.96	0.94
7	Flowers	510	465	434	0.93	0.85
8	Girl	262	242	206	0.85	0.78
9	People	330	234	171	0.73	0.52
10	Sea	301	284	234	0.82	0.78
11	Sky	815	768	732	0.95	0.90
12	Snow	256	242	221	0.91	0.86
13	Sun	290	278	259	0.93	0.89
14	Tree	331	301	285	0.94	0.86
15	Water	641	589	536	0.91	0.84

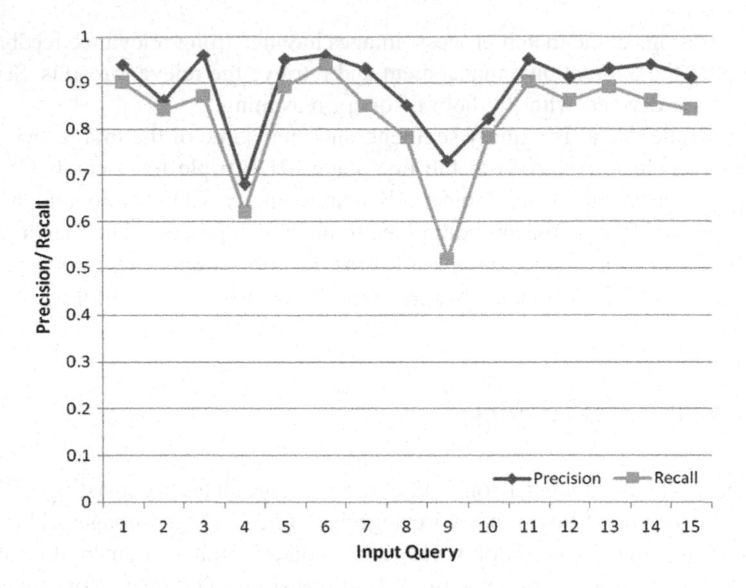

Fig. 2 Precision and recall of input queries

input queries. For building, people, sea input query results are obtained with low precision and recall, the reason being, most of this query relevant image database tags have not matched with domain ontology. In the case of other input queries, our results prove the effectiveness of the proposed approach. Figure 2 shows the precision and recall curves of input queries.

5 Conclusion

This paper proposes a Region based semantic image retrieval using ontology and very significantly improves the precision and recall of image retrieval by enhancing image annotations. The method applies both search space and semantic gap reduction. Another merit of this method is sharing of information among image seekers with the support of ontology framework. The experimental results vouch for the efficacy of our approach to retrieve query relevant results. In the future, research related to word sense and personal name disambiguation will be taken up in the verification of image tags.

References

1. M.R. Naphade and T.S. Huang, Extracting semantics from audio-visual content: the final frontier in multimedia retrieval, IEEE Trans. on Neural Networks, vol. 13, no. 4, pp. 793–810 (2002).
2. A.W.M. Smeulders, M. Worring, S. Santini, A. Gupta, R. Jain, Content-based image retrieval at the end of the early years, IEEE Trans. Pattern Anal. Mach. Intell. 22 (12) 1349–1380 (2000).
3. X.S. Zhou, T.S. Huang, CBIR: from low-level features to high level semantics, in: Proceedings of the SPIE, Image and Video Communication and Processing, vol. 3974, San Jose, CA, pp. 426–431(2000).
4. Ying Liu, XIN Chen, Chengcui Zhang, and Alan Sprague, Semantic Clustering for region based Image Retrieval, Ninth IEEE International Symposium on Multimedia, pp. 167–172 (2007).
5. J. Li, N.M. Allinson, Relevance feedback in content-based image retrieval: a survey, in: Handbook on Neural Information Processing, Springer, Berlin Heidelberg, pp. 433–469 (2013).
6. Roung Shiunn Wu, Wen Hsien Hsu, A Semantic Image Retrieval Frame work based on Ontology and Naïve Bayesian Inference, International Journal of Multimedia Technology, p. 36–43(2012).
7. J. Lehmann, R. Isele, M. Jakob, A. Jentzsch, D. Kontokostas, P. N.Mendes, S. Hellmann, M. Morsey, P. van Kleef, S. Auer *et al.*, Dbpedia–a large-scale, multilingual knowledge base extracted fromwikipedia, *Semantic Web Journal*, (2013).
8. RC. Veltkamp, M. Tanase, Content-Based Image Retrieval Systems: A Survey, rapport no UU-CS-2000-34, (2000).
9. R. Datta, J. Li, JZ. Wang, Content-based image retrieval: approaches and trends of the new age, in: Proceedings of the 7th ACM SIGMM International Workshop on Multimedia Information Retrieval, pp. 253–262(2005).
10. M.S. Lew, N. Sebe, C. Djeraba, R. Jain, Content-based multimedia information retrieval: state of the art and challenges, ACM Trans. Multim. Comput., Commun., pp. 1–19(2006).
11. R. Datta, D. Joshi, J. Li, J. Wang, Image retrieval: ideas, influences, and trends of the new age, ACM Comput. Surv. (CSUR) 40 (2)(2008).
12. R. Brunelli, O. Mich, Image retrieval by examples, IEEE Trans. Multim. 2 (3), pp. 164–171 (2000).
13. C.M. Bishop, Pattern Recognition and Machine Learning, Springer, (2006).
14. W. Niblack, R. Barber, W. Equitz, M. Flickner, E. Glasman, D. Pektovic, P. Yanker, C. Faloutsos, G. Taubin, The QBIC project: Querying images by content using color, texture and shape, in: Proceedings of the SPIE Storage and Retrieval for Image and Video Databases, San Jose, CA, (1994).
15. J.R. Smith, S.F. Chang, VisualSEEk: a fully automated content-based image query system, in: Proceedings of the Forth ACM International Conference on Multimedia '96, Boston, MA, (1996).
16. J.Z. Wang, J. Li, G. Wiederhold, SIMPLIcity: semantics-sensitive integrated matching for picture libraries, IEEE Trans. Pattern Anal. Mach. Intell., pp. 947–963(2001).
17. B.S. Manjunath and W.Y. Ma, "Texture features for browsing and retrieval of image data," IEEE Trans. on Pattern Analysis and Machine Intelligence, vol. 18, no. 8, pp. 837–842 (1996).
18. W.Y. Ma and B.S. Manjunath, "NeTra: A Toolbox for Navigating Large Image Databases," Multimedia Systems, vol. 7, no. 3, pp. 184–198(1999).
19. T.S. Huang, S. Mehrotra, and K. Ramchandran, "Multimedia analysis and retrieval system (mars) project," in Proc of 33rd Annual Clinic on Library Application of Data Processing—Digital Image Access and Retrieval (1996).

20. A. Pentland, R. Picard, and S. Sclaroff, "Photobook: Content-based manipulation of image databases," Int. J. Computer Vision, vol. 18, no. 3, pp. 233–254(1996).
21. J.R. Smith and S.-F. Chang, "Visualseek: A fully automated content-based image query system," in ACM Multimedia, pp. 87–98(1996).
22. I. Kompatsiaris, E. Triantafillou, and M. G. Strintzis, "Region-Based Color Image Indexing and Retrieval," in Proc. IEEE International Conference on Image Processing, Thessaloniki, Greece, (2001).
23. J.R. Smith, S.F. Chang, Visually searching the Web for content, IEEE Multim., pp. 12–20 (1997).
24. C. Carson, S. Belongie, H. Greenspan, J. Malik, Blobworld: image segmentation using expectation-maximization and its application to image querying, IEEE Trans. Pattern Anal. Mach. Intell., 1026–1038(2002).
25. S. Sclaroff, M. LaCascia, S. Sethi, L. Taycher, Unifying textual and visual cues for content-based image retrieval on the World Wide Web, Comp. Vis. Image Understand, pp. 86–98(1999).
26. Ning RUAN, Ning HUANG, Wen Hong, Semantic based image retrieval in remote sensing archieve: An Ontology approach, IEEE, pp. 2888–2891(2006).
27. Sohail Sarwara, Zia Ul Qayyum, Saqib Majeed, Ontology Based Image Retrieval Framework using Qualitative Semantic Image Descriptions, ICKBIIES, Elsevier, pp. 285–294(2013).
28. Anuja Khodaskar, Siddarth Ladhake, New-Fangled Alignment of Ontologies for Content based Semantic Image Retrieval, ICCC, Elsevier, pp. 298–303(2015).
29. Madiha Liaqat, Sharifullah Khan, Muhammad Majid, Fuzzy Ontology based Model for Image Retrieval, Mobile web and Intelligent Information Systems, LNCS, Vol. 9847, pp. 108–120, Springer (2016).

Validation of Lehman Laws of Growth and Familiarity for Open Source Java Databases

Arvinder Kaur and Vidhi Vig

Abstract Lehman's laws of software evolution have been widely researched and validated but there exists very few studies that verified these laws for databases in open source. Database evolution jeopardize the semantical and syntactical cogency of an applications, but, is their incremental augmentation restrained by the growth and familiarity? To verify this, the current study explores the properties of growth for database evolution by analyzing Lehman's fifth and sixth law of software evolution: Law of Conservation of Familiarity and Continuous Growth on three open source Java databases spread across 63 releases for 17774 number of bugs. The study found that Lehman's laws of growth and familiarity applies on Open Source Java databases also and laws were validated by all the datasets.

Keywords Open source · Databases · Lehman laws of software evolution · Bugs

1 Introduction

Lehman laws of software evolution [1–4] were defined and redefined from late 1960s to early 1990s. Lehman laid three Laws of software evolution

1. Law of Continuing Change
2. Law of Increasing Entropy
3. Law of Statistically Smooth Growth.

A. Kaur · V. Vig (✉)
USICT, Guru Gobind Singh Indraprastha University, Sec 16-C Dwarka, New Delhi, India
e-mail: vidhi.ipu@gmail.com

A. Kaur
e-mail: arvinderkaurtakkar@yahoo.com

S.C. Satapathy et al. (eds.), *Computer Communication, Networking and Internet Security*, Lecture Notes in Networks and Systems 5,
DOI 10.1007/978-981-10-3226-4_43

In 1978, modification of earlier laws and redesigning resulted in five laws, three of which were the old ones only. The new laws introduced were Law of Invariant Work Rate and Law of Incremented Growth.

Though, the first three laws remained same, only their definitions changed. These laws were validated specifically on E (embedded) type of systems which involved human interception. Unlike the other two types of programs (S (specified) type and P (problem solving) type), E type programs showed evolutionary pattern and changed as a result of change in its environment. Laws 3–5 experienced a new version in year 1980 and were stated as:

3. Law of fundamental law of program evolution
4. Law of conservation of organizational stability
5. Law of conservation of familiarity

Continuous uncertainty in the software led to reformulations of laws in 1996. The laws now increased from 5 to 8 and since then, are used as it is till date by the researchers to prove or refute the laws for their study. The reason behind revision of these laws was change in development and maintenance standards of the software, with time. Thought of global users coming together to work and discuss software was totally unimaginable until internet came into existence. Lehman totally abandoned Free Libre Softwares while formulating the earlier laws resulting in formulation and reformulation of laws as the time progressed. Law VIII is one such example, where libre softwares strongly hold and progress.

1. Law of Continuing Change
2. Law of Increasing Complexity
3. Law of Self-Regulation
4. Law of Conservation of Organizational Stability
5. Law of Conservation of Familiarity
6. Law of Continuing Growth
7. Law of Declining Quality
8. Law of Feedback Systems

Databases resembles a lot to E-type systems as they revolve around a community (developers and users) that solve real world problems (use of queries) [5] therefore, they must apply to Laws of Software evolution also. Evolution of databases has been rarely studied for its entire lifecycle in spite of the fact that alteration in schema of databases may lead an application to crash or behave abnormally [5]. In fact, no study till date has explored the incremental growth and familiarity of these databases for Open Source Java applications. The current study therefore specifically explores the Law of Continuous Growth and Conservation of Familiarity for three open source Java databases spread across 63 releases for 17774 number of bugs.

The related work is presented in Sect. 2 followed by Data Collection Methodology in Sect. 3. Validation of Law of Evolution for Growth Attributes of Databases is presented in Sect. 5 and Conclusion in Sect. 5.

2 Related Work

Lehman laws of software evolution have been studied and validated widely. Many researchers verified these laws on closed source and industrial projects. Lehman himself proposed these laws for closed source applications. [6] in their study gave an elaborate description on categorization of projects in S, P and E type. [7] on the other hand, explained the methodology to perform empirical research in this field for software evolution. Consequently, these laws were explored on different platforms and programming languages by various studies [8].

A schema evolution jeopardize the semantical and syntactical validity of the related applications and tremendously affect the users as well as developers [5]. Despite this fact, only four studies have verified evolution on databases [5, 6, 9, 10]. [6, 9, 10] studied only the statistical attributes of the evolution and fail to render details of formal mechanism of database evolution. [5] on the other hand studied evolution on open source databases and found that the evolution of databases augmented in controlled manner. [11, 12] analyzed schema evolution for databases too.

The current study on the other hand, focuses on growth and change brought in open source Java databases and verifies fifth law, Law of Conservation of familiarity and sixth law, Law of Continuous Growth only. Till date no study has explored these laws individually and specifically for Java databases in open source. Moreover, this study verifies these laws for every major and minor releases in details unlike [5] study, who just explored the ripples brought in by releases as a whole.

3 Data Collection Methodology

The all-volunteer Apache Server Foundation (ASF) develops, stewards, and incubates more than 350 Open Source projects and initiatives that cover a wide range of technologies [13]. Apache Server Foundation was mined for database projects whose bugs were tracked under Jira [14] repository to maintain uniformity of data. These projects were then checked for their status. Projects under status 'Active' were kept in the pool for selection and randomly three projects were chosen for the study.

These datasets were then mined for their artefacts and used in the study. It must be noted that Versions with status 'Released' and Bugs with status 'Closed and Resolved' were considered for the study in order to avoid spurious results. This process is described briefly in Fig. 1 and the datasets thus collected are presented in Table 1 given.

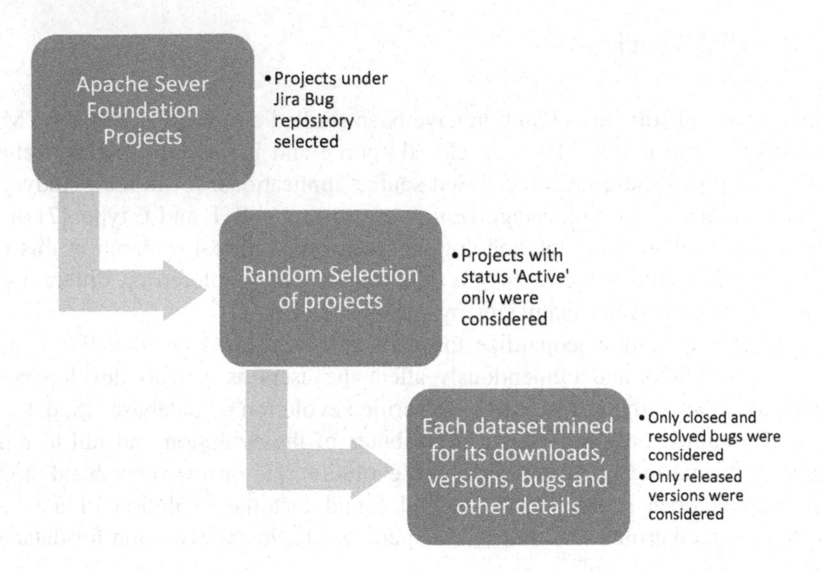

Fig. 1 Data collection process

Table 1 Datasets selected for the study

Datasets	URL	Category	Brief description	Date of first release	Date of last release	No of versions
Pig	www.pig.apache.org	Hadoop, database	Platform for analyzing large data sets on Hadoop*	29/10/2007	06/06/2015	22
Hive	www.hive.apache.org	Hadoop database	Data warehouse software facilitates querying and managing large datasets residing in distributed storage*	30/04/2009	15/02/2016	16
Zookeeper	www.zookeeper.apache.org	Database	Distributed computing platform*	13/11/2007	02/09/2015	25

*Information gathered from https://projects.apache.org/ *Note* The artefacts were last updated in May 2016

4 Validation of Law of Evolution for Growth Attributes of Databases

In Jira repository, every issues was categorized into five categories: Bug, Wish, Task, Improvement and New feature [14]. The current study explored these issues for the amount of change brought by each one of them. It was observed that bugs contributed to more than 60% of change in all the datasets. Figure 2 given below presents the details of the issues of datasets selected for the study.

To further statistically validate this behavior, single factor ANOVA was applied. To verify this the following null and alternative hypothesis were laid.

H_0: All the issues brings in equal change.
H_0: At least one issue (Bugs) brings in more change than others.

The null hypothesis states that all the issues bring in equal change in the application but since, p value is less than 0.05 (Table 2), we can reject the null hypothesis and support the alternate hypothesis. Consequently, the study mined and verified bugs for each version in order to identify their trend major and minor versions. The study gathered details of 17774 bugs in 3 open source Java databases for 63 versions.

Law of Conservation of Familiarity: This law proposes that system's incremental growth tends to remain statistically invariant or to decline. This is so, because the developers need to understand the program's source code and behavior. A corollary is often presented, stating that releases that introduce many changes will be

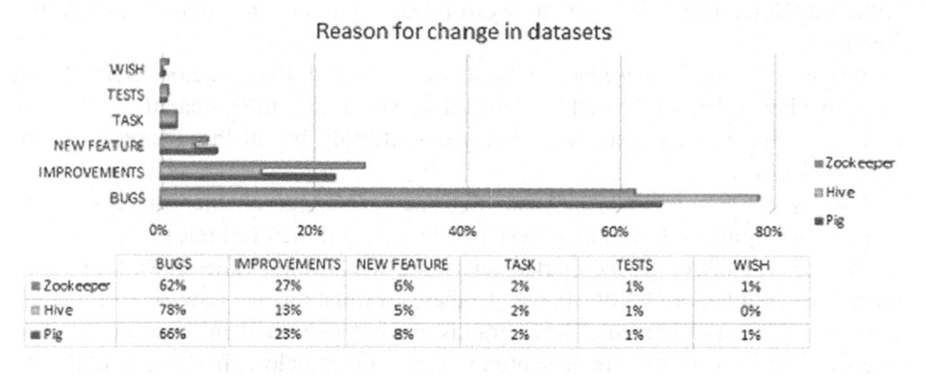

	BUGS	IMPROVEMENTS	NEW FEATURE	TASK	TESTS	WISH
Zookeeper	62%	27%	6%	2%	1%	1%
Hive	78%	13%	5%	2%	1%	0%
Pig	66%	23%	8%	2%	1%	1%

Fig. 2 Graph displaying various reasons for change in datasets

Table 2 Result of single factor ANOVA

Source of Variation	SS	Df	Ms	F	P-value	F crit
Between Groups	7484626	5	1496925	3.74666	0.028252	3.105875
Within Groups	4794431	12	399535.9			
Total	12279057	17				

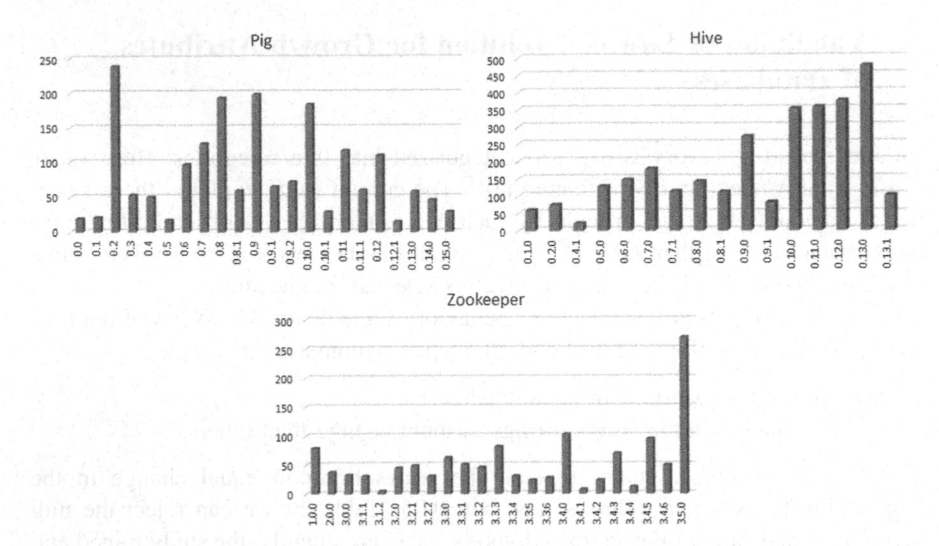

Fig. 3 Graphs displaying the distribution of bugs in Pig, Hive and Zookeeper dataset with Bugs on x axis and versions on y axis

followed by smaller releases that correct problems introduced in the prior release, or restructure the software to make it easier to maintain [1–4].

To further understand the behavior of changes every datasets was mined for its each and every versions. Since, it has already been observed that bugs were the biggest reason for change (Fig. 2), bugs for every version of the datasets were collected. Note: The study uses the word release and version synonymously in the chapter.

While analyzing of these bugs, a trend was observed. Every major version had a huge number of bugs followed by their minor versions with comparatively smaller number of bugs. This trend was observed uniformly by all the datasets and are presented in Fig. 3.

Law of continuous growth: Line of Code (LOC) for all the versions of the datasets were gathered from their very first releases till their last release before May 2016 (because after this we started analysing the results). The study found and increasing trend in LOC of all the databases validating the Law of Continuous Growth for all the datasets. These graphs are not presented in the chapter since graphs of change in LOC are presented in Fig. 4 given below. Increase in LOC can be observed from these graphs also.

Interestingly, the study found that change in LOC (presented in Fig. 4) of the major versions were far more greater than the change in LOC of the minor versions. Further investigations revealed that every major version in the selected dataset either brought a 'New Feature' or an 'Improvement' resulting in a sudden increase in LOC. This sudden increase in LOC further brought an increase in bugs (in major versions) also. The minor versions, on the other hand, were maintenance releases or releases launched after unit testing and rectification of bugs.

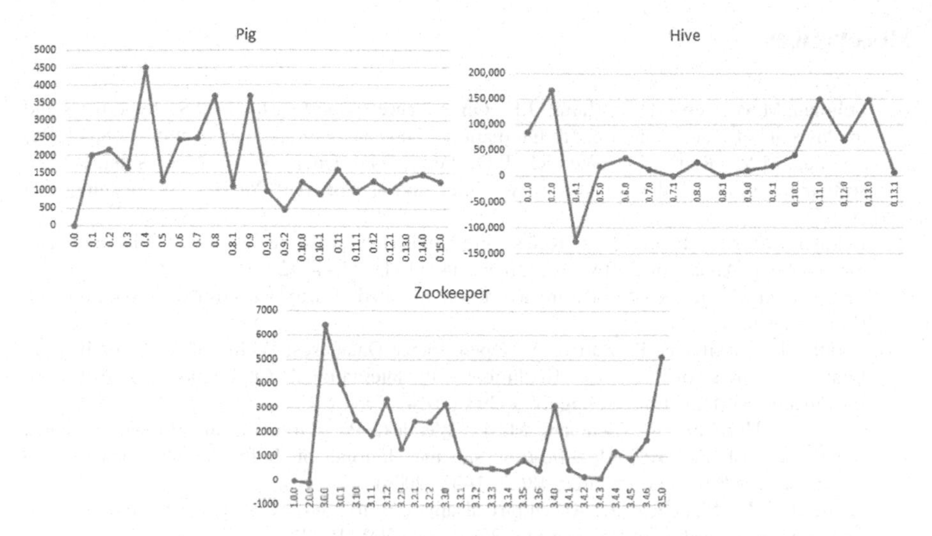

Fig. 4 Change in LOC of each version of selected datasets

Hence, it can be stated after the analysis that Law of Continuous growth and Conservation of Familiarity hold for the selected datasets. It can also be observed that the releases (Versions) that bring maximum change are the "major versions" of the software system and these releases are often followed by the "minor versions" that bring smaller changes wherein bugs are the leading factor of change.

5 Conclusion

The study explored the laws concerning the growth attributes of database evolution which is an indispensable and almost untouched field of software evolution. It was observed that the databases continuously grow and expand periodically in the system validating the Law of Continuous Growth in the system. The study discovered that bugs were unanimously the biggest reason for change in the system and should be paid due attention while validation of Laws of Software Evolution. Law of Conservation of Familiarity was also validated by all the datasets and it was found that releases that bring maximum change are the "major versions" and these releases are often followed by the "minor versions" that bring smaller changes wherein bugs are the leading factor of change.

Further research is going on to explore the generalizability of these results on all the Open Source Java databases. Other laws of evolution are also under study for this domain.

References

1. Lehman, M.M., Perry, D.E., Ramil, J.F.: On evidence supporting the FEAST hypothesis and the laws of software evolution, 5th International Software Metrics Symposium, 84–88 (1998)
2. Lehman, M.M., Ramil, J.F., Wernick, P.D., Perry, D.E., Turski, W.M.: Metrics and laws of software evolution—the nineties view, International Software Metrics Symposium, 0–3 (1997)
3. Lehman, M.M., Ramil, J.F.: Rules and tools for software evolution planning and management, Annals of Software Engineering, 11 (1), 15–44 (2001)
4. Lehman, M.M.: Laws of software evolution revisited, European Workshop on Software Process Technology (1996)
5. Skoulis, I., Vassiliadis, P., Zarras, A.: Open-Source Databases: Within, Outside, or Beyond Lehman's Laws of Software Evolution?. In International Conference on Advanced Information Systems Engineering, 379–393 (2014)
6. Cook, S., Harrison, R., Lehman, M.M., Wernick, P.: Evolution in software systems: foundations of the SPE classification scheme. Journal of Software Maintenance and Evolution: Research and Practice, 18(1), 1–35 (2006)
7. Kemerer, C.F., Slaughter, S.: An empirical approach to studying software evolution, IEEE Transactions on Software Engineering, 25(4), 493–509 (1999)
8. Kaur, A., Vig, V.: Mining software repositories for empirical validation of laws of software evolution for Java projects, International Journal of Computational Systems Engineering, 3, 155–173 (2016)
9. Papastefanatos, G., Vassiliadis, P., Simitsis, A., Vassiliou, Y.,: Metrics for the prediction of evolution impact in etl ecosystems: A case study. Journal on Data Semantics, 1(2), 75–97 (2012)
10. Sjøberg, D.: Quantifying schema evolution. Information and Software Technology, 35(1), 35–44 (1993)
11. Manousis, P., Panos V., Apostolos Z., George, P.: Schema Evolution for Databases and Data Warehouses, In European Business Intelligence Summer School, 1–31 (2015)
12. Cleve, A., Maxime, G., Loup, M., Jerome, M., Jens, W.: Understanding database schema evolution: A case study, Science of Computer Programming 113–121 (2015)
13. Apache Server Foundation, http://www.apache.org
14. Jira, https://www.atlassian.com/software/jira

Genetic Algorithm and Particle Swarm Optimization: Analysis and Remedial Suggestions

Hari Mohan Pandey

Abstract A comprehensive comparison of two powerful evolutionary computational algorithms: Genetic Algorithm and Particle Swarm Optimization have been presented in this paper. Both the algorithms have the global exploration capability; is being applied to the difficult optimization problems. The operators of each algorithm greatly contribute to the success have been reviewed, focusing on how they affect the searching in the problem space. The rationale of conducting this study is: to bring additional insights into how these algorithms work, and suggest remedies, if incorporated, improves the performance.

Keywords Bio-inspired algorithm · Crossover · Mutation · Genetic algorithm · Particle swarm optimization · Nature inspired algorithm

1 Introduction

There exist four most popular evolutionary algorithms (EA) are: Genetic Algorithm (GA), Evolutionary Programming (EP), Evolutionary Strategy (ES) and Genetic Programming (GP). These EA are powerful global optimization algorithm, has strength to explore the problem space adequately in their own way, have been applied to solve many complex problems, and was difficult to solve employing the traditional approach. Therefore, no question about the computational capabilities, these algorithms hold, but still there exists few problems, where these algorithms do not show a good behavior as expected, motivated researchers to develop new algorithms. The outcome of this is the Particle Swarm Optimization (PSO) algorithm was proposed by Eberharth and Kennedy [1].

This paper reports a comprehensive analysis of two algorithms: GA and PSO. The focus is given to the operators used in these algorithms. The GA, is of course,

H.M. Pandey (✉)
Department of Computer Science & Engineering, Amity University,
Sector-125, Noida, Uttar Pradesh, India
e-mail: profharimohanpandey@gmail.com

S.C. Satapathy et al. (eds.), *Computer Communication, Networking and Internet Security*, Lecture Notes in Networks and Systems 5,
DOI 10.1007/978-981-10-3226-4_44

not a new algorithm, has shown its capabilities, and was developed by Holland [2]. The performance of the GA affects by the basic armory: crossover, mutation and selection it uses. The author has conducted this study and comparison with the PSO. It is worth to mention that the term GA refers the basic binary version of a GA used in this paper. The author also wants to mention that there exist several literatures were proposed talks about the basic configuration of the GA and suggest an amount of different modification to be made to improve the performance. Hence, the aim of this paper is not to compare GA and PSO rather the objective is to present a declarative statement about the better performance of one or the other.

The author assumes that the audience of this paper has the basic understanding of both the algorithms, therefore the basic details about the GA and PSO has not been provided, rather a comprehensive analysis and discussion is reported, pay much to the researchers. The analysis and discussion reported in this paper are the author's experience gained during the computational experimentations.

2 Analysis: GA and PSO

The PSO uses particle to represent the population analogous to the GA's chromosome. A particle is used to represent a candidate solution in the PSO very similar to the chromosome used in the GA. The strength of the searching for the PSO and GA is directly proportional to the fitness. The exploration and search for the higher fitness value depends on the reproduction operators and selection in the GA, whereas it can be achieved by utilizing the particle profit from the discoveries and previous experience of the particles. The PSO algorithm incorporates various different schemes (global best (*gbest*) and local best (*lbest*)), particles use to collaborate. The *gbest* conceptually connects particles in the swarm to one another, whilst *lbest* form some sort of neighborhood of individual particle comparing itself with some fixed number of its neighbors.

The *i*th particle in *j* dimensional search space of the swarm can be represented as: $x_i = (x_{i1} \ldots x_{ij})$. The velocity of the particle x_i can be shown as: $v_i = (v_{i1}, v_{i2} \ldots v_{ij})$. The best position of previous visit of the particle is represented as: $x_{ij}^{\#} = (x_{i1}^{\#}, x_{i2}^{\#} \ldots x_{ij}^{\#})$. If all the particles are considered as the neighbors and is the best solution found so far, then velocity and new position of the particle is determined using Eqs. (1) and (2) respectively.

$$v_{ij}(t+1) = wv_{ij}(t) + c_1 r_1 (x_{ij}^{\#}(t) - x_{ij}(t)) + c_2 r_2 (x_{ij}^{*}(t) - x_{ij}(t)) \quad (1)$$

$$x_{ij}(t+1) = x_{ij}(t) + v_{ij}(t+1) \quad (2)$$

where, r_1 and r_2 are random numbers maintain the diversity of the population, uniformly distributed between [0, 1]. These numbers are scaled by acceleration coefficients and, where $0 \leq c1 \leq c2 \leq 2$ [1], and w is an inertia weight plays an important role contributes in balancing the global and local search [2], thereby maintains the trade-off between global and local exploration abilities of the particles (flying points). The following convention is used "higher" w facilitates the global exploration, whereas lower w tends to support local exploration, helps in fine-tune the searching.

The crossover operator plays a significant role in the performance of the GA. The crossover operator is mainly an exploration operator being utilized to explore the problem space and direct the searching towards the global optimum. The effect of the crossover operator varies in the GA's run following the convention "at the beginning chromosomes are usually randomized, hence the effect of the crossover is high and significant, whilst towards the end the effect of the crossover should be less due to smaller structure of the chromosomes, often converged". This discussion concludes that the crossover probability should be higher at the beginning, whereas should be low at the end of the run. This discussion also motivated researchers and scientist to develop dynamic algorithms, showed the capability to vary the crossover probability according to the requirement during a run employing some heuristics. The PSO algorithm does not support the crossover. However, it is achieved due to the particle acceleration. Each particle in the PSO stochastically accelerated towards its own former best position using Eqs. (1) and (2). The PSO algorithm has reflected a geometric feel; it may, in fact, show the nature of the recombination operator as in the ES.

Mutation showed the opposite effect with respect to the crossover during a run of a GA following the convention "mutation operator shows the lower level effect near the beginning of a run whilst more near the end". It happens at beginning and initially generated population is found random and therefore flipping a bit at this stage and at an early stage may not be very effective as flipping the bits near the end during a GA's run, when the population is about the convergence. If a researcher is planning to vary the mutation probability, then the recommendation is "use relatively lower mutation probability in the beginning, increase towards the end". The theoretical study—have been conducted concludes that "GA search can reach to any point in the problem space using the mutation operator". The theoretical study proves that the GA is theoretically ergodic, is not true in a practical sense (as GA runs in an iterative manner, which is not fixed). The EP is proven a true ergodic because it gives some sort of guarantee of finite probability of reaching to the better solution via a jump/generation.

The PSO shows the behavior somewhere between the EP and GA. The PSO cannot reach to any point in the problem space in one jump/iteration; however, it is possible in the beginning of the execution, provided sufficiently large velocity (Vmax). The working of the PSO depends on the survival of the particles in a generative manner, can go anywhere in the problem space, given sufficient generations. This discussion shows the ergodic nature of the PSO, stronger than the GA.

In PSO, each particle travels some distance using velocity; hence the PSO mutation behavior is directional with a certain momentum built-in, provided "*usebest*" is active. The mutation operator in the GA is assumed to be omnidirectional indicates that any bit in an individual is allowed to be flipped. The interesting fact about the GA's mutation is "GA support a bit-position based mutation, which affects the directionality". The mutation operator of the EP is also omnidirectional, which shows the mutation effect on the parameter-by-parameter basis. The PSO algorithm evaluates the difference between the *pbest* and the present location shows the same flavor as discussed for the EP, however the Vmax is alike for all the parameters.

3 Discussion

The above analysis leads to the number of ideas have come out for the GA and PSO. In fact, it is clear that the performance of the GA is largely affected by the crossover, mutation and selection. On the other hand, it can be seen that the PSO is an EA that does not support these operators, but these operator's services are achieved by the particle motion maintains a momentum. The w, c_1, c_2 and Vmax, contributes much to the success of the PSO. The recent work with the PSO presented the advocacy of the dynamic PSO, where these parameter values can be manipulated according to the requirement of the computational system. The PSO is an effective optimization algorithm used to solve several problems [3]. But few limitations were observed during the implementation of the PSO are: the swarm may prematurely converge [3, 4] and the PSO has stochastic approach and performance is largely dependent on the problem. The PSO using a GA was suggested takes the merits of the PSO and GA respectively. The PSO and GA both are used to solve optimization problems through evolutionary strategy. In [1] modifications were made in the PSO using the GA to improve the performance and reach to the global optimum. The GA and PSO both are used to find the optimal or the near optimal solutions. But, unfortunately these algorithms suffers with the problem of the premature convergence is a situation when he diversity of the population/swarm decreases over the generations leads to an undesirable convergence. Pandey et al. [5] has presented a comprehensive review of the various approaches were proposed to address premature convergence in the GA. Premalatha and Natarajan [6] have discussed a hybrid PSO and GA based approach for the global optimization presents the advocacy of the GA and PSO operators for handling the premature convergence. Hence, these two algorithms complement each others. The parameter quantification of these algorithms is an effective helps in alleviating the risk of the premature convergence [7]. The GA is based on the survival of the fittest concludes that selection of the population in the next generation considerably affects the performance of the GA. There exist several selection methods were proposed in GA. Pandey et al. [8] has evaluated the performance of selection techniques and presented a statistical test results using a travelling salesman problem.

4 Conclusions

This research successfully meets what it aimed for, i.e. a comprehensive analysis of two most popular algorithms: GA and PSO. The GA and PSO offer a ways to solve complex optimization problems. Few facts reported in this paper have been already discovered, but was not reported in a systematic and organized way. The main contribution of this paper is: it reports the different factors affect the working of the GA and PSO. The GA uses crossover, mutation and selection (parent selection and survival selection) guide the searching, whilst the PSO on the other hand perform the same task in a different way using factors: location, motion, inertia weight and acceleration coefficients. The PSO is the only algorithm does not support selection is achieved using ancestry path of the particles.

This paper has shown some aspects of the GA and PSO are really important to study and more such studies are required to understand the better working of both the algorithms. An obvious outcome of this study would be the development of the new improved algorithm in the near future. In this paper, the author critically evaluates the effects of the parameters affects the working of both the algorithm, hence the author is confident that this study will be advantageous for other researchers applying the GA and PSO.

References

1. Eberhart, Russ C., and James Kennedy. "A new optimizer using particle swarm theory." *Proceedings of the sixth international symposium on micro machine and human science*. Vol. 1. 1995.
2. Goldberg, David E., and John H. Holland. "Genetic algorithms and machine learning." *Machine learning* 3.2 (1988): 95–99.
3. Shi, Yuhui, and Russell Eberhart. "A modified particle swarm optimizer." Evolutionary Computation, Proceedings, 1998. IEEE World Congress on Computational Intelligence. The 1998 IEEE International Conference on. IEEE, 1998.
4. Van den Bergh, Frans, and Andries Petrus Engelbrecht. "Cooperative learning in neural networks using particle swarm optimizers." *South African Computer Journal* 26 (2000): p-84.
5. Pandey, Hari Mohan, Ankit Chaudhary, and Deepti Mehrotra. "A comparative review of approaches to prevent premature convergence in GA." *Applied Soft Computing* 24 (2014): 1047–1077.
6. Premalatha, K., and A. M. Natarajan. "Hybrid PSO and GA for global maximization." *Int. J. Open Problems Compt. Math* 2.4 (2009): 597–608.
7. Pandey, Hari Mohan. "Parameters Quantification of Genetic Algorithm." *Information Systems Design and Intelligent Applications*. Springer India, 2016. 711–719.
8. Pandey, Hari Mohan, et al. "Evaluation of Genetic Algorithm's Selection Methods." *Information Systems Design and Intelligent Applications*. Springer India, 2016. 731–738.

A Dual Phase Probabilistic Model for Dermatology Classification

Shweta and Sangeeta Rani

Abstract Medical data processing is one of the most required and critical application areas. The work requires the expert information processing with each stage. The data collection to classification must be defined with specified rule formulation and observations. In this paper, a specialized skin disease processing method is defined for Dermatology Disease. The proposed work model is divided in three main stages. In first stage, the data processing and analysis is applied under statistical parameters. In second stage, these parameters are observed with probabilistic measures to generate the predictive cell structure. In the final stage, the Bayesian network is applied to perform the disease classification. The implementation of work is applied on three different sample sets. The implementation results show that the method has provided the highly accurate results.

Keywords Dermatology · Skin disease · Classifier · Bayesian net

1 Introduction

Medical data processing is an evidence based analysis applied in different forms and on different data sets individually as well as collectively. The patient history, disease symptoms, disease history, environment, patient details etc. are some common databases which are been used in an integrated form to perform disease prediction. The evidence based analysis is applied separately for different diseases under expert supervision. Other than this, the diagnosis information, methods and processes can be applied on recorded data for individually patient and relatively for individual visit to the physician. The secondary information includes the social,

Shweta (✉) · S. Rani
Department of Computer Science & Engineering, Amity University,
Noida, Uttar Pradesh, India
e-mail: shweta.bhutani98@gmail.com

S. Rani
e-mail: schugh@amity.edu

S.C. Satapathy et al. (eds.), *Computer Communication, Networking and Internet Security*, Lecture Notes in Networks and Systems 5,
DOI 10.1007/978-981-10-3226-4_45

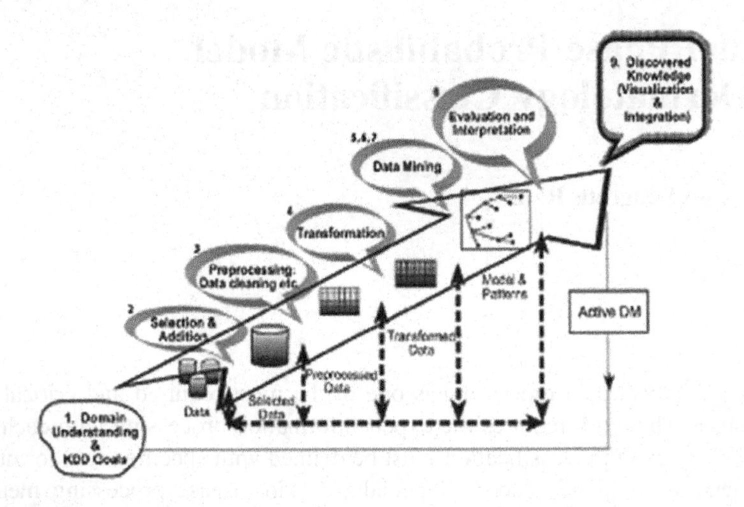

Fig. 1 KDD process stages

economic and demographic information processing. The clinical information processing and the constraints specific knowledge discovery are required to obtain the accurate prediction of disease as well as to identify the appropriate diagnosis respective to disease. A series of data mining operations are required to perform effective knowledge processing. KDD (Knowledge Discovery in Database) defines each stage of data processing with some constraints and procedures. These processes are applied in series under observation to predict the disease. These process specific stages of KDD are shown in Fig. 1.

The KDD is a scientific process which requires an intelligent and vast knowledge base. In this paper, a cell feature based probabilistic method is presented for disease classification. In this section, the basic medical data processing and relative criticalities are discussed. The basic KDD process model with associated components is also discussed. In Sect. 2, the work defined by earlier researchers is discussed. In Sect. 3, the proposed work model is presented and discussed. In Sect. 4, the results and observations of the work is presented. In Sect. 5, the conclusion drawn from work is presented.

2 Related Work

Medical data processing is one of the critical mining areas to identify the disease. The disease prediction requires more intelligent work to identify the disease. These algorithms are disease specific and analysis on featured symptoms. Different researchers provided the contribution on disease identification and classification. Some of the work provided by earlier researchers is discussed in this section. Weitschek et al. [1] provided the identification of various data mining issues relative to medical data processing was described. Author identified the data processing challenges and its requirement observation to identify the structural and unstructured observations for disease prediction. Author analyzed the complex associated tasks to generate the comparative observation to generate the frequent task under disease prediction.

Rajkumar et al. [2] has provided a work on dermatology detection under kernel space specification using neural network modeling. Author used the transformed feature space to intelligent computation to identify the dermatology disease class. The method signifies that the work model obtained the high degree association and classification with promising results. Rahman et al. [3] provided the integration of feature based neural network for identification of dermatology disease. Author defined correlation filter based feature selection method to achieve accuracy of disease recognition in parametric evaluation method. Author used the dimension reduction model to gain accurate results with lesser efforts. Azzini et al. [4] used the problem resolution model using neural network method. Author defined a hybrid method using neural network integration to the genetic approach to improve the prediction vector. Author used the statistical evaluation at different layers and in neuron formation to improve the classification results. Zhang et al. [5] provided a work on hybrid genetic algorithm to perform rule specific classification. Author applied the rule pruning to optimize the generation process and compute the effective results for Iris and Dermatology databases. Another work on multi-class classification using Cuckoo [6] search algorithm was provided for dermatology disease detection. Author applied the fuzzy featured rules for statistical characterization under confusion matrix, accuracy and sensitivity measures to identify the disease class. Author applied the receiver curve characterization for disease prediction. Barreto et al. [7] provided a work on statistical measures to provide the disease dermatology detection. Author applied the featured analysis sing statistical method and applied the rule formation based on multivariate observation. The LDA integrated classifier is applied to improve the accuracy of recognition method.

Feature processing is the basic process applied on databases to improve the recognition accuracy. The feature processing includes feature selection, feature generation and pruning methods. One such feature selection method was proposed by Cesar et al. [8] using fuzzy entropy measures. Author used the mean classification method for feature selection under uncertainty consideration.

Some of the researchers applied the disease mining for images. A skin image based markov [9] model implementation will be applied for disease classification based on feature analysis. Author applied the dimension specific analysis on imbalanced dataset to identify the skin lesion and identified the mining association rules. The rule filtration is defined using PCA and SVD approach for disease classification using standard features. Dhinagar et al. [10] provided the work on skin disease detection using skin images. Author applied the feature extraction and expert system processing to differentiate the skin features to generate significant results. The relative skin lesion processing under different feature vector was provided to identify the disease type. Rojas et al. [11] used a work on geometric feature using PCA method to generate the fractal dimension and area ratio analysis. Author used the polarity feature analysis to observe different tissues to identify the disease time using segmentation methods. Maglogiannis et al. [12] used the vision features for lesion classification using feature data processing methods. Author applied the segmentation to generate the textural features and border detection to generate the statistical features. Author applied the corresponding finding to identify the diagnostic of the defined disease.

Researchers provided the work on some other common diseases including the skin disease, breast cancer, heart disease etc. Fidelis et al. [13] used the genetic as the feature classifier to discover the rules for disease prediction under rule condition analysis. Author applied the method for identification of skin and breast cancer detection. The results observation shows the model has provided the significant derivation of rules. Alfisahrin et al. [14] applied the mining technique and optimization for liver disease classification. Author used the probabilistic measure using naive bays method and decision tree for attribute formation and identify the disease probability based on featured evaluation. Ranganatha et al. [15] defined an algorithmic model for heart disease detection. Author used the medical information analysis to identify the disease occurrence probability and predict the chances of disease under conditional observation on different associated methods. In this section, the earlier work for dermatology and other disease is described. In next section, the proposed model for dermatology detection is presented with relative experimentation.

3 Proposed Probabilistic Model

Dermatology is medicine term to provide the solution to skin, hair problems and relative disease. It provides the impact analysis of allergy, environmental effect on skin, hair, scalp etc. If the symptoms are known, it is easy for a doctor to identify the disease occurrence. But to provide the digitization of medical system, it is required to identify the disease through some real time application. This project is effort in same direction. In this work, a probabilistic method is provided for dermatology disease prediction. The work is designed as a real time graphical interface that will load the data and perform the characteristics observation on data of different attributes. In this work Bayesian network is used for probabilistic measure of disease and implementation is done using weka integrated java environment. The data is collected from web source and processed under cell driven probabilistic analysis. The basic flow of work is shown in Fig. 2. The implementation results of the work are discussed in next section.

(A) Probabilistic Classifier

The feature formulation with probabilistic behavior is here identified to formulate the rule. The conditional feature specific estimation is applied for disease identification. A contribution driven probabilistic estimation is here applied for condition

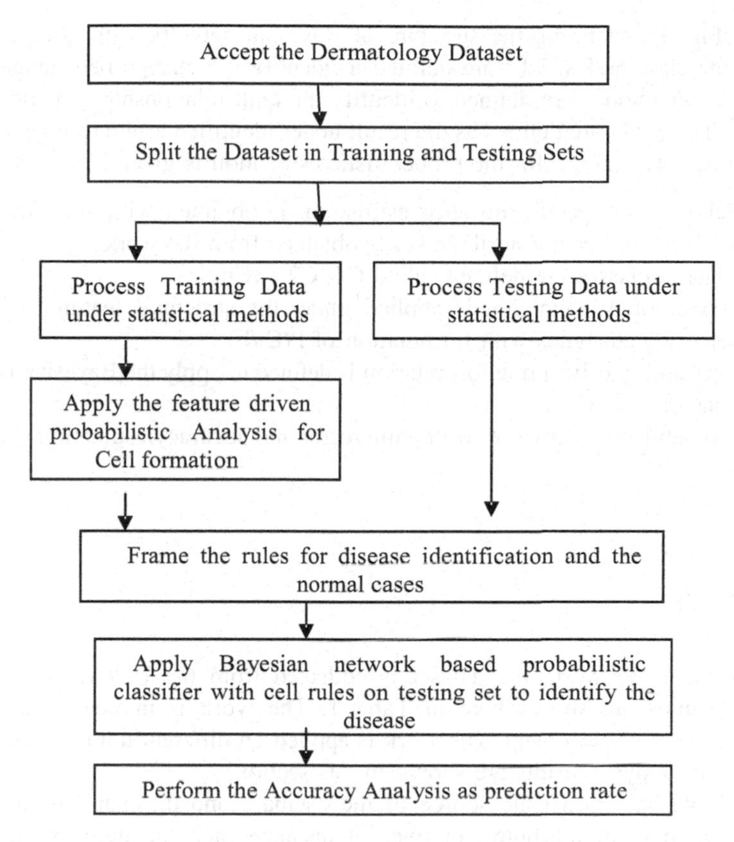

Fig. 2 Proposed model

probability generation. This observed attribute value is considered to provide the structure specific disease identification. The probabilistic classifier is shown in Fig. 3.

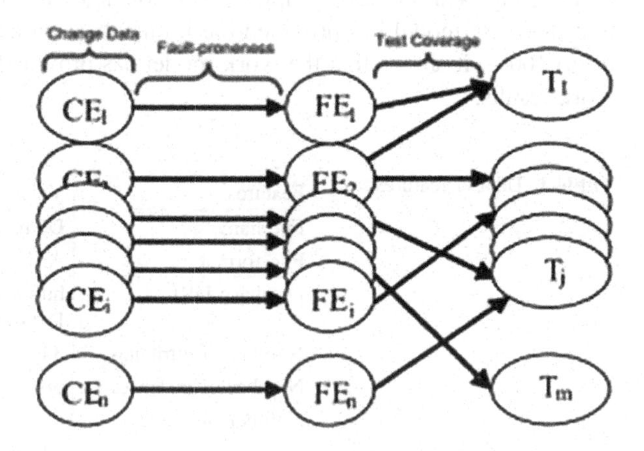

Fig. 3 Bayesian network model

Here Fig. 3 is defining the structure of Bayesian network with three different levels. The class nodes (CE) are defined to identify the change percentage. Fault Proneness (FE) nodes are defined to identify the fault relationship or nodes in the network. Target (T) are defined as the result nodes identified as the result class. The algorithmic specification for the probabilistic estimation is given here under

- The class label specific instance estimation is obtained with specification of class label. The feature attribute set is obtained from the work.
- The dataset classes are defined called C1, C2 …Cm.
- The conditional estimation is applied under the structural feature estimation between two attributes with formulation of P(Ci/F).
- The probability driven rule formulation is defined to apply the Bayesian network formulation.
- The probabilistic estimation with gain vector and accuracy estimation is applied in this work.

4 Results

To implement the work, the dataset is collected from the UCI repository. The dataset features are shown here in Table 1. The work is implemented in java integrated weka environment. The work is applied on different training and testing sets and the analysis results are shown in this section.

The Table has shown the source of the database and the major properties in terms of number of attributes, number of instances and the number of disease classes. The experimentation based results obtained from the work are described here under. The work is applied on three different training and testing sets. The description of the datasets is given here in Table 2.

The implementation results are here taken in the form of accuracy of the classification process. The accuracy results are here shown in Fig. 4.

Here Fig. 4 shows the accuracy results obtained for the work. The Figure shows that the work model has provided the minimum accuracy of 97.8% and maximum up to 100%. It shows that the work model has provided the effective and accurate work results.

Table 1 Dataset features

Features	Values
File name	Dermatology.arff
File format	Arff
Database URL	http://repository.seasr.org/Datasets/UCI/arff/dermatology.arff
Number of attributes	34
Number of instances	366
Number of classes	6

Table 2 Testing set description

Testing	Training set size	Testing set size
Test 1	300	66
Test 2	366	131
Test 3	251	91

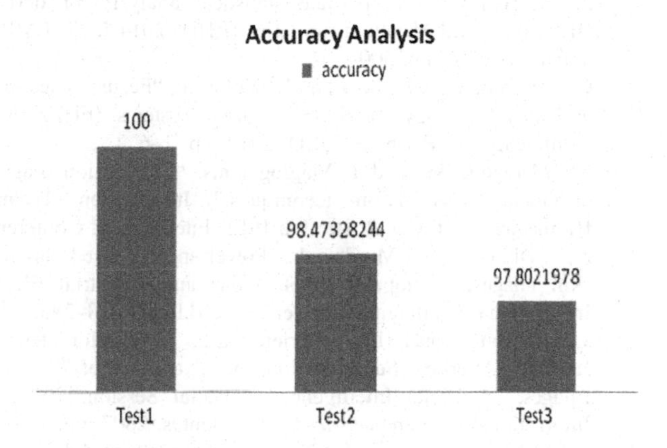

Fig. 4 Accuracy analysis

5 Conclusion

In this paper, a cell driven probabilistic measure is provided for classification of dermatology disease. The work model has performed the statistical analysis and probabilistic observation on training data to form a cell structure. Based on this cell structure, the testing data process is implied under bayesian classifier. The work is applied on three different sample sets. The implementation results shows that the method has provided the higher accuracy over 97%.

References

1. E. Weitschek, G. Felici and P. Bertolazzi, "Clinical Data Mining: Problems, Pitfalls and Solutions," Database and Expert Systems Applications (DEXA), 2013 24th International Workshop on, Los Alamitos, CA, 2013, pp. 90–94.
2. N. Rajkumar and P. Jaganathan, "A new RBF kernel based learning method applied to multiclass dermatology diseases classification," Information & Communication Technologies (ICT), 2013 IEEE Conference on, JeJu Island, 2013, pp. 551–556.
3. S. Abdul-Rahman, Ahmad Khairil Norhan, M. Yusoff, A. Mohamed and S. Mutalib, "Dermatology diagnosis with feature selection methods and artificial neural network," Biomedical Engineering and Sciences (IECBES), 2012 IEEE EMBS Conference on, Langkawi, 2012, pp. 371–376.
4. A. Azzini and S. Marrara, "Dermatology Disease Classification via Novel Evolutionary Artificial Neural Network," Database and Expert Systems Applications, 2007. DEXA '07. 18th International Workshop on, Regensburg, 2007, pp. 148–152.

5. Zhongyang Xiong, Yufang Zhang, Lei Zhang and Shujie Niu, "A Parallel Classification Algorithm Based on Hybrid Genetic Algorithm," Intelligent Control and Automation, 2006. WCICA 2006. The Sixth World Congress on, Dalian, 2006, pp. 3237–3240.
6. S. Chakravarty and P. Mohapatra, "Multi-class classification using Cuckoo Search based hybrid network," 2015 IEEE Power, Communication and Information Technology Conference (PCITC), Bhubaneswar, India, 2015, pp. 953–960.
7. A. S. Barreto, "Multivariate statistical analysis for dermatological disease diagnosis," Biomedical and Health Informatics (BHI), 2014 IEEE-EMBS International Conference on, Valencia, 2014, pp. 500–504.
8. C. Iyakaremye, P. Luukka and D. Koloseni, "Feature selection using Yu's similarity measure and fuzzy entropy measures," Fuzzy Systems (FUZZ-IEEE), 2012 IEEE International Conference on, Brisbane, QLD, 2012, pp. 1–6.
9. M. Maragoudakis and I. Maglogiannis, "Skin lesion diagnosis from images using novel ensemble classification techniques," Information Technology and Applications in Biomedicine (ITAB), 2010 10th IEEE International Conference on, Corfu, 2010, pp. 1–5.
10. N. J. Dhinagar and M. Celenk, "Power spectra based classification of cancerous nevoscope skin images," Computer Applications and Industrial Electronics (ICCAIE), 2011 IEEE International Conference on, Penang, 2011, pp. 278–283.
11. J. C. Riaño-Rojas, F. A. Prieto-Ortiz, L. J. Morantes, E. Sánchez-Camperos and F. Jaramillo-Ayerbe, "Segmentation and Extraction of Morphologic Features from Capillary Images," Artificial Intelligence - Special Session, 2007. MICAI 2007. Sixth Mexican International Conference on, Aguascallentes, 2007, pp. 148–159.
12. I. Maglogiannis and C. N. Doukas, "Overview of Advanced Computer Vision Systems for Skin Lesions Characterization," in IEEE Transactions on Information Technology in Biomedicine, vol. 13, no. 5, pp. 721–733, Sept. 2009.
13. M. V. Fidelis, H. S. Lopes and A. A. Freitas, "Discovering comprehensible classification rules with a genetic algorithm," Evolutionary Computation, 2000. Proceedings of the 2000 Congress on, La Jolla, CA, 2000, pp. 805–810 vol. 1.
14. S. N. N. Alfisahrin and T. Mantoro, "Data Mining Techniques for Optimization of Liver Disease Classification," Advanced Computer Science Applications and Technologies (ACSAT), 2013 International Conference on, Kuching, 2013, pp. 379–384.
15. S. Ranganatha, H. R. P. Raj, C. Anusha and S. K. Vinay, "Medical data mining and analysis for heart disease dataset using classification techniques," Research & Technology in the Coming Decades (CRT 2013), National Conference on Challenges in, Ujire, 2013, pp. 1–5.

Performance Evaluation of Reproduction Operators in Genetic Algorithm

Hari Mohan Pandey and Nidhi Jain

Abstract The performance of a GA largely depends on its parameters: crossover, mutation and selection. There exist many crossover and mutation operators are proposed. The primary interest of this paper is to investigate the effectiveness of the various reproduction operators. The conceptual characteristics of the combination of reproduction operators in the context of Travelling Salesman Problem (TSP) are discussed. Extensive experiments are conducted to compare the performance of 3-crossovers and 3-mutation operators. The computational experiments are performed and the results are collected. Statistical tests are conducted that demonstrate the superiority of 2-point cut crossover and swap mutation operators combination.

Keywords Crossover · Genetic algorithm · Mutation · Reproduction operators etc.

1 Introduction

Evolutionary algorithms (EAs) are looking at techniques that are propelled from Darwin's hypothesis of normal choice and survival of the fittest in the biological world. EA is a meta-heuristic optimization algorithm is based on the biological evolution, such as reproduction, crossover, mutation and selection. The main independently developed EAs are: Evolutionary Strategies (ESs), Evolutionary Programming (EP), Genetic Algorithm (GA) [1] and Genetic Programming (GP). EA differs from other traditional optimization methods since they incorporate a pursuit from a mass of arrangements and not from a solitary point. A major obstacle in EAs is slow convergence rate. This became more prominent when the optimized

H.M. Pandey (✉) · N. Jain
Department of Computer Science & Engineering, Amity University,
Sector-125, Noida, Uttar Pradesh, India
e-mail: profharimohanpandey@gmail.com

N. Jain
e-mail: nidhijain1794@gmail.com

S.C. Satapathy et al. (eds.), *Computer Communication, Networking and Internet Security*, Lecture Notes in Networks and Systems 5,
DOI 10.1007/978-981-10-3226-4_46

became more complex and numerically intensive to solve. GA is based on the evolutionary system [2–4] of human was proposed by Holland. It has successfully been applied to the fields of optimization, machine learning, neural networks, fuzzy logic and various other sectors. The GA method depends on the Darwinian guideline of survival of the fittest [5–7].

GA is used successfully in different optimization problems [8]. GA's performance depends upon the encoding technique and selection of GA's operators [9]. There exist various GA's operators have been suggested. There exist several crossover operators have been proposed to permutation presentations, are used in a large number of combinatorial optimization problems. In this direction, Traveling Salesman Problem (TSP) is one of the most studied problems. TSP is known as the classical combinatorial optimization problem. The basic concept of TSP is to find the shortest closed tour that connects a number of cities in a region. TSP is easy to understand, but extremely hard to solve, perhaps because it is classified as non-polynomial (NP)-complete problems, meaning that the amount of computation required increases exponentially with the number of cities [10, 11].

This study would give a reason to the researchers of various levels to conduct more suitable research in this domain to identify the suitable combinations of the reproduction operators, which might lead GA towards the global optimum solution of various other similar problems.

The rest of the paper is organized as follows: Section 2 discusses the GA and the reproduction operators adapted to conduct this study. The experimental results, analysis and discussion are reported in Sect. 3. The conclusion is drawn in Sect. 4.

2 Genetic Algorithm and the Reproduction Operators Adapted

The process of a GA starts with providing the basic essential details such as number and location (coordinates) of the cities, the initial population size, maximum number of generations, the mutation and crossover rates and the tournament size. The fitness of each of the chromosomes in the initial population is determined using a fitness function and the next generation is evolved through the reproduction operator. The selection operator embodies the principle of survival of the fittest. In each generation, the fittest individual is *selected* [14]. There exist many different selection techniques that can be used [12–14]. In our case, a tournament selection is used that works on a tournament size. It works on some sort of tournament is being applied to a number of randomly selected individuals and then one wins goes further for reproduction operation. The concept of the crossover operation is analogous to biological crossovers from where the GA has been devised. The crossover operator plays a significant role in the exploration of the search space. We have adapted 3-crossover methods: 1-point cut crossover (C1), 2-point cut crossover (C2) and ordered crossover (C3) for the comparison. Mutation prevents local search and tremendously increases the probability of finding the global optima. It actually prevents premature convergence. The mutation techniques used for the comparison are: Swap mutation (M1), Insert Mutation (M2), Inversion Mutation (M3).

3 Computational Experiment, Analysis and Observations

Extensive computational experiments are performed creating the reproduction operators combination. All the computational experiments are conducted using Java on Intel Core i5-3230 M CPU @ 2.60 GHz and 4 GB RAM with Window 7 Ultimate. The TSP problem is considered to perform the experiments. The minimum distance route constituting all the cities is considered as an optimization criteria. Extensive parameters tuning is done apply a Taguchi signal to noise ratio and an orthogonal array based approach [15]. The tuned parameter's values are: population size: 100, chromosome size: 200, tournament size: 5, crossover rate: 0.2, mutation rate: 0.015, maximum generation: 100 and number of cities: 20.

Figure 1a represents the location of the different cities is taken as an input. The dots show the cities and X and Y axis represents the coordinates of each city. Figure 1b represents the best optimal tour obtained. From the results depicted in Table 1, it can be seen that he optimal solution to the TSP is 580.9, which is obtained from the C3M1 reproduction operator combination. The reproduction operator combinations C1M2 yields the worst result. Also from Table 2, the means and standard deviations of C3M1 reproduction operator combination again seems to be the optimal solution of TSP. In order to reach to an exact conclusion, the authors conducted statistical tests. The hypotheses are formed considering "there is not a significant difference in the performance at 5% confidence interval".

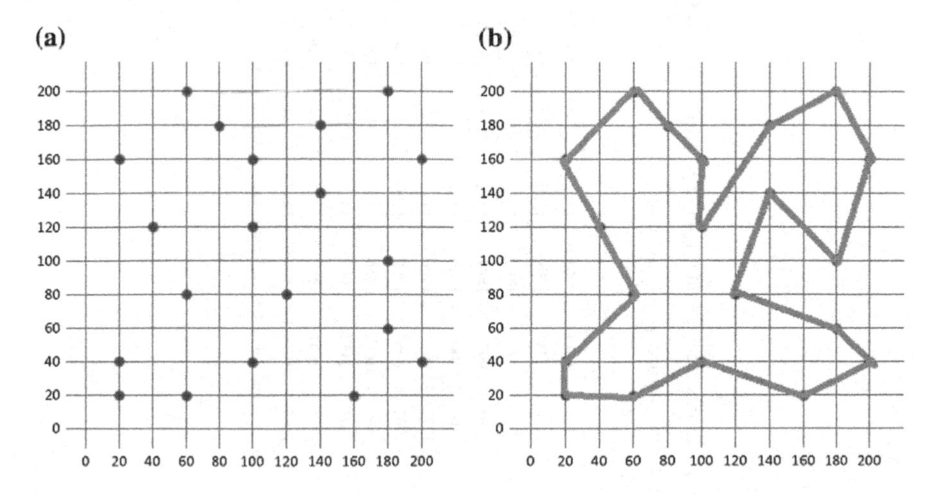

Fig. 1 **a** Location of cities (X and Y axes represents the coordinate of the cities) and **b** optimal solution

Table 1 Results received after applying various reproduction operator combinations

Exp. No.	Reproduction operator combinations																	
	C1M1		C1M2		C1M3		C2M1		C2M2		C2M3		C3M1		C3M2		C3M3	
	I	F	I	F	I	F	I	F	I	F	I	F	I	F	I	F	I	F
1	1689	1014	1903	1312	1670	1513	1924	1378	1794	1257	2023	432	1909	643	1891	1046	1966	983
2	1874	873	1799	1550	1899	1336	1903	1129	1816	1215	1887	736	1910	489	1872	1255	1648	1093
3	1940	999	1897	1349	1907	1351	1948	1267	1873	1171	1861	682	1969	633	1870	1068	1902	1007
4	1793	997	1881	1258	1851	1361	1668	1057	1852	1383	1870	467	1759	492	1900	1025	1828	1226
5	1809	1006	1746	1461	1928	1320	1812	1263	1945	1238	1850	564	2021	676	1736	858	1713	1112
6	1772	991	1900	1410	1852	1630	1890	1272	1661	1280	1775	787	1900	567	1767	1161	1705	1207
7	1962	1060	1992	1390	1863	1243	1890	1102	1901	1089	1777	703	1914	587	1931	1112	1633	1066
8	1890	1042	2056	1425	2010	1302	1902	1143	1944	1402	1790	841	1797	702	1808	1196	1732	1291
9	1975	1154	1900	1395	1735	1261	1909	1385	1755	1142	1861	518	1789	377	1979	1045	1890	950
10	1948	1051	1885	1587	1732	1310	1887	1358	1872	1012	1905	682	1874	643	1710	905	1867	1287
Average	1865.2	1018.7	1895.9	**1413.7**	1844.7	1362.7	1873.3	1235.4	1841.3	1218.9	1859.9	641.2	1884.2	**580.9**	1846.4	1067.1	1788.4	1122.2

I: initial value, F: Final value, C1: Ordered crossover, C2: 1-point crossover, C3: 2-point crossover, M1: Swap mutation, M2: Insert mutation, M3: Inverse mutation

$$H_0 : \mu_1 = \mu_2 = \mu_3 = \mu_4 = \mu_5 = \mu_6 = \mu_7 = \mu_8 = \mu_9$$
$$H_A : \textit{At least one of the combinations showing different results.}$$

The F-test is based on the ANOVA, is conducted. The samples were drawn from each reproduction operator combinations. The descriptive analysis is depicted in Table 2 and Fig. 2 graphically reflects the mean for the nine reproduction operator combinations. The graph shows the nine reproduction operator combinations

Table 2 Descriptive analysis

	N	Mean	Std. deviation	Std. error	95% Confidence interval for mean		Minimum	Maximum
					Lower bound	Upper bound		
C1M1	10	1018.70	70.424	22.270	968.32	1069.08	873	1154
C1M2	10	1413.70	100.373	31.741	1341.90	1485.50	1258	1587
C1M3	10	1362.70	119.186	37.690	1277.44	1447.96	1243	1630
C2M1	10	1235.40	120.336	38.054	1149.32	1321.48	1057	1385
C2M2	10	1218.90	121.979	38.573	1131.64	1306.16	1012	1402
C2M3	10	641.20	138.468	43.788	542.15	740.25	432	841
C3M2	10	580.90	101.265	32.023	508.46	653.34	377	702
C3M2	10	1067.10	122.683	38.796	979.34	1154.86	858	1255
C3M3	10	1122.20	124.737	39.445	1032.97	1211.43	950	1291
Total	90	1073.42	297.641	31.374	1011.08	1135.76	377	1630

C1: Ordered crossover, C2: 1-point crossover, C3: 2-point crossover, M1: Swap mutation, M2: Insert mutation, M3: Inverse mutation

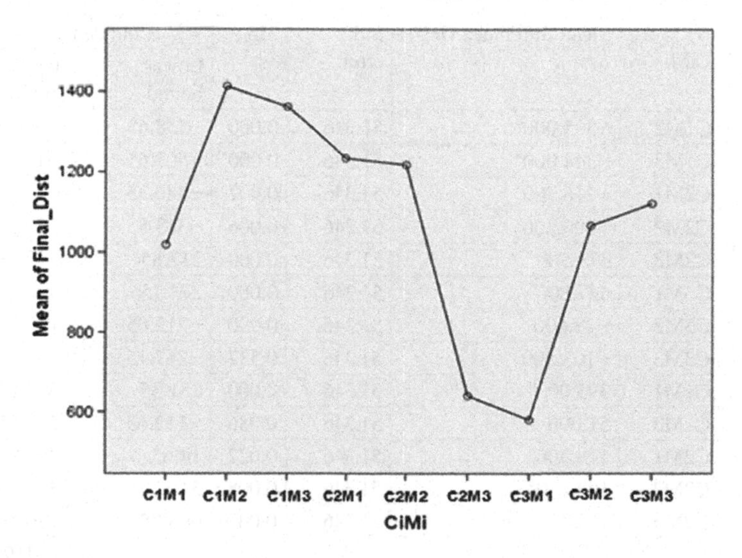

Fig. 2 Estimated marginal mean plot for the reproduction operator combinations

Table 3 Main ANOVA table

	Sum of squares	df	Mean square	F	Sig.
Between groups	6816791.356	8	852098.919	64.641	.000
Within groups	1067738.600	81	13181.958		
Total	7884529.956	89			

(C1M1, C1M2, C1M3, C2M1, C2M2, C2M3, C3M1, C3M2 and C3M3) on X-axis. Y-axis shows the estimated marginal mean of distance. Figure 2 indicates the superiority of C3M1 in achieving the minimum travel distance for TSP. Also, it is noticed that C1M2's performance is the worst. Table 3 shows the results received from the F-test. The determined significance value (p-value) = 0.000 < 0.05 concludes that H_0 could be rejected. But this result does not provide any symptoms of which reproduction operator combinations are responsible for it. Therefore, the authors have applied multiple comparison tests (Posthoc test).

The post hoc tests compares one single combination results with other combinations' results. The Tukey test is applied for further information investigation. The process maintains the alpha values till their intended values as long as the statistical model assumptions are met. The 1st column "(I) CiMi" and 2nd column "(J) CiMi" represents reproduction operators combination. The "mean difference (I-J)" represents the mean differences of (I) combinations and (J) combinations respectively taken from Table 2. The significant difference can be verified by comparing the p-value present in 5th column of Table 4. It can be seen that the significance value of C1M1 with C1M2, C1M3, C2M1, C2M2, C2M3, and C3M1 is less than 0.05, so

Table 4 Multiple comparison test results

(I) CiMi	(J) CiMi	Mean difference (I−J)	Std. error	Sig.	95% Confidence interval	
					Lower bound	Upper bound
C1M1	C1M2	−395.000[a]	51.346	0.000	−558.65	−231.35
	C1M3	−344.000[a]	51.346	0.000	−507.65	−180.35
	C2M1	−216.700[a]	51.346	0.002	−380.35	−53.05
	C2M2	−200.200[a]	51.346	0.006	−363.85	−36.55
	C2M3	377.500[a]	51.346	0.000	213.85	541.15
	C3M1	437.800[a]	51.346	0.000	274.15	601.45
	C3M2	−48.400	51.346	0.990	−212.05	115.25
	C3M3	−103.500	51.346	0.537	−267.15	60.15
C1M2	C1M1	395.000[a]	51.346	0.000	231.35	558.65
	C1M3	51.000	51.346	0.986	−112.65	214.65
	C2M1	178.300[a]	51.346	0.022	14.65	341.95
	C2M2	194.800[a]	51.346	0.008	31.15	358.45
	C2M3	772.500[a]	51.346	0.000	608.85	936.15

(continued)

Table 4 (continued)

(I) CiMi	(J) CiMi	Mean difference (I−J)	Std. error	Sig.	95% Confidence interval	
					Lower bound	Upper bound
	C3M1	832.800[a]	51.346		669.15	996.45
	C3M2	346.600[a]	51.346	0.000	182.95	510.25
	C3M3	291.500[a]	51.346	0.000	127.85	455.15
C1M3	C1M1	344.000[a]	51.346	0.000	180.35	507.65
	C1M2	−51.000	51.346	0.986	−214.65	112.65
	C2M1	127.300	51.346	0.258	−36.35	290.95
	C2M2	143.800	51.346	0.131	−19.85	307.45
	C2M3	721.500[a]	51.346	0.000	557.85	885.15
	C3M1	781.800[a]	51.346	0.000	618.15	945.45
	C3M2	295.600[a]	51.346	0.000	131.95	459.25
	C3M3	240.500[a]	51.346	0.000	76.85	404.15
C2M1	C1M1	216.700[a]	51.346	0.002	53.05	380.35
	C1M2	−178.300[a]	51.346	0.022	−341.95	−14.65
	C1M3	−127.300	51.346	0.258	−290.95	36.35
	C2M2	16.500	51.346	1.000	−147.15	180.15
	C2M3	594.200[a]	51.346	0.000	430.55	757.85
	C3M1	654.500[a]	51.346	0.000	490.85	818.15
	C3M2	168.300[a]	51.346	0.039	4.65	331.95
	C3M3	113.200	51.346	0.413	−50.45	276.85
C2M2	C1M1	200.200[a]	51.346	0.006	36.55	363.85
	C1M2	−194.800[a]	51.346	0.008	−358.45	−31.15
	C1M3	−143.800	51.346	0.131	−307.45	19.85
	C2M1	−16.500	51.346	1.000	−180.15	147.15
	C2M3	577.700[a]	51.346	0.000	414.05	741.35
	C3M1	638.000[a]	51.346	0.000	474.35	801.65
	C3M2	151.800	51.346	0.091	−11.85	315.45
	C3M3	96.700	51.346	0.627	−66.95	260.35
C2M3	C1M1	−377.500[a]	51.346	0.000	−541.15	−213.85
	C1M2	−772.500[a]	51.346	0.000	−936.15	−608.85
	C1M3	−721.500[a]	51.346	0.000	−885.15	−557.85
	C2M1	−594.200[a]	51.346	0.000	−757.85	−430.55
	C2M2	−577.700[a]	51.346	0.000	−741.35	−414.05
	C3M1	60.300	51.346	0.960	−103.35	223.95
	C3M2	−425.900[a]	51.346	0.000	−589.55	−262.25
	C3M3	−481.000[a]	51.346	0.000	−644.65	−317.35
C3M1	C1M1	−437.800[a]	51.346	0.000	−601.45	−274.15
	C1M2	−832.800[a]	51.346	0.000	−996.45	−669.15
	C1M3	−781.800[a]	51.346	0.000	−945.45	−618.15

(continued)

Table 4 (continued)

(I) CiMi	(J) CiMi	Mean difference (I−J)	Std. error	Sig.	95% Confidence interval	
					Lower bound	Upper bound
	C2M1	−654.500[a]	51.346		−818.15	−490.85
	C2M2	−638.000[a]	51.346	0.000	−801.65	−474.35
	C2M3	−60.300	51.346	0.960	−223.95	103.35
	C3M2	−486.200[a]	51.346	0.000	−649.85	−322.55
	C3M3	−541.300[a]	51.346	0.000	−704.95	−377.65
C3M2	C1M1	48.400	51.346	0.990	−115.25	212.05
	C1M2	−346.600[a]	51.346	0.000	−510.25	−182.95
	C1M3	−295.600[a]	51.346	0.000	−459.25	−131.95
	C2M1	−168.300[a]	51.346	0.039	−331.95	−4.65
	C2M2	−151.800	51.346	0.091	−315.45	11.85
	C2M3	425.900[a]	51.346	0.000	262.25	589.55
	C3M1	486.200[a]	51.346	0.000	322.55	649.85
	C3M3	−55.100	51.346	0.976	−218.75	108.55
C3M3	C1M1	103.500	51.346	0.537	−60.15	267.15
	C1M2	−291.500[a]	51.346	0.000	−455.15	−127.85
	C1M3	−240.500[a]	51.346	0.000	−404.15	−76.85
	C2M1	−113.200	51.346	0.413	−276.85	50.45
	C2M2	−96.700	51.346	0.627	−260.35	66.95
	C2M3	481.000[a]	51.346	0.000	317.35	644.65
	C3M1	541.300[a]	51.346	0.000	377.65	704.95
	C3M2	55.100	51.346	0.976	−108.55	218.75

[a]The mean difference is significant at the 0.05 level

Table 5 Homogeneous subset table for cost function

CiMi	N	Subset for alpha = 0.05					
		1	2	3	4	5	6
C3M1	10	580.90					
C2M3	10	641.20					
C1M1	10		1018.70				
C3M1	10		1067.10	1067.10			
C3M3	10		1122.20	1122.20	1122.20		
C2M2	10			1218.90	1218.90	1218.90	
C2M1	10				1235.40	1235.40	
C1M3	10					1362.70	1362.70
C1M2	10						1413.70
Sig.		0.960	0.537	0.091	0.413	0.131	0.986

Means for groups in homogeneous subsets are displayed
a. Uses Harmonic Mean Sample Size = 10.000

the mean of the C1M1 is significantly different for these combinations and not for C3M2, C3M3. We conducted homogeneity test to verify the group similarity is shown in Table 5. The mean values listed under subset 1 with C3M1 and C2M3 have the optimal mean values, but are not reliably statistically different from each other, also they have significant values greater than 0.05. In subset 3, C3M2, C3M3 and C2M2 are not reliably statically different from each other, but having significant value as 0.091 which is lesser than 0.05.

4 Conclusions

In this paper, we have implemented nine different combinations of crossover and mutation operators in order to test the influence of the recombination of the genetic search process when applied to the TSP. The obtained results for the random locations as a test of the TSP show high performance creating the shortest paths. The best known result for the TSP instance, which was obtained by using 2-point crossover and swap mutation. The F-test is conducted to verify the hypothesis and the results conclude that null hypothesis could be rejected. This result indicates that one or the other reproduction operators are showing a different result than the others. To reach to an exact conclusion, we applied post hoc test (Tukey test). The results of the Tukey test conclude that the operator combination C3M1 has more ability to explore the search space more than any other combination. Also, we found that the performance of C2M3 is slightly is similar to C3M1. From the simulation results, we concluded that not all the combinations of reproduction operators give higher performance for a GA, but it can be achieved employing an appropriate reproduction operator combination.

References

1. Koenig, Andreas C. "A study of mutation methods for evolutionary algorithms." University of Missouri-Rolla (2002).
2. Lin, Wen-Yang, Wen-Yung Lee, and Tzung-Pei Hong. "Adapting crossover and mutation rates in genetic algorithms." J. Inf. Sci. Eng. 19.5 (2003): 889–903.
3. Srinivas, Mandavilli, and Lalit M. Patnaik. "Adaptive probabilities of crossover and mutation in genetic algorithms." IEEE Transactions on Systems, Man, and Cybernetics 24.4 (1994): 656–667.
4. De Jong, Kenneth. "Adaptive system design: a genetic approach." IEEE Transactions on Systems, Man, and Cybernetics 10.9 (1980): 566–574.
5. Kaya, Yılmaz, and Murat Uyar. "A novel crossover operator for genetic algorithms: ring crossover." arXiv preprint arXiv:1105.0355 (2011).
6. Pandey, Hari Mohan, Ankit Chaudhary, and Deepti Mehrotra. "A comparative review of approaches to prevent premature convergence in GA." Applied Soft Computing 24 (2014): 1047–1077.

7. Pandey, Hari Mohan, Anurag Dixit, and Deepti Mehrotra. "Genetic algorithms: concepts, issues and a case study of grammar induction." Proceedings of the CUBE International Information Technology Conference. ACM, 2012.
8. Holland J. H. "Genetic algorithms." Scientific American 267.1 (1992): 66–72.
9. Magalhaes-Mendes, Jorge. "A comparative study of crossover operators for genetic algorithms to solve the job shop scheduling problem." WSEAS transactions on computers 12.4 (2013): 164–173.
10. Noraini, Mohd Razali, and John Geraghty. "Genetic algorithm performance with different selection strategies in solving TSP." (2011).
11. Grefenstette, John, et al. "Genetic algorithms for the traveling salesman problem." Proceedings of the first International Conference on Genetic Algorithms and their Applications. Lawrence Erlbaum, New Jersey (160–168), 1985.
12. Shukla, Anupriya, Hari Mohan Pandey, and Deepti Mehrotra. "Comparative review of selection techniques in genetic algorithm." Futuristic Trends on Computational Analysis and Knowledge Management (ABLAZE), 2015 International Conference on. IEEE, 2015.
13. Pandey, Hari Mohan. "Performance Evaluation of Selection Methods of Genetic Algorithm and Network Security Concerns." Procedia Computer Science 78 (2016): 13–18.
14. Pandey, Hari Mohan, et al. "Evaluation of Genetic Algorithm's Selection Methods." Information Systems Design and Intelligent Applications. Springer India, 2016. 731–738.
15. Pandey, Hari Mohan. "Parameters Quantification of Genetic Algorithm." Information Systems Design and Intelligent Applications. Springer India, 2016. 711–719.

Rolling Circle Algorithm for Routing Along the Boundaries of Wireless Sensor Networks

Rajesh Sharma, Lalit Kumar Awasthi and Naveen Chauhan

Abstract Geographic routing has emerged as one of the most suitable strategies for routing in Wireless Sensor Networks (WSNs). Greedy forwarding (GF) is an efficient form of geographic routing in which a packet is progressively pushed closest to the destination in each hop. The presence of communication voids may lead GF to fail at dead-ends. The dead-end situation is usually handled by using methods like flooding, face routing, and routing along network boundaries. Existing schemes for handling void problems are either too inefficient or are rely on some unrealistic assumptions like unit disk graph model of connectivity. In this chapter, a scheme called Rolling Circle Algorithm (RCA) is presented for recovering from dead-ends by routing packets along the boundary of the WSN until GF can resume again. The proposed technique is based on a variant of alpha-shapes method of detecting boundaries of a point cloud. The value of the parameter alpha used to detect boundary nodes adapts according to the local topology and location information of 1-hop neighbors of the forwarding node.

Keywords Wireless sensor networks · Geographic routing · Perimeter routing · Rolling-circle algorithm

1 Introduction

Geographic routing (GR) has emerged as a suitable routing strategy for WSNs due to its scalability, localized operation, small state per node, and robustness. GR is based on two basic assumptions: (1) every node is aware of its own and its

R. Sharma (✉) · L.K. Awasthi · N. Chauhan
National Institute of Technology, Hamirpur, India
e-mail: rajesh.nitham@gmail.com

L.K. Awasthi
e-mail: lalit@nith.ac.in

N. Chauhan
e-mail: naveenchauhan.nith@gmail.com

S.C. Satapathy et al. (eds.), *Computer Communication, Networking and Internet Security*, Lecture Notes in Networks and Systems 5,
DOI 10.1007/978-981-10-3226-4_47

immediate neighbors' locations, and (2) the source node of the message is informed of the location of the destination. In GR, a node utilizes the information about geographic locations of itself, its neighbors and the destination nodes to choose the next appropriate hop. *Greedy forwarding* (GF) is a simpler and efficient form of geographic routing with worst-case complexity of $\Omega(d^2)$ [1], where d is the Euclidean distance between the source and the destination of the packet. In GF, a packet is progressively forwarded closest to the destination in each hop. Only limitation of GF is that it suffers from *local maximum* problem that arises when none of the neighbors of the forwarding node can yield any positive progress towards the destination. A local maximum also called a *dead-end* always lies on the boundary of some *communication void* in the network arising due to random deployment of nodes, irregular terrain topography, obstacles, and other sources of obstruction to radio waves. Hence, dead-end problem is often referred to as *void problem* also. GF guarantees loop-free operation and yields a nearly optimal route in densely deployed networks [1]. Many GR protocols (Compass Routing-II [2], GPSR [3], GFG [4], GOAFR [5], GAR [6], CBG [7]) leverage the efficiency of GF by using it as default routing method and switching to *recovery mode* when dead-end problem surfaces. The recovery from dead-end situation is done by switching to some stand-by reliable routing method which usually is inefficient as compared to GF. The recovery mode continues until the packet is delivered or GF is feasible again. Various strategies for handling void problem in geographic routing can be broadly classified as [8] flooding-based, heuristic, planar graphs-based, spanning-tree based and geometric.

Flooding is not suitable for resource-constrained WSNs due to inherent problems of implosion, duplicate messages, and resource-blindness. However, restricted forms of flooding like 1-hop flooding [9] has been exercised for handling voids in WSNs.

Heuristic recovery strategies rely on some intuitive notions and are generally not supported by sound theoretical analysis. For example, in [10], a dead-end node actively explores its neighborhood by increasing its transmission power until the destination or some node that can make a positive progress towards the destination falls within its transmission range. Energy consumption and interference caused by increasing the transmission power are the main drawbacks of active exploration.

Planar graph-based recovery strategies traverse the faces of a planarized graph using "right-hand rule" and switching faces progressively towards destination. Graph planarization structures like Relative Neighborhood Graph (RNG) and Gabriel Graph (GG) assume Unit Disk Graph (UDG) model for network connectivity. UDG model assumes perfect circular and same transmission ranges for all nodes and is often criticized [11] for being too idealistic to model the real behavior of wireless radios. Cross link detection protocol (CLDP) [11] could planarize any arbitrary graph without UDG assumption, but incurs high overheads due to multiple probing of each edge.

Spanning tree based techniques construct a spanning tree in the WSN graph when a dead-end is encountered during GF. The packet is either flooded [12] or forwarded towards the direction in the spanning tree nearer to the destination [13].

Geometric based methods use geometric aspects of the network to avoid or overcome the dead-end situation.

Alpha-shapes [14] is a well-known concept of computational geometry used to approximate the shape rendered by a point cloud in space. Rolling-Circle Algorithm (RCA) is based on a variant of localized alpha-shapes to determine the boundary of the WSN for the purpose of recovering from the dead-end problem. RCA is a geometric solution based on actual connectivity graph and locations of 1-hop neighbors of the nodes.

2 Related Work

"Compass Routing II" [2] was the first routing strategy based on face routing (FR). In FR, the faces of the planarized network graph are traversed using "right hand rule". If S is the source and D is the destination of a message, FR switches to the adjacent face of the planar graph at an edge that crosses the line $\overline{SD}$ at a point closer to D than the point where the current face was entered. In worst case, FR is no better than basic flooding [5] and takes Ω(n) steps before arriving at the destination [1], where n is the number of nodes in the network. Adaptive Face Routing (AFR) [5] confines the search area of FR by a bounding ellipse (with S and D as two foci and $1.2 \times |\overline{SD}|$ as major axis) during face traversals. The execution cost of AFR is bounded above by square of the cost of the optimal path. GOAFR [5] uses GF in conjunction with AFR as recovery strategy as an end-to-end geographic routing protocol.

GFG [4], GPSR [3], Compass Routing [2], GAR [6], CBG [7], GOAFR [5] begin to operate in greedy mode and switch to FR when a local maximum is encountered. The routing mode switches back to greedy mode as soon as the packet reaches at a node that is closer to the destination than the dead-end that initiated the recovery. Removal of edges during planarization may cause detours while routing. The detours may extend up to stretch factors of $\Theta(\sqrt{n})$ for GG and $\Theta(n)$ for RNG [15].

GDSTR [13] progressively constructs hull trees by aggregating convex hulls formed by the immediate neighbors and other hull subtrees. The stuck packets are routed either downward or upward in the hull-tree depending upon the location of the destination.

BoundHole [16] was one of the early algorithms to use geometric aspects of the network to identify holes. It uses a rule called "tent rule" to identify holes around a dead-end node. A path pruning algorithm to shorten the detours in recovery schemes is presented in [17]. Fayed et al. [18] proposed a method for detecting boundaries of a WSN using α-shapes. Greedy Anti-void Routing (GAR) [6] and Contention-based Georouting (CBG) [7] algorithms handle void region problem by routing packets along the boundary of the WSN detected using variants of α-shapes. GAR combines GF with Rolling-ball UDG boundary Traversal (RUT) to achieve end-to-end geographic routing. RUT uses a rolling ball of radius $r/2$

(r is transmission range of the node) starting from the dead-end node in counter-clockwise direction. The sequence of nodes swept by such rolling ball constitutes the boundary of the void region. The Rotational Sweep (RS) algorithm presented in [7] is a contention-based form of RUT. In RS, the forwarder node broadcasts a Request to Send (RTS) message, all neighboring nodes called candidates, schedule to respond by a Clear to Send (CTS) message after a certain delay. The DATA packet is forwarded to the first candidate that replies with the CTS. Other candidates then cancel their scheduled CTS transmissions. The delay is a function of the relative locations of the forwarder, the candidate and the previous node. The delay function can be geometrically visualized as a rotating semicircle of radius $r/2$ where r is the radius of the unit disk. A severe drawback of GAR and RS is that these algorithms rely on the assumption of UDG model. A slight deviation in transmission ranges of nodes may lead to unexpected behavior of these schemes.

3 System Model and Terminology

The sensor nodes are randomly deployed in an Euclidean plane $\mathbb{R}^2$ and each node is aware of its geographic location in the form of its Cartesian coordinates (x, y). An undirected graph $G(V, E)$ models a WSN, where V, the set of vertices represent nodes and E, the set of edges represent the wireless links between nodes. A node v connected to node u through a direct wireless link is called a neighbor of u. The set of neighbors of a node u is represented by $\mathcal{N}(u)$. $|uv|$ denotes the Euclidean distance between nodes u and v. The links between nodes are symmetric i.e. $u \in \mathcal{N}(v) \rightarrow v \in \mathcal{N}(u)$. There is no global addressing scheme for nodes and a node is identified by its location only. There is only one static sink node in the WSN and all nodes are informed of its location.

α-Shape: Let S be a set of points in a plane and $\alpha \geq 0$, a point $p \in S$ is an *α-extreme* if it is situated on the boundary of a disc of radius $1/\alpha$ and no other points of S exists inside that disc. Two neighboring α-extremes p and q existing on the boundary of the same disc are called *α-neighbors*. The set of all α-extremes and edges connecting them constitute an *α-shape*.

4 Rolling Circle Algorithm

RCA is based on a localized variant of α-shapes to determine the boundary of the WSN when GF fails. The stuck packet is then routed along the boundary nodes until a node closer to the destination than the node that initiated the recovery is found, where RCA terminates and GF resumes. In contrast to predetermined fixed value of the parameter α, the size of the α-disc is adjusted according to the local topology and geometric properties of the forwarding node. The proposed method is described below by taking an example of Fig. 1.

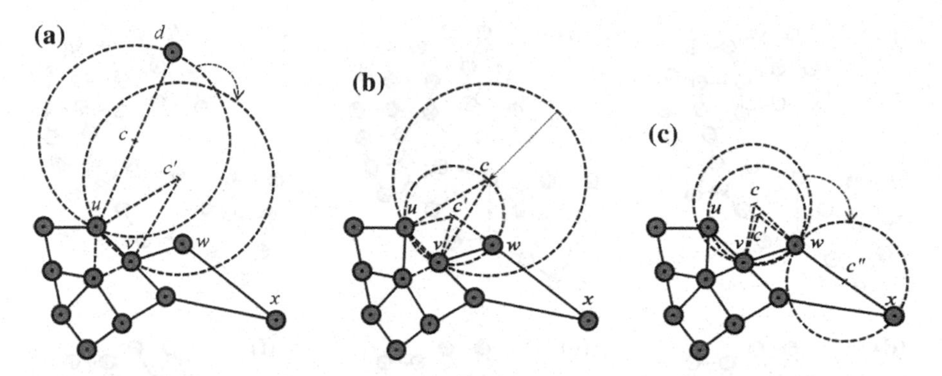

Fig. 1 Steps in RCA: **a** Clock-wise sweep, **b** Deflating the circle, **c** Inflating the circle

Initialization: The dead-end node u towards destination d initiates the perimeter mode with the mid-point of the line joining u and d (point c) as the initial center and $2/|ud|$ as the initial value of α i.e. $|ud|/2$ as the initial radius of the rolling circle.

Termination: The perimeter mode terminates in either of the following conditions:

(a) RCA terminates and GF resumes at a node that is closer to d than u.
(b) Network partition is detected when a packet returns back to the dead-end node that had initiated the recovery, in this case the packet is dropped.

Radius adjustment: When a packet is received by a boundary node v from the node u in perimeter mode, some of the neighbors of v (w in Fig. 1b) may lie inside the rolling circle due to larger initial value of its radius. In this case, the circle is deflated so as to make it a circumcircle of the $\Delta\, uvw$. The center of the new circle is computed as the intersection point c' of the perpendicular bisector lines of uv and vw. The radius of the new circle is set as $|\overline{vc'}|$. Similarly, some of the neighbors may not ever get swapped by the rolling circle (neighbor x of w in Fig. 1c) due to smaller value of the rolling circle radius. In such a case, the circle is inflated up to the radius $|\overline{wx}|/2$ where x is the farthest neighbor of w. The new center of the circle passing through v and w having radius $|\overline{wx}|/2$ is computed.

Rolling: The forwarding node $u(x_u, y_u)$ determines the sweep angles for all its neighbors. The sweep angle for a neighbor $v(x_v, y_v)$ is the minimum angle by which the rolling circle $c(x_c, y_c)$ is to be rotated about the forwarding node u so that the node v is swept by the circle. The node having minimum sweep angle is chosen as the next boundary node. The new center of the rolling circle $c'(x_c', y_c')$ and sweep angle $\delta\theta_v$ are computed as:

$$x' = \frac{x_1 + x_2}{2} + \frac{y_2 - y_1}{|\overline{uv}|}\sqrt{|\overline{uc}|^2 - \left(\frac{|\overline{uv}|}{2}\right)^2} \tag{1}$$

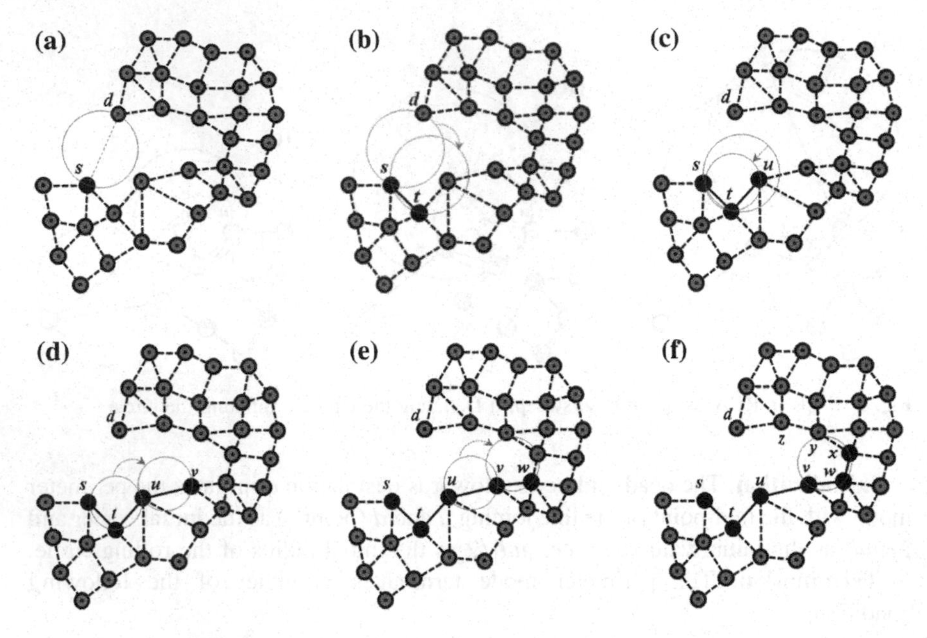

Fig. 2 Working of RCA

$$y' = \frac{y_1 + y_2}{2} - \frac{x_2 - x_1}{|\overline{uv}|}\sqrt{|\overline{uc}|^2 - \left(\frac{|\overline{uv}|}{2}\right)^2} \quad (2)$$

$$\theta_{v1} = \tan^{-1}\frac{y_1 - y}{x_1 - x}, \quad \theta_{v2} = \tan^{-1}\frac{y_1 - y'}{x_1 - x'} \quad \text{and} \quad \delta\theta_v = \theta_{v1} - \theta_{v2} \quad (3)$$

Example A complete example of RCA is demonstrated in Fig. 2.

Node *s* is a dead-end towards *d* and initiates the RCA with $|\overline{sd}|$ as the diameter of the rolling circle. The neighbor *t* has the minimum sweep angle and is selected as the next boundary node. The node *u* lies inside the circle, hence the circle is deflated so that it becomes the circumcircle of the triangle *stu*. The node *u* is selected as the next boundary node. The nodes *v*, *w* and *x* are subsequently selected as next boundary nodes. Node *x* finds a neighbor *y* that is closer to *d* than *s*, hence recovery mode is terminated and greedy mode resumes at *x*. The packet is now greedily forwarded to *d* through nodes *y* and *z*.

Algorithm: Rolling Circle Algorithm

Inputs: *Current node* **u**, *packet* **p**, *destination* **d**

Data Structures and Notations:

seq_no : Sequence number of the boundary node; (x, y) : Coordinates of the node, *getAngle(u,v)*: Anti-clockwise angle between the line *uv* and positive x-axis; *getCenter(u,v,w)* : Center of the circle passing through points *u*, *v* and *w*; $|\overline{uv}|$: Euclidean distance between u and v; $\mathcal{N}(u)$: Set of neighbors of u.

Fields encapsulated into the packet:

mode: routing mode (GREEDY or PERIMETER); *prev_node*: Previous node of the boundary; *init_node*: Node initiating the perimeter mode; *rc_center*: Center of the currently rolling circle; *seq_no*: Sequence no. of the previous boundary node.

	begin
	/* Terminate if network partition is detected*/
1.	$\textbf{\textit{if}}\ p.mode = \textbf{\textit{PERIMETER}} \wedge p.init\_node = u\ \textbf{\textit{then exit}}$
2.	$c \leftarrow \boldsymbol{p.rc\_center}$
3.	$i \leftarrow \boldsymbol{p.init\_node}$
	/*Resume greedy forwarding if feasible*/
4.	$U \leftarrow \{w: w \in \mathcal{N}(u) \wedge |\overline{wd}| < |\overline{\iota d}|\}$
5.	$\textbf{\textit{if}}\ U \neq \emptyset\ \textbf{\textit{then}}$
6.	$p.mode \leftarrow \textbf{\textit{GREEDY}}$
7.	$next\_hop \leftarrow w: w \in U \wedge |\overline{wd}| = \min_{v \in U} |\overline{vd}|$
8.	$\textbf{\textit{send}}(p, next\_hop)$
9.	$\textbf{\textit{else if}}\ p.mode = \textbf{\textit{GREEDY}}$
10.	$V \leftarrow \{w: w \in \mathcal{N}(u) \wedge |\overline{wd}| < |\overline{ud}|\}$
11.	$\textbf{\textit{if}}\ V \neq \emptyset\ \textbf{\textit{then}}$
12.	$next\_hop \leftarrow w: w \in V \wedge |\overline{wd}| = \min_{v \in V} |\overline{vd}|$
13.	$\textbf{\textit{send}}(p, next\_hop)$
14.	***else***
15.	$p.mode \leftarrow \textbf{\textit{PERIMETER}}$
16.	$p.init\_node = u$
17.	$p.rc\_center = d$
18.	$p.seq\_no = 0$
19.	***end if***
20.	***end if***
	/* Deflate the circle if required*/
21.	$\textbf{\textit{while}}\ \exists w\ (w \in adj(u) \wedge |\overline{wc}| < |\overline{uc}|)$
22.	$c \leftarrow \textbf{\textit{getCenter}}(u, p.prev\_node, |\overline{wc}|)$
23.	***end while***
	/* Inflate the circle if required*/
24.	$\textbf{\textit{while}}\ \exists w\ (w \in adj(u) \wedge |\overline{uc}| < 2|\overline{uw}|)$
25.	$c \leftarrow \textbf{\textit{getCenter}}\left(u, p.prev\_node, \frac{|\overline{uw}|}{2}\right)$
26.	***end while***
	/* Roll the circle to get next boundary node*/

27.	$\theta_1 \leftarrow \boldsymbol{getAngle}(u, c)$
28.	$l \leftarrow \sqrt{(u.x - c.x)^2 + (u.y - c.y)^2}$
29.	$\boldsymbol{for\ each}\ w \in adj(u)$
30.	$\quad c_w \leftarrow \boldsymbol{getCenter}(u, w, l)$
31.	$\quad \theta_2 \leftarrow \boldsymbol{getAngle}(u, c_w)$
32.	$\quad \boldsymbol{\delta\theta}_w \leftarrow \theta_1 - \theta_2$
33.	$\boldsymbol{end\ for}$
34.	$next\_hop \leftarrow w{:}\ \delta\theta_w = \min_{x \in U} \delta\theta_x$
35.	$p.rc\_center \leftarrow c_w$
36.	$p.prev\_node \leftarrow u$
37.	$u.seq\_no \leftarrow p.seq\_no$
38.	$p.seq\_no \leftarrow p.seq\_no + 1$
39.	$\boldsymbol{send}(p, next\_hop)$
	$\boldsymbol{end}$

5 Simulation Results

Simulation of RCA is carried out in OMNeT++ [32]. The parameters used in the simulation are mentioned in the Table 1.

The performance metric used in this study is *hop stretch factor* (HSF)—the ratio of the number of hops on the route between two nodes to the number of hops in the optimal path. The comparison of HSF achieved by GPSR [19], GAR [12] and Greedy-RCA combination under varying node densities in different compositions of greedily routable region (GRR) and dead-end regions is carried out in the simulation. RNG graph structure is utilized to implement planarization for GPSR. In simulation, each node sends one packet to the sink at a random instant of time. We take the average of the number of hops for each such end-to-end routing to

Table 1 Simulation parameters

Parameter	Value
Deployment terrain	1000 m × 1000 m
Number of nodes	Ranging from 100 to 1000
Node placement	Random uniform
Node density	Uniform distribution 1 per 900 m^2
Application	Constant packet rate (CPR) @ 1 pkt/s
Payload size	16 bytes
Data rate	250 Kbps
Radio range	60 m
Simulation rounds	50
Average node degree	4–30

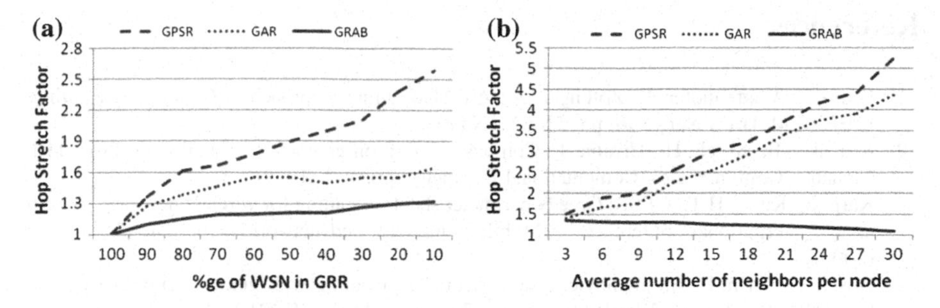

Fig. 3 **a** Effect of GRR composition on HSF. **b** Effect of node degrees on HSF

compute hop stretch factors. For the purpose of our simulation, we define node density in terms of average of the degree of all nodes in WSN.

In a void-free deployment case (Fig. 3a), the HSF achieved by each of GPSR, GAR and GRCA approaches 1 as all packets are routed using greedy mode only. GPSR shows a steep increase in HSF with decreasing portion of WSN in GRR. Though GAR and GRCA identify the same boundary nodes in an UDG model, the performance of GRCA gets better due to path pruning based on the sequence numbers during recovery. The effect of node densities on HSF is presented in Fig. 3b. In simulation 50% of WSN is taken as GRR and 50% as dead-end region. Interestingly the HSF for both GPSR and GAR increase with increase in density, because the number of hops in the shortest path decrease with density and number of faces in the planarized graph (in case of GPSR) and number of boundary nodes (in case of GAR) increase with node density. On the contrary, the performance of GRCA gets better with increase in node densities. This is due to the fact that GRCA utilizes sequence number of boundary nodes for GF during recovery.

6 Conclusion

The schemes used to recover from dead-end problems are generally inefficient when compared with GF. Recovery schemes based on the concept of alpha-shapes provide a localized and efficient way to route the stuck packets along the boundary of the void-region. A major drawback of existing alpha-shapes based schemes is the underlying assumption of UDG model of connectivity. By adapting the parameter alpha according to the local topology and geometric aspects, the rolling circle algorithm could efficiently detect the boundary nodes of WSN and without UDG assumption. Once the boundary nodes are identified, the dead-end situation is handled by routing packets along the boundary nodes until GF is feasible again. The pruning of the path along the boundary of the WSN further shortens the path during recovery. Simulation results have manifested the efficiency of RCA in reducing hop stretch factor under different simulation setups.

References

1. Kuhn, F., Wattenhofer, R, Zollinger, A.: An Algorithmic Approach to Geographic Routing. IEEE/ACM Trans Netw. 16, pp. 51–62. (2008).
2. Kranakis, E., Singh, H., Urrutia, J.: Compass routing on geometric networks. In: Proc. 11th Canadian Conference on Computational Geometry. pp. 1–4 (1999).
3. Karp, B., Kung, H.T.: GPSR: Greedy perimeter stateless routing for wireless networks. In: 6th annual international conference on Mobile computing and networking pp. 243–254. ACM (2000).
4. Bose, P., Morin, P., Stojmenović, I. and Urrutia, J.: Routing with guaranteed delivery in ad hoc wireless networks. Wireless networks. 7(6), pp. 609–616 (2001).
5. Kuhn, F., Wattenhofer, R. and Zollinger, A.: Asymptotically optimal geometric mobile ad-hoc routing. In: 6th international workshop on Discrete algorithms and methods for mobile computing and communications. pp. 24–33. ACM (2002).
6. Liu, W.J. and Feng, K.T.: Greedy routing with anti-void traversal for wireless sensor networks. IEEE Transactions on Mobile Computing, 8(7), pp. 910–922 (2009).
7. Rührup, S. and Stojmenovi, I.: Contention-based georouting with guaranteed delivery, minimal communication overhead, and shorter paths in wireless sensor networks. In: IEEE Int. Symp. on Parallel & Distributed Processing (IPDPS), pp. 1–9. IEEE (2010).
8. Chen, D. and Varshney, P.K.: A survey of void handling techniques for geographic routing in wireless networks. IEEE Communications Surveys & Tutorials, 9(1), pp. 50–67 (2007).
9. Stojmenovic, I. and Lin, X.: Loop-free hybrid single-path/flooding routing algorithms with guaranteed delivery for wireless networks. IEEE Transactions on Parallel and Distributed Systems, 12(10), pp. 1023–1032 (2001).
10. Chen, D., Deng, J. and Varshney, P.K.: On the forwarding area of contention-based geographic forwarding for ad hoc and sensor networks. In: SECON (Sept-2005) pp. 130–141 (2005).
11. Kim, Y.J., Govindan, R., Karp, B. and Shenker, S.: Geographic routing made practical. In: 2nd conference on Symposium on Networked Systems Design & Implementation-Volume 2, (May-2005) pp. 217–230. USENIX Association (2005).
12. Radhakrishnan, S., Racherla, G., Sekharan, C.N., Rao, N.S. and Batsell, S.G.: DST-A routing protocol for ad hoc networks using distributed spanning trees. In: Wireless Communications and Networking Conference (WCNC-1999), pp. 1543–1547. IEEE. (1999).
13. Leong, B., Liskov, B. and Morris, R.: Geographic Routing Without Planarization. In: NSDI (May-2006), Vol. 6, pp. 25–25 (2006).
14. Edelsbrunner, H., Kirkpatrick, D. and Seidel, R.: On the shape of a set of points in the plane. IEEE Transactions on information theory, 29(4), pp. 551–559 (1983).
15. Bose, P., Devroye, L., Evans, W. and Kirkpatrick, D.: On the spanning ratio of gabriel graphs and β-skeletons. In: Latin American Symposium on Theoretical Informatics (April-2002), pp. 479–493, Springer, Berlin Heidelberg (2002).
16. Fang, Q., Gao, J. and Guibas, L.J.: Locating and bypassing holes in sensor networks. Mobile networks and Applications, 11(2), pp. 187–200 (2006).
17. Ma, X., Sun, M.T., Zhao, G. and Liu, X.: An efficient path pruning algorithm for geographical routing in wireless networks. IEEE Transactions on Vehicular Technology, 57(4), pp. 2474–2488 (2008).
18. Fayed, M. and Mouftah, H.T.: Localised alpha-shape computations for boundary recognition in sensor networks. Ad Hoc Networks, 7(6), pp. 1259–1269 (2009).
19. Rührup, S. and Stojmenovi, I.: Contention-based georouting with guaranteed delivery, minimal communication overhead, and shorter paths in wireless sensor networks. In: IEEE Int. Symp. on Parallel & Distributed Processing (IPDPS-2010), pp. 1–9. IEEE (2010).

Improving Shared Cache Performance Using Variation of Bit Set Insertion Policy

Dilbag Singh and Dupinder Kaur

Abstract To satisfy the need for ever increasing demand of processing speed, arises need of significant change in the processor design from single to multi-core. Multi-core architecture design allows the increase in computing power of chips without increasing their complexity. Traditional processors employ a dedicated cache, the introduction of multi-core platform considers the shared use of cache memory. To increase the hit ratio an effective cache replacement policy needs to be developed. In this paper, a new scheme for block substitution in cache is proposed namely: Bit Set Insertion in random manner. The research methodology used in this research is experimental. The performance of the proposed algorithm is computed on INTEL Core Duo processor. Hit Ratio is computed by keeping the access patterns of string constant and compared with previously known policy for thrashing pattern in shared cache in multi-core environment.

Keywords Shared cache · Multi-core · Replacement · Thrashing

1 Introduction

For moving from single to multi-core processors system architects have several options of organizing the system but the best possible one is the cache performance improvement. In some organization architecture choose to keep a dedicated cache

D. Singh (✉) · D. Kaur
Department of Computer Science and Applications,
Chaudhary Devi Lal University, Sirsa, HR, India
e-mail: dscdlu7@gmail.com

D. Kaur
e-mail: dupinder.pahwa@gmail.com

S.C. Satapathy et al. (eds.), *Computer Communication, Networking and Internet Security*, Lecture Notes in Networks and Systems 5,
DOI 10.1007/978-981-10-3226-4_48

for each core while other explores the shared last level cache among different cores for better resource utilization and performance [1]. INTEL CORE Duo processor is the first processor which introduce the concept of sharing of cache into multi-core environment using its INTEL Smart Cache. This architecture allows both cores to have access to entire 2 MB last level cache to have in order to reduce resource underutilization. This fetch logic improves the system efficiency by prefetching data in L2 cache even before cache request is made [2].

In order to make best utilization of shared cache for the overall system performance, the page replacement policy must be effective. Page Replacement Policy decides the swapping out of a cache block and its replacement in order to fulfill the requirement of incoming request [3]. A replacement policies with fixed size cache is required to increase the overall efficiency by increasing hit rate. Thus the objective of this paper is to suggest a variation of Bit Set Insertion Policy with Random implementation for shared cache architecture by assigning a status bit k with each data block. In case the requested block is already present within cache then its status bit is set to 1 and a hit is increased by one. In case of cache miss, the cache block having $k = 0$ is searched if it exists then the page is placed at that location and K is set to 1. In case if all the cache blocks are already set to 1 then 50% of the blocks are vacated on the random basis.

2 Objectives of the Study

Poor page replacement policy can have adverse effect on the system performance. So, an efficient page replacement policy is needed for improving the cache utilization. Keeping this aspect in mind the present has been carried by taking the following objectives:

- To study the performance measures of the multi core processors.
- To propose the method to increase the performance of cache.
- To compare the hit ratio of the new policy with the LRU.

3 Bit Set Insertion Policy (BSIP) with Random Implementation

In this paper hit ratio is computed on dual core processor system by introducing a variation of Bit Set Insertion Policy with Random Implementation. The main concern of this policy is to make hit by setting the status bit of the cache line to ensure its usability for future. This policy overcomes the drawback of old policy BSIP with LRU implementation.

In BSIP with Random implementation an additional status bit K with each cache line having default value 0 is used. In case of a cache Hit, the bit K for that cache line is set to 1 which implies that this line may again get hit in future. To illustrate this concept initially cache is considered with A, B, C, T, S tag having k = 0 for each. Three cases are considered for a new data request:

Case 1: If a request is made for data existing with cache, its k bit is set to 1 and it is considered to be hit (Table 1).

Case 2: If a request for the data not available with cache, then it is replaced with first block having k = 0 by traversing the cache blocks and its bit k is set to 1. If the cache is full and all k = 1 then 50% of the cache blocks are vacated and new data is placed at the first vacant block (Table 2).

Table 1 Illustration of BSIP hit

Tag Bits	Status Bit(K)
A	0
B	0
C	0
T	0
S	0

BSIP HIT →

Tag Bits	Status Bit(K)
A	0
B	1
C	0
T	0
S	0

Table 2 Illustration of BSIP Miss

Tag Bits	Status Bit(K)
A	1
B	0
C	1
T	1
S	0

BSIP MISS →

Tag Bits	Status Bit(K)
A	1
D	1
C	1
T	1
S	0

4 Proposed Algorithm

The aim of this algorithm is to make efficient utilization of the cache in case of thrashing access pattern. The proposed algorithm for BSIP with Random policy is as under: In this algorithm, the policy is designed on the basis on presence and absence of data in the cache block. For presence K value is set to 1 and for absence it is 0.

In case of occurrence of miss the allocation of block is made by two different ways. Firstly, if for any of the status bit k = 0 then the required block is placed at that location. Secondly if k = 1 for all cache blocks then 50% of the cache is vacated by calling a subroutine generating the block number on random basis. After vacating the blocks, the current data is placed at first location having k = 0 and status bit for that particular block is to 1.

Algorithm: Bit Set Insertion Policy with Random implementation

```
1. Tag← random request of new cache block
2. k← Status bit for each cache line
3. cache=0
4. flag=0
5. Hit= 0
6. while cache < total_cache_length do
7. if tag = = cache_tag then
8. Set k for that cache line
9. flag=1 , Hit=Hit+1 and return the cache
10. break
11. end if
12. cache = cache + 1
13. end while
14. else if (flag= =0) then
15. cache = 0
16. while cache < total_cache_length do
17. if (k = =1) then
18. cache = cache + 1
19. end if
20. use a random function to make 50% of cache empty.
21. goto 24
22. break
23. if (k= =0)
24. Insert the cache block at that line. Set k=1,c=1 and then return the cache
25. end if
26. end if
27. end while
```

5 Output Evaluation

A simulator has been prepared for analyzing the performance of the proposed technique by setting-up the hypothetical conditions. For simulating the problem INTEL Core Duo processor is assumed by placing the default data in shared cache as A, B, C, D and initially their status bit k is set to 0.

The new string access pattern is remained to be same for two policies as shown above.

The Figs. 1 and 2 shows the snapshot of the hit ratio for the Bit set insertion policy with LRU. As illustrated above hit ratio in BSIP with LRU is 0.375.

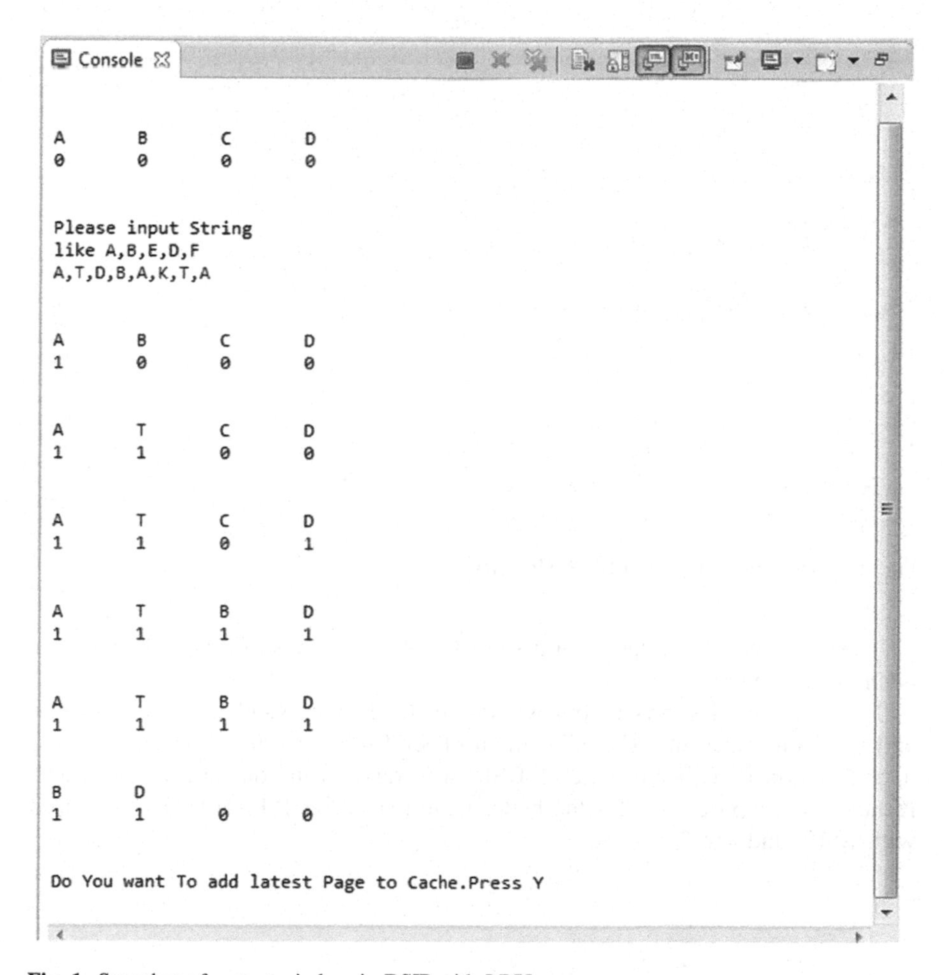

Fig. 1 Snapshot of output window in BSIP with LRU

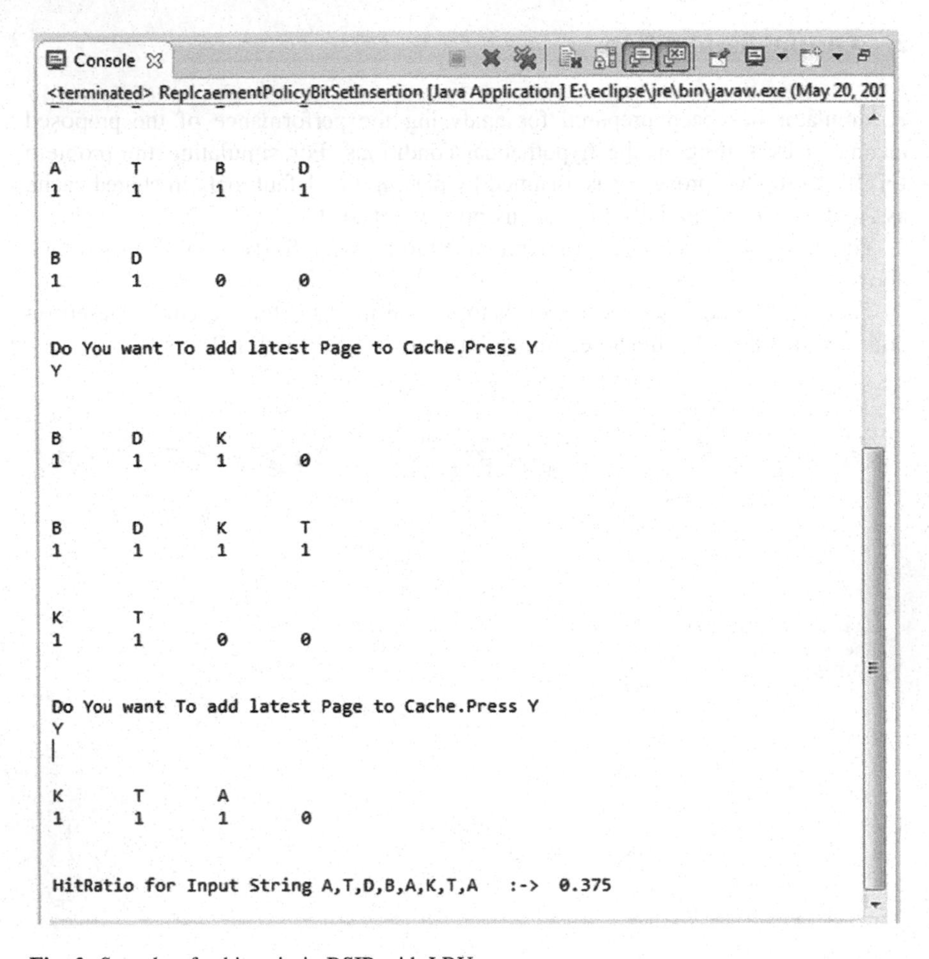

Fig. 2 Snapshot for hit ratio in BSIP with LRU

The access pattern of the string is same for BSIP with random and output for this is shown as under:

Figures 3 and 4 shows output window of BSIP with Random implementation and hit ration comes out to be 0.5 while for BSIP LRU it was 0.37. Hence, it is clear that hit ration is higher in case of BSIP with random in comparison with LRU. Hence, it is proposed that for the better performance BSIP LRU may be replaced with BSIP random.

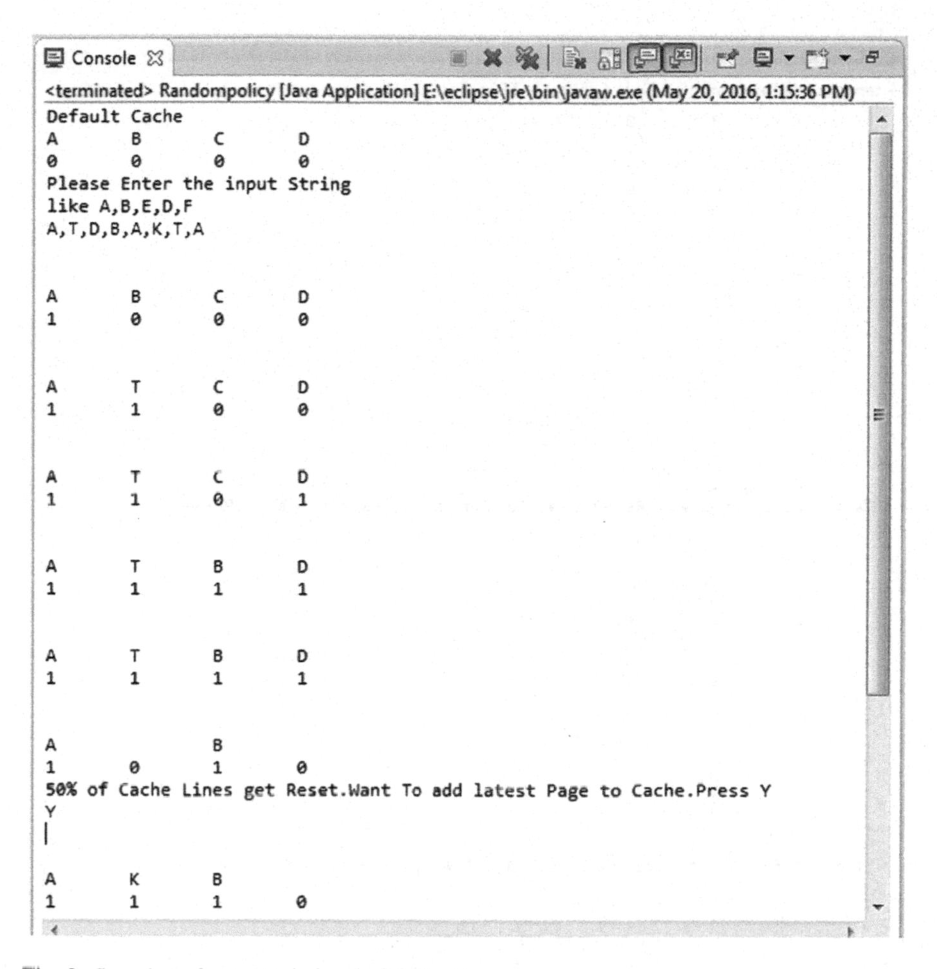

Fig. 3 Snapshot of output window in BSIP with random

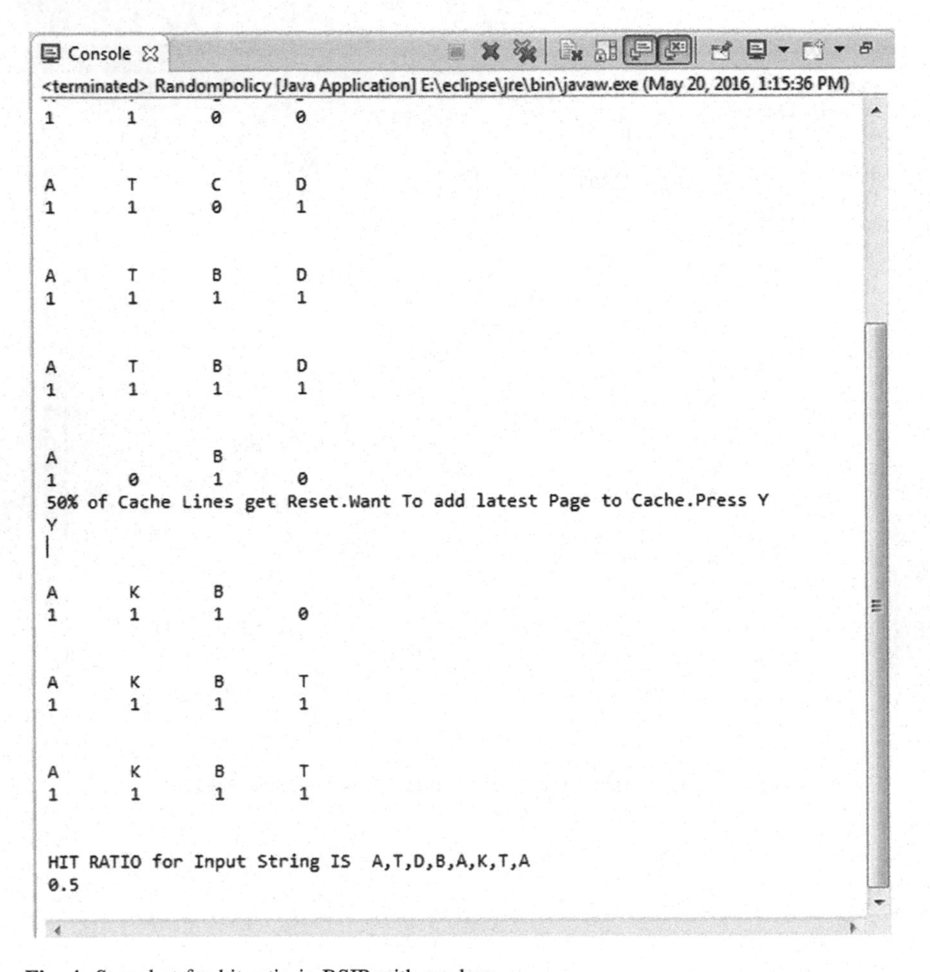

Fig. 4 Snapshot for hit ratio in BSIP with random

Table 3 Hit ratio comparison for BSIP with LRU and random policy

Access patters of string	Hit ratio of BSIP with LRU policy	Hit ratio of BSIP with random policy
A, K, L, T, B, Y, N, M, R, A	0.1	0.2
A, B, D, C, E, A, B, T, H	0.4	0.5
A, T, D, B, A, K, T, A	0.325	0.5
R, B, C, T, A, D, T, B, R	0.33	0.4
A, B, T, R, C, A, K, T, D	0.2	0.33
A, B, D, A, C, T, H, T	0.75	0.75
A, B, D, H, C, A, H, B	0.5	0.625
A, B, H, T, D, C, B, A, H, T	0.2	0.3

6 Conclusion

In this paper a new cache block replacement technique namely: BSIP with random implementation is proposed. This technique states that if requested block is already present in cache then its status bit is set to 1 and hit is increased by one. In case of cache miss, the cache block having k = 0 is searched if it exists then the page is placed at that location and K is set to 1. In case if all the cache blocks are already set to 1 then 50% of the blocks are vacated on the random basis.

The process described above resulted in improved the efficiency and processing speed of multi core system as it increases the hit ratio. In this paper comparison is also made between BSIP with LRU and BSIP with Random by carrying out execution on INTEL Core DUO processor. Hit ratio is calculated for both policies by keeping access patterns of string constant and following measurements are made.

Table 3 shows the output of the simulator for eight different access patterns for eight strings for both the policies. First string is taken as A, K, L, T, B, Y, N, M, R, A and hit ratio comes out to be 0.1 for BSIP with LRU whereas it is 0.2 for BSIP with random policy. For second string A, B, D, C, E, A, B, T, H the hit ratio is found 0.4 for BSIP LRU while for BSIP with random is 0.5. Hit ratio for string A, T, D, B, A, K, T, A in BSIP LRU is 0.325 and for BSIP with random it is 0.5. For string A, B, T, R, C, A, K, T, D hit ratio for BSIP LRU is 0.2 and for BSIP with random is 0.33. For A, B, D, A, C, T, H, T the hit ratio for BSIP LRU and BSIP with random comes out 0.75 and is same for both. For string A, B, D, H, C, A, H, B hit ratio for BSIP LRU is 0.5 while for BSIP with random is 0.625. Hit ratio for string A, B, H, T, D, C, B, A, H, T for BSIP LRU is found as 0.2 but for BSIP with random is 0.3.

The above output of the simulator shows that BSIP with Random implementation gives higher Hit rate as compared to BSIP with LRU. The greater hit Ration means the better efficiency of the techniques. Hence, the BSIP-Random implementation is the better approach in comparison with the BSIP-LRU. Hence it is recommended that the designer must use BSIP-Random approach for page replacement so as to enhance the throughput of the system.

References

1. Tian Tian, “Software Techniques for shared-cache Multi-core systems” https://software.intel.com/enus/articles/software-techniques-for-shared-cache-multi-core-systems, (2007).
2. Tian Tian, “Effective Use of the Shared Cache in Multi-core architectures” http://www.drdobbs.com/parallel/effective-use-of-the-shared-cache-in-mul/196902836, (2012).
3. Khushboo Patidar, Dr. Richa Gupta.: A Taxonomy of Cache Replacement Algorithms. International Journal of New Technologies in Science and Engineering, vol. 2, Issue 3, (2015).

Phrase Based Web Document Clustering: An Indexing Approach

Amit Prakash Singh, Shalini Srivastava and Sanjib Kumar Sahu

Abstract Clustering documents within a cluster are mostly based on single term analysis over present document set and naming one of them can vector space model. In view to get improved results, more explanatory features are to be included. These informative features could be like phrases and the weights they hold for a specific document. Importance of document clustering can be explained with the need of categorization of documents, making sets of search engines results, building taxonomy of documents, and many more. This paper is going to present few such important pats of document clustering. Two algorithms famous in this field will be discussed and an analytical study of the results presented by famous search engine and the tool based on this algorithm will be talked about. The model will show efficient and less relevant results of a search engine and an improved document similarity.

Keywords Web mining · Document similarity · Phrase-based indexing · Document clustering · Document structure · Document Index Graph · Phrase matching

A.P. Singh (✉)
University School of Information and Communication Technology,
Guru Gobind Singh Indraprastha University, Dwarka, Delhi, India
e-mail: amit@ipu.ac.in

S. Srivastava
Integrated Institute of Technology, Dwarka, Delhi, India
e-mail: shln.srivastava@gmail.com

S.K. Sahu
Department of Computer Science and Application, Utkal University,
Bhubaneshwar, Odisha, India
e-mail: sahu_sanjib@rediffmail.com

S.C. Satapathy et al. (eds.), *Computer Communication, Networking and Internet Security*, Lecture Notes in Networks and Systems 5,
DOI 10.1007/978-981-10-3226-4_49

1 Introduction

There have been many researches going on in struggle to keep up with the ever so increasing data over the World Wide Web. These researches have been focusing on finding the ways to organize this vast information as into deal with them in an easy way. Accuracy of the results along with the efficiency of these search engines is what we desire. Web pages over the internet are present in a format given by HTML which leads us to go for text data mining techniques. For mining information from this formatted text different data mining techniques can be used and they help us for finding the meaning stored within the text. Clustering use these mining results for identifying the documents to group them into intra cluster or inter cluster similarity [1]. Web mining is same as text mining when applied over a web page.

When we say Web mining, there comes the categorization of these techniques given by Kosala and Blockeel [2] segregated the web mining into three parts based on:

(1) Structure, (2) Usage, and (3) Content. This paper will be dealing with the web content mining. The need of document clustering comes from application like clusters of documents retrieved presenting structured and comprehensible results to the user (e.g., [3]), collecting documents (e.g., digital libraries) and the most famous IR from set of clusters.

Clustering technique focuses on:

1. Representing data in a model,
2. Generating a similarity measure,
3. Designing a cluster model, and
4. An algorithm for clustering web pages using above data model and the similarity measure.

Going further with this, we all know that Vector Space Model works with web pages as vector of the terms (words) appearing in all the web page sets. It weights (usually term-frequencies) them based on the frequency of their occurrence.

As said by Hammoud and Kamel [1] in their work, clustering should not be just based on single term but on phrases. Phrase based means that the rather that finding similarity between documents based on single term frequencies; we should rather use matching phrases. This phrase based clustering is yet to researched with more efforts still it may provide efficient results in comparison to single term based similarity. Work done by Zamir et al. [4–6] has proposed Suffix Tree Clustering (STC) for phrase-based document clustering. This includes the use of a "**tree**" (a Compact tree) structure which carries common suffixes between documents. These suffixes are the base for clustering documents within a same cluster using a connected-component graph algorithm.

According to the authors, it takes nlogn performance and provides quality clusters. In this paper, we propose a system for Web clustering based on a part of STC and the other of DIG (Document Index Graph) as the base. Both of these algorithms are phrase based and work together to overcome each other's drawback.

System works as

1. Restructuring of web pages in different levels of importance a part holds.
2. Then clusters are built using a phrase based indexing model called as Document Index Graph (DIG) in view to capture individual sentences in a web page, in place of acting over single terms. It implements the cluster level of programming using graph theory and uses graph's property for matching any- length phrase in a web page to those present in previous web pages.
3. Then a similarity measure is used for assigning scores of similarity in between web pages.
4. Finally web pages will be clustered.

Incorporating these components has shown better results comparative to search engine we chose. The overall system design is illustrated in Fig. 1.

As illustrated in the figure above, DIG has given us the platform for creating clusters based on a group of significant terms known as phrase. This model is indexed based which calculates the match between pages using a new similarity methodology which uses phrases too. This method is a club between the single term based similarity and phrase based similarity. Similarity, which is given by this measure, has proven to be providing more effective clustering quality as it is very less prompt to useless terms for giving poor results.

The proposed web page clustering efforts for improving the quality of cluster based on pair-wise web page matching within each cluster for providing a maximized similarity in every cluster. It has been observed that quality of clusters obtained by this method is improved over previously used clustering techniques. The improvement is over 25% from other used techniques.

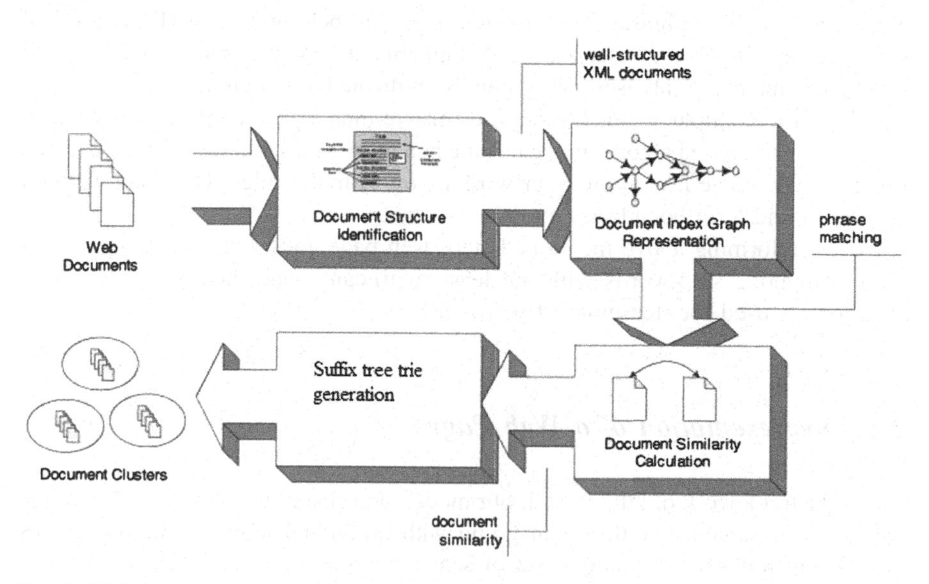

Fig. 1 Web document clustering system design [1]

The rest of this paper is organized as follows: Sect. 2 provides an analysis along with presentation of significant features gives view of presents and analysis of the important features of these structured Web documents [7]. Section 3 introduces the Document Index Graph model. Section 4 presents the phrase based similarity measure. Section 5 discusses the experimental results. Finally, we conclude and discuss future work in the last section.

2 Structure of a Webpage

Web pages are known for their semi structured format. This states that the HTML tags are used to designate different parts of the page. Since this language specifies the framework of the page and is very user friendly and this feature of its makes it standout from other languages for designing web documents. Even after being user friendly and easy to understand, yet we can easily differentiate between the key parts within the page using the HTML tags. Finding more informative parts within the page and assigning them a level of significance is what this tool will be using. Different levels of significance will be based on the appearance of the terms or phrases.

The proposed system will be analyzing HTML page and then formalizes it according to the need of DIG based algorithm. There will be three levels of significance as; HIGH, MEDIUM, and LOW used by Hammoud and Kamel in their work [1].

Structuring the web page according to this concept gives us a method for finding similarity between two pages. As an explanation to this we can go with the scenario where the matching phrases between two pages are belonging to a HIGH level of significance will give greater value of similarity as compared to LOW level of significant matching phrases. This can be indicated by arguing with match in headings, for example, is much more informative than a phrase match in body text.

The indexed based algorithm locates the boundary of a sentence with the use of a finite state machine lexical analyzer working on heuristic rules. This same method is used to find term boundaries too.

After performing above mentioned steps web page goes with a cleaning step in view to remove stop-words with the least significance and then Porter Stemmer algorithm is used for stemming of words [8].

2.1 Representation of a Web Page

Using the framework of DIG model, our model too is using the representation of the web page as sentences rather than going with individual terms of the page. This model constitutes a web page as set of sentences consisting for set of terms.

Going with the working of Vector Space Model, we will represent a page as vector of sentences:

$$D_i = \{s_{ij} : j = 1, \ldots p_i\},$$
$$S_{ij} = \{t_{ijk} : k = 1, \ldots, l_{ij}, w_{ij}\}$$

where

d_i is document i,
s_{ij} is sentence j in document i,
p_i is the number of sentences in document i,
t_{ijk} is term k of sentence s_{ij},
l_{ij} is the length of sentence s_{ij}, and
w_{ij} is the weight associated with sentence $s_{ij.}$

This above representation maps the original web page into a more formal layout by breaking it into set of sentences. Then specific weights are given to each based on their significance level. Here we ignore the count of sentences occurrence. We will consider this frequency of sentences while going for matching phrases.

3 Document Index Graph

Because of one of the major drawbacks of vector space model i.e. ignoring the relationship between the terms of a sentence has pushed this technique a little too far in the queue. Even VSD ignore the structure of a sentence too which does not satisfy the prime requirements of the current search engines need. On the other hand, our system is using the hybrid of DIG and STC which are the most famous phrase based techniques. The first one indexes the pages and also keeps the structure of the sentence safe which overcomes the drawback of VSD. It maintains level of significance of actual sentence as well (Fig. 2).

3.1 Structure of the Web Page Based on DIG

DIG is a digraph with

$$G = (V,\ E)$$

in which

V: Collection of nodes $\{v1, v2, \ldots, vn\}$, representing unique terms within a web page; and

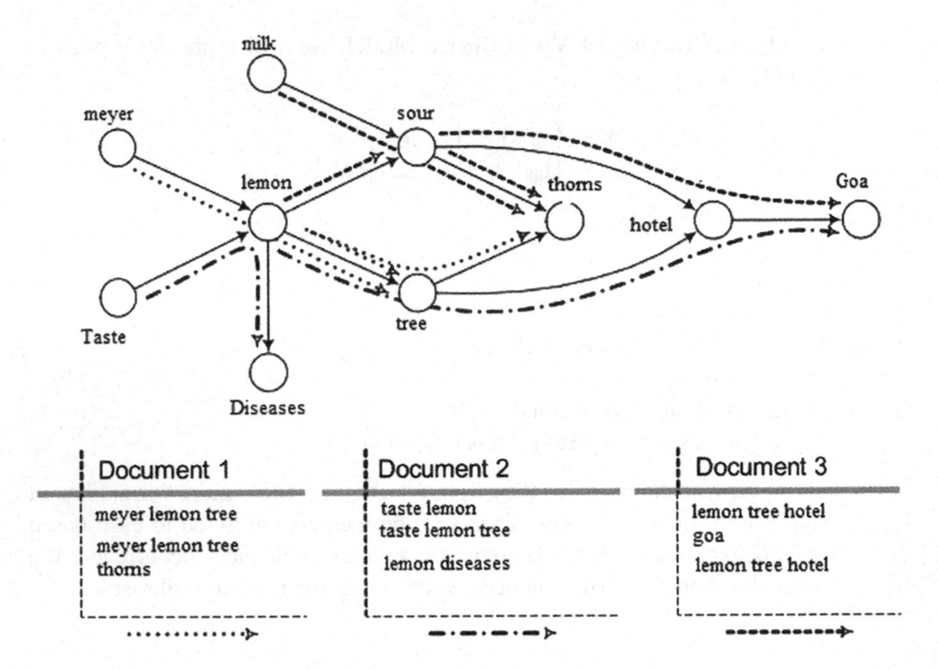

Fig. 2 Layout of Document Index Graph

E: Collection of *edges* $\{e1, e2, \ldots, em\}$, representing links between two consecutive terms in an ordered pair (vi, vj).

Our digraph will have an edge from one node to another only if the word a_i appears consecutively with the word a_j. Number of nodes present in the graph will be the exact count of unique words within the page. Thus it overcomes the drawback of STC by removing redundant data. Each node within the graph carries information regarding pages these terms or words has appeared in and maintains path information too. Edges between each node and in whole graph are responsible for maintaining sentence structure. Thus DIG maintains an inverted list of pages it has considered so far. The list is inverted as sentence boundaries do affect the direction of graph.

To justify this let's assume that a phrase of say k words of a page have a sequence of words as $\{a_1, a_2, a_3 \ldots, a_m\}$. A phrase or sentence is represented by the path from a_1 to a_m as $(a_1, a_2), (a_2, a_3), \ldots (a_{m-1}, a_m)$ edges in the graph.

In place of following the method of DIG for storing information in each vertex, we will be using method of STC for doing that. Sentences are uniquely identified through this information. If there are sentences who are sharing phrases, there will be shared vertices between them showing similarity of words. A table is maintained containing number of pages. With every entry, it stores the frequency of occurrence of a word within that document. While storing these words, it stores them based on level of significance of their occurrence within the page (Fig. 3).

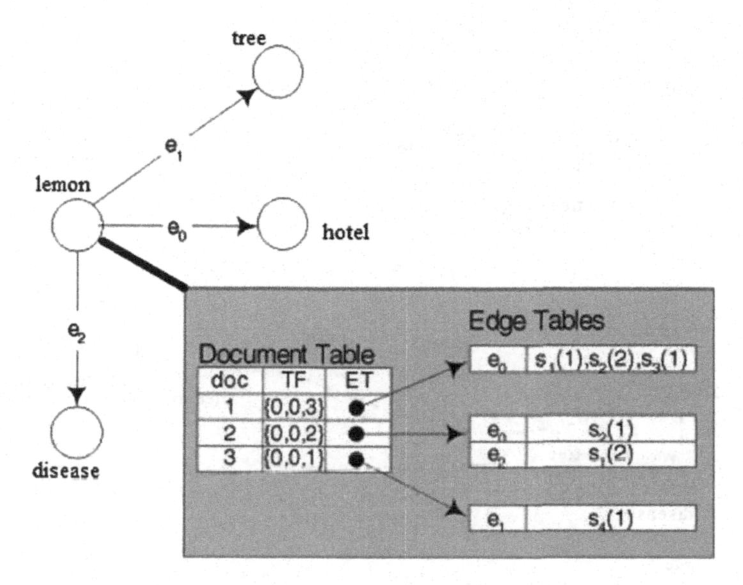

Fig. 3 Example of structure designed of storing the words

This table is useful for calculating actual similarity along with level of significance. As it's a graph that is actually storing all these words, it also stores the number of incoming or outgoing edges too. Storing information in this manner tells us the path within the sentence. This figure illustrates the structure we got from our design.

To better illustrate the graph structure, Fig. 4 presents an example.

This graph shows how many sentences are there in the page with overlapping between them along with the successive words coming in the page are put into the graph as nodes attached with edges. Thus, it is mapping the paths within a sentence. Dotted lines are symbolizing phrases from page 1, dash-dotted lines are showing phrases belonging to page 2, and dashed lines are there to show sentences page 3. With every occurrence of any phrase within same page, frequency counter of words within it are increased accordingly. In real world storing very unique word will the major issue we need to work on.

3.2 Graph Construction

This is an incremental approach. Whenever a new page enters into the corpus, it is looked in a sequential manner and necessary updates are done within the graph. With every new coming term, nodes and respective changes to edges are done. If there is no fresh entry done within the table, our storage requirement gets satisfied. Graph enters stability with such situation when such pages are scanned from the

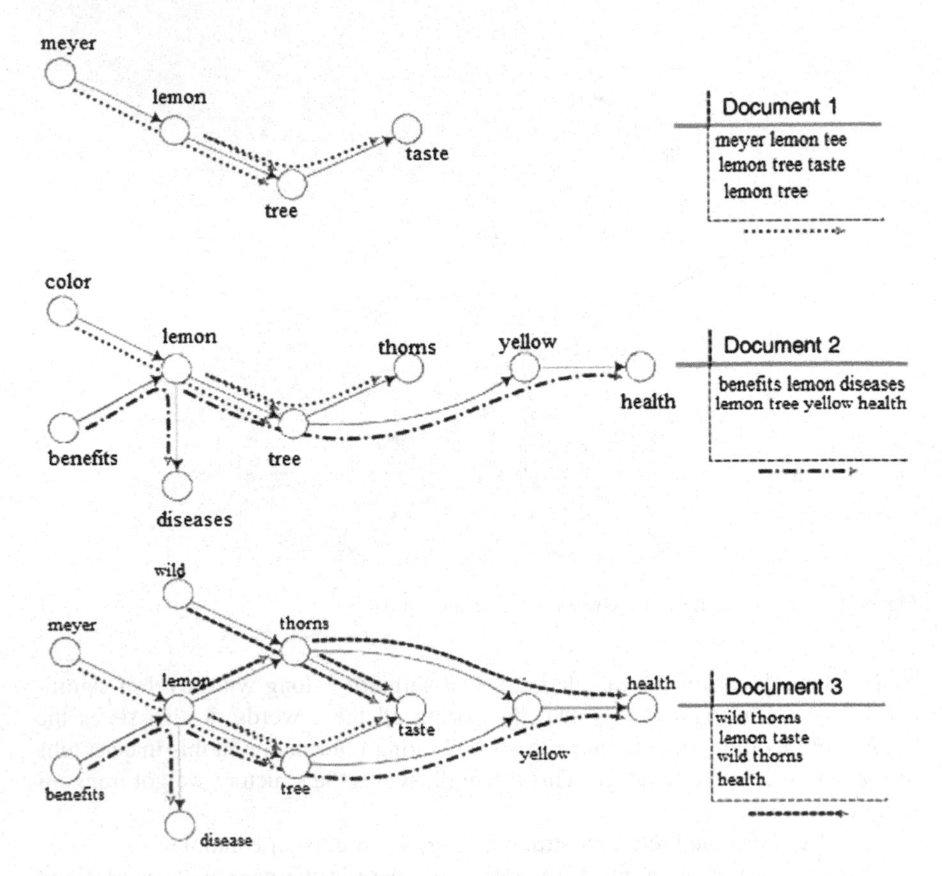

Fig. 4 Incremental construction of DIG

database completely on a specific search keyword. The operation we are left with after this is of updating the phrase structure in view to house new sentences introduced. This situation creates a critical situation because this needs to inspect a new page introduced for finding those terms which were already present in the graph, in place of going with each node. DIG does this efficiently. Going with our previous example, it illustrates three pages (Fig. 4).

In this extended example an incremental construction procedure is followed. In this, new nodes are added along with new edges incrementally. DIG is responsible for providing complete information about every sentence between documents and thus supports an efficient information retrieval. In place of providing aimed information for the search keyword entered, it goes for providing information about the degree of overlap among a pair of pages, which further gives us degree of similarity between them.

3.3 *Finding Matching Sentences*

Our tool finds matching phrases using DIG only. Its algorithm follows an incremental approach for creating graph. It starts with individual page from the database. It processes every sentence with its boundary from individual page in line 1. After processing each page, it looks for matching phrases with a confined context. It stores the length of sentences too which further be used for similarity calculation. It repeats all those steps we talked about in paper before for process terms within the web page. For every new sentence, its first word is stored with a hash table consultation. If the term is already there, then we start from that node of the graph or else add a new node and start over.

A list L is maintained which contains entry for every previous page sharing a phrase with the current page P. Page identifiers are maintained for every page tool has countered with and the list L is updated with extending phrases. And if nothing as such appears then, it just updates the respective vertex of graph reflecting new phrase path. As soon as the whole page is dealt with, L will have listed matching phrases between (a_{m-1}, a_m). This L will be given as output with list of matching phrases and everything necessary about these phrases. This algorithm takes in roughly $O(m)$ time, where m is the number of words in page P.

4 Calculating Similarity

Context similarity leads to document similarity and this phrase based document clustering actually provides results based on this. The similarity measure used by Hammoud and Kamel in their work calculates similarity for matching phrases rather than individual term. It's an advanced version of general similarity measure we usually use. This method is somewhat near to the work of Isaacs et al. [9] created a pair-wise probabilistic web page matching measure based on Information Theory. This method was based on VSD. This measures the similarity as a function of four components:

- Count of matching phrases in document S.
- Length of matching phrases $(l_i : i = 1, 2 \ldots S)$.
- Number of matching phrases in matching documents $(n_{i1} \text{ and } n_{i2} : i = 1, 2, \ldots S)$, and
- Significance levels (*weight*) of matching phrases in two matching web pages $(wt_{i1} \text{ and } wt_{i2} : i = 1, 2 \ldots S)$.

Hammoud and kamel [1] gave us with the following similarity measure and we have used this only for calculating similarity between the pages.

$$\mathrm{Sim}(s_1, s_2) = \frac{\sqrt{\sum_{i=1}^{P} [g(li), \{f_{i1} w_{i1} + f_{i2} w_{i2}\}]^2}}{\sum_j |8_{i1}| \cdot w_{i1} + \sum_k |8_{k2}| \cdot w_{i2}} \quad (1)$$

This technique has produced best results. Normalizing of two web pages is to be done through above Eq. (1) is a mandatory step as to enable comparison between similarities between documents.

5 Experimental Results

To test our hypothesis, we created a tool based on DIG and somewhat STC and then calculated similarity between the pages from our database. The experimental setup consists of first 50 links given by one of the famous search engine Yahoo.com on a specific search keyword **"Lemon Tree"**. These were the results given by Yahoo.com for the search keyword as "Lemon Tree". Out tool has given nil web pages as result because none of these 50 urls were actually giving the context as **"Lemon Tree"**.

They were just the random results focusing on LemonTree Hotel. In order to provide proof of our tool efficiency, we have calculated F-measure that is made up of the *Precision* and *Recall* concepts of Information Retrieval literature. Cluster j's precision and recall with respect to any class *i* are defined as:

$$P = \mathrm{Precision}(i, j) = \frac{N_{ij}}{N_i} \quad (2)$$

$$R = \mathrm{Recall}(i, j) = \frac{N_{ij}}{N_i} \quad (3)$$

We have taken **Entropy** as the second measure, which assigns a value of **"goodness"** a cluster is composed of. It gives the homogeneity of a cluster and thus higher the value it has the more homogeneous the cluster is and vice versa. Cluster **j's** entropy is calculated using following formula:

$$E_c = \sum_{J=1}^{m} (N_j * E_j) \quad (4)$$

The idea is to maximize the value of F-measure and minimize value of entropy in order to achieve high quality clusters. Table 1 shows an improvement in clustering because of our tool. It has achieved a blend factor value in between 70 and 80% (Fig. 5).

Table 1 Phrase based clustering improvement

	F-measure	Entropy
Rank algorithm	0.904	0.103
DIG	0.919	0.141

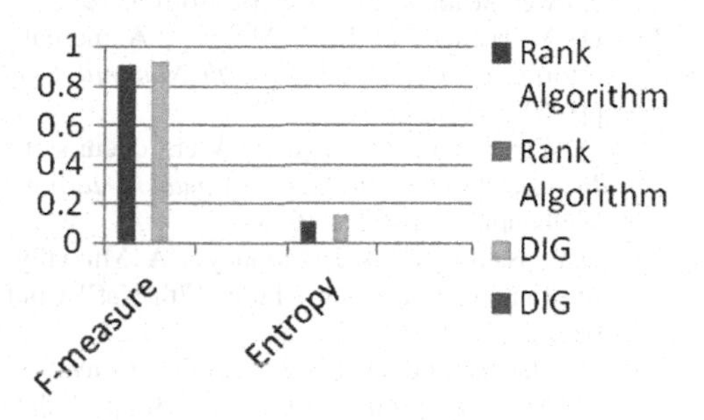

Fig. 5 DIG performance

6 Conclusion and Future Research

In this system the main idea is of using phrase based clustering for enhancing results of search engines used now days. This can be a great achievement if the storage space criteria can be handled with better care. Considering context of every sentence has given us with far more relevant results as compared to current rank based algorithms on which most of the real time search engines are based. If this can be done with other languages too then it's a major scope of our paper.

References

1. K.M. Hammouda and M.S. Kamel, "Efficient Phrase-Based Document Indexing for Web Document Clustering," IEEE Trans.Knowledge and Data Eng., vol. 16, no. 10, pp. 1279–1296, Oct 2004.
2. A. K. Jain and R. C. Dubes. *Algorithms for Clustering Data*. Prentice Hall, Englewood Cliffs, N.J., 1988.
3. Hung Chim and Xiaotie Deng, "Efficient Phrase-Based Document Similarity for Clustering" IEEE Trans. Knowledge and Data Eng., Vol. 20, No. 9, pp. 1217–1228, Sep 2008.
4. O. Zamir, O. Etzioni, O. Madanim, and R. M. Karp. Fast and intuitive clustering of web documents. In *Proceedings of the 3rd International Conference on Knowledge Discovery and Data Mining*, pages 287–290, Newport Beach, CA, August 1997. AAAI.
5. O. Zamir, O. Etzioni, O. Madanim, and R.M. Karp, "Fast and Intuitive Clustering of Web Documents," Proc. Third Int'l Conf. Knowledge Discovery and Data Mining, pp. 287–290, Aug. 1997.
6. R. Kosala and H. Blockeel, "Web Mining Research: A Survey," ACM SIGKDD Explorations Newsletter, vol. 2, no. 1, pp. 1–15, 2000.
7. M. Yamamoto and K.W. Church, "Using Suffix Arrays to Compute Term Frequency and Document Frequency for All Substrings in a Corpus," Computational Linguistics, vol. 27, no. 1, pp. 1–30, 2001.

8. J. D. Isaacs and J. A. Aslam. Investigating measures for pairwise document similarity. Technical Report PCS-TR99-357, Dartmouth College, Computer Science, Hanover, NH, June 1999.
9. M. Steinbach, G. Karypis, and V. Kumar. A comparison of document clustering techniques. *KDD-2000 Workshop on Text Mining*, August 2000.
10. K. Cios, W. Pedrycs, and R. Swiniarski. *Data Mining Methods for Knowledge Discovery*. Kluwer Academic Publishers, Boston, 1998.
11. U. Y. Nahm and R. J. Mooney. A mutually beneficial integration of data mining and information extraction. In *17th National Conference on Artificial Intelligence (AAAI-00)*, pp. 627–632, 2000.
12. O. Zamir and O. Etzioni. Web document clustering: A feasibility demonstration. In *Proceedings of the 21st Annual International ACM SIGIR Conference*, pp. 46–54, Melbourne, Australia, 1998.
13. U.Y. Nahm and R.J. Mooney, "A Mutually Beneficial Integration of Data Mining and Information Extraction," Proc. 17th Nat'l Conf. Artificial Intelligence (AAAI-00), pp. 627–632, 2000.
14. U. Manber and G. Myers, "Suffix Arrays: A New Method for OnLine String Searches," SIAM J. Computing, vol. 22, no. 5, pp. 935–948, 1993.
15. R. Baeza-Yates and B. Ribeiro-Neto, Modern Information Retrieval. Addison Wesley, 1999.

Effectiveness of FPA in Sparse Data Modelling and Optimization

R.S. Umamaheswara Raju, V. Ramachandra Raju and R. Ramesh

Abstract The work is to develop an intelligent model to predict the surface roughness and increase the productivity, as surface roughness estimation is a complex task and cannot be easily done for a given cutting parameters due to the complexity of the machining. In order to predict the surface roughness from cutting parameters, a forward mapping predicted model using flower pollination algorithm (FPA) is developed. While predicting surface roughness the FPA model showed a better performance than the existing regression technique, SVM with competitively a minimum percentile of error.

Keywords Cutting parameters · Surface roughness · FPA

1 Introduction

Manufactured products are generally quantified for dimensional accuracy and surface roughness. In metal cutting to achieve the required surface roughness in the first trail itself is a challenge and it depends mainly on the selection of cutting parameters like speed, feed, and depth of cut (DOC). Manufacturing industries for selection of cutting parameters mainly hinge on cutting tool manufacturers manuals or on the experts or on operator's expertise. Intelligent Machining centers require proper cutting parameters selection. The elements are to be modeled and compensated in real machining condition Ramesh et al. [1].

Turning operation was performed on Ti–6Al–4 V alloy [2]. Prediction of surface roughness was done using Multi regression technique in two stages, in the first stage, the surface roughness is predicted from tool vibrations. Due to the error

R.S. Umamaheswara Raju (✉) · R. Ramesh
Department of Mechanical Engineering, MVGRCE, Vizianagaram, AP, India
e-mail: maheshraju@mvgrce.edu.in

V. Ramachandra Raju
RGUKT (IIIT), AP, India

S.C. Satapathy et al. (eds.), *Computer Communication, Networking and Internet Security*, Lecture Notes in Networks and Systems 5,
DOI 10.1007/978-981-10-3226-4_50

percentile being more close to 24 percentile found not accurate enough. In the second stage added cutting parameters predicted roughness error percentile is low as 8%. The boring operation was performed on SS material [3]. Several parameters like nose radius, cutting speed, feed and volume of material removed are used for prediction of roughness, tool wear, and workpiece vibration. ANN model predicted the values with the least error of 4.5 percentile. Turning of Inconel 718 super alloy is done in [4] optimizing residual stress and surface roughness are studied. Mean errors of GA values are given as inputs to ANN model in turn used for the prediction of surface roughness and residual stress. Finally for optimizing cutting parameters for better roughness and stress GONNS technique is used. Capabilities of ANN techniques are evaluated for prediction of surface roughness by [5].

Reviewed several other works and proved that in order to have a better prediction model need to determine the number of layers and nodes in hidden layers. Turning operations are performed on Al–11.3Si–2Cu–0.4 Fe die casting alloy [6], the effect of different additives and cutting parameters on the surface roughness are studied. Dry turning operations are performed on stainless steel in [7], A Taguchi technique is used for getting optimized cutting conditions. Milling operations are performed on Hadfield steel material [8]. Machining productivity with low-cost high quality with high material removal was studied in [9]. Cutting parameters effects on surface roughness and material removal are analyzed using statistical investigation based on analysis of variance (ANOVA).

A second order regression model is formulated to predict the responses in turning and to minimize the responses Taguchi technique is used for estimation of roughness [10]. Surface roughness prediction model in turning operation is explored using Response Surface Methodology [11]. Cutting parameters and tool vibrations are used for prediction of surface roughness. Improving the surface quality by optimizing several parameters to achieve better roughness is studied using RSM for integrally bladed rotors [12]. Firstly for prediction of optimal roughness a single factor is considered. Secondly, optimal factors are derived from the RSM models and optimal parameters enhanced the cutting quality [13]. The analytical parameters are compared with experimental data and found the model to be significant. Prediction of surface roughness using relevance vector machine in grinding operation is done [14]. Series of grinding operation are performed to validate the RVM predicted model and found the trend to be very close. Intelligent systems are used for the prediction of roughness in end milling [15]. Three types of models are used for prediction namely radial basis function neural networks, adaptive neuro-fuzzy inference systems, and genetically evolved fuzzy inference systems. RBFN showed considerably good results.

Lou et al. [16] used the regression technique to study the effect of variation of operating parameters on surface finish. In 84 total experimental data sets, 60 are used for training the regression model. The 24 remaining sets are used for testing the model. The regression model could predict the surface roughness with an average accuracy of 90.03%. The regression equation developed is of order first and is given below.

$$Y_i = 22.9468 + 10.9357X_{2i} - 0.004274\,X_{1i}X_{2i} + 0.674909X_{1i}X_{3i} - 69.7679X_{2i}X_{3i} \quad (1a)$$

A system is proposed for surface roughness estimation y Ramesh et al. [17]. SVM tool is used to handle multi-dimensional operating parameters for prediction of surface roughness. The data given by [16] is used for prediction of roughness using SVM.

The past appraised works basically foreseen several regression tools and numerical methods which fundamentally stuck in local minima Chakravarthy [18, 19], the main problems lies in the convergence characteristics which hinge mainly on the population given. If the population is adequate the prediction will be rational. As a substitute to these, evolutionary computing tools like GA, PSO, TLBO and FPA are used. Generally, such tools won't depend on the initial population to solve, generally selects random weights to solve engineering problems. In this chapter, a nature inspired FPA which mimics the pollination process in flowering plants is presented for the prediction of surface roughness from cutting parameters. Random weights are selected by weighted sum approach used to solve multi-objective problems. This algorithm is evolutionary computing tool developed by Yang [20] in the year 2013 for multi-objective problems in several disciplines. In the current work FPA used to predict the surface roughness.

2 Flower Pollination Algorithm

Every spices strive to protect and facilitate perpetuation of their own spices to increase their population without any change in their breed. Pollination is the art of transportation of pollen grains from one plant male anther to another female stigma. There are basically two different types of pollinators which help in pollination namely biotic and abiotic. The plants that need pollination from animals are usually bright in color and have a strong odor to attract bees. The relationship between the flower and regular visiting bees gets developed and this mutualism is called as floral constancy Amaya Marquez [21]. Another way is by the wind which collects pollens from one plant and transport to others globally.

Evaluation of cost fetches an individual I from Eq. 5. Training data 1 is entered from Eq. 2a which are SFD for a corresponding Y_1. The fitness function as in Eq. 4 is the difference between the predicted and the actual measured value D_1. Similarly for the individual 1 training data 2 is entered for corresponding fitness D_2. Likewise training data 3, 4…t, are entered, where t is 24 in present case where differences are depicted as D_3, D_4…D_t respectively. The average of all the fitness differences D_1, D_2…D_t is the fitness average F_1 for the individual 1. In this case, the individuals selected are 1, 2…N, where N is 20. The fitness averages attained are F_1, F_2…F_N which is an input to FPA for further prediction. The flow and working of FPA is shown the Fig. 1 flow chart.

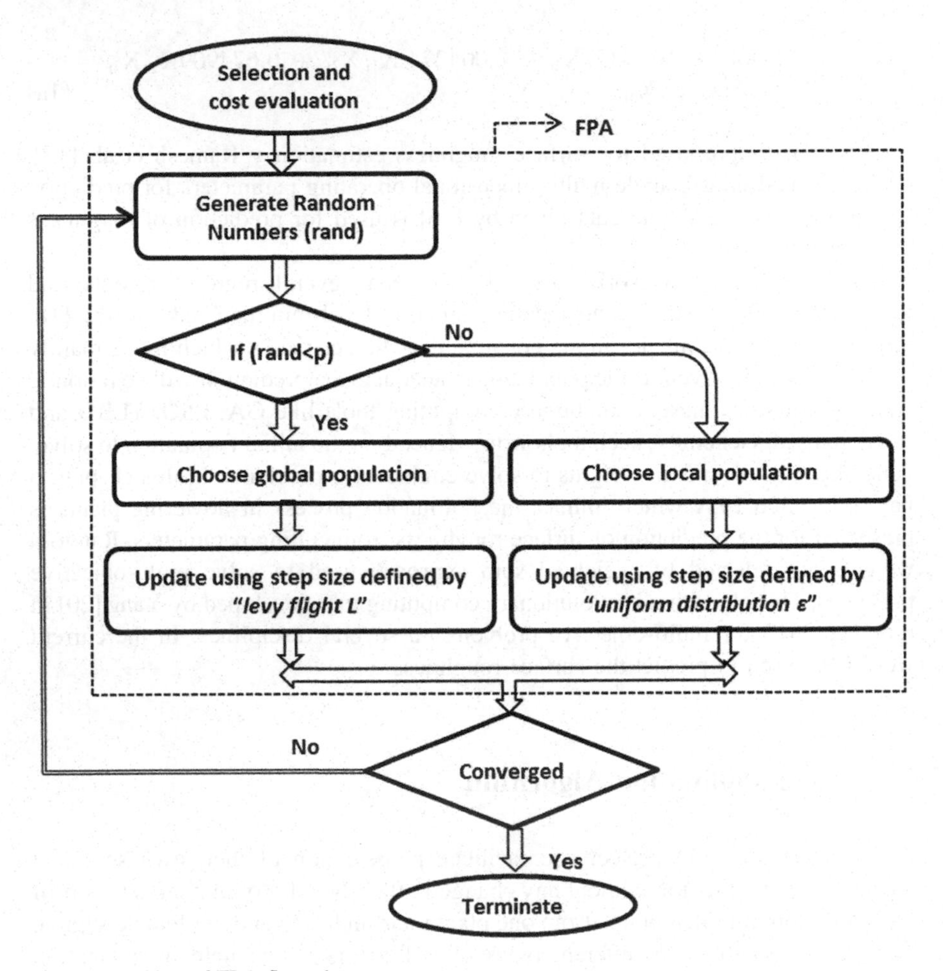

Fig. 1 Working of FPA flow chart

In FPA optimization pollen is considered as a solution, pollen is fundamentally selected from the cost evaluation. Then FPA generates the random variables the initial population A, B…. H. FPA checks the generated random variables whether they are less than or greater than P, where P is 0.8 in this case as specified by Yang [19] which is the best value for all engineering problems solved by FPA. If P is less than 0.8 in FPA the probability switching happens. Yang has given preference for the global population where the pollens jump from initial point globally for solution, based on step size as defined by levy flight L [22]. The mathematical representation of flower constancy [23] is given as follows.

$$x_i^{k+1} = x_i^k + L\left(x_i^k - g^*\right) \tag{1b}$$

The pollen is considered as "i" solution, the solution vector is given as X_i and whole flower pollination is considered as solution space. Probability switching is done by choice for self-pollination or else for global pollination. Resulted in fittest (g*) fruitful pollination and reproductions with either of pollination technique. The floral constancy equation is shown in (1b). L is levy flight as mentioned above.

3 Experimental Work

A comprehensive analysis and evaluation of the performance of several models are done in this section. In this work, a comparison of FPA model is done to Lou et al. [16] multi regression model and to SVM Model developed by Ramesh et al. [17] for the same experimental data. Lou [16] conducted end milling operation on CNC vertical machining center and aluminum being the workpiece material. Four spindle speeds ranging 750, 1000, 1250 and 1500 rpm, seven feed rates ranging 6, 9, 12, 15, 18, 21 and 24 in. per min and three depth of cuts (DOC) 0.01, 0.03 and 0.05 in. are selected for experimentation. Totally 84 different experiments were conducted out of which 24 are randomly selected for testing and remaining 60 data are used for training the models. The surface roughness (R_a) of the components is measured using stylus-based profilometer.

4 Fitness Formulation for Prediction Problem

4.1 Implementation Function

In this section, the objective function formulation is presented. Empirical 1st order polynomial equation favorable for surface roughness estimation is selected.

The FPA the starting point is population initialization. The training data "T" is a vector of 3 × t dimension where "t" is a number of trial experiments performed. Each row is one set of experimental data whereas each set constitutes parameters of the corresponding row numbered trial. Similarly, the measured value in each trial is a scalar of t-dimensions which is also used as a reference training data "R". The representation of the T and R matrices is as given in 2a and 2b.

Training Data:

$$T = \begin{bmatrix} X_{1,1} & X_{1,2} & X_{1,3} \\ X_{2,1} & X_{2,2} & X_{2,3} \\ - & - & - \\ - & - & - \\ X_{t,1} & X_{t,2} & X_{t,3} \end{bmatrix} \tag{2a}$$

$$R = \begin{bmatrix} R_1 \\ R_2 \\ - \\ - \\ R_t \end{bmatrix} \tag{2b}$$

where $X_{1,1}$, $X_{1,2,}$ and $X_{1,3}$ are the experimental training data like Speed1, Feed1, and Depth of cut1 respectively and the corresponding surface roughness being R_1, the second set of training data being $X_{2,1}$, $X_{2,2,}$ and $X_{2,3}$ and the corresponding surface roughness being R_2, likewise wise totally "t" number of training data taken for the model development and "t" is equal to 24 in our case.

4.2 Formulation of Objective Function for Prediction Model

Speed, Feed, and depth of cut are one of the main elements which contribute to the surface roughness as said by many researchers. As there is no fixed and yet unknown relationship between the cutting parameters and roughness and is dynamic which changes from machine to machine. Similarly as in Eq. (1a) even in this model the same type of first order nonlinear prediction model is developed as shown.

$$Y(A, B, \ldots H) = A + BX_1 + CX_2 + DX_3 + EX_1X_2 + FX_2X_3 + GX_1X_3 + HX_1X_2X_3 \tag{3}$$

where X_1, X_2, X_3 are the speed, feed and depth of cut respectively and Y is corresponding surface roughness. The constants A = 2.3832, B = 0.008753, C = −0.00084, D = −1, E = −0.00000419, F = −0.00297, G = 0.000617, H = 0.00000231,

Fitness equation:

$$f = |Y - R_m| \tag{4}$$

where Y is predicted roughness and Rm are measured in trial experiments.

Initial population:

Totally population size is N individuals and N-dimensional solution space is provided for every individual for achieving a solution. The number of design variables selected is N and each and every solution will be in a set of the N-dimensional vector which is given below.

$$\begin{bmatrix} A_1 & B_1 & C_1 & D_1 & E_1 & F_1 & G_1 & H_1 \\ A_2 & B_2 & C_2 & D_2 & E_2 & F_2 & G_2 & H_2 \\ - & - & - & - & - & - & - & - \\ - & - & - & - & - & - & - & - \\ A_N & B_N & C_N & D_N & E_N & F_N & G_N & H_N \end{bmatrix} \tag{5}$$

where N = 20 for this model.

5 Comparison of Performance

The setting of cutting parameters is a trivial task, as it depends on several operational conditions of the machine tools. FPA proposed in this work seems to be a better computational tool in forward mapping prediction of surface roughness. The performance comparison is made in between FPA and several other models are shown in Table 1.

The after analyzing the data present in Table 1 in bold, in SPSS they are 7 better-predicted values, in SVM there are 6 better-predicted values and FPA had 10 better-predicted values. Even after analyzing the worst model having more than 10% of error FPA had less number 4 at SL. no 8, 16, 18, 19, whereas the other models had 8 worst in SPSS and 6 worst in SVM. FPA proved to be a better model than SPSS and SVM.

Table 1 Comparison of SPSS, SVM and FPA predicted surface roughness (Ra)

S. No	Actual Ra	SPSS Ra	SVM Ra	FPA Ra	SPSS % error	SVM % error	FPA % error
1	2.18	2.89	2.92	2.76	32.56	33.94	**26.74**
2	1.42	2.03	1.81	1.50	42.95	27.46	**5.98**
3	3.66	2.86	3.39	3.23	21.85	**7.37**	11.6
4	3.73	3.54	3.71	3.47	5.09	**0.53**	6.70
5	2.44	2.46	2.35	2.34	**0.81**	3.68	3.8
6	2.34	2.12	2.42	2.45	9.40	**3.41**	4.8
7	3.15	3.19	3.22	3.11	1.26	2.22	**1.2**
8	4.37	3.81	3.98	3.53	12.8	**8.92**	19.1
9	2.21	2.47	2.42	2.37	11.7	9.50	**7.3**
10	4.75	5	4.72	4.847	5.26	**0.63**	2.06
11	2.11	2.28	2.3	2.15	8.05	9	**2.16**
12	3.18	3.39	4	3.63	6.60	25.7	14.3
13	3.07	2.94	2.95	3.12	4.23	3.90	**1.86**
14	2.69	2.66	2.79	2.61	**1.11**	3.71	2.9
15	3.61	3.37	3.46	3.33	6.6	**4.15**	7.72
16	2.39	2.19	2.3	1.97	8.3	**3.76**	17.2
17	2.59	2.65	2.52	2.59	2.31	2.70	**0.004**
18	2.52	2.29	2.27	2.21	9.1	9.92	12.09
19	2.39	2.14	2.11	2.18	10.4	11.7	**8.69**
20	2.08	2.1	2.11	1.93	0.96	1.44	6.80
21	4.52	4.46	4.71	4.44	1.32	4.20	1.6
22	1.42	1.73	1.51	1.50	21.8	6.33	**5.92**
23	2.85	3.13	3.21	3.04	9.82	12.63	**6.91**
24	1.78	2.11	2.07	1.74	18.5	16.29	**2.17**

6 Conclusion

An intelligent system is proposed in this work, such model prescribed above can use to build intelligence in open architectural machines to achieve the required surface finish. A comparison is made to the several models like SPSS multi-regression, SVM and proposed FPA model. FPA prediction is proved to be a better model than the former models. Developing such automated systems for open architectural machine tools in estimating the cutting parameters, so that a number of trial and error experiments, skilled labor for setting parameters, reduces production time and finally cost effective. The above said work can be made into an intelligent support system for operators in estimating the cutting parameters.

References

1. R. Ramesh, S. Jyothirmai, K. Lavanya. Intelligent automation of design and manufacturing in machine tools using an open architecture motion controller. Journal of Manufacturing Systems 32 (2013) 248–259.
2. Vikas Upadhyay, P.K. Jain, N.K. Mehta. "In-process prediction of surface roughness in turning of Ti–6Al–4 V alloy using cutting parameters and vibration signals". Measurement 46 (2013) 154–160.
3. K. Venkata Rao, B.S.N. Murthy, N. Mohan Rao. "Prediction of cutting tool wear, surface roughness and vibration of workpiece in the boring of AISI 316 steel with artificial neural network". Measurement 51 (2014) 63–70.
4. Farshid Jafarian, Hossein Amirabadi & Mehdi Fattahi. "Improving surface integrity in finish machining of Inconel 718 alloy using intelligent systems". Int J Adv Manuf Technol (2014) 71:817–827.
5. Azlan Mohd Zain, Habibollah Haron, Safian Sharif. "Prediction of surface roughness, in the end, milling machining using Artificial Neural Network". Expert Systems with Applications 37 (2010) 1755–1768.
6. Mohsen Marani Barzani, Erfan Zalnezhad, Ahmed A.D. Sarhan, Saeed Farahany, Singh Ramesh. "Fuzzy logic based model for predicting surface roughness of machined Al–Si–Cu–Fe die casting alloy using different additives-turning" Measurement 61 (2015) 150–161.
7. D. Philip Selvaraj, P. Chandramohan, M. Mohanraj. "Optimization of surface roughness, cutting force and tool wear of nitrogen alloyed duplex stainless steel in a dry turning process using Taguchi method". Measurement 49 (2014) 205–215.
8. Turgay Kıvak. "Optimization of surface roughness and flank wear using the Taguchi method in milling of Hadfield steel with PVD and CVD coated inserts". Measurement 50 (2014) 19–28.
9. Lakhdar Bouzid, Smail Boutabba, Mohamed Athmane Yallese, Salim Belhadi, Francois Girardin. "Simultaneous optimization of surface roughness and material removal rate for turning of X20Cr13 stainless steel". Int J Adv Manuf Technol, 14 June 2014.
10. Guojun Zhang, Jian Li, Yuan Chen, Yu Huang, Xinyu Shao, Mingzhen Li. "Prediction of surface roughness in end face milling based on Gaussian process regression and cause analysis considering tool vibration". Int J Adv Manuf Technol, 21 August 2014.
11. Zahia Hessainia, Ahmed Belbah, Mohamed Athmane Yallese, Tarek Mabrouki, Jean-François Rigal. "On the prediction of surface roughness in the hard turning based on cutting parameters and tool vibrations". Measurement 46 (2013) 1671–1681.

12. Tao Zhao, Yaoyao Shi, Xiaojun Lin, Jihao Duan, Pengcheng Sun & Jun Zhang. "Surface roughness prediction and parameters optimization in grinding and polishing process for IBR of aero-engine". Int J Adv Manuf Technol, 08 June 2014.
13. A. Bougharriou, W. Bouzid, K. Sai. "Analytical modeling of surface profile in turning and burnishing". Int J Adv Manuf Technol, 26 July 2014.
14. Jianliang Guo. "Surface roughness prediction by combining static and dynamic features in cylindrical traverse grinding". Int J Adv Manuf Technol, 16 August 2014.
15. Abdel Badie Sharkawy, Mahmoud A. El-Sharief, M-Emad S. Soliman. "Surface roughness prediction in end milling process using intelligent systems". Int. J. Mach. Learn. & Cyber. (2014) 5:135–150.
16. Dr. Mike S. Lou, Dr. Joseph C. Chen, Dr. Caleb M. Li. "Surface Roughness Prediction Technique For CNC End-Milling". Journal of Industrial Technology,Volume 15, 1999, Number 1.
17. R. Ramesh, K. S. Ravi Kumar, G. Anil. "Automated intelligent manufacturing system for surface finish control in CNC milling using support vector machines". Int J Adv Manuf Technol (2009) 42:1103–1117.
18. V.S.S.S. Chakravarthy, S.R. Chowdary Paladuga, M. Rao Prithvi. "Synthesis of circular array antenna for side lobe level and aperture size control using flower pollination algorithm. International journal of antennas and propagation", volume 2015, article ID 819712, 9 pages.
19. V.S.S.S. Chakravarthy, P. M. Rao. "On the convergence characteristics of flower pollination algorithm for circular array synthesis", in proceedings of 2nd international conference on electronics and communication systems 9 ICECS'150, PP. 485–489, IEEE, Feb 2015.
20. Yang, X. S. (2012), Flower pollination algorithm for global optimization, in Unconventional Computation and Natural Computation, Lecture Notes in Computer Science, Vol. 7445, pp. 240–249.
21. M. Aaya-Marquez, "Floral constancy in bees: a revision of theories and a comparison with others pollinators", Revista Colombian de Entomologia, vol 35, no 2, 2009.
22. I. Pavlyukevich, "Levy flights, non-local search and simulated annealing," journal of computational physics, vol. 226, no. 2, pp. 1830–1844, 2007.
23. S. Lukasik and P.A. Kowalski, "Study of flower pollination algorithm for continuous optimization," Advances in intelligent systems and computing, vol. 332, pp. 451–459, 2015.

Voice Recognition Based on Vector Quantization Using LBG

D. Nagajyothi and P. Siddaiah

Abstract The process which recognizes the speaker based on the information present in the speech is called Voice recognition. This can be used to many applications like identification, voice dialling, tele-shopping, voice based access services, information services, tele-banking, security control of confidential information. The variation of Speaker exists in speech signals because of different resonances of the vocal tract. MFCC is the technique to exploit the differences of the speech signal. Similarly, the technique of Vector Quantization (VQ) emerged as useful tool. In this chapter, the VQ is employed for efficient creating the extracted feature vector. The acoustic vectors extracted from input speech of a speaker and provide a set of training vectors. LBG algorithm is used for clustering a set of L training vectors into a set of M codebook vectors.

Keywords Speech processing · Vector quantization · LBG algorithm · MFCC

1 Introduction

The technique of Speech processing is one of the significant area of interest with signal processing engineers. The process depends on the speech perception of the recognizer which greatly influenced by the intelligence of the system. Considering the advantages and to achieve the final goal several industries are providing huge support and efforts. Analysis of the voice signal involves in feature extraction of the signal which is later processed to eliminate the redundant data. This enables to

D. Nagajyothi
ECE Department, Vardhaman College of Engineering, Shamshabad, Telangana, India
e-mail: Nagajyothi1998@gmail.com

P. Siddaiah (✉)
Department of ECE, University College of Engineering and Technology, Acharya Nagarjuna University, Guntur, India
e-mail: Siddaiah-p@yahoo.com

S.C. Satapathy et al. (eds.), *Computer Communication, Networking and Internet Security*, Lecture Notes in Networks and Systems 5,
DOI 10.1007/978-981-10-3226-4_51

capture useful information which can be further used for processing. This step typically involves in converting speech signal to a set of statistical data [1–5] which is considered as front end processing. Sometimes there is chance of ignoring useful information leading information loss. For the development of this system we follow some properties like separating sub word class, speaker variability, minimisation of degradation effect on speech signals. These set of properties are used to facilitate a fine analysis of speech signal. They typically involve in exploring the acoustic properties of the signal which are useful for further analysis in the processing. The process of feature extraction typically corresponds to meaningful depiction of the speech signal. This is often referred as parameterization.

Initial step of voice recognition involves in preserving those components of audio signal which are useful in identifying the linguistic content. This serves speech recognition and helps to differentiate people by their voice. This also favours automated recognition of the speaker. Typically the entire process is divided into two blocks. First step is speaker identification (SI) and the next step is speaker verification (SV). The technique is to identify which of the training speakers made the test utterance. Finally the system output referred as the name of the training speaker or its ID. In otherwise case is what a possible rejection if the utterance has been made by an unidentified speaker. In the due course of time, it is evident that engineers are practicing several varieties of methodologies with different states of success. The Speech recognition process can be broadly divided into two phases like the training phase and the testing phase.

2 Structure of Voice Recognition System

The voice recognition system mimics human auditory system which refers to a matching problem which involves in comparing the input speech signal with the one in brain or computing machine.

Typical block diagram of the existing system is as shown in Fig. 1. The technique involves in processing training and testing/recognition [4]. Vector

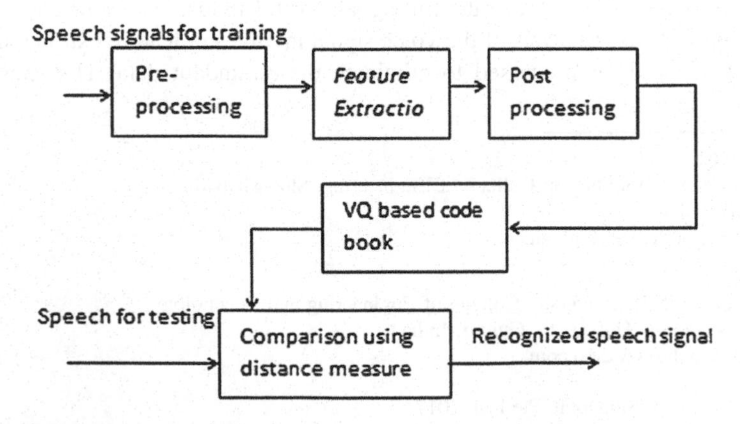

Fig. 1 Structure of voice recognition system

Quantization refers to mapping of vectors from one vector space to several number of regions. Each region is referred as a cluster depicted by its center called a codeword. On the line the term codebook refers to collection of all such code words. Several algorithms like LBG and K-means are proposed for mapping or clustering the feature vectors [6].

3 Feature Vector Representation

Extracted statistical information of a speech signal is referred as speech signal feature vector [7]. The major steps for speech signal feature extraction (SFE) are framing, pre-emphasis, windowing, DFT, mel-scaling, logarithm and DCT. Simple block diagram representing the process is as shown in Fig. 2. Preliminary stage of SFE involves in boosting high frequency energies [6–10]. It is to be noted that there is more energy associated with spectrum of a voice signal like vowels. This change in energy through the frequencies is due to the nature of the glottal pulse. The stage of boosting the high frequencies allows more amount of the energy available for acoustic model which improves detection accuracy. Pre-emphasis is performed using simple filter which is a 1st order high-pass filter. When the input is x[n], $0.9 \leq \alpha \leq 1.0$, the filter equation is given as

$$y[n] = x[n] - \alpha * x[n-1]. \tag{1}$$

A running window facilitates for the analysis of a small concentrated portion of the image. It is possible to extract the spectral features of the signal encapsulated in the window and arrive at an assumption of the signal for further analysis. With in certain limits the window assumes non-zero value while, elsewhere it is 0 [10].

The characterization of the window is dealt using three parameters. Size of the window, its offset with successive windows and shape of the window are these three parameters [7]. The extraction of the signal can be done by multiplying the

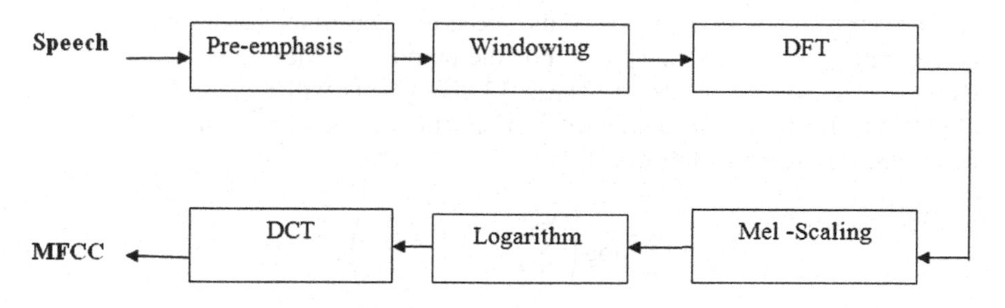

Fig. 2 Block diagram of feature extraction

model of the signal at time n say s[n], with the corresponding window component at time n, w[n]:

$$\mathrm{Y[n]} = \mathrm{w[n]} * \mathrm{s[n]}. \tag{2}$$

The hamming window model for such applications is modelled as follows.

$$\mathrm{W(n)} = 0.54 - 0.46 * \cos(2\pi \mathrm{n}/(\mathrm{N}-1)); 0 \le \mathrm{n} \le \mathrm{N}-1. \tag{3}$$

The next step involves in extracting spectral information for modelled window signal [8]. For this reason the version of fourier technique referred as discrete Fourier transformation (DFT) is applied using the following equation.

$$X[k] = \sum_{n=0}^{N-1} x(n)e^{-j2\pi nk/N}. \tag{4}$$

In DFT, the version known as Fast Fourier Transformation (FFT) is employed in this work. The FFT of a signal provides the information about the energy distribution at each frequency band. Human hearing, has varying sensitivity to frequency bands. It is less sensitive at frequencies above 1000 Hz. The Mel frequency (m) is computed using the raw acoustic frequency as follows:

$$\mathrm{Mel(f)} = 1127 * \ln(1 + \mathrm{f}/700). \tag{5}$$

The corresponding spectrum of the Mel-Ceptrum is as shown in Fig. 3. A bank of filters are employed in MFCC computation. The role of these banks is to collect energy from each filter. Like this 10 filters are operated with a band gap of 100 Hz, while the remaining banks are spread logarithmically over 1000 Hz. During MFCC computation, this intuition is implemented by creating a bank of filters which collect energy from each frequency band, with 10 filters spaced linearly below 1000 Hz, and the remaining filters spread logarithmically above 1000 Hz. The log of each of the Mel spectrum values are taken as shown in Fig. 6. Using log, the feature estimates are made less sensitive to variations in input.

The cepstrum refers to the log of the spectrum. Cepstrum is as a significant way of separating the source and filter. For the purpose of MFCC extraction, the 12 Cepstral values are considered. These 12 coefficients will represent information about vocal tract filter, separated from information about glottal source. The model of Ceptrum is given as follows.

$$c[n] = \sum_{n=0}^{N-1} \log\left(\sum_{n=0}^{N-1} x(n)e^{-j2\pi nk/N}\right) e^{j2\pi nk/N}. \tag{6}$$

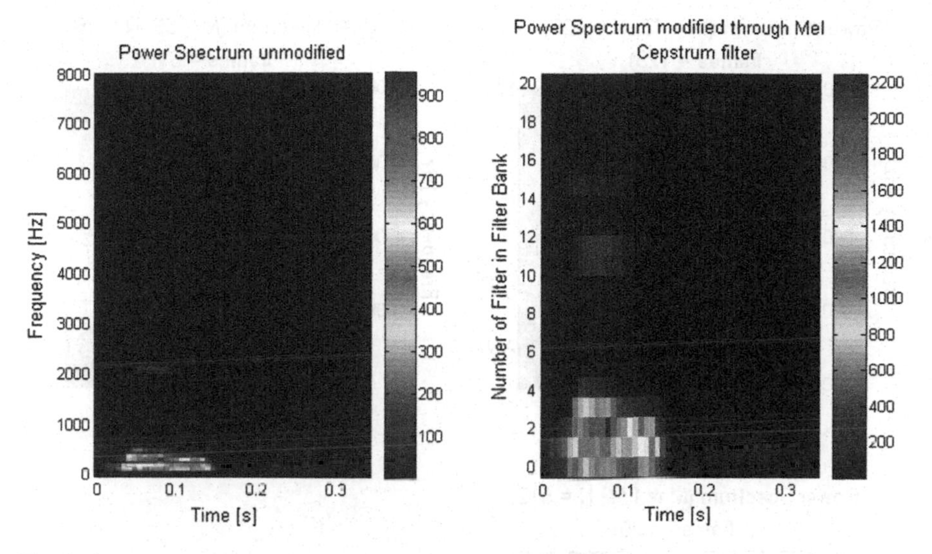

Fig. 3 Spectrum of Mel-cepstrum for single frame

The cepstrum can also be defined as inverse DCT of the log magnitude of the DFT of a signal. On time scale, the parameter energy is computed by adding the power of all the samples in the frame. Accordingly, the representation of asignal 'x' over a sample in range $(t_1 - t_2)$ is given as

$$\text{Energy} = \sum_{t=t_1}^{t=t_2} x^2(t)$$

The change in Cepstral are added over a time in order to accomplish the fact that the speech signal is varying from frame to frame. The final power spectrum is as shown in Fig. 4.

This is done adding for each of the 13 features a delta or velocity feature, and a double delta or acceleration feature. Computing the difference between the frames mathematically contributes to evaluation of deltas. Thus the delta value d(t) for a particular Cepstral value c(t) can b estimated as

$$\mathrm{D(t)} = [\mathrm{c(t+1)} - \mathrm{c(t-1)}]/2. \tag{7}$$

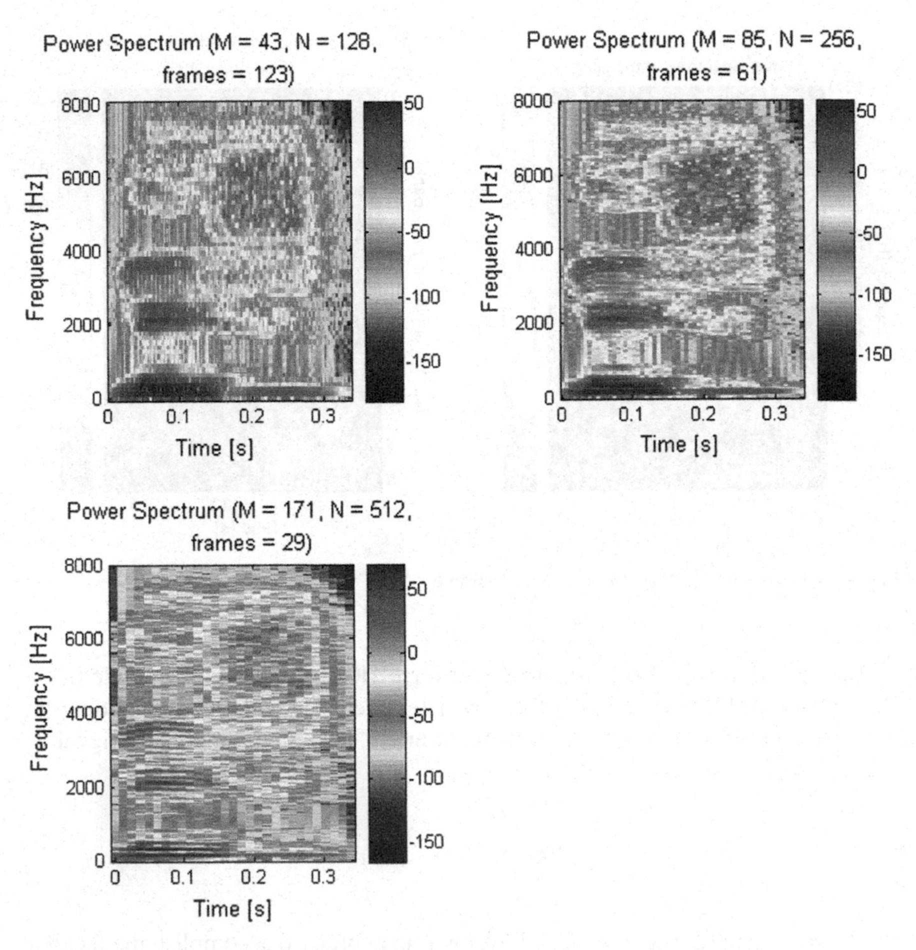

Fig. 4 Power spectrum of frames of the signal

4 Vector Quantization

Pattern recognition is a constituent factor that evolved as a problem associated with speech recognition [11]. The aim of pattern recognition involves in classifying objects of interest into several categories or classes [6]. Patterns are objects of interest. In our speech recognition techniques, these patterns are the acoustic vectors which are statistical data representing the speech signal using some of the techniques discussed above [12]. Each class corresponds to individual vocal data. The classification procedure employed in this work is also referred to as feature matching [13]. Supervised pattern recognition [14] refers to set of pattern which are already known. This is exactly the case considered as during the training session every speaker is assigned with an ID. However, the remaining patterns are employed to test the classification algorithm. These patterns are collectively

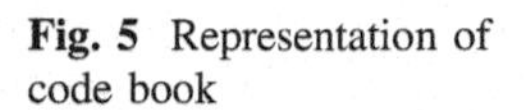

Fig. 5 Representation of code book

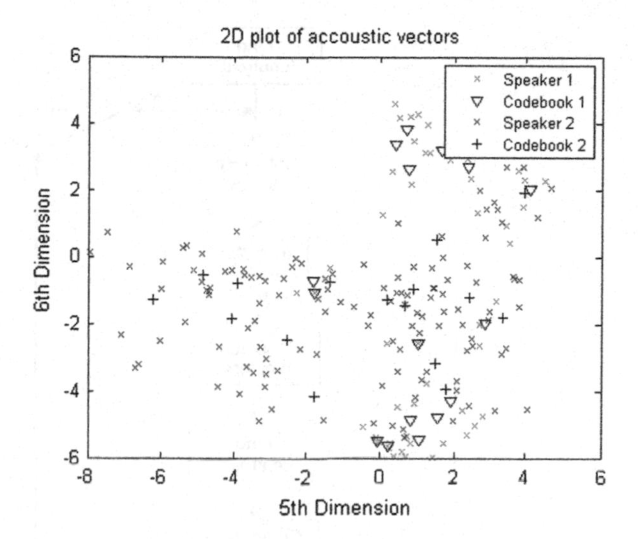

referred to as the test set. Evaluation of the performance of the technique is possible if the correct classes of the individual patterns in the test set are also known. The codebook is based on this principle.

Several matching techniques are proposed in speaker recognition like Dynamic Time Warping (DTW), Hidden Markov Modeling (HMM), and Vector Quantization (VQ) [3]. In this chapter, the VQ approach is employed. The employed method has ease of implementation and high accuracy. The corresponding codebook is as shown in Fig. 5.

5 Algorithm Approach

After the completion of enrolment session, the feature vectors extracted from input of a speaker and which provides a set of training vectors [9, 11, 14]. In this discussion the building of a code book for a specific person/speaker from the training vectors has done as shown in Fig. 6.

LBG algorithm is employed for generating the codebook [10]. The method typically involves in clustering from a set of training vectors and mapped to a set of codebook vectors [15]. The following steps describe the algorithm flow [2].

a. Frame a one-vector codebook (CB) in order to create a centroid.
b. Enhance the size of CB according to the splitting principle.
c. Perform Nearest-Neighbour Search.
d. Update the centroid.

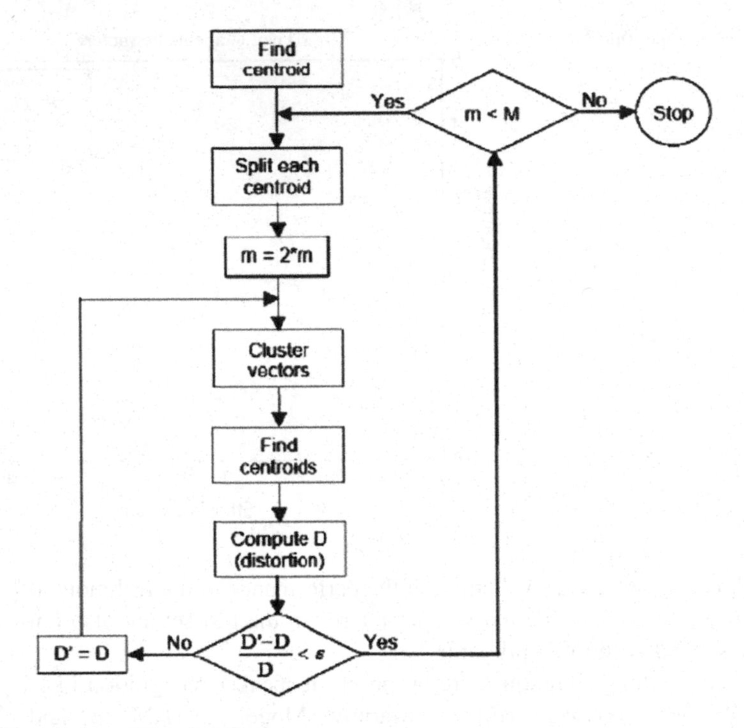

Fig. 6 Description of the algorithm steps

e. Iteration1: repeat the procedure of steps 3 and step 4 until the average distance less than present threshold [16].
f. Iteration 2: repeat steps 2, 3 and 4 until a codebook is designed.

6 Conclusion and Future Scope

Implementation of a speaker identification/Voice recognition system has been performed effectively. The SV system consists of two modules. First module is an Enrolment or training the system which involves in developing a database of existing or available speakers. The second module is testing the model using same speakers in order to validate the efficiency in identifying the words in the speech. This typically validates the system proposed. In the training phase, each and every speaker has to provide the speech samples to build a reference models. It broadly divided into two significant parts. Initially the first part typically involves in processing every individuals voice input further which is condensed and abstracted the features of the corresponding vocal-tracts. The next part corresponds to putting each individuals data consolidated and constitute a resultant matrix. During testing phase feature extraction and matching techniques are considered. Feature extraction is the process of extracting a small amount of data from the voice signal. Feature matching involves the actual procedure to identify the known speakers speech.

References

1. Kekre, H.B., Vaishali, Kulkarni: Speaker Identification by using Vector Quantization. Intl J of Engg Sc and Tech, vol. 2, pp - 1325–1331 (2010)
2. Linde, Y., Buzo, A., Gray, R.: An algorithm for vector quantizer design. IEEE Trans on Comm, vol. 28, pp - 84–95 (1980)
3. Rita Singh, Bhiksha Raj, Stern, R.M: Automatic Generation of Sub word Units for Speech Recognition Systems. IEEE Trans on speech and audio processing. Vol. 10 (2002)
4. Jesus Savage, Carlos Rivera, Vanessa Aguilar, "Isolated Word Speech Recognition Using Vector Quantization Techniques and Artificial Neural Networks", Facultad de Ingenieria, Departamento de Ingeniería en Computación, University of Mexico, UNAM, Mexico City C. P. 04510, Mexico (1994)
5. Rabiner. L. R, Juang, B.H.: Fundamentals of Speech Recognition, Prentice-Hall, Englewood Cliffs, N.J (1993)
6. Tzu-Chuen Lu, Ching-Yun Chang, "A Survey of VQ Codebook Generation", Journal of Information Hidingand Multimedia Signal Processing, Ubiquitous International, Vol. 1, No. 3, pp. 190–203, (2010)
7. Furui. S, Speaker-independent isolated word recognition using dynamic features of speech spectrum, IEEE Trans on Acoustic, Speech, Signal Processing, Vol. 34, pp. 52–59, (1986)
8. Furui. S.: An overview of speaker recognition technology, ESCA Workshop on Automatic SpeakerRecognition, Identification and Verification, pp. 1–9, (1994)
9. Soong, F. E., Rosenberg, A. E., Juang, B. H.: A vector quantization approach to speaker recognition, AT & TTechnical Journal, Vol. 66, No. 2, pp. 14–26, (1987)
10. Wiploa J. G., Rabiner L. R.: A Modified K-Means Clustering Algorithm for Use in Isolated Word Recognition, IEEE Trans on Acoustics, Speech and Signal Processing, Vol. 33 No.3, pp. 587–594, (1985)
11. Marcel R. Ackermann, Johannes Blomer, Christian Sohler,: Clustering for Metric and Nonmetric DistanceMeasures, ACM Transactions on Algorithms, Vol. 6, No. 4, Article 59, pp. 1–26, (2010)
12. Shih-Ming Pan, Kuo-Sheng Cheng,: An evolution-based tabu search approach to codebook design, PatternRecognition, Vol. 40, No. 2, pp. 476–491, (2007)
13. Chin-Chen Chang, Yu-Chiang Li, Jun-Bin Yeh,: Fast codebook search algorithms based on tree-structuredvector quantization, Pattern Recognition Letters, Vol. 27, No. 10, pp. 1077–1086, (2006)
14. Buzo, Gray, A.H., Gray, R.M., Markel, J. D.,: Speech coding based upon vector quantization, IEEE Trans on Acoustics, Speech and Signal Processing, Vol. 28, No. 5, pp. 562–574, (1980)
15. Linde, Y., Buzo, A., Gray, R. M.,: An algorithm for vector quantizer design, IEEE Transactions on Communications, Vol. 28, No.1, pp. 84–95 (1980)
16. Modha, D., Spangler. S.,: Feature weighting in k-means clustering. Machine Learning, Vol. 52, No.3, pp. 217–237, (2003)

Improved In-System Debugging of High Level Synthesis Generated FPGA Circuits

A. Murali, K. Hari Kishore, A. Trinadha, Saswat Tripathy and P. Dhanunjaya Rao

Abstract The new approaches for in-system debug of High-Level Synthesis generated hardware have been proposed. With these approaches the source level events are observed in the final hardware by using Event Observability Ports (EOP). And also Event Observaility Buffers (EOB) is proposed for tracing events with the use of EOPs. In EOBs the data storage is done in cycle-by-cycle manner on the basis of EOB-by-EOB. This approach leads to the loss of different event timing relationships information stored in various trace buffers. These event relationships are recovered by the proposed two methods in this chapter. The observed results in this chapter will describe the effectiveness and feasibilities of EOB trace strategy.

Keywords Event observability ports (EOP) · Event observability of buffers (EOB) · Trace buffer · Field programmable gate array · Design verifications

1 Introduction

Now a day the advanced High-Level-Synthesis tools have gained very much attractiveness in the field of industries and academics [1]. Using these HLS tools the design productivity has been boost-up by 5 times [2]. The design engineers have

A. Murali (✉) · K. Hari Kishore
Department of ECE, KLEF, K.L University, Green Fields, Guntur, AP, India
e-mail: amurali3@gmail.com

K. Hari Kishore
e-mail: kakarla.harikishore@klunivrsity.in

A. Murali · S. Tripathy · P. Dhanunjaya Rao
Department of ECE, Raghu Engineering College, Visakhapatnam, AP, India
e-mail: saswat.tripathy@gmail.com

P. Dhanunjaya Rao
e-mail: dhanupaila.dp@gmail.com

A. Trinadha
Department of ECE, Chinthalapudi Engineering College, Guntur, AP, India
e-mail: akula.trinadha@gmail.com

S.C. Satapathy et al. (eds.), *Computer Communication, Networking and Internet Security*, Lecture Notes in Networks and Systems 5,
DOI 10.1007/978-981-10-3226-4_52

observed that the productivity have been boosted due to reduction in the simulation activity timing achieved by the software based verification of design specifications. The loss of productivity can be obtained once the bug is identified when the HLS design is operating at high speed. The user needs to first analyze the complete circuit in detail before going for debugging can overcome the productivity loss. The drawback of this is only due to lack of effectiveness of source level debugging in present HLS tools. To achieve the effectiveness and integrated debugging must be done at the source level where the source is the only approach to modify and fix the bug. To debug at the source level for improving the productivity HLS takes the first place.

The debugging of designs based on FPGA is done by using a standard Embedded Logic Analyzers (ELA). The net-list signals are used to configure the ELAs for trace buffer connection. While the active execution of a design signals are captured on a cycle-by-cycle basis. The captured data on ELAs are limited by the data present in an on chip memory. If necessary the in-system debugging of HLS design on FPGA can also be done by ELAs. The RTL code generated by HLS tools are used for configuring the ELA for recording the correct information in required time. This procedure may not be known to all the users. Thus for optimizing the trace buffers the ELAs cycle-by-cycle approach may not be that much flexible for using high level knowledge even the HLS tool provides the high level knowledge.

In this chapter the new approaches have been proposed for the optimized in-system debugging of FPGA designs generated by HLS. This approach overcomes the drawback of capturing the relevant information in a limited memory. The information available in the HLS tool is utilized by the instrumentation process to configure the trace buffers for flexibility in resource usage and memory efficiency. Thus the new approach gives an advantage to the user no need of getting familiar with low-level organization of FPGA circuit instead it depends upon the high level information provided by the HLS tool.

2 Background

Currently the available HLS tools take a high level language (Java and C) based sequential software input specification as an input and generates the relevant hardware. Usually the sequential software specification contains the information of control flow and how the events are updated. By analyzing the event updates we can easily debug the software when it goes wrong. These event updates and control flow is utilized by the user to analyze the debugging of final hardware as of done in software.

Generally the HLS compiler follows a design flow which is described as. Initially the standard front-end compiler is used to compile the sequential input specification to an intermediate representation. The IR contains the information of required operation for implementing the input specification in software. Assembly language based format is used to represent the IR. In general the two level control flow diagrams is used to represent the IR in which higher level blocks and lower level blocks are considered. The control flow decision operation terminates the basic blocks to execute the next basic block. Once the scheduler receives the IR, the operations are assigned from the CFD to the entire control steps involved in a state transition graph (STG) [3]. As the scheduling completes the operations related to functional unit and registers are assigned from the STG. Depending on scheduling and binding results RTL code will be generated. The higher level modules instance the generated RTL code in order to follow the vendor tool flow to generate the bit-stream for FPGA dumping.

3 Observability Ports

3.1 Event Observability Ports

The proposed HLS designs [4, 5] built with Event Observability Ports (EOP) to observe the update of states and events of control flow. The data signals and event signals are present in the EOPs to observe the completed operations and to validate the data signal corresponding to the event operation. The advantages of the EOP are adding ports to the design after scheduling and bounding.

Identifying the data signals and events which are relevant is needed to observe initially before design is instrument with EOP. To complete this task STG is used. An STG is a state transition Graph with a collection of control states with different state transition with respect to each state. State transition is happened with various transition conditions. A set of operations are contained in a control state to control the execution of each state. The process involved in generation of RTL code uses STG flow based on structured circuit model like FSMs.

3.2 Tracing Event Observability Ports

Initially the EOPs are developed to trace the state update events and the control flow efficiently. To trace the EOPs, Event Observability Buffers (EOBs) have been proposed. Mostly the EOBs are used due to its small size and it contains data storage enable input and data input. Because of its small size it can be allocated on an EOP-by-EOP basis. The storage depth of EOBs is configured with respect to the RAR of the EOPs event signal. If the RAR of the EOPs event signal is high, the depth of the EOB can be configured to high. The EOPs event signal connected to the data storage enable input of EOB will store the event data signal when an actual event is occurred. These trace buffers also have an advantage after place and route for the incremental insertion [6, 7]. When the occurring of events are invalid then the time spent for storing is waste so ELAs are used to prevent this by tracing both event and data signals on an EOP-by-EOP basis. For all trace buffer inputs these ELAs will provide with equal storage depth. Hence, the storage depth cannot be adjusted by the ELAs on input basis.

4 Timing and Event Order Recovery

The EOB approach is mainly proposed to identify the events timing relationship when it is lost. Earlier the ELA approach has been used to recover this lost due to its monolithic trace buffer. Due to small size of the EOBs are adopted to recover the problem of timing relationship with event reference trace. This approach traces EOPs event signals using single EOB by connecting EOPs event signals to the EOBs data input.

The data storage enable input of EOB is created by ORing the same event signals. This will filters the data storage when no events are occurred during cycles. With replacement of previous data storage enable signal with a constantly asserted enable signal can leads to creation of cycle accurate reference trace.

Once EOP tracing completes, the corresponding EOB event recovery can be done by using event reference trace. The recovery is done when the match occurs with assertion of event reference trace and the last value written into the corresponding EOB [8]. For every sample of event reference trace this step should be repeated. In Fig. 1 it observed that how the event recovery is done based on the cycle accurate event reference trace. From the Fig. 1 it is observed that there are two event signals represents as cycle accurate event reference trace contents. By examining the final sample 4 event contents recovery algorithm starts. The data corresponding to each event should be stored in the corresponding event EOBs, but from the Fig. 1 it is observed that the data for event 2 should be stored in the event 2 EOB but it has been stored in event 4 EOB. Event trace sample number is labeled to this data by event reference trace. This process will be repeated by all the steps on previous trace sample until it reaches the beginning of event trace. This process is continued until the recording of two trace buffers stops.

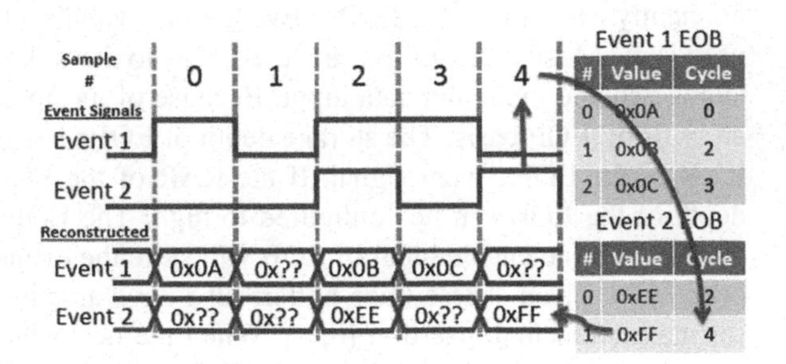

Fig. 1 Event reference trace based timing and event order recovery

5 Experimental Results

5.1 Tool Setup

Based on C-language each benchmark has been programmed and by using VIVADO HLS tool it is compiled to RTL [9]. For Fully pipelined design VIVADO HLS tool is targeted for ZynQ 7020 device by adjusting the minimum clock period to 10 ns and for partially pipelined and un-pipelined designs the minimum clock period is set to 25 ns. The capturing of events could be reduced with lower clock rates assigned for pipelined and un-pipelined designs may result in un-proper recovery. The HLS PIPELINE directives will be added to the each partially pipelined and fully pipelined design for enabling the pipelining.

The desired tracing is done by incrementing the design with EOPs with corresponding state update events and control flow. This involves four steps to instrument the HLS benchmarks. First, to examine the C-code to obtain the set of events which are involved in result of better design executions. Second, events in the RTL and C-code mapping have been obtained. Third, synthesis of a design is performed by XST (14.6) and generated EDIF net-list. Finally, top level ports of EDIF net-list have been connected to the EOPs. The EOPs are added by searching the required signals and then routing these signals to the top-level ports. The clock and start signal ports are also created with respect to the EOP signals. A keep attribute will be added to the RTL source code if any required signal is not found and the design will be re-synthesized to generate the EDIF net-list.

5.2 Results

For each test design it performs two sets of experiments. In first set the keep attribute is used to maintain the same name from RTL to the EDIF net-list. The efficiency of various trace strategies which are used to trace the EOPs have been tested by remaining experiments. The Table 1 describes the Baseline design statistics. The results observed in Figs. 2 and 3 is simplified to the design statistics in Table 1.

From the Figs. 2 and 3 it is observed that the name preserving done by keep attribute has very little effect on usage of LUTs or on the minimum clock period. Due to many of the signals are already existed with different names before keep

Table 1 Baseline design resource utilization

Design	REGs	LUTs	Min. Clk. period (ns)
Un-pipelined	358	332	18.738
Partially-pipelined	817	1908	19.886
Fully-pipelined	922	696	8.726

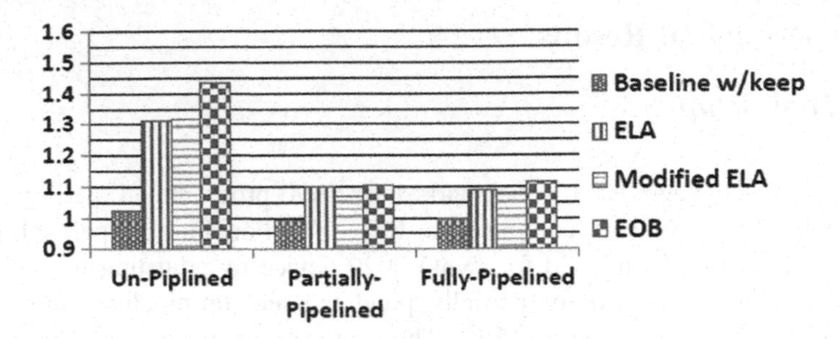

Fig. 2 The normalized LUT for each experiment to the baseline

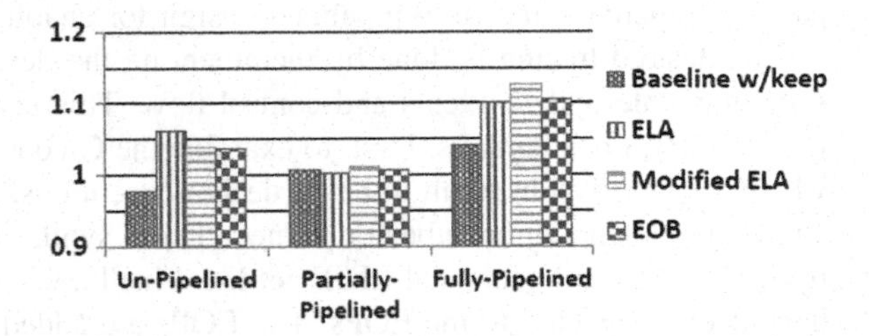

Fig. 3 The normalized minimum clock period to the baseline

attribute was assigned. It is observed that the overall effect of the keep attribute is around less than 5%.

The three various trace strategies have been used to trace the EOPs in remaining experiments. In first strategy is the cycle by cycle ELA strategy which is a standard strategy in which all the EOP data event signals are directly connected to the inputs of EOB. The EOBs are forced to store all EOP signals regardless of whether the event had occurred or not, his is done only because of connecting the EOBs enable signals to the high. In the second strategy instead of tying high to the EOBs enable signals, performing OR logic on event signals which prevents the storing of data in memory whether the event occurred or not. Final strategy has an advantage of storing the data when the events are actually occurred. In this strategy all the data and event signals are directly connected to the corresponding EOB enable and input signals.

6 Conclusion

This chapter has presented an improved in-system debugging of high level synthesis generated FPGA circuits have been presented. The proposed method improved the observability of source-level events by instrumenting the HLS generated with EOPs. In this proposed system the EOBs are used to capture the events when it is actually occurred. Based on the requirements of source-level events the depth of these EOB can be adjusted to store the captured events. From the obtained results it is observed that the usage of EOBs gives better results when compared to the ELAs. To recover the order and timing relationships of events we have also proposed two new approaches. Additionally, to configure the depth of EOB efficiently we have introduced a concept of RAR in this chapter.

References

1. J. Cong, L. Bin, S. Neuendorffer, J. Noguera, K. Vissers, and Z. Zhiru.: High-level synthesis for fpgas from prototyping to deployment. Computer-Aided Design of Integrated Circuits and Systems, IEEE Transactions. vol. 30, no. 4, 473–491 (2011)
2. K. Rupnow, L. Yun, L. Yinan, and C. Deming.: A study of high-level synthesis Promises and challenges, in ASIC (ASICON), 2011 IEEE 9th International Conference on, Conference Proceedings, pp. 1102–1105 (2011)
3. J. Cong and Z. Zhang.: An efficient and versatile scheduling algorithm based on sdc formulation. pp. 433–438 (2006)
4. K. S. Hemmert, J. L. Tripp, B. L. Hutchings, and P. A. Jackson.: Source level debugger for the sea cucumber synthesizing compiler in Field- Programmable Custom Computing Machines, 2003. FCCM 2003. 11th Annual IEEE Symposium on, Conference Proceedings, pp. 228–237 (2003)
5. K. S. Hemmert, “Source level debugging of circuits synthesized from high level language descriptions. (2004)
6. E. Hung and S. J. Wilton, “Incremental trace-buffer insertion for fpga debug.(2003)
7. J. Keeley.: An incremental trace-based debug system for field programmable gate-arrays. (2013)
8. J. Curreri, G. Stitt, and A. D. George.: High-level synthesis techniques for in-circuit assertion-based verification in Parallel & Distributed Processing, Workshops and PhD Forum (IPDPSW), 2010 IEEE International Symposium on, Conference Proceedings, pp. 1–8 (2010)
9. Xilinx.: Vivado design suite user guide: High-level synthesis. vol. UG902–Version 2013.4 (2013)

Design of Cascaded Hybrid Interleaver for Fast Turbo Coding and Decoding

M. Rajani Devi, K. Ramanjaneyulu and B.T. Krishna

Abstract Turbo code in an attempt to realize a technique that approaches the theoretic limit utilizes a crucial component, interleaver, which converts burst errors caused by impulsive noise to simple errors. But the processing of data through the interleaver incurs time as well as memory. Both time and memory complexities are directly related to the total number of information bits. The time required for interleaver processing is so high because of which some applications omits the complete interleavers. But the consequent schemes suffer from high bit error rate. In this chapter a novel hybrid two stage interleaving scheme was proposed which reduces the time required for interleaving processing while maintaining the BER criteria up to the levels obtained with block or 3GPP interleavers.

Keywords Interleaver · Turbo code · Block interleaver · 3GPP interleaver

1 Introduction

The main function of a communication system is to transmit information from the source to the destination with sufficient reliability. In the last two decades, there has been an explosion of interest in the transmission of digital information mainly due to its low cost, simplicity, higher reliability and possibility of transmission of many services in digital forms [1].

M. Rajani Devi (✉)
JNTU Kakinada, East Gdavari District, Kakinada, AP, India
e-mail: rajimerigala@gmail.com

K. Ramanjaneyulu
PVP Siddhartha Institute of Technology, Kanuru, Krishna District, Vijayawada, AP, India
e-mail: kongara.raman@gmail.com

B.T. Krishna
JNTU, Vizianagaram, AP, India
e-mail: tkbattula@gmail.com

S.C. Satapathy et al. (eds.), *Computer Communication, Networking and Internet Security*, Lecture Notes in Networks and Systems 5,
DOI 10.1007/978-981-10-3226-4_53

Theoretically, Shannon stated that the maximum rate of transmitted signal or capacity of a channel over Additive White Gaussian Noise (AWGN), with an arbitrarily low bit error rate depends on the Signal to Noise Ratio (SNR) and the bandwidth of the system (*W*), according to [2].

Instead of *S/N*, the channel capacity can be represented based on the signal to noise ratio per information bit (E_b/N_0). Considering the relationship between SNR and E_b/N_0, and the channel capacity (with value R).

$$\frac{S}{N} = \frac{E_b}{N_o} X \frac{R}{W} \tag{1}$$

$$\frac{C}{W} = \log_2\left(1 + \frac{E_b}{N_o} \cdot \frac{R}{W}\right) \tag{2}$$

In the case of an infinite channel bandwidth $\left(W \to \infty, \frac{C}{W} \to 0\right)$ the Shannon bound is defined by:

$$\frac{E_b}{N_o} = \frac{1}{\log_2 e} = 0.693 \tag{3}$$

In order to achieve this bound, i.e. $\frac{E_b}{N_o} = -1.59\,\text{dB}$ value, it would be necessary to use a code with such long length that encoding and decoding would be practically impossible. However, the most significant step in obtaining this target, was by Forney, who found that a long code length could be achieved by concatenation of two simple component codes with short lengths linked by an interleaver [3].

Conventionally, a turbo code is analyzed as a block code by using a block interleaver and terminating RSC encoders to a known state at the end of each data block. Codes with this structure are generally decoded using an iterative decoding technique. Depending on the number of iterations, the decoding procedure can approach the Maximum Likelihood decoding. Similarly to linear block codes, the probability of error for turbo codes in an AWGN channel is upper bounded [4].

Analysis indicates that turbo codes with block interleavers often have a relatively low value with high number of multiplicities which results in insufficient code performance in the low to high signal to noise ratios. In fact, the error will not decrease proportionally with increments of signal to noise ratio. This phenomenon is called "error floor".

One of the most effective solutions to reduce the error floor is utilization of a suitable interleaver compatible with the structure of constituent RSC encoders. The best performance of turbo code is attained by permuting the input bit streams randomly to the interleaver of the different memories [5]. However, in most applications it is not advisable, when the input bit stream length is large [6]. A number of studies were undertaken on the importance of interleaver on the turbo code performance [7–9] and a number of interleavers are presented in the literature with diverse features and complexities [10–12]. The rest of the chapter is structured as follows. The next two sections deal with block and 3GPP interleavers respectively. Section 4 presents the proposed hybrid interleaver. Section 5 presents the simulation performance of the proposed scheme. Section 6 concludes the chapter.

2 Block Interleaver

A block interleaver accepts a set of symbols and rearranges them, without repeating or omitting any of the symbols in the set. The number of symbols in each set is fixed for a given interleaver. The interleavers' operation on a set of symbols is independent of its operation on all other sets of symbols. Block interleavers can be classified into mainly three categories. They are Matrix interleaver, Random interleaver and Algebraic interleaver.

The Random interleaver rearranges the elements of its input vector using a random permutation. The random interleaver accepts a column vector input signal. The number of elements parameter indicates how many numbers are in the input vector. The Initial seed parameter initializes the random number generator that the interleaver uses to determine the permutation. The interleaver is predictable for a given seed, but different seeds produce different permutations. The Random interleaver chooses a permutation table randomly using the **initial seed** parameter that user provide. By using the same **initial seed** value in the corresponding random de-interleaver, one can restore the permuted symbols to their original ordering. If the de-interleaver and the interleaver have the same value for initial seed, then the two are inverses of each other.

3 3GPP Interleaver

The input bits to the internal interleaver Turbo code are indicated by $x_1, x_2, x_3, \ldots x_K$, where K is any integer bit number and takes one value of $K \leq 5114$. The relation between the input bits to the Turbo code interleaver and the channel coding is defined by.

3.1 *Bits-Input to Rectangular Matrix with Padding*

The turbo code internal interleaver input bit sequence $x_1, x_2, x_3, \ldots x_K$ is presented as the rectangular matrix as follows.

1. The rectangular matrix rows, R, is such that

$$R = \begin{cases} 5, & if(K \leq 159) \\ 10, & if(160 \leq K \leq 200)\, or\, (481 \leq K \leq 530) \\ 20, & if(K = any\, other\, value) \end{cases}$$

The rows of rectangular matrix are 0, 1 … R − 1 from top to bottom.

2. The prime number is determined, which is to be used in the intrapermutation, p, and the rectangular matrix columns, C, is as follows

   if $(481 \leq K \leq 530)$ then $p = 53$ and $C = p$.
   else
   the minimum p is evaluated form the Table 2, and is given by
   $K \leq Rx(p+1)$, and C is determined, where

$$C = \begin{cases} p-1, & \mathit{if}\; K \leq R.(p-1) \\ p, & \mathit{if}\; R.(p-1) < K \leq R.p \\ p+1, & \mathit{if}\; R.p < K \end{cases}$$

   end if

   The rectangular matrix columns are numbered from left to right as 0, 1, …, C − 1.

3. The input sequence $x_1, x_2, x_3, \ldots\ldots, x_K$ is written row by row into the rectangular matrix of order $R \times C$, with y_1 in column 0 of row 0:

$$\begin{bmatrix} y_1 & y_2 & y_3 & \cdots\; y_C \\ y_{(C+1)} & y_{(C+2)} & y_{(C+3)} & \cdots\; y_{2C} \\ \vdots & \vdots & \vdots & \vdots \\ y_{((R-1)C+1)} & y_{((R-1)C+2)} & y_{((R-1)C+3)} & \cdots\; y_{RXC} \end{bmatrix}$$

where $y_k = x_k$ for k = 1, 2, …, K and if R × C > K, the dummy bits are padded such that $y_k = 0$ or 1, for k = K + 1, K + 2, …, R × C. These dummy bits are pruned away from the output of the rectangular matrix after intra-row and inter-row permutations.

4 Hybrid Interleaver

In the previous sections block and 3GPP interleavers are described. Now let us take another look at block interleavers. Consider matrix interleaver for the time and memory complexity analysis. Assume the incoming flow of data bits are framed as MXN array. The values of M and N depend on the expected size of burst error. The memory requirement at the interleaver stage will be MN bits (assuming bits as reference). But overriding of bits on filling and emptying can't be allowed, hence a memory of 2MN bits is required; MN bits of memory when writing the bits on MN array and separate MN bits which will be filled by reading from previous MN array.

Similar memory requirement is needed at the de-interleaver as well. Hence a total of 4MN bits of memory is required for MN bits of data. The time complexity is crucial thing in fast communications. Some of the practical installations omit the interleaving and de-interleaving in order to reduce the latency introduced by the interleaver pair. The minimum end-to-end delay due to the block interleaver is (2MN − 2M + 2). The other interleaver considered is 3GPP interleaver which is basically the improved matrix based block interleaver suggested by Third Generation Partnership Project. The time and memory constraints are close to that of standard block interleaver. When the number input frames is high then the complexity of both the interleaving schemes becomes similar because in 3GPP interleaver in addition to standard matrix writing and reading there are some additional operations which are somehow scalar operations.

Now if the total number of bits in input frame is split into subgroups and if the individual groups are processed using the normal interleavers the time complexity can be reduced. Let the input frame is split into 4 groups. So the total number of input bits to individual interleaver becomes MN/4. They can organized using M/2 × N/2 matrix. Hence the time complexity of individual interleaver becomes (MN/2 − M + 2). The total time complexity becomes 4 * (MN/2 – M + 2). But when the total input bits are split into four groups and applied to separate interleavers, the four groups of input bits are concentrated in that region only. That means the first quarter of bits is placed again in the first quarter only, similarly the remaining quarters. Hence a second stage of interleaving is proposed in which these quarters are interleaved based on a direct assignment without using another interleaver. Hence this scheme can be regarded as a two stage interleaving scheme shown in Fig. 1.

In this chapter block interleavers are used and any other interleavers and any other combinations can also be considered. The structure of the proposed interleaver is shown in Fig. 1. From the above discussion it is clear that by splitting the input frame into four parts the execution time will be reduced greatly. The Table 1 gives the comparison between the above interleavers theoretically.

5 Simulation Results

In this section the simulation results of the turbo codes with four varieties of interleavers are presented. The four varieties are block interleaver, 3GPP interleaver, no interleaver and the proposed hybrid interleaver. The input frame sizes are varied from 4 to 1024 and in this section input frames sizes 16, 64, 256 and 1024 bits are considered. First of all consider input frame size of 16. Note that the decoding algorithm used throughout this work is Max-log-MAP algorithm. The BER versus E_b/N_o performance for input frame size of 16 bits with different interleavers is shown below in Figs. 2 and 3.

The performance of turbo code with different interleavers seems to be almost similar, the simulation time is also in the same lines ranging from 8.28 s in the case of no interleaver to 14.71 s in the case of hybrid interleaver. The purpose of interleaver is to reorder the information bits so that the signal to noise ratio is high as the interleaver converts the burst error caused by impulsive noise to simple

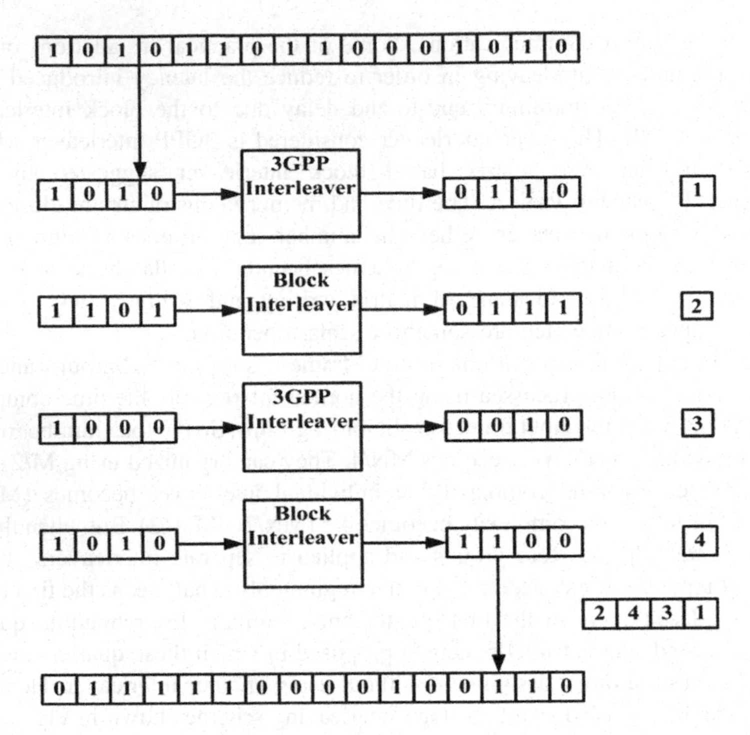

Fig. 1 Block diagram of hybrid interleaving scheme

Table 1 Theoretical time calculations for MN = 1024

MN	M	N	Block interleaver	Hybrid interleaver
1024	2	512	2046	2048
	4	256	2042	2040
	8	128	2034	2024
	16	64	2018	1992
	32	32	1986	1928
	64	16	1922	1800
	128	8	1794	1544
	256	4	1538	1032
	512	2	1026	8

errors. Hence use of interleaver is mandatory. But to check the performance the no interleaver case is considered. In the above case where the frame length is 16 bits, the BER performance of turbo code with no interleaver is similar to that of other cases as the chance of occurrence of impulsive noise increases as the length increases, hence the effect of impulsive noise in no interleaver case prevails when the frame length is high. Now consider the frame length to be 1024 bits. The BER versus E_b/N_o performance was given in the Figs. 4 and 5.

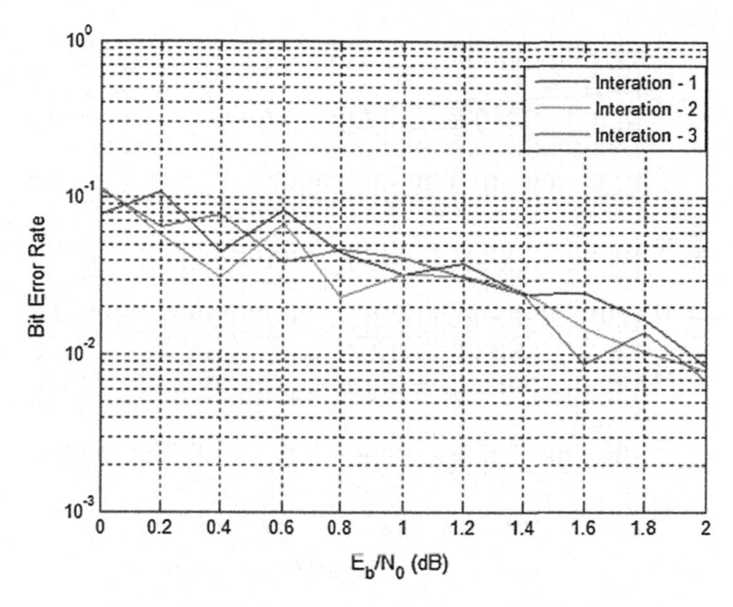

Fig. 2 BER versus E_b/N_o, 3GPP interleaver, input frame size = 16 bits

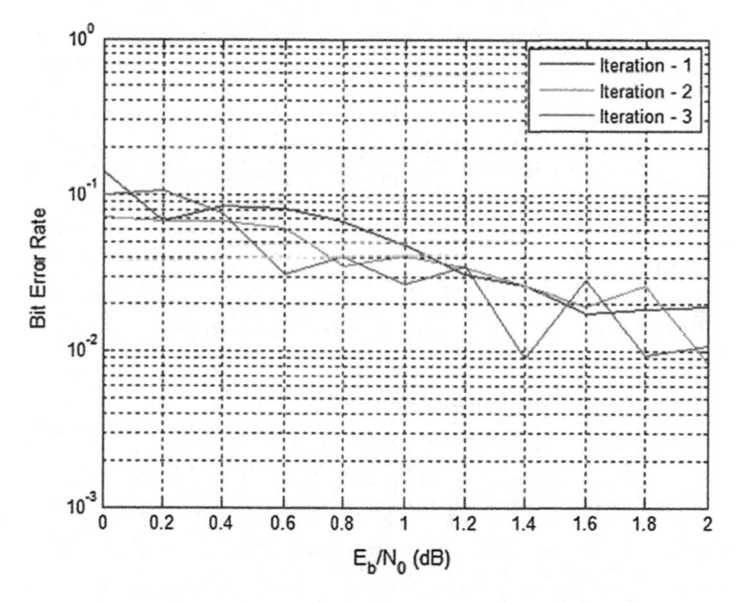

Fig. 3 BER versus E_b/N_o, hybrid interleaver, input frame size = 16 bits

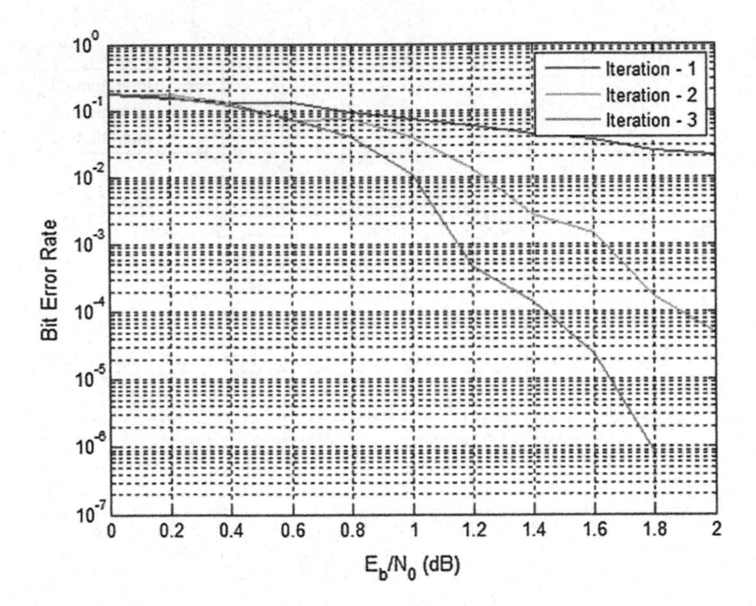

Fig. 4 BER versus E_b/N_o, 3GPP interleaver, input frame size = 1024 bits

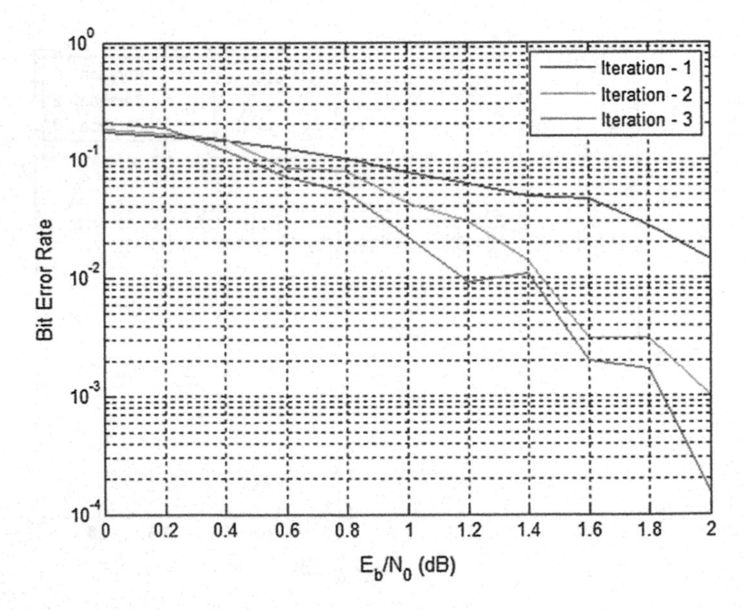

Fig. 5 BER versus E_b/N_o, hybrid interleaver, input frame size = 1024 bits

Table 2 Time required for interleaving and de-interleaving in seconds

	16-bits	64-bits	256-bits	1024-bits
No interleaver	8.2872	6.6354	10.6015	38.0655
Block interleaver	13.556	19.755	102.8677	1105.413
3GPP interleaver	11.8857	14.5764	753.506	4305.971
Hybrid interleaver	14.7156	11.4243	24.1712	148.2161

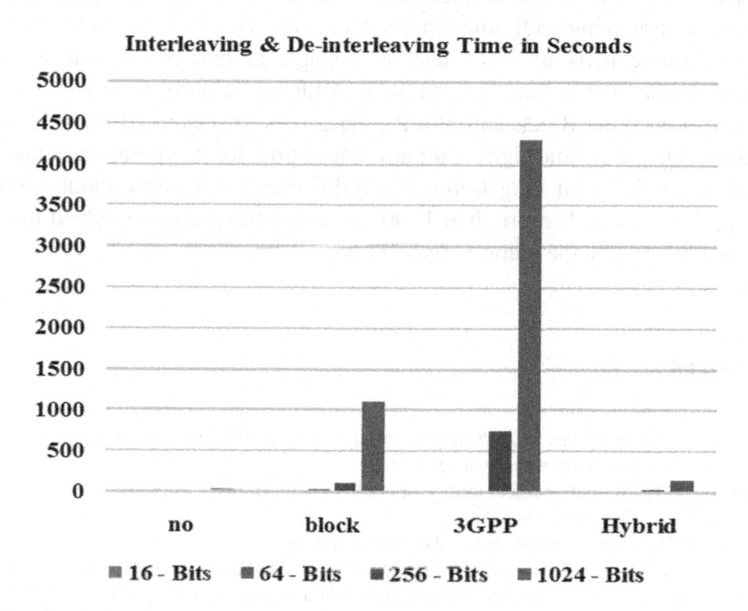

Fig. 6 Interleaving and de-interleaving time in seconds

As thought in the last paragraph the BER with no interleaver was very high compared to all other cases, i.e., with interleaver. The presence of interleaver has lowered the BER. Now the question is how much does this interleaving effect on the total time of execution? The interleaving has a serious effect on the total time required to transfer the data. In some applications in order to enable fast communication interleaver is dropped owing to high error rate. The interleaving scheme proposed in this chapter lowers the time required for coding and decoding. Table 2 gives the time required for completion of interleaving and de-interleaving in turbo codes. The Fig. 6 plots the data given in the Table 2. From the figure it is evident that the time required in case of hybrid interleaver is comparable to that of no interleaver case.

6 Conclusion

In this chapter an attempt has been made to reduce the time required for turbo coding and decoding in the presence of interleaving. Absence of interleaver, block interleaver and 3GPP interleavers are considered. A novel hybrid two stage interleaver was presented. The first stage of the interleaver deals with splitting of input frame and interleaving individual parts there off. The second stage deals interleaving complete parts of first stage as a single atomic block. The absence of interleaver presents a serious escalation in BER particularly for long frames and incurs very low time. Block and 3GPP interleavers lower the BER because of the interleaving but take some significant amount of time for its operation to take place. At times, for a 1024 bit long frame the 3GPP interleaver takes about 4300 s and block interleaver spends more than 1100 s whereas the proposed hybrid interleaver takes about 150 s for the same length frame.

References

1. Benedetto. S., Bieglieri. E.,: Principles of digital transmission with wireless applications. *Kluwer academic/Plenum publishers*, (1999)
2. Shannon. C.E,: A mathematical theory of communication. *Bell System Technical Journal*, vol. 27, pp. 379–423(1948)
3. Forney. G.D,: Concatenated codes. *MIT Press* (1966)
4. Seghers. J,: On the free distance of turbo codes and related product codes. *Swiss Federal Institute of Technology, Zurich, Switzerland, Diploma project 6613*, vol. 42, (1995)
5. Dolinar. S, Divsalar. D,: Weight distributions for turbo codes using random and nonrandom permutations. *TDA Progress Report*, vol.15, pp. 56–65, (1995)
6. Sun. J, Takeshita. O.Y,: Interleavers for turbo codes using permutation polynomials over integer rings. *IEEE Transactions on Information Theory*, no. 1, pp. 101–119, (2005)
7. Atousa H. S. Mohammadi, Weihua Zhuang,:Variance of the Turbo Code Performance Bound Over the Interleavers. IEEE Transactions on Information Theory, Vol. 48, (2002)
8. Prabhavati D. Bahirgonde, Shantanu K. Dixit,: BER Analysis of Turbo Code Interleaver. International Journal of Computer Applications (0975–8887) Volume 126 – No.14, (2015)
9. Maan A. Kousa,: Performance of Turbo codes with Matrix Interleavers. The Arabian Journal for Science and Engineering, Volume 28, Number 2B, (2003)
10. Ching-Lung Chi,: Quadratic Permutation Polynomial Interleavers for LTE Turbo Coding. International Journal of Future Computer and Communication, Vol. 2, No. 4, (2013).
11. Arya Mazumdar, Chaturvedi. A.K, Adrish Banerjee,: Construction of Turbo Code Interleavers from 3-regular Hamiltonian Graphs. IEEE Communications Letters, Vol. 10, No. 4, (2006)
12. Jing Sun, Takeshita. O.Y,: Interleavers for Turbo Codes Using PermutationPolynomials Over Integer Rings. IEEE Transactions On Information Theory, Vol. 51, No. 1, (2005)

Multi User Authentication in Reliable Data Storage in Cloud

Sk. Yakoob, V. Krishna Reddy and C. Dastagiraiah

Abstract Information out sourcing is the fundamental centering term in distributed computing. More information proprietors accomplishes in outsource their information into cloud servers. In any case, touchy data ought to be secured before outsourcing for protection necessities, which obsoletes data utilization like catchphrase based papers recuperation. A safe multi-catchphrase evaluated seek arrangement over secured thinking information, which in the meantime encourages intense update capacities like expulsion and position of records. In particular, the vector space outline and the broadly utilized TF–IDF configuration are blended in the list development and inquiry. But the ability of searching encryption is main challenge in secure multi keyword search with user activities of deletion, revoking of users with multiple user access control in cloud computing. In this paper we propose to BGKM (Broadcast Group Key Management) schema for multi user access control in user activities like data insertion, deletion and in user revocation in real time cloud application development. We evaluate the security of our BGKM plan and evaluate it with the current BGKM techniques which are mostly used for data accessing in real time cloud computing for achieving real time applications in cloud. Our experimental results show efficient data extraction from multi user access in searching of data from cloud servers.

Keywords Cloud computing · BGKM · Secure multi key word search · Quality based encryption

Sk. Yakoob (✉) · V. Krishna Reddy · C. Dastagiraiah
Department of CSE, K L University, Vijayawada, India
e-mail: yakoob_cs2004@yahoo.co.in

V. Krishna Reddy
e-mail: vkrishnareddy@kluniversity.in

C. Dastagiraiah
e-mail: dattu5052172@gmail.com

S.C. Satapathy et al. (eds.), *Computer Communication, Networking and Internet Security*, Lecture Notes in Networks and Systems 5,
DOI 10.1007/978-981-10-3226-4_54

1 Introduction

Distributed computing has been considered as new of business IT offices, which can mastermind tremendous wellspring of handling, stockpiling and programs, and permit clients to appreciate mainstream, helpful and on-interest system access to a circulated offer of configurable preparing sources with extraordinary execution and minimal financial cost. By considering a portion of the components in distributed computing outsourcing delicate data, for example, messages individual record subtle elements, money related organization points of interest, to remote disjoins being security concerns in cloud information stockpiling. The cloud administration suppliers (CSPs) that keep the data for clients may get to clients' fragile data without authorization [1, 2]. A general way to deal with ensure the data protection is to secure the data before outsourcing. Retrieveable security methods empower the client to store the secured information to the thinking and perform catchphrase and key expression search for over figure content area. So far, numerous works have been suggested under different risk models to achieve various look for performance, such as single keyword and key phrase look for, likeness look for, multi-keyword boolean look for, rated look for, multi-keyword rated look for, etc. Among them, multi-keyword rated look for accomplishes more and more attention for its practical usefulness. As shown in the above search data from user search content. The csearch requirements is as shown in Fig. 1.

Consider a server that sends information to a gathering of clients in a multicast/broadcast session through an open correspondence channel (as shown in Fig. 2). To guarantee information privacy, the server offers a mystery gathering key K with all gathering individuals and encodes the show information utilizing a symmetric encryption calculation with K as the encryption key [2, 3]. This procedure is called upgrade or re-keying. The procedure to keep up, circulate and upgrade the gathering keys is called gathering key administration. In this document, we recommend a new BGKM plan which, to the best of our information, is the first provably protected BGKM plan [4]. Our new plan is versatile, effective and

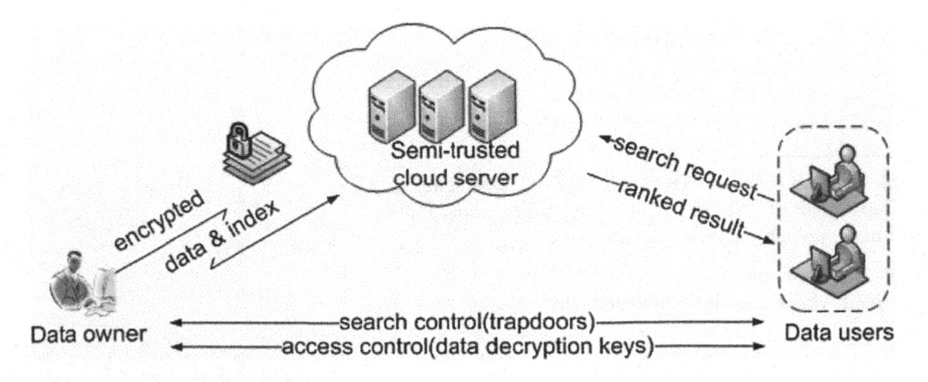

Fig. 1 Cloud data search retrieval based on user search requirements

protected. It keeps the use of protected personal interaction programs little by not demanding any private communications when rekeying occurs either among the team associates or between the key server and a persisting team participant.

The remainder of this paper organised as follows. Section 1 describes Related work for out sourcing from cloud servers. Section 2 describes Background approach for searching content of user in real time cloud data storage. Section 3 describes multi user access control in data outsourcing in cloud. Section 4 Describes procedure for searching content in cloud based on multi user process. Section 5 describes conclusion for multi user access in real time cloud applications.

2 BGKM Security Procedure with Architecture

In this area, we officially determine a transmitted team key control plan and its protection, and recommend a new team key control plan which allows any legitimate participant in the group which keeps an personal registration symbol to obtain a typical team key [4, 5].

Definition 1 (BGKM): Braodcast Group Key Management (BGKM) is consisting of two entities: (1) a key server (Svr), and (2) group members (Usrs), a chronic transmitted channel from Svr to all Usrs, an ephemeral personal channel3 between Svr and each personal Usr, and the following phases:

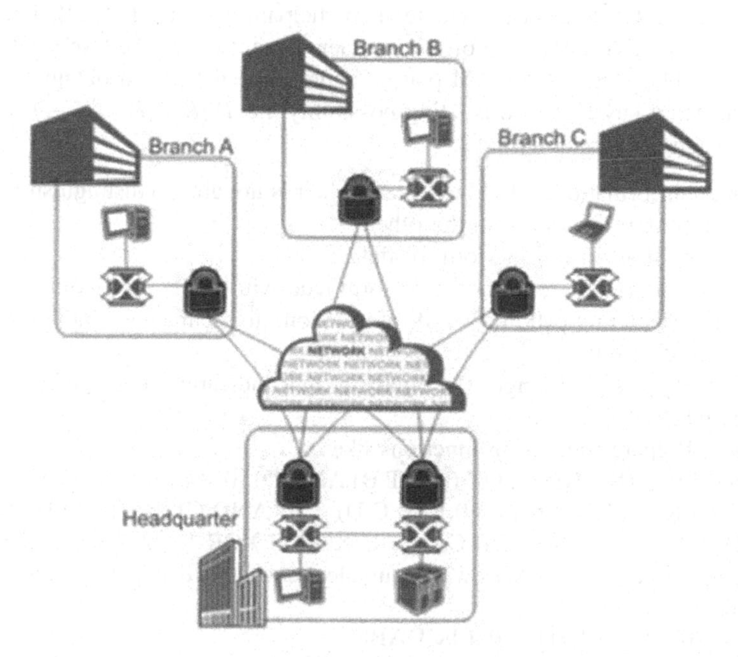

Fig. 2 Advanced key distribution in cloud server environment

ParamGen Svr requires as feedback a protection parameter k and results a set of community parameters Param, such as the sector KS of possible key principles [6, 7].

TkDeliv Svr delivers each Usr an personal registration symbol (IST) through a personal route.

KeyGen Svr selects a distributed team key K $ KS. In accordance with the ISTs of Usrs, Svr computes a set of principles PubInfo. Svr keeps K key, and shows through the transmitted channel PubInfo to all team associates Usr.

KeyDer Usr uses its IST and PubInfo to estimate the distributed team key K. Update When the distributed team K can no more be used (e.g., when there is a modify of group characteristics such as be a part of and leaving of team users), Svr produces new team key "K" and "PubInfo", then shows the new PubInfo to the team. Each Usr uses its IST and the new "PubInfo" to estimate the new distributed team key "K". We contact the program after the Update phase a new "session". The Upgrade stage is also known as a rekeying stage [8].

3 BGKM with SHAMIR

A BGKM plan should allow a real team participant to obtain the distributed team key, and prevent anyone outside the team from doing so. Officially discussing, a BGKM plan should fulfill the following protection qualities. It must be appropriate, audio, key concealing, and forward/backward key defending [9, 10].

(1) Correct: Let Usr be a present team participant with an IST [9]. Let K and PubInfo be Svr's outcome of the KeyGen stage. Let "K" be Usr's outcome of the KeyDer stage. A BGKM plan is appropriate if Usr can obtain the appropriate team key K with frustrating possibility, i.e. $P_r[K = K'] \geq 1 - f(k)$ where f is negligible function for k.

With the information of PubInfo, the attacker is not able to distinguish one of its selected important factors from the other.

Input: Sending files in the form of data.

Output: Encrypted form/Decrypted form data with buffer size requirements.

Step 1: Appending padding bit of information, divide message into 64 bits with multiples of 512 bits.

Step 2: Append the length (In binary format indicating length of the original message into 64 bit).

Step 3: Prepare processing functions like

f(t;B,C,D) = (B AND C) OR ((NOT B) AND D) (0 <= t <= 19) f(t;B,C,D) = B XOR C XOR D (20 <= t <= 39) f(t;B,C,D) = (B AND C) OR (B AND D) OR (C AND D) (40 <= t <= 59) f(t;B,C,D) = B XOR C XOR D (60 <= t <= 79).

Step 4: Initiate buffers sizes with equivalent constants depending on the number of words:

H0 = 0x67452301 H1 = 0xEFCDAB89

H2 = 0x98BADCFE H3 = 0x10325476
H4 = 0xC3D2E1F0
Step 5: Processing Message in 512 bit blocks: K(0), K(1),…, K(79): 80 Processing Constant Words H0, H1, H2, H3, H4, H5: 5 Word buffers with initial values.

Algorithm 1: Shamir Algorithm for generate multiple keys with different signatures.

(2) **Sound:** Let Usr be a person without a legitimate IST. A BGKM is sound if the likelihood that Usr can get the right gathering key K by substituting the IST with a worth value that is definitely not one of the legitimate ISTs and afterward taking after the key induction stage KeyDer is immaterial.
(3) **Key concealing**: A BGKM is key concealing if given PubInfo, any gathering which does not have a substantial IST can't recognize the genuine gathering key from an arbitrarily picked esteem in the key-space KS with non-negligible likelihood.
(4) **Forward/in reverse key ensuring**: Algorithm 1 organizes as follows: Input as files and getting output as encrypted and decrypted format for accessing permitted files in cloud. In step 1 convert original text to formatted text with respect to binary numbers for padding to arrangement of all the bits in required format. In step 2, assign each bit length to 64 bit arrangement for uploaded text from files. After assign bit length to all the text preset in uploaded files using step 3 we perform X-OR operations between each assigned padding length for generating different hash functions with respect to its primary verification for bit length in cloud data storage. Initiates equivalent primary number verification with hash function in step 4 then it will generates different keys for sharing multiple users at a single file sharing in cloud data sharing. After sharing these keys data will be encrypted and decrypted file in cloud [11–13].

4 Experimental Evaluation

In this section we analyze the computational performance of ACV-BGKM. We imitate the KeyGen stage at Svr and the KeyDer stage at Usrs. In the research, we differ both the dimension the actual primary area Fq and the dimension the number of Usrs, and evaluate the Svr-side and Usr-side calculations time [12, 13]. To highlight on the mathematics functions, we do not depend plenty of here we are at hashing functions in the research. The rule is published in the Magma scripting language, and uses Magma's inner collection for limited area mathematics and fixing linear systems (Fig. 3).

The research was conducted on a machine running GNU/Linux kernel edition 2.6.9 with a Double Primary AMD Opteron (TM) Processer 2200 MHz and 16 G bytes storage. Only one processor was used for calculations. The following

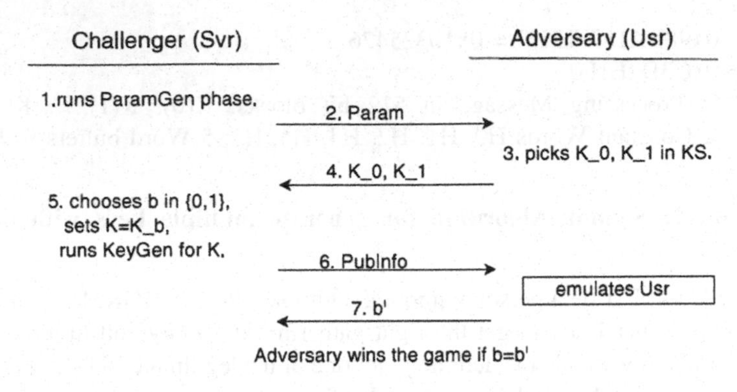

Fig. 3 The attacker activity for BKGM's key concealing residence

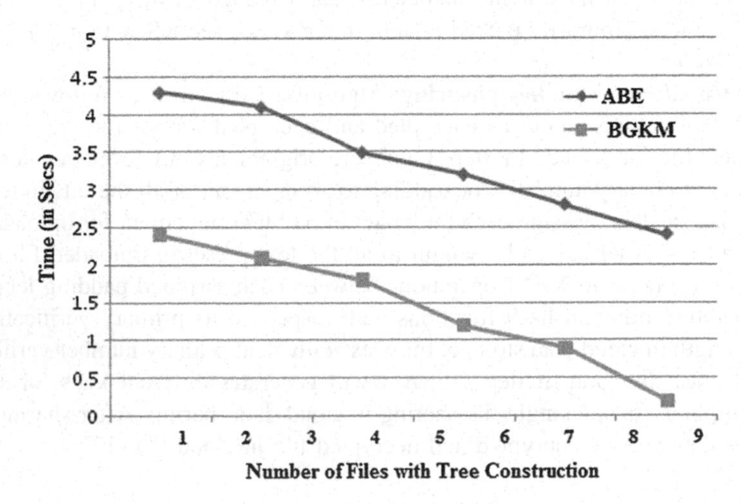

Fig. 4 Comparison of Svr in ABE and Usr in BGKM in term of time with respect to tree generation

diagrams tell performance of broadcast group key management with access control aprtunuties. Figure 4 reviews the ACV-BGKM operating time at Svr and Usr for team dimensions 600, 800, 1000 and 1200, and with the dimension the primary area which range from 64 to 128 pieces [5, 14]. The operating time is averaged over 20 versions. As proven in the determine, the common calculations time improves in common as the dimension the primary area improves. The real operating time relies on the prime field that is selected and the way area mathematics is conducted in Magma.

Figure 5 reviews the ACV-BGKM operating time at Svr and Usr for set area measures (in bits) 64, 80, 96 and 112, with the dimension the team which range from 100 to 2000 associates. The running time is averaged over 20 versions. Further performance gains can be carried out when the primary variety q is selected

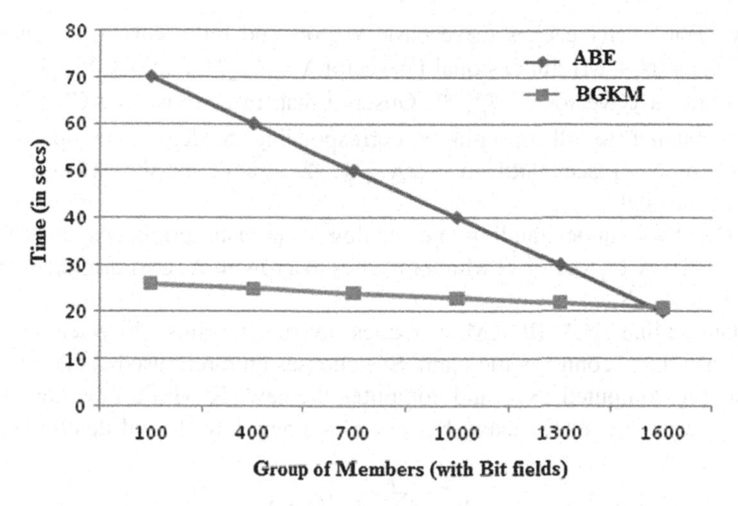

Fig. 5 Computational time efficiency in application process of ABE and BGKM in access control in usability

to be in a unique type, e.g., a common Mersenne primary (Salinas prime), for which quick area mathematics in F_q is available [5].

We lay the base towards a quicker method of ACV-BGKM, known as FACVBGKM, at the price of extra area, pre-computation and an catalog. It follows a baby step-giant-step (BSGS) rekey procedure where irregular massive actions are conducted analogous to the ACV-BGKM plan [2]. However, the amortized computational and interaction price is reduced by the release of regular small actions. Due to area restrictions, we only describe the changes to the ACV-BGKM method below.

Protocol (FACV-BGKM): FACV-BGKM performs under identical circumstances as ACV-BGKM. ParamGen Svr chooses N" = N + M where M $\geq$ N. For the highest possible protection and minimum amortized price, it is suggested to set M = N.

TkDeliv Svr assigns an index $i(1 \leq i \leq N)$ chosen consistently at unique, to each of the n current customers. Svr selects N ISTs and delivers an IST and corresponding catalog to each customer. The staying N − n precomputed ISTs are used for rekeying when new customers be a part of the team [12].

KeyGen Svr makes an $N \times (N \times M)F_q - matrix$ A where for a given $i(1 \leq i \leq N)$.

$$a_{i,j} = \left\{ \begin{array}{cc} 1 & if\, i = j \\ 0 & if\, 1 \leq j \leq N\, and\, i \neq j \\ H(ist_i || Z_j) & if\, N < j \leq N + M \end{array} \right\}$$

Like in ACV-BGKM method, Svr determines the zero area of A with a set of its M basis vectors, and chooses an accessibility management vector Y as one of the primary vectors. Svr caches these basic vectors and represents Y as "used." Svr constructs an (N + M)-dimensional Fq-vector $X = (\sum_{i=1}^{n} K.e_i^T) + Y$ where e_i is the ith standard basis vector of F_q^{N+M}. Observe that, in contrast to ACVBGKM, the key is included to all the places corresponding to legitimate spiders. Like, ACVBGKM, Svr places PubInfo = &X, (z1, z2… zM)', and shows PubInfo via the broadcast channel.

KeyDer Usri, understanding the catalog i and isti, originates the (N + M)-dimensional row Fq-vector vi which matches to a row in A. Usri originates the team key as K = vi · X.

Update Unlike ACV-BGKM, Svr does not run the finish KeyGen stage again [9]. If a new Usrs connects the team, Svr chooses an rarely used catalog t and is it from the pre-computed ISTs and computes the new ^X with a new key ^K. If an present User results in the team, Svr chooses a new key ^K and determines a new

$$\hat{X} = (\sum_{j=1}^{n} K.e_i^T) + Y$$

Where ^Y is an "unused" foundation vector which is among the pre-computed set in KeyGen stage. Svr marks ^Y as "used", and shows only ^X while maintaining the other community details the same. We contact these functions a "baby-step rekey" since it only needs time O(N) in comparison to O(N3) in ACV-BGKM. A finish KeyGen (i.e. "giant-step rekey") eventually O(N3) needs to be performed every M Up-dates since otherwise a team participant who has been legitimate for the last M classes can restore the zero area of A, thus the matrix A itself. A giant-step rekey also needs to be conducted with a resized matrix A before M updates, if the variety of connects exceeds N − n after the present giant-step rekey to provide new customers [7, 13].

As described above, the KeyGen price is amortized to acquire a plan quicker than the ACVBGKM scheme. Due to area restrictions, we bypass the security/performance analysis of the FACV-BGKM plan from this document. We observe that it is an exciting start analysis problem to choose the maximum M and N principles based on the program situation.

5 Conclusion

It could be a critical yet troublesome future work to outline a vivacious retrievable security arrange for whose redesigning capacity can be finished by thinking server just, in the mean time orchestrating the capacity to bolster multi-watchword evaluated seek. Moreover, as the a large portion of works about retrievable security, our arrangement for the most part perspectives the procedure from the thinking server. Really, there are numerous safe troubles in a multi-client arrangement. We have proposed another BGKM arrangement ACV-BGKM which is overseen by a trusted key server, and permits any true blue client in the group to procure a circulated group key all alone from transmitted group subtle elements. The plan reduces the use of personal peer-to-peer communication programs, and only uses a transmitted route to provide new rekeying messages when the team key needs to be modified. The interaction expense is straight line with the number of customers in the team. The plan uses only effective hash functions and straight line geometry over finite areas in calculations, and does not require any security plan.

References

1. Zhihua Xia, Xingming Sun,: A Secure and Dynamic Multi-keyword Ranked Search Scheme over Encrypted Cloud Data. IEEE trans on parallel and distributed systems, vol. 99, pp: 340–352, (2015).
2. Ren. K., Wang. C., Wang. Q.,: Security challenges for the public cloud. *IEEE Internet Computing*, vol. 16, pp. 69–73, (2012).
3. Kamara. S., Lauter. K,,: Cryptographic cloud storage. *Financial Cryptography and Data Security*. Springer, pp. 136–149 (2010).
4. Gentry. C.: "A fully homomorphic encryption scheme," Ph.D. dissertation, Stanford University, (2009).
5. Wang. C, Ren. K, Yu. S, K. M. R. Urs.: Achieving usable and privacy-assured similarity search over outsourced cloud data. *INFOCOM, 2012 Proceedings IEEE*. IEEE, pp. 451–459 (2012).
6. Cao. N, Wang. C., Li. M, Ren. K., Lou. W.,: Privacy-preserving multi-keyword ranked search over encrypted cloud data in *IEEE INFOCOM*, pp. 829–837, April (2011).
7. Sun. W., Wang. B., Cao. N., Li. M, Lou. W, Hou. Y.T., Li. H,: Privacy-preserving multi-keyword text search in the cloud supporting similarity-based ranking. in *Proceedings of the 8th ACM SIGSAC symposium on Information, computer and communications security*. ACM, pp. 71–82 (2013).
8. Orencik. C, Kantarcioglu. M., Savas. E.: A practical and secure multi-keyword search method over encrypted cloud data. in *Cloud Computing (CLOUD), 2013 IEEE Sixth International Conference on*. IEEE, pp. 390–397 (2013).
9. Zhang. W., Xiao. S., Lin. Y., Zhou. T., Zhou. S.: Secure ranked multi-keyword search for multiple data owners in cloud computing. in *Dependable Systems and Networks (DSN), 2014 44th Annual IEEE/IFIP International Conference on*. IEEE, pp. 276–286 (2014).
10. Kamara. S., Papamanthou. C., Roeder. T.: Dynamic searchable symmetric encryption. in *Proceedings of the 2012 ACM conference on Computer and communications security*. ACM, pp. 965–976, (2012).
11. Kamara. S, Papamanthou. C.: Parallel and dynamic searchable symmetric encryption in *Financial Cryptography and Data Security*. Springer, pp. 258–274 (2013).
12. Cash. D, Jarecki. S, Jutla. C, Krawczyk. H., Rosu. M. C., Steiner. M.: Highly-scalable searchable symmetric encryption with support for boolean queries. in *Advances in Cryptology– CRYPTO(2013)*. Springer, pp. 353–373 (2013).
13. Yu, Wang. S., C., Ren, K., & Lou, W.,: Achieving secure, scalable, and fine-grained data access control in cloud computing. In *INFOCOM, (2010) Proceedings IEEE*, pp 1–9, (2010).
14. Shang. N., Nabeel. M., Paci. F., Bertino. E.: A privacy-preserving approach to policy-based content dissemination. in *ICDE'10: Proceedings of the 2010 IEEE 26th International Conference on Data Engineering*, 1–33, (2010).

Bio-Inspired Approach for Energy Aware Cluster Head Selection in Wireless Sensor Networks

D. Rajendra Prasad, P.V. Naganjaneyulu and K. Satya Prasad

Abstract Since 2000, the usage of Wireless Sensor Networks (WSNs) in the military and civilian applications is increased rapidly. WSNs consist of low cost hardware which acts as multi-functional sensor nodes and are linked to the base station to efficiently aggregate and transmit the data. However, the limitation of sensor nodes is limited battery power. Efficient cluster head (CH) selection manages the WSNs in reducing the overall energy consumption. The cluster head combinations the data from the neighbouring nodes and transmits to the base station. The selection of cluster head in WSN is a Non-deterministic Polynomial-hard problem. In this paper, a bio inspired approach called firefly algorithm is used for selection of CH in WSNs. The synchronous firefly algorithm selects the optimal CH node among all nodes participated in the network, which also increases the performance of the network. The experimental results show that the proposed algorithm has effective CH selection procedure when compared to the other existing algorithms. The proposed algorithm reduced the packet loss ratio by 8.32% and it improved the reduction of energy consumption of overall network.

Keywords Wireless sensor networks · Cluster head · Firefly algorithm · Energy

---

D. Rajendra Prasad (✉)
Department of ECE, St. Ann's College of Engineering & Technology,
Chirala, India
e-mail: rp.devathoti@gmail.com

P.V. Naganjaneyulu
Department of ECE, MVR College of Engineering & Technology,
Vijayawada, India
e-mail: pvnaganjaneyulu@gmail.com

K. Satya Prasad
Department of ECE, Jawaharlal Nehru Technological University,
Kakinada, India
e-mail: prasad_kodati@gmail.com

S.C. Satapathy et al. (eds.), *Computer Communication, Networking and Internet Security*, Lecture Notes in Networks and Systems 5,
DOI 10.1007/978-981-10-3226-4_55

1 Introduction

Since 2000, WSNs [1] are majorly incorporated into the field of military applications as well as civilian applications. The major usage of WSNs is in biomedical applications, target tracking, habit monitoring, building management systems, surveillance and disaster management [2]. WSNs have minimum storage capacity and limited battery power and it is not possible to replace the battery in the WSNs. Hence, the utilization of energy in the WSNs is very crucial and decides the network performance. Therefore, energy is concerned as the major issue in WSNs. There designing energy efficient protocol is more important in WSNs [3]. In general, the working procedure of WSNs is to collect the information from the sensor nodes and sending it to the base station. Figure 1 shows the complete hierarchy of working procedure of WSNs.

In WSNs, for optimizing the nodes energy, some CH nodes are selected and these CH nodes aggregate data from the neighbouring nodes. The CH node sends the aggregated data to the BS, which reduces the network overhead and leads to the reduction of energy consumption.

LEACH is one of the efficient protocols which randomize the selection of CH nodes among the neighbouring sensor nodes. The basic principle of LEACH protocol is selection of CH node periodically. The CH node is organized with two stages. One is selection of cluster nodes and another one is data communication between the cluster nodes. The cluster head formation is done based on the remaining energy in the nodes. At the cluster formation, each node is assigned with a random number and is compared with the threshold value μ(n). If the number is less than the threshold value μ(n), then the node is selected as CH node, otherwise it

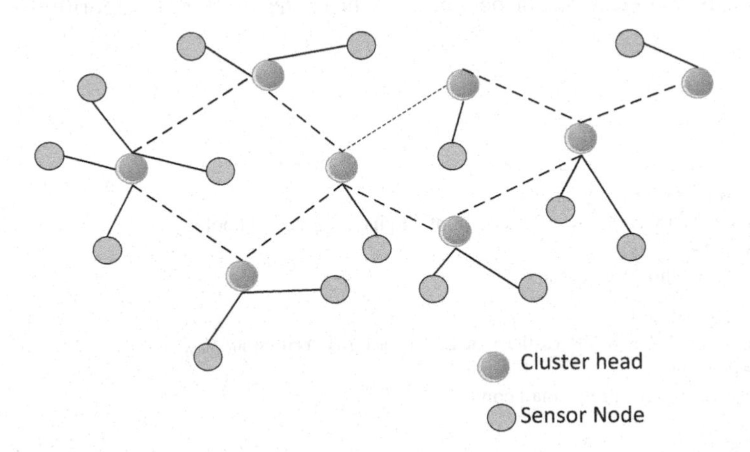

Fig. 1 Wireless sensor networks

is selected as the normal sensor node in the cluster. The threshold value for the cluster is given in (1)

$$\mu(n) = \begin{cases} \frac{\alpha}{1-\alpha\times(L \bmod (1/\alpha)} & if\, n \in N \\ 0 & if\, n \notin N \end{cases} \quad (1)$$

Here, n is the current node, N is the set of nodes in the cluster, α is the ratio of CH nodes in the cluster, L represents the number of rounds.

The LEACH protocol completely depends on the probability model for selecting the CH nodes. If the CH nodes are nearer to each other, it leads to high energy consumption, so this LEACH protocol is failed to address this issue. To overcome the discussed issue, many heuristic algorithms were proposed such as particle swarm optimization (PSO), genetic algorithm (GA) and ant colony optimization algorithm (ACO). In this paper we are addressing the energy minimization problem by the firefly algorithm. The firefly algorithm works with principle of intensity of light produced by the fireflies. The major advantage of this algorithm is the avoidance of multiple CH nodes.

2 Related Work

In WSNs, Energy consumption is treated as the major issue. To overcome this problem, many studies have been represented by the researchers. In [4], the authors proposed energy aware cluster based optimization mechanism. This mechanism is concentrated on reducing the cluster heads and it achieved 16.3% reduction in overall energy consumption of the network. They managed to reduce the delay between the sensor nodes. In [5], the authors presented the survey on energy consumption factors and the existing approach to deal with the energy consumption in WSNs. They made the comparative analysis on all existing algorithms and presented the pros and cons of all algorithms with respect to different parameters. The authors in [6] proposed the prediction based energy efficient clustering algorithm for minimizing the energy consumption at the broadcasting the information from the cluster head. They made an attempt to improve the overall performance by considering the factors such as delay and throughput. In [7], Beacan et al., presented the model for minimizing energy consumption based on the distance of the nodes form the base station. The bio inspired approaches for CH selection and cluster development in the WSNs had gained popularity in recent years. The approaches such as particle swarm optimization genetic algorithm and ant colony optimization algorithms are extremely popular in this category. In [8], kuila et al., proposed the CH selection mechanism by using the multi-objective function. The mechanism identifies the CH node energy consumption and end to end delay between nodes for forwarding the packets. In the proposed model, each node is assigned as particle in the algorithm. In [9], the author combined both LEACH and PSO algorithm for finding the suitable CHs and sensor nodes with in the cluster. So, this paper is concentrated on employing the firefly algorithm as the optimization technique for finding the CH node within the cluster.

3 Problem Formulation

The problem formulation for the WSNs are carried out using the graph G = (V, E), where V is the vertices and E is the edges and each edge is associated with some weight W that are associated with some parameters. The objective function for the firefly algorithm is given in Eq. 2.

$$\text{Problem: } \min F(x) = \sum_{i=1}^{N} W_i x_i$$

$$\text{Subjected to } x_i = \begin{cases} 1 & \textit{if edge } E_i \textit{ is selected} \\ 0 & \textit{otherwise} \end{cases} \tag{2}$$

The objective function f(x) is organized with three parameters, end to end delay, packet delivery ratio and energy consumption. The objective function is to minimize the stated parameters.

4 Energy Consumption Model for WSNs

The energy consumption of the two nodes which are in the communication is shown in Fig. 2.

The transmission energy for Q bit packet for a distance of d between two nodes is given by Eq. 3.

$$E_T(Q,d) = \begin{cases} QE_e + Q\varepsilon_f d^2, & \textit{where } d < d_0 \\ QE_e + Q\varepsilon_a d^4, & \textit{where } d > d_0 \end{cases} \tag{3}$$

$$d_0 = \sqrt{\frac{\varepsilon_f}{\varepsilon_a}} \tag{4}$$

The receiving energy for Q bit packet is given by Eq. 5.

$$E_R(Q) = QE_e \tag{5}$$

Fig. 2 Energy consumption model for two nodes

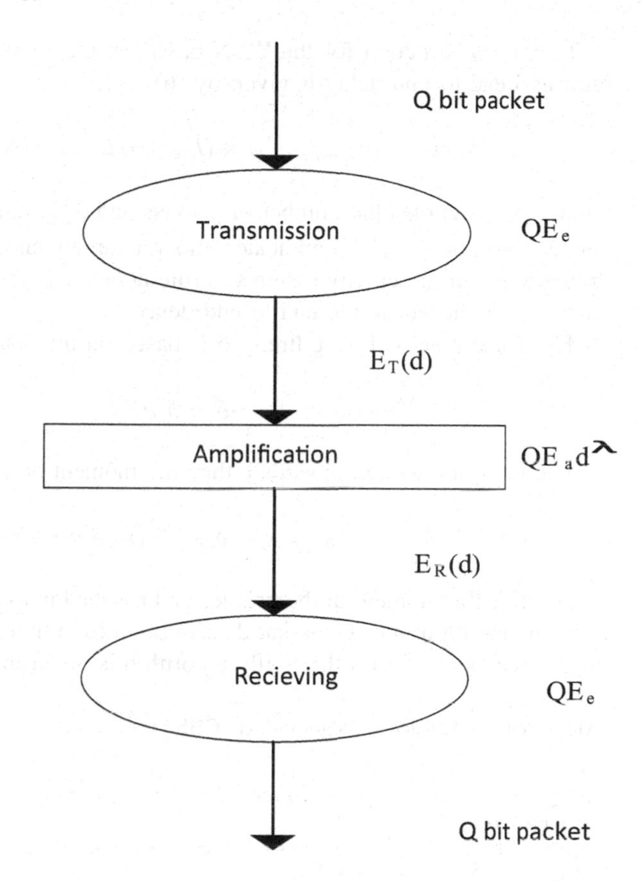

where Q is the number of bits E_T is the transmission energy and E_R is the receiving energy and d is the distance between the nodes. E_e, E_f, E_u are the coefficients of the energy consumption model.

5 Firefly Algorithm for CH Selection

Due to the interesting behaviour of the fireflies, in late 2007 and 2008, Cambridge University developed firefly algorithm. The firefly algorithm has the following properties

- Fireflies attracted to each other without considering the gender, as all the fireflies are unisex.
- The firefly can be attracted by the other firefly based on the intensity of the light. The value of the intensity of light is high means there is a chance of attracted by the other firefly.
- The objective function defines the light intensity of the firefly.

The fitness function for the WSN based on the energy calculation, packet loss ratio and end to end delay is given by (6)

$$f(x) = \{(\rho_{drop}/\rho_{total}) \times (E_{rem}(i)/E_{initial})\}/(\exp^{-e_{delay}/e_{allow}}) \tag{6}$$

where, ρ_{drop} denotes the number of packets and ρ_{total} indicates the total number of packets assigned. $E_{rem}(i)$ indicates the remaining energy at node i and $E_{initial}$ indicates the initial energy assigned to the node i, e_{delay} is current end to end delay and e_{allow} is the allowable end to end delay.

The attractiveness of the firefly δ is based on the distance 'd' as given in (7).

$$\delta = \delta_0 e^{-\vartheta d^2} \tag{7}$$

The firefly i attracted towards j, then the moment of the firefly is given in Eq. 8.

$$\kappa_i = \kappa_i + \delta_0 e^{-\vartheta d^2}(\kappa_i - \kappa_j) + r\omega \tag{8}$$

where, K is the moment of the fireflies and r is the randomized variable and ω is the uniform distribution or Gaussian distribution of the time. The algorithm for cluster head selection based on the firefly algorithm is given in Algorithm 1.

Algorithm 1: CH selection algorithm (FFCHSA)

```
Begin
Step 1: Initialize the nodes with the parameters /*Initial population
of fireflies*/
Step 2: Evaluate the fitness function of each node within the cluster
based on the Eq. 6.
Step 3: Sort the nodes based on the fitness function and find the current
best
Step 4: Compare the fitness values of the nodes
If (δj > δi)
Make the δj as the current best node
Step 5: Assign all the nodes to the δj
Step 6: Evaluate the distance d and update the attractiveness
Step 7: Go to step 2 and repeat the process until all the nodes in the
cluster completes its fitness function
Step 8: Select the best node as the cluster head
End
```

6 Simulation Environment

To evaluate the proposed FFCHSA algorithm, we consider the other existing algorithms such as Genetic Algorithm [10] and Particle Swarm Optimization algorithm [11]. The experimental setup is carried by using the NS2 simulator.

Table 1 Simulation parameters

Parameter	Value
Area size	1250 m × 1250 m
Number of nodes	100
Traffic type	CBR
Transmission range of the node	250 m
Maximum transmission power	0.28 W
Electronics energy	50 nJ/bit
Amplifier energy	100 pJ/bit
Initial energy of the node	40 J
Simulation time	500 s

To illustrate the mechanism of proposed algorithm, we consider the network of size 1250 m × 1250 m with 100 nodes randomly distributed. If the energy of the node is down, then that node is treated as the dead node. The average node degree is taken as 3–5 nodes and the traffic type for the network is taken constant bit rate (CBR) with 256 byte data packet. The mobility rate of the node is considered as 0–15 m/s. The complete simulation parameters are given in Table 1.

6.1 Results Evaluation

The simulation is carried out with NS-2 Simulator. The performance of the proposed algorithm is compared with other existing algorithms. The parameters used for the performance evaluation are packet loss, energy consumption, end to end delay and the frequency of change in cluster head. Figure 3 shows the packet loss of the proposed FFCHSA algorithm with the other existing algorithms. The

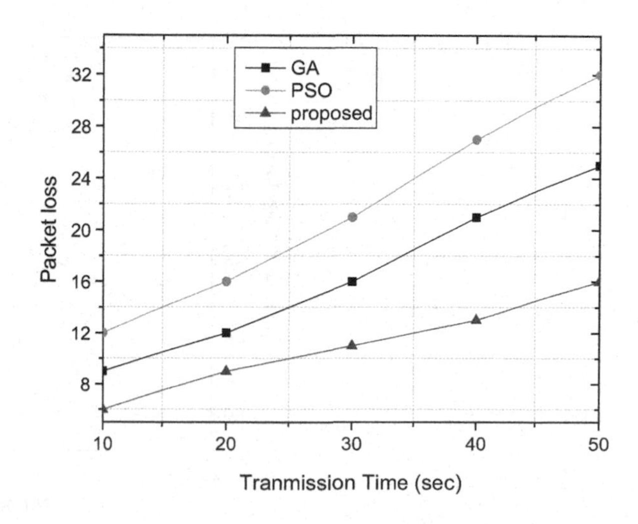

Fig. 3 Packet loss versus transmission time

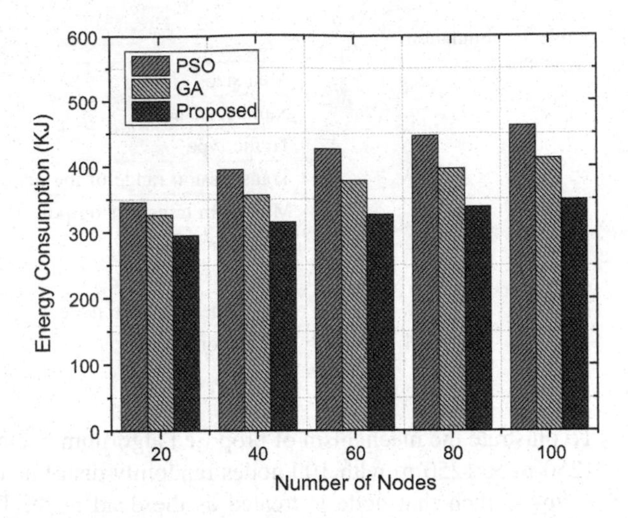

Fig. 4 Comparison of energy consumption

proposed algorithm recorded the less packet loss ratio by 8.32% when compared to the GA and PSO algorithms. Figure 4 represents the energy consumption of three algorithms. The energy consumption of proposed algorithm is recorded as 20% less while comparing with the other two algorithms. This is due to the effective utilization of nodes and reduction of number of CH nodes in the FFCHSA algorithm. The plot in the Fig. 5 represents the end to end delay between the sensor nodes of the proposed FFCHSA, GA and PSO algorithm. It is identified that the proposed FFCHSA algorithm recorded the minimum end to end delay and it improves the performance of the overall network.

Figure 6 shows the comparison results of changing frequency of cluster heads. If the cluster head is stable, the energy consumption of the network is minimal otherwise, the energy consumption drastically increases with the change of cluster

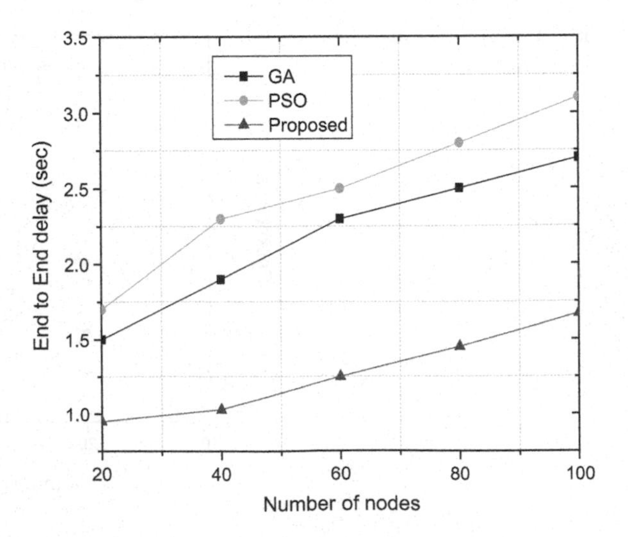

Fig. 5 Comparison of end to end delay

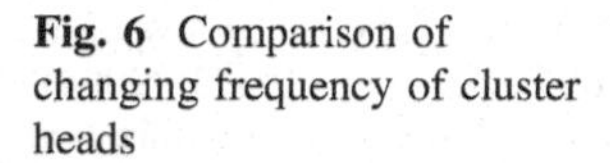
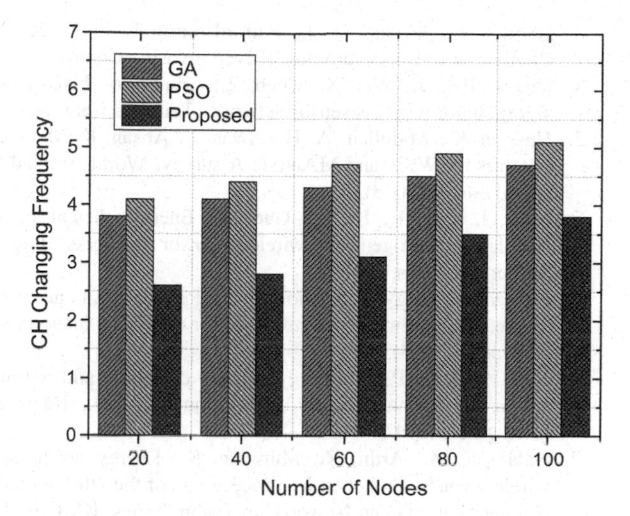

Fig. 6 Comparison of changing frequency of cluster heads

heads. In Fig. 6, it is observed that the proposed FFCHSA algorithm recorded the minimal value for the frequency of changing the cluster heads when compared to the GA and PSO algorithms.

7 Conclusion

In this paper, we are concentrated on developing the bio-inspired approach for the selection of CH node within the cluster. The cluster head selection is carried out using the firefly algorithms (FFCHSA). The algorithm collects the parameters of energy consumption, packet drop ratio and end to end delay of the nodes which are participating in the cluster and calculates the fitness function of each and every node. The fitness function decides the cluster head selection within the cluster and finally groups all the nodes with the cluster head. The proposed algorithm proved its efficiency under different criteria's such as reduction in packet drop ratio, reduction in energy consumption of the network and minimization in end to end delay. The overall performance of the network is improved using the proposed algorithm.

References

1. Hoang Duc Chinh, et al.: Real-time implementation of a harmony search algorithm-based clustering protocol for energy-efficient wireless sensor networks, IEEE Transactions on Industrial Informatics, pp. 774–783, (2014).
2. Yick, J., Mukherjee, B., Ghosal, D.: Analysis of a prediction based mobility adaptive tracking algorithm, In: Proceedings of the 2nd International Conference on Broadband Networks (BROADNETS '05), pp. 809–816, (2005).

3. Lorincz, K., Malan, D. J., Fulford-Jones T. R. F. et al.: Sensor emergency response: Challenges and opportunities, IEEE Pervasive Computing, vol. 3, no. 4, pp. 16–23, (2004).
4. Hu, S., Han, J., Wei, X., Chen, Z.: Amulti-hop heterogeneous cluster-based optimization algorithm for wireless sensor networks, Wireless Networks, vol. 21, no. 1, pp. 57–65, (2015).
5. Hussain, K., Abdullah, A. H., Awan, F. Ahsan, K. M., Hussain, A.: Cluster head election schemes for WSN and MANET: A survey, World Applied Sciences Journal, vol. 23, no. 5, pp. 611–620, (2013).
6. Peng, J., Liu, T., Li, H., Guo, B.: Energy-efficient prediction clustering algorithm for multilevel heterogeneous wireless sensor networks, International Journal of Distributed Sensor Networks, (2013).
7. Bencan, G., Tingyao, J., Shouzhi, X., Peng, C.: An energy heterogeneous clustering scheme to avoid energy holes in wireless sensor networks, International Journal of Distributed Sensor Networks, (2013).
8. Kuila, P., Jana, P. K. :Energy efficient clustering and routing algorithms for wireless sensor networks: Particle swarm optimization approach, Engineering Applications of Artificial Intelligence, (2014).
9. Natarajan, M., Arthi, R., Murugan, K.: Energy aware optimal cluster head selection in wireless sensor networks, in: Proceedings of the 4th International Conference on Computing, Communications and Networking Technologies (ICCCNT '13), pp. 1–4, IEEE, July (2013).
10. Peirav Ali, Habib Rajabi Mashhadi, Hamed Javadi, S. :An optimal energy-efficient clustering method in wireless sensor networks using multi-objective genetic algorithm, International Journal of Communication Systems, pp. 114–126, (2013).
11. Singh, Buddha, and Daya Krishan Lobiyal.: A novel energy-aware cluster head selection based on particle swarm optimization for wireless sensor networks, Human-Centric Computing and Information Sciences, (2012).

Performance of Slotted Hexagonal Patch for Wireless Applications

K. Madhu Sudhana Rao and M.V.S. Prasad

Abstract Slots transform the conventional geometry into multi-resonant structures. These multi resonant patterns on conducting patch contribute to multiple band characteristics in patch antennas. Gaps between multiple resonant bands can be referred to band rejection characteristics. In this paper, such a multiband and band rejection characteristics of a hexagonal patch antenna with a semi hexagonal slot line is reported. A thorough parametric analysis is carried out to arrive at the radiation characteristics of the antenna. The antenna is simulated in CST Microwave Studio. The analysis of the antenna is carried out using several reports like return loss and VSWR. Ultra wide band (UWB) characteristics are evident from the analysis.

Keywords Slot antenna · Hexagonal · Multi resonant

1 Introduction

Since the release of band 3.1–10.6 GHz by FCC as unlicensed spectrum for commercial and civilian applications. The world witnessed a boom in wireless application. This led to a several congestion in the UWB due to heavy traffic

K. Madhu Sudhana Rao (✉)
Department of ECE, ANU College of Engineering & Technology, Guntur, AP, India
e-mail: sudanrao65@gmail.com

K. Madhu Sudhana Rao
Department of ECE, KKR & KSR Institute of Technology & Sciences, Vinjanampadu, Guntur, AP, India

M.V.S. Prasad
Department of ECE, RVR & JC College of Engineering, Chowdavaram, Guntur, AP, India
e-mail: mvs_prasad67@yahoo.co.in

S.C. Satapathy et al. (eds.), *Computer Communication, Networking and Internet Security*, Lecture Notes in Networks and Systems 5,
DOI 10.1007/978-981-10-3226-4_56

wireless personal applications [1, 2]. In such situations the role of the antenna is vital for interference from transmission and reception of electromagnetic signals. Non-conventional shaped geometry and slot antennas are often considered for such application basing on this strategy a wide variety of antennas like triangular, circular, hexagonal and in some cases fractal shaped antennas are proposed for multi-band and wideband characteristics [3]. Electromagnetic Band Gap (EBG) structures and Frequency Selective Surfaces (FSS) are proposed which are capable of changing the radiation pattern as well as enhancing in the gain. Among these, the FSS possess low profile but provide better gain enhancement of antenna [4]. It is proved that with FSS it possible to improve the bandwidth [5–7]. The other techniques include designing arrays using conventional and meta-heuristic techniques [8].

In this paper the technique of combining hexagonal geometry and slotted strip along four sides of hexagonal patch leaving the other two sides is considered. The radiation characteristics are studied with parametric analysis on the simulated geometry using CST Microwave Tool. The rest of the paper is organized as follows. The geometry of the proposed antenna is given in Sect. 2. Results pertaining to the parametric analysis are given in Sect. 3 and Conclusion of the work is presented in Sect. 4.

2 Geometry of the Proposed Antenna

Initially, for proper study and investigation of the resonance characteristics of the hexagonal antenna with slot it is desirous to study the radiation characteristics of the hexagonal patch antenna. This process allows to investigate the impact of the slot. Accordingly a hexagonal patch antenna is designed and simulated as shown in Fig. 1. Simulated reports in terms of return loss and VSWR of the hexagonal

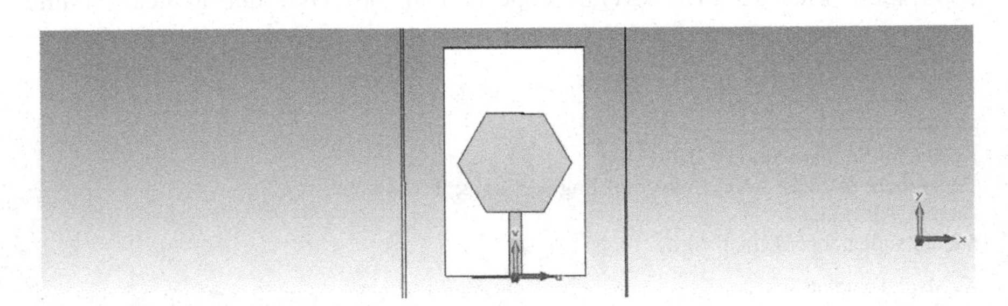

Fig. 1 Geometry of hexagonal patch antenna

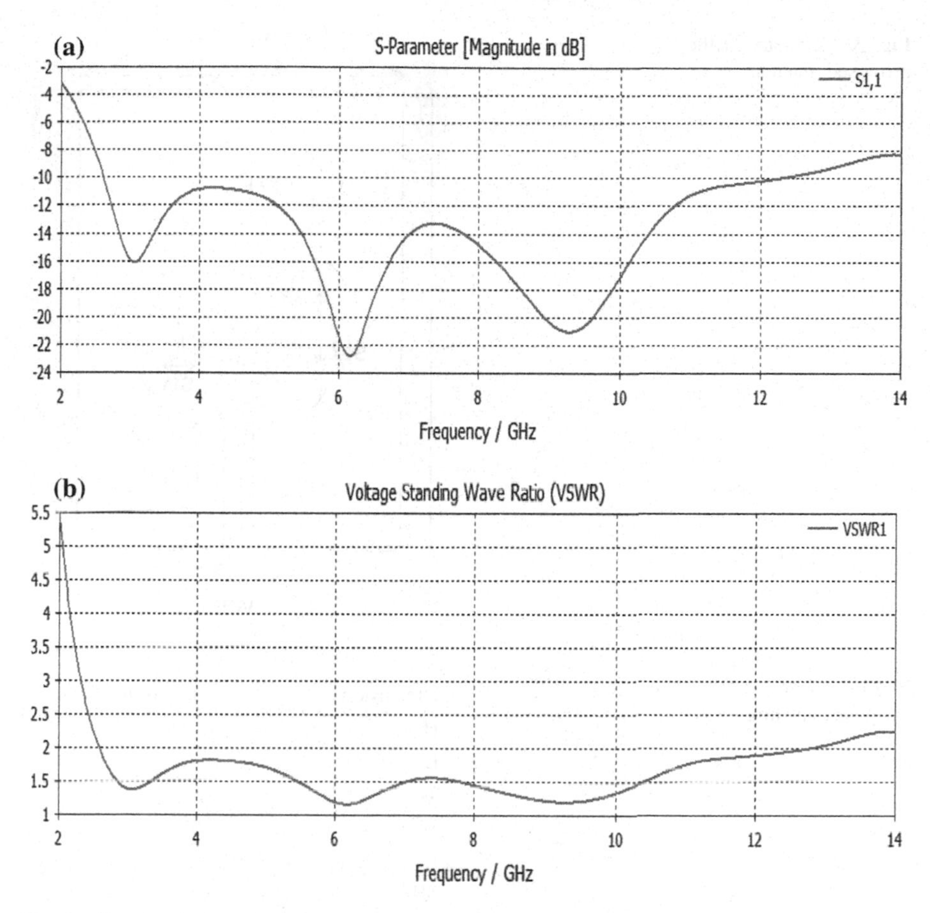

Fig. 2 Simulated plots of **a** return loss and **b** VSWR

antenna are presented in Fig. 2a, b. The UWB characteristics of the antenna are evident from the return loss plot where the bandwidth of the antenna is extending from 3 to 12 GHz covering the entire UWB range.

The geometry of the proposed antenna is as shown in Fig. 3. The typical geometry consists of hexagonal patch of side 13 mm. A strip line running along four inner sides of hexagon patch tilted by 45° is etched. The microstrip line feed arrangement is made by running a micro strip line from the slot less side of the hexagonal patch to the substrate edge. The ground arrangement is made under the substrate which is FR4.

The dimensions of the proposed patch antenna are listed in the Table 1. The substrate used is FR4. The choice of FR4 is due to the reason that it economically viable and easily available in the commercial market of laminates.

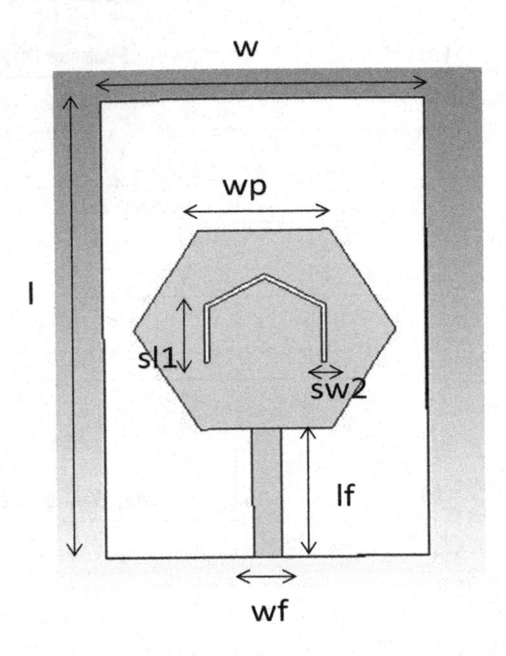

Fig. 3 Geometry of the proposed antenna

Table 1 Dimensions of the proposed antenna

S. no	Dimensions	Value (mm)
1	H	1.6
2	L	52
3	W	32
4	Wf	2.8
5	Lf	14.5
6	Sl1	6.5
7	Sw2	0.5
8	Sw1	7.2
9	Wp	13
10	Tp	0.01
11	Lg	14
12	g	0.5

3 Parametric Analysis of the Antenna

3.1 Tuning Parameters

A thorough parametric analysis is performed on the antenna. The physical parameters like the length of the slotted strip and width of slotted strip are chosen for parametric analysis. The reason for this selection is due to the implicit fact that the resonant characteristics are greatly effected by the slotted section in the hexagonal patch. The parameter 'Sl1' is varied from 5.5 to 7.5 mm in three steps while the other parameters Sw2 is varied over a range 0.3–0.7 mm in three steps. In the following subsections, all the results pertaining to this sweep are presented.

3.2 *Case1: Varying 'Sl1'*

Radiation Parameters like return loss and VSWR are taken over a frequency range of 2–14 GHz. The resultant plots of S11 and VSWR are as shown in Fig. 4a, b. It is evident from the reports that the effect of varying 'Sl1' is minor on the radiation characteristics of the hexagonal patch with slotted strip. The operating band extends from 2 to 13 GHz with notch band at 3.1–4.1 GHz. The band remains an altered for change in Sl1.

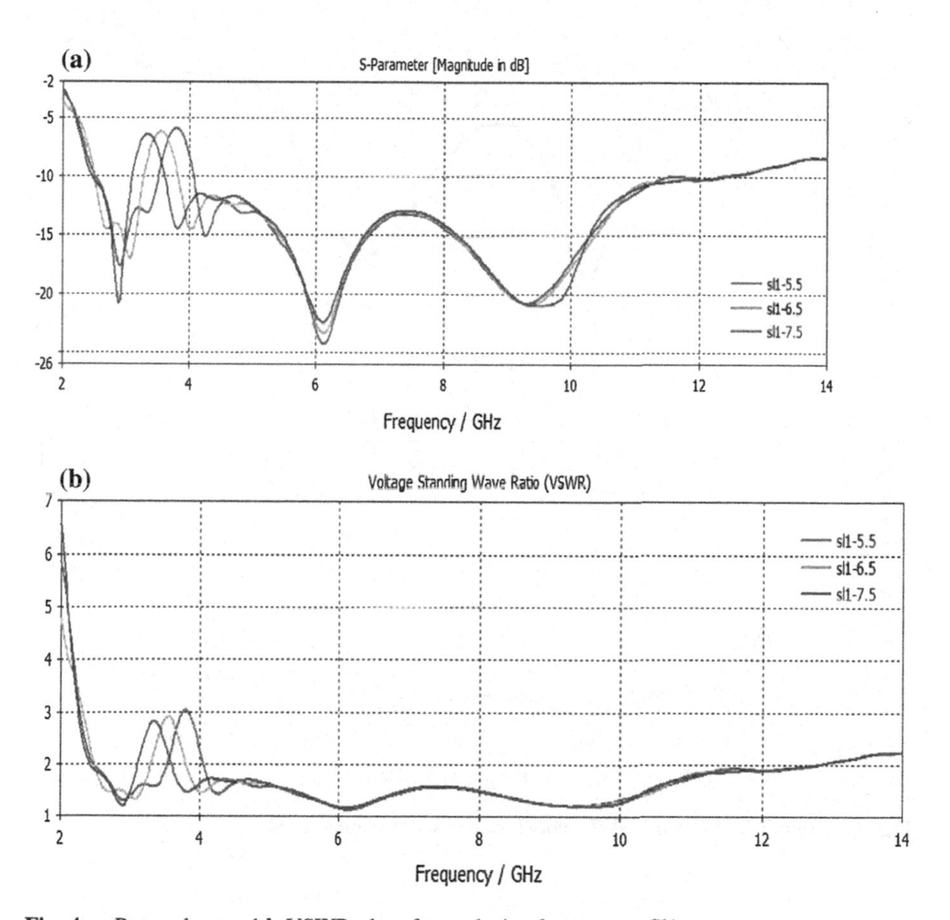

Fig. 4 **a** Return loss and **b** VSWR plots for analysis of parameter Sl1

3.3 *Case 2: Varying 'Sw2'*

The return loss and VSWR plots for varying Sw2 parameter over a set of values (0.3, 0.5 and 0.7 mm) are presented in Fig. 5a, b respectively. The effect of the parameter on the radiation and the bandwidth are evident from these plots that there is not much change with Sw2 variation. However, a small deviation in the notch band is observed.

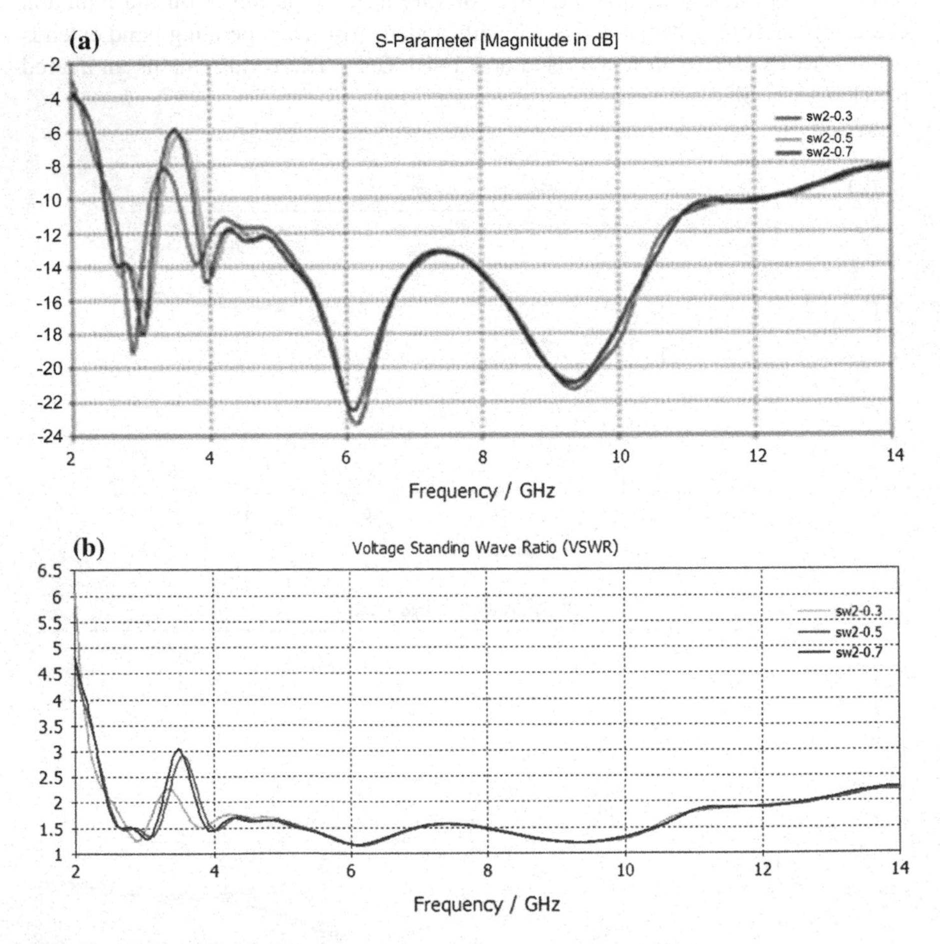

Fig. 5 **a** Return loss and **b** VSWR plots for analysis of parameter Sl1

4 Conclusion

A slotted strip resembling an incomplete hexagonal shape is generated on a hexagonal patch rotated by 45° is studied for its radiation characteristics. The resultant geometry exhibited excellent UWB features. A notch band is produced to avoid interference due to WLAN application. However, the parametric analysis states that more than the length of the side 'Sl1', the notch band width is effected by variation in 'Sw2'. Measurement carried out on the fabricated prototype would be a good slope of future work.

References

1. Kumar, G., K. P. Ray: Broadband Microstrip Antennas, Artech House Publishers, London (2003).
2. Bahl, I. J., P. Bhartia.: Microstrip Antennas, Artech House, Dedham, MA (1980).
3. Chowdary, P. S. R., Prasad, A. M., Rao, P. M, Anguera, J.: Design and performance study of sierpinski fractal based patch antennas for multiband and miniaturization characteristics, Wireless Personal Communications, 83, 1713–1730, (2015).
4. Pirhadi, A., Bahrami, H., Nasri, J.: Wideband high directive aperture coupled microstrip antenna design by using a FSS superstrate layer. IEEE Trans Antennas Propagate, 60, 2101–2106, (2012).
5. Chen, H. Y.,Y. Tao, Y.:Bandwidth enhancement of a U-slot patch antenna using dual-band frequency-selective surface with double rectangular ring elements. *Microwave Opt. Technol. Lett.*, 53, 1547–1553 (2011).
6. Monavar, F. M., Komjani, N.: Bandwidth enhancement of microstrip patch antenna using Jerusalem cross-shaped frequency selective surfaces by invasive weed optimization approach, Prog In Elect magnet Research, 121, 103–120, (2011).
7. Chaimool, S., Chung, K. L., Akkaraekthalin, P.: Simultaneous gain and bandwidths enhancement of a single-feed circularly polarized microstrip patch antenna using a meta material reactive surface. Prog In Elect magnet Research B, 22, 23–37 (2010).
8. Vedula, V. S. S. S. C., Chowdary, P. S. R., Rao, P. M.: Synthesis of Circular Array Antenna for Sidelobe Level and Aperture Size Control Using Flower Pollination Algorithm. Int J. Antennas and Propag, vol. 2015, Article ID 819712, 9 pages, (2015). doi:10.1155/2015/819712.

On the Notch Band Characteristics of Ring Antenna

Anjaneyulu Katuru and Sudhakar Alapati

Abstract Notch characteristics are often required in wide band antennas in order to reject the frequencies that may lead to huge interference. There are several varieties of geometrics which accomplish this process. In this paper, such a notch band antenna is proposed with a circular ring shaped patch. The simulated antenna excellent notch characteristics at selected frequencies. The analysis of the antenna is carried out using several simulated reports like return loss, VSWR and Radiation Patterns. The simulation has been performed using CST.

Keywords Notch band · Ring antenna · Wide band · CST

1 Introduction

With the widespread personal communication systems, the available spectrum is congested. Such a situation is a challenging task for antenna engineers to design radiating system for that does not involve any interference with the existing wireless systems [1–3]. This case is significantly achieved by designing antennas that exhibit band rejection characteristics such band rejection features are known as notch band

A. Katuru (✉)
Department of ECE, ANU College of Engineering & Technology,
Guntur, AP, India
e-mail: anjikaturu@gmail.com

S. Alapati
Department of ECE, RVR & JC College of Engineering,
Guntur, AP, India
e-mail: alapati_sudhakar@yahoo.com

S.C. Satapathy et al. (eds.), *Computer Communication, Networking and Internet Security*, Lecture Notes in Networks and Systems 5,
DOI 10.1007/978-981-10-3226-4_57

characteristics as they project high reflection coefficients over rejected frequencies in order to provide poor radiation. Slot antennas, obtained by etching a portion of the radiating patch are often employed for generating multi resonant geometry. A U-shaped slot is etched on a resonant patch to reject WLAN frequencies in [4, 5]. Similarly, C-shaped slots are generated to reject frequencies used by dedicated short range devices [6]. In order to avoid destructive interference, the choice of fractals also accomplished the task with its inherent multi-resonant features. Fractals possess radiation characteristics at certain bands of frequencies which are spaced by notch bands [7]. Antenna arrays are the other appropriate multiple antenna geometry for enhanced gain characteristics keeping the bandwidth constant [8, 9].

In this paper, one such an attempt of designing UWB antennas with notch band characteristics is proposed geometry is simulated in finite element method based electromagnetic modelling tool known as Computer Simulation Technology (CST). The generated radiation parameters revealed the UWB characteristics of the antenna. Notch characteristics are evident with the inclusion of annular ring.

Further, the paper is organized as follows. The geometrical description of the proposed antenna is given in Sects. 2 and 3. Results pertaining to the simulation based experimentation are given in Sect. 4. Conclusion of the work is provided in Sect. 5.

2 Conventional Circular Patch Antenna

Unlike regular rectangular patch, a circular patch is considered, for its obvious advantages with geometrical structure. The structure of circular patch seems to be similar from an angle on the plane of the patch. Also, it is quite an easy task to provide a strip feed. The location of the feed has a wide freedom of choice along the circumference of the patch. Considering the above discussion, a circular patch is as shown in Fig. 1a. The geometry of the partial ground is given as shown in Fig. 1b. A narrow strip connecting the edge of patch to the substrate end is laid to facilitate for microstrip feed. The dimensions of the circular patch are given in Table 1. The UWB characteristics extending from 2 to 10 GHz are evident from the return loss plot which is as shown in Fig. 2a, b.

3 Proposed Circular Ring Patch Antennas

The geometry of the proposed ring slot antenna is obtained by simply etching a circular shaped area at centre and outside the ring. The ring strip runs around a circular slot. With a radius of r1 while the radius of outer circle constituting the ring is r2. The resultant geometry is as shown in Fig. 3. A rectangular strip of dimensions ‘Sl’ and ‘Sw’ is left on side opposite to the strip feed. The strip is projected towards inner area of circle. This projected section is very useful in modifying the

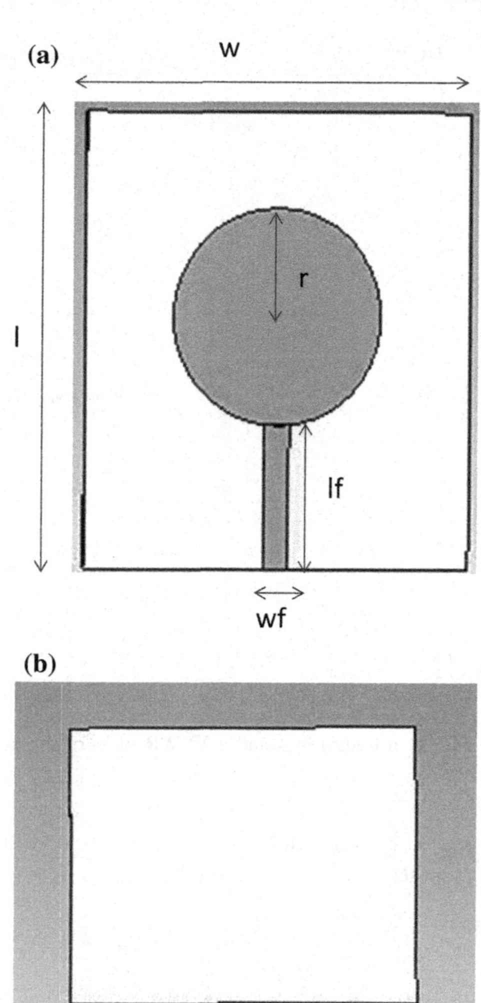

Fig. 1 Geometry of **a** conventional circular patch antenna and **b** ground plane

Table 1 Dimensions of the conventional patch antenna

Dimension	Value in mm
W	50
L	57
r	13.5
wf	3
lf	18.5
gndl	17.7
gndw	50

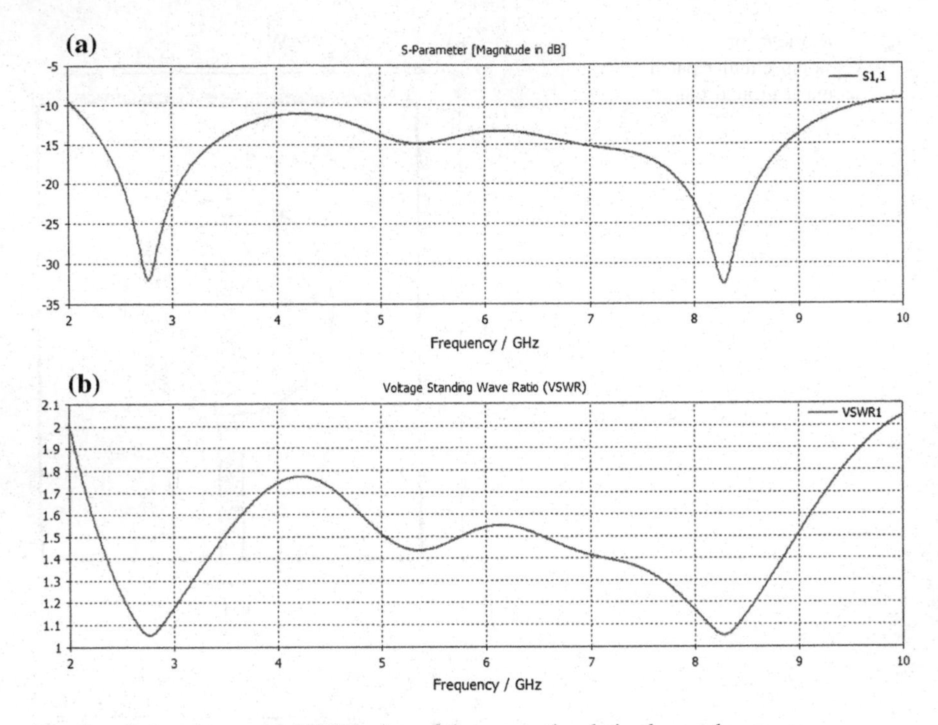

Fig. 2 **a** Return loss and **b** VSWR plots of the conventional circular patch antenna

Fig. 3 Geometry of the proposed antenna

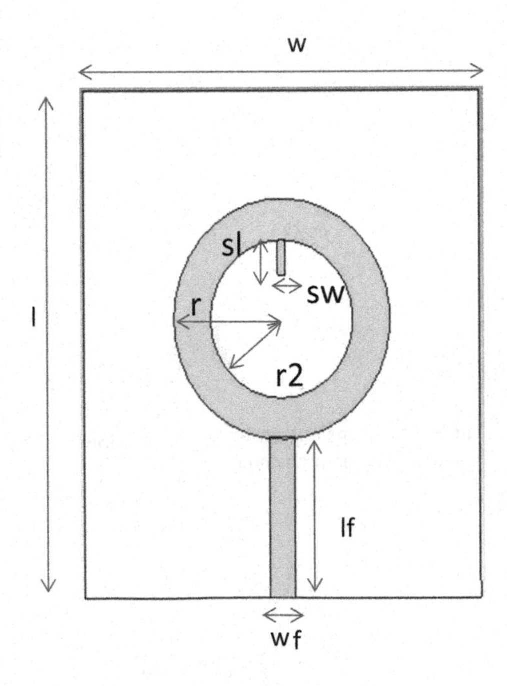

Table 2 Dimensions of proposed antenna

Dimension	Length (mm)
l	57
r	13.5
r2	9
wf	3
lf	18.5
gndl	17.7
gndw	50
w	50
sl	4
sw	1

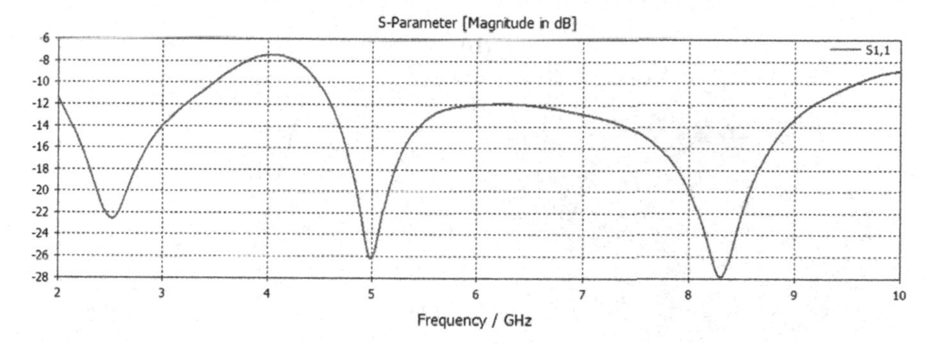

Fig. 4 Return loss plot of proposed antenna

current distribution along the strip line. The dimensions of the proposed antenna are given in Table 2.

4 Results and Discussion

The radiation characteristics of the proposed geometry in the previous section are given in this section. The analysis of the simulated geometry is done using generated simulation reports. The resonant frequencies are identified from the return loss plot as shown in Fig. 4. These frequencies are also visible in the corresponding VSWR plot as shown in Fig. 5.

Further, the study of the radiation characteristics is also performed using the 2D and 3D radiation pattern plots as shown in Fig. 6a, b. The pattern resembles that of a patch antenna. The radiated energy distribution along both azimuthal and elevation angles is studied using 3D plot. When compared with the resonant characteristics of simple circular patch the bandwidth of the ring antenna appears to be broken over a band of frequencies between 2 and 5 GHz. This significantly

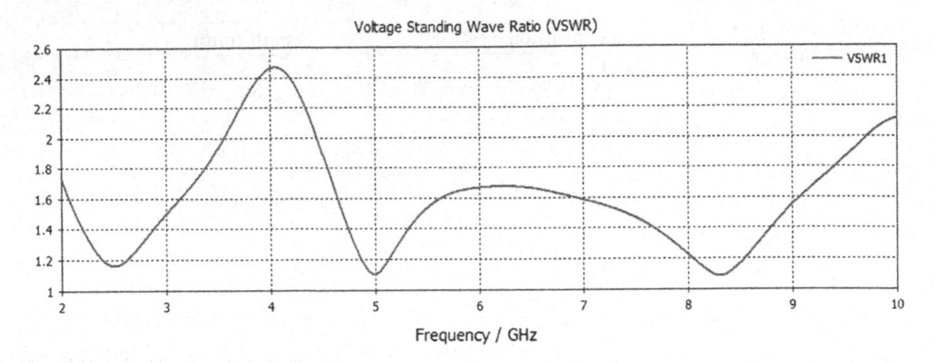

Fig. 5 VSWR plot of proposed antenna

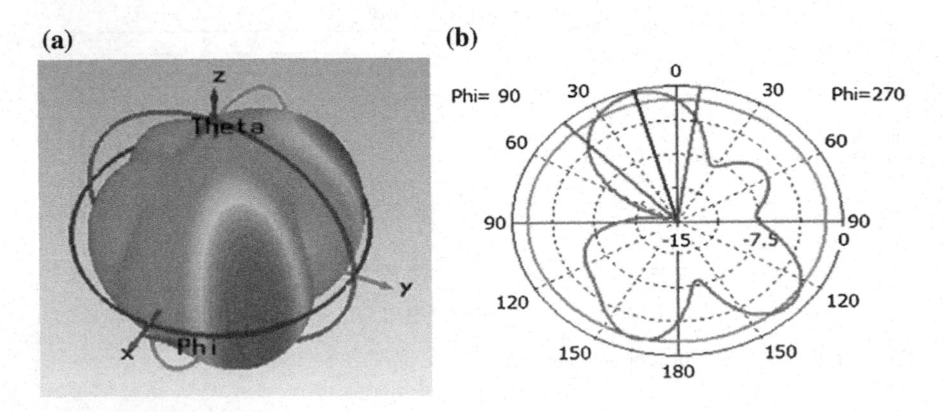

Fig. 6 Radiation pattern plots **a** 3D and **b** 2D

contribute to the notch band characteristics. The accomplishment to notch characteristics is made possible with etched circular slot at the centre along with the inner small rectangular projection.

5 Conclusion

Another technique of generating notch band characteristics by using a circular slot and a rectangular section of inner projection is successfully simulated. The effect of the circular slot is evident on the radiation characteristics of a circular patch is evident with reported return loss and VSWR plots. Further the study of the current distribution and its effect on validation characteristics would be good slope of future work. The fabrication of the proposed antenna and measurements carried on it would be a better mode to validate the antenna geometry.

References

1. First Report and Order Revision of Part 15 of the Commission's Rule Regarding UWB Transmission System FCC 02–48, (2002).
2. Kumar, G., Ray, K. P., Broadband Microstrip Antennas, Artech House, (2003).
3. Mishra, S. K., Mukherjee, J.: WLAN band-notched printed U-shape UWB antenna. In: International Conference on Signal Processing and Communications (SPCOM), pp. 1–5, (2012).
4. Kunwal, M., Bharadwaj, G., Sharma, M. M.: Compact U-shaped Monopole Antenna for UWB Applications with WLAN Band Notched, Communication Systems and Network Technologies (CSNT). In: 2015 Fifth International Conference on, pp. 28–32, (2015).
5. Agrawall, N., Kumar, G., Ray, K.: Wide-band planar monopole antennas, IEEE Trans. Antennas Propagat., 46, pp. 294–295 (1998).
6. Li, F., Ren, L. S., Zhao, G., Jiao, Y. C.: Compact triple-band monopole antenna with C-shaped and S-shaped meander strips for wlan/wimax applications. Progr In Elect magnet Research Lett, 15, 107–116 (2010).
7. Chowdary, P. S. R., Prasad, A. M., Rao, P. M., Anguera, J.: Design and performance study of sierpinski fractal based patch antennas for multiband and miniaturization characteristics, Wireless Personal Communications, 83, 1713–1730, (2015).
8. Vedula, V. S. S. S. C, Chowdary, P. S. R., Rao, P. M.: Synthesis of Circular Array Antenna for Sidelobe Level and Aperture Size Control Using Flower Pollination Algorithm. Int J. Antennas and Propag, vol. 2015, Article ID 819712, 9 pages, (2015).
9. Chakravarthy. V, Mallikarjuna Rao. P,: Implementation of TLBO for circular array synthesis. Int. J. of Adv Tech and Innovat Research, 7(4), 351–355, (2015).

Design of a Compact Wideband Bow-Tie Dielectric Stacked Patch Antenna for Ku-Band Spectrum

Venkateswara Rao Tumati, Rajesh Kolli and Anilkumar Tirunagari

Abstract The present communication gives the design of a compact patch antenna module to be functional from 12 to 18 GHz EM spectrum which is given the name Ku-band. The design of the antenna focuses the dielectric stacking approach. Base substrate is chosen FR4 dielectric and it consists of microstrip feed line and a partial ground plane on its top and bottom respectively, a dielectric material along with patch having bow-tie shape is stacked on the base substrate. The proposed antenna is designed and simulated using ANSYS HFSS. The FEM based simulations proven the −10 dB return loss criteria and the VSWR below 2.0 are satisfied for the proposed model under 12.07–19.11 GHz band. The geometrical aspects of this model ensures the wideband functionality of the antenna with a peak gain of 6.23 dB.

Keywords Dielectric stacking · Bow-tie · Ku-band · HFSS

1 Introduction

The antennas are playing a major key role in transmitting/receiving the signals in terms of EM radiation and also known as a transducer. Further, since their inception the microstrip patch antennas are being expanding their roots in many fields and state their presence. Though the MSAs are having several advantages such as low profile, conformal, light weight and less expensive and also suffers from narrow bandwidth and low gain like demerits [1]. This led to an innovative research in the

V.R. Tumati (✉) · R. Kolli · A. Tirunagari
Department of Electronics and Communication Engineering,
SIR C. R. Reddy College of Engineering, Eluru, India
e-mail: tumati01@gmail.com

R. Kolli
e-mail: raajesh.kolli@gmail.com

A. Tirunagari
e-mail: anilthebetter@gmail.com

S.C. Satapathy et al. (eds.), *Computer Communication, Networking and Internet Security*, Lecture Notes in Networks and Systems 5,
DOI 10.1007/978-981-10-3226-4_58

field of antenna engineering. The microwave spectrum of C-band, X-band, Ku, Ka bands are having applications in satellite communications, TV broadcasting, BSS (Broadcasting Satellite Service), etc. Ku-band is having a very wide spectrum from 12 to 18 GHz after the C and X-bands. The government and research organizations are encouraging the novel antenna models for the Ku-band activities. The power of the signal in Ku-band has no power limitations when compared to C-band for interference avoiding with communication systems that are terrestrial based. The antennas are significantly smaller dimension at Ku-band which can contribute for the smaller earth stations for VSAT terminals. Ku-band downlink frequencies has the range from 11.7 to 12.7 GHz for fixed and broadcast satellite services (FSS and BSS).

In the recent study, fractals are suggested for multiband, broadband and miniaturization characteristics [2]. Stacking of the patch antennas reduces the need of accommodating the radiating elements in a coplanar configuration. This approach minimizes the size of the antennas and can lead to implement the compact designs with good features such as wide operating band and improvement in gain. Various types of stacking procedures were so far proposed and demonstrated in literature. Some of them are presented here. In [3], stacked modified patch antenna was presented capable of wide operating band from 2 to 10 GHz and contains two patches. Effect of various substrates has been explained. A design with two-layers of patches with E-shape were stacked were reported in [4] is exhibiting around 10% of impedance bandwidth at 2.4 GHz. In [5], the analysis of creating a notch structure in a microstrip patch antenna is presented. A microstrip antenna with triangular slot defected ground structure is reported for obtaining bandwidth enhancements is discussed in [6]. The stacked patch antennas have contributed the domain of designing such UWB antennas in the antenna literature. A MPA with probe feeding was proposed in [7] with entrenched two unsymmetrical opposite rectangular slots in the rectangular patch and thicker dielectric was used for obtaining wideband for UWB wireless applications. In [8], a DRA with compact size is presented for wideband applications in which they have employed two dielectric resonators with cylindrical shape that are located asymmetrically w.r.t the aperture slot and operates from 9.62 to 12.9 GHz. The reduction in antenna beam width is attained with this configuration. A patch antenna for Ku-band has been proposed in [9] with a defected ground structure for obtaining wide bandwidth 15.27–16.51 GHz and at the resonant frequency gain, directivity are reported as 4.45 dB, 5.17 dBi respectively.

In this letter a compact antenna is designed for Ku-band applications. Two dielectric materials are used in the proposed model one is base substrate and patch dielectric. The feed line and ground plane are metallized on both sides of base substrate at its top and bottom. Patch dielectric is stacked on the base substrate and the patch is layered on patch substrate. This design approach is different and is designed and simulated using 3D EM computer aided engineering tool called HFSS. The study of this antenna is discussed in the following section.

2 Stacked Dielectric Patch Antenna Design

The design of patch antenna is discussed here. The FR4 dielectric substrate with relative permittivity 4.4 and 0.02 loss tangent is used to support the metallization layers. The proposed antenna is designed using the stacking approach. The substrate is square shaped with dimensions 20 mm × 20 mm × 1.6 mm. The electric signal is allowed to fed to the antenna using the microstrip line feeding matched to the standard 50 Ω characteristic impedance. The width '*Wf*' of the feed line is calculated using the relation given in the [1]. In this design the patch element does not lie on the same plane rather than it is kept at a height '*hdr*'. The rectangular patch of *Lp* × *Wp* size is modified by removing the two isosceles triangular notches (having height '*c*') from the upper and lower half of the rectangular patch. This forms the bow-tie shape patch which characterizes the wideband operation of the antenna. Further, the architecture is loaded with another dielectric material upon thc substrate which takes the form of the bow-tie as that of the patch. The Fig. 1 and Table 1 show the clear idea of the geometrical structure and parameters.

The feed line provides the excitation to the patch by means of electromagnetic coupling by the including EM fields in the stacked dielectric material. The ground plane is etched partially on the substrate's bottom. The above mentioned

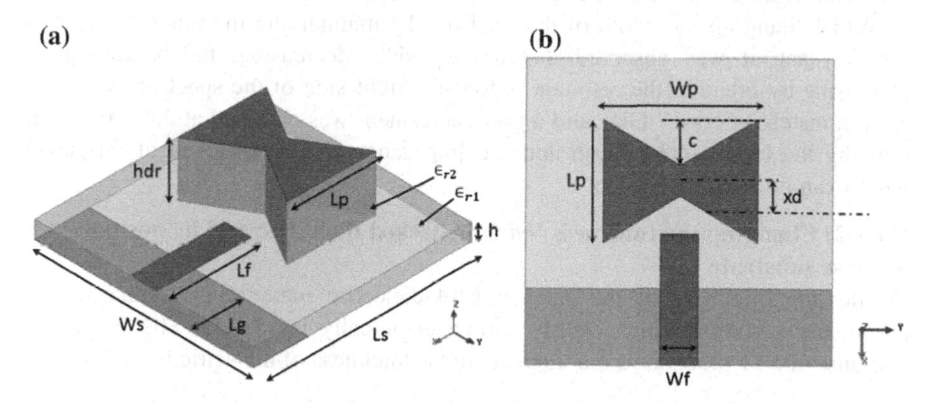

Fig. 1 Architecture of stacked dielectric patch antenna **a** 3D view **b** 2D view

Table 1 Geometrical parameters of the proposed antenna

Parameter	Value in mm	Parameter	Value in mm
Ls	20	*Wf*	2.4
Ws	20	*Lg*	5
h	1.6	*c*	3
Lp	8	*xd*	2
Wp	10	*hdr*	4.8
Lf	8.5		

geometrical aspects ensures the antenna to be a compact one. The present design contains higher degree of freedom such that we can optimize the antenna using various parameters to obtain required performance. The subsequent section is discussed with the parametric variation of the antenna parameters to achieve to desired antenna performance.

3 Parametric Study

In this section the feed line dimensions, change of stacked dielectric material (relative permittivity variation) and other parameters are varied and this made a clear study and optimization of the antenna for required characteristics.

Case 1: Variation of the length '*Lf*' and width '*Wf*' of the feed line
The length of the feed line is parametrically analyzed to study the antenna frequency domain characteristics with its change. The length is swept from 8 to 15.5 mm with a step change of 0.5 mm. As shown in Fig. 2a the bandwidth decreases and the lower cut-off frequency shifts left as the '*Lf*' increases from its mid value to the higher lengths. The decrease in '*Lf*' value results the bandwidth increases. But the value 8.5 mm gives the better result at which the −10 dB return loss bandwidth criteria and the level of return loss is seems to be better.

While changing the width of the feed line by maintaining the other parameters are constant, it was observed that as the width decreasing, the bandwidth is increasing by creating the resonances towards right side of the spectrum which is approximately at 16–17 GHz and a good agreement was obtained at 2.4 mm width meeting the bandwidth criteria and the impedance is well matched at this level which can be seen in Fig. 2b.

Case 2: Changing the thickness '*hdr*' of stacked dielectric and its position '*xd*' on base substrate
Further the thickness of the stacked FR4 dielectric material was subjected to parametric variation which clearly shown graphically in Fig. 3a. The frequency response of S11 plot shows that increasing the thickness of dielectric is leading to a

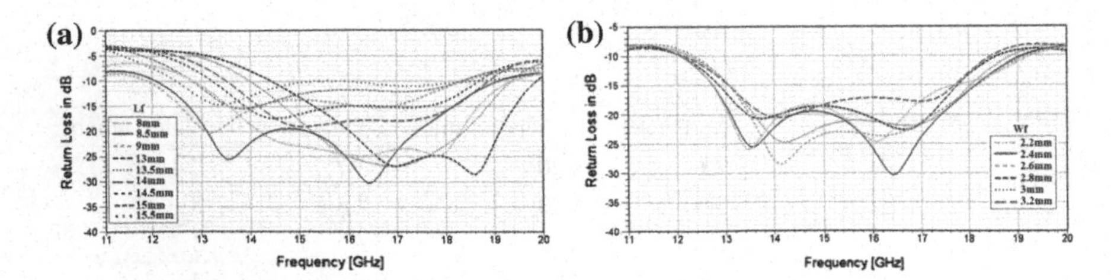

Fig. 2 Simulated return loss characteristics of the antenna with parametric variation of **a** '*Lf*' and **b** '*Wf*' parameters

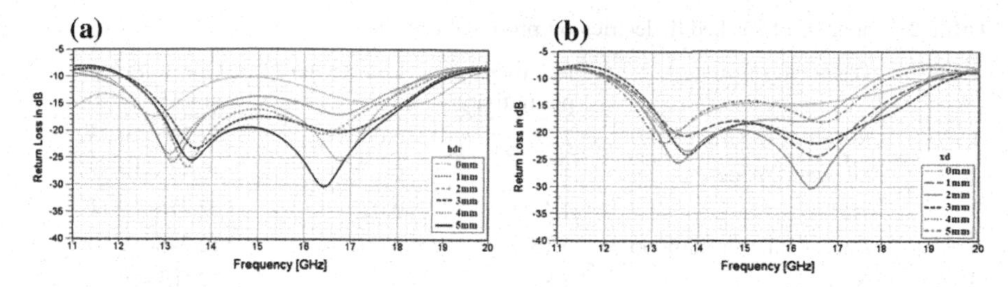

Fig. 3 Return loss variations of the antenna by varying **a** '*hdr*' and **b** '*xd*' parameters

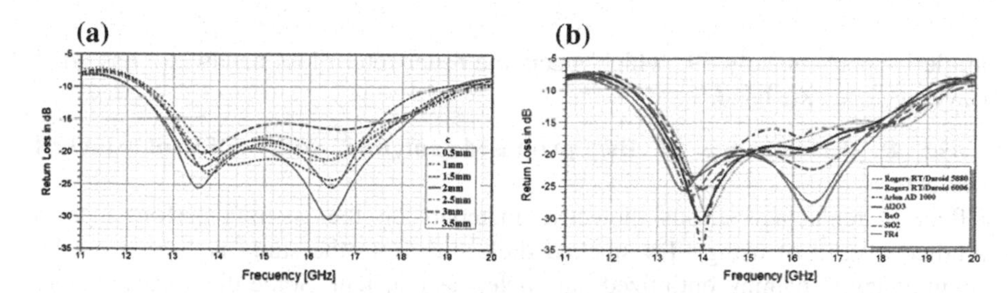

Fig. 4 Return loss characteristics for **a** variation in '*c*' and **b** for different stacked dielectric materials

return loss below −15 dB, thus dielectric thickness increase will create the deeper resonances also maintains considerable bandwidth with decreasing the Q-factor of the antenna. The position of the stacked dielectric is allowed to change linearly from the center of the substrate with a vertical shift. It was identified that the position of the dielectric effects the spectral characteristics of the antenna and it is effecting the return loss level maintaining around −15 dB at lower values of '*xd*'. Figure 3b demonstrates that the resonance frequency is slightly shifts towards the right side while the position moved further upside. At 2 mm upside from the substrate center the geometry gives the good reflection loss less than −20 dB at its two resonant frequencies. This can be a better value at which the stacked dielectric can be placed.

Case 3: Changing the height '*c*' triangular notch cut in the patch and dielectric

The triangular notch is having a critical parameter called its height '*c*'. This indirectly varies the bow-tie angle at which two of its edges are excited by the feed line. Keeping all other parameters unchanged, this parameter is swept from 0.5 to 4.5 mm. From Fig. 4a, at lower values of '*c*' the angle subtended at the vertex of the triangular cut is more and the angle becomes less when the '*c*' value getting increased. Further decreasing this angle by increasing the '*c*' value the bandwidth reduces and disappearing the other resonant frequencies. At 3 mm the corresponding

Table 2 Variation of stacked dielectrics on proposed antenna

S. no	Dielectric substrate	Relative permittivity	Loss tangent	Operating band in GHz
1	Rogers RT/Duroid 5880	2.2	0.0009	12.235–19.232
2	Silicon Dioxide	4	0	12.443–19.527
3	FR4	4.4	0.02	12.076–19.112
4	Rogers RT/Duroid 6006	6.15	0.0019	12.246–19.109
5	BeO	6.8	0	12.607–19.427
6	Al_2O_3 Ceramic	9.8	0	12.411–18.868
7	Arlon AD 1000	10.2	0.0023	12.607–19.226

angle is approximately 59° which meets the better return loss bandwidth has good coverage over Ku-band.

Case 4: Performance of the proposed antenna with different stacked dielectrics
Effect of replacing various dielectric materials on the antenna performance is studied instead of using FR4 stacked dielectric. Here the study is carried out by maintaining remaining optimized parameters as constant. Some dielectrics namely Rogers RT/Duroid 5880, Al_2O_3, BeO, FR4, Silicon Dioxide, Rogers RT/Duroid 6006, Arlon AD 1000 are used in this study. The reflection loss behavior is shown in Fig. 4b operating bands of the proposed antenna at various stacked dielectrics are tabulated in Table 2.

4 Simulation Results

The HFSS EM simulation results of the proposed module is mentioned in both network parameters based characteristics and that of radiated field observations.

4.1 Simulated Return Loss and VSWR Characteristics of the Proposed Antenna

The proposed antenna performance according to its reflection loss characteristics is can be understood from Fig. 5a. The simulation of the proposed antenna using FEM based HFSS shows the −10 dB return loss bandwidth from 12.07 to 19.11 GHz with a bandwidth of 7.03 GHz and consisting of two resonances at 13.5 and 16.4 GHz where the return loss is noted as −25.51 dB and −30.37 dB respectively. The voltage standing wave characteristics shown in Fig. 5b has the lower cutoff frequency starting

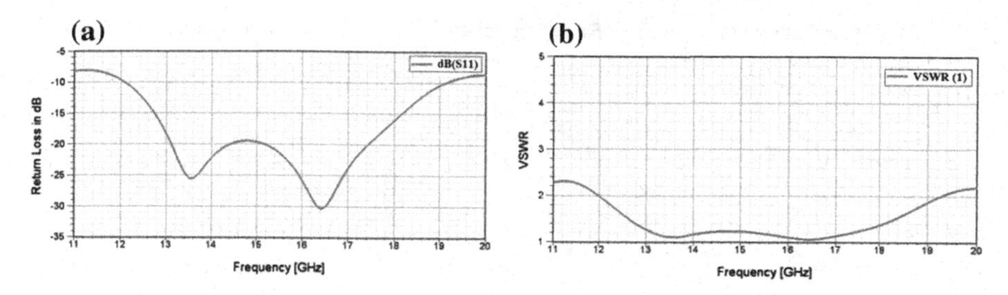

Fig. 5 Simulated **a** Return loss **b** VSWR characteristics of the proposed antenna

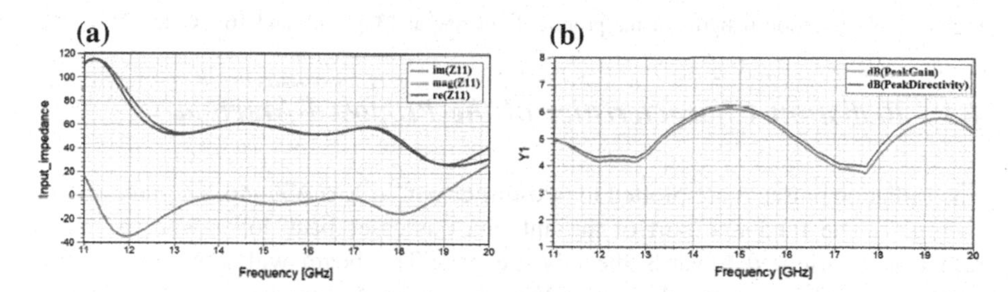

Fig. 6 Simulated **a** Input impedance **b** Gain versus frequency characteristics of the proposed antenna

from 11.97 GHz and the higher cutoff frequency limited up to 19.27 GHz with a VSWR $\leq$ 2.0 bandwidth of 7.29 GHz. At the resonant frequencies of 13.6 and 16.4 GHz the VSWR values are maintained as 1.11 and 1.06 respectively. The magnitude of input impedance of its port swings between 50 and 60 Ω showing good matching with the characteristic impedance within Ku-band which can be seen from Fig. 6a.

4.2 Simulated Gain Versus Frequency Characteristics of the Proposed Antenna

The gain and directivity versus frequency performance of the proposed antenna is shown in Fig. 6b. On the vertical axis the peak gain and the peak directivity are taken and plotted across the frequency scale on horizontal axis. The peak gain of the antenna swings from a minimum value of 3.74 dB to a maximum value of 6.11 dB with an average maintaining gain of 4.93 dB. Whereas the peak directivity varying from 3.99 to 6.23 dB with an average directivity of 5.08 dB within its operating band. At mid-frequency of operating band, the gain is more.

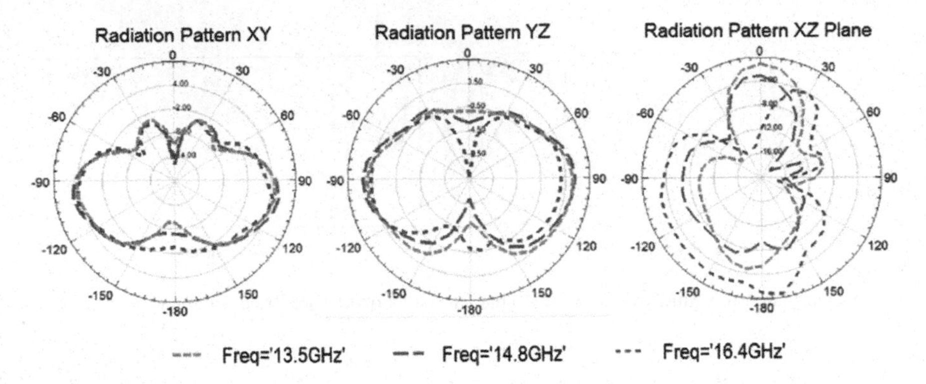

Fig. 7 **a** 2D radiation patterns of the proposed antenna at 13.5, 14.8 and 16.4 GHz

4.3 Radiation Characteristics of the Proposed Antenna

The radiation patterns are plotted in two-dimension in XY, YZ and XZ plane shown in Fig. 7. The feed line part of the antenna possesses back lobes whereas major radiation is oriented towards the bow-tie part. The beam width of the radiation pattern varies from 40 to 48° in its XY plane within the operating band and that of YZ plane it has the elevation angular coverage of 82–110° to the coordinates specified in the geometry of the antenna.

5 Conclusion

A compact stacked dielectric patch antenna designed considering two layers of dielectrics having an overall dimensions of 20 × 20 × 6.6 mm^3. The radiating patch is placed on the second dielectric layer makes this a different one. The patch is electromagnetically coupled to the feed line excitation through the EM fields induced within the dielectric. Simulation results are showing the −10 dB impedance bandwidth from 12.07 to 19.11 GHz which covers most of the Ku-band also satisfying the criteria VSWR below 2.0 with minimum VSWR of 1.06. Also a maximum gain is obtained 6.23 dB and bidirectional radiation pattern is obtained. Further the effect of various dielectrics in place of stacked dielectric is studied. The proposed model can be best suited for satellite phones and VSAT terminals.

Acknowledgements Authors would like to express their sincere gratitude to the Research Centre of Dept. of ECE and management of SIR C. R. Reddy College of Engineering, Eluru for providing the licensed 3D EM simulation software ANSYS HFSS ver17 and their esteemed guidance and support to the work.

References

1. Huang, Y., Boyle, K.: Antennas: from theory to practice. John Wiley & Sons, (2008).
2. Chowdary, P. S. R., Prasad, A. M., Rao, P. M., Anguera, J.: Design and performance study of Sierpinski fractal based patch antennas for multiband and miniaturization characteristics, In: Wireless Personal Communications, vol. 83 (3), 1713–1730. (2015).
3. Pathak, H. M., Pankaj C.: Stacked Modified Patch Antenna Comparison for Various Dielectric Material. In: International Conference on Machine Intelligence and Research Advancement (ICMIRA), pp. 162–166. Katra (2013).
4. Devraj, V., Ajayan, K. K., Baiju, M. R.: A Novel Optimization Technique for a Stacked Patch Antenna. In: Asia-Pacific Microwave Conference, pp. 1–4. Bangkok (2007).
5. Venkateswara Rao, T., Sudhakar, A., Padma Raju, K.: Analysis of Notch in Microstrip Antenna. In: International Journal of Advanced Electrical and Electronics Engineering, pp. 15–19. (2015).
6. Venkateswara Rao, T., Sudhakar, A., Padma Raju, K.: Design of Microstrip Antenna with Notch Filter for UWB Application. In: International Journal of Electronics & Communication Technology, vol. 7(1), pp. 30–33. (2015).
7. Elkorany, A. S., Elhalafawy, S. M., Shahid, S., Gentili, G. G.: UWB integrated microstrip patch antenna with unsymmetrical opposite slots. In: IEEE-APS Topical Conference on Antennas and Propagation in Wireless Communications (APWC), pp. 426–429. Turin (2015).
8. Majeed, A. H., Abdullah, A. S., Elmegri, F., Sayidmarie, K. H., Abd-Alhameed, R. A., Noras, J. M.: Aperture-Coupled Asymmetric Dielectric Resonators Antenna for Wideband Applications. In: IEEE Antennas and Wireless Propagation Letters, vol. 13, pp. 927–930. IEEE, (2014).
9. Bhadouria, A. S., Kumar, M.: Wide Ku-Band Microstrip patch antenna using defected patch and ground. In: International Conference on Advances in Engineering and Technology Research (ICAETR), pp. 1–5. Unnao (2014).

Design of Scan Cell for System on Chip Scan Based Debugging Applications

A. Murali, K. Hari Kishore, G. Vijaya Padma, L. Srikanth and R.H. Gopalkrishna

Abstract This paper presents a scan based debugging technique approach for a SOC with multiple clock domains. In this paper, the existing techniques are used for debugging the SOC has been described and the limitations of existing debugging approach have also explained. To overcome the limitations of the existing debugging techniques, new scan cell approach has been proposed. This debugging technique also supports online debugging and it holds the present status and shifts previous state. The IEEE 1500 test wrapper has been optimized to minimum area with the proposed debugging technique.

Keywords System-on-chip (SOC) debugging · Design-For-Debug (DFD) · Scan-based debugging · Scan design · Online debugging

1 Introduction

Nowadays, the chip sizes have been reduced to smaller and smaller with the semiconductor technologies. The smaller chip usages have been grown higher and the debugging effort for small chip's increased due to huge complexity integrated on a single chip. A run stop debugging method has been used by the debug

A. Murali (✉) · K. Hari Kishore · G. Vijaya Padma
Department of ECE, KLEF, K.L University, Green Fields, Guntur, AP, India
e-mail: amurali3@gmail.com

K. Hari Kishore
e-mail: kakarla.harikishore@klunivrsity.in

G. Vijaya Padma
e-mail: gujjuvijayapadma@gmail.com

A. Murali · L. Srikanth · R.H. Gopalkrishna
Department of ECE, RAGHU Engineering College, Visakhapatnam, AP, India
e-mail: srikanth451@gmail.com

R.H. Gopalkrishna
e-mail: harigopalakrishnarudra@gmail.com

S.C. Satapathy et al. (eds.), *Computer Communication, Networking and Internet Security*, Lecture Notes in Networks and Systems 5,
DOI 10.1007/978-981-10-3226-4_59

engineers to debug these chips. The run-stop debugging methodology includes the states as Run, Stop, Resume Operations and Scan Dump as in [1]. The improvements of observability and reduced debugging time have been achieved with storage of internal data/signals traces which leads to increase in area [2].

Most of the advance SOC's have multiple clock domains. The logic core present inside the SOC communicates with different interfaces developed in SOC using an interconnect bus which operates either with the same clock or different clocks. SOC's are designed with multiple clock domains, to which multiple interfaces triggers and if the execution is stopped it returns the ambiguity due to the multiple clocks stops at different time instances [3–5]. The uncorrelated clocks results in complexity in debugging. These complexities have been reduced to some extent with the use of online debugging, transaction and message level debugging approaches.

Generally, the Design-For-Debug (DfD) techniques implements the on-chip and off-chip debugging support logics. DfD technique optimizes the IEEE standard equipped test infrastructure like 1149.1, 1500 wrapper and the test access mechanism. In an embedded core, each core consists of one or more scan chains and it is wrapped with the use of the 1500 (IEEE standard) wrapper for testing embedded core. The test infrastructures while in debug are restricted to debugging limitations like clock phase relationship inconsistency and data invalidation [6], which may occur in Cores under debug (CUD) conditions results in stop and restart of debugging. To overcome these problems, the paper presented with a design of scan cell architecture which consists of a couple of flip-flops and control logic which improves the debugging speed. These limitations of the existing system are addressed by the proposed scan cell for performing debug operations, which holds the current and previous state.

The proposed scan cell design involves many works to support the debugging and testing which uses an additional flip-flop or latch for holding the state by performing the scan shift. The author DasGupta presented [7] a online testing with the use of Stable-Shift Register Latch (SSRL) which consists of the shift register latch (SRL), these are used for the Level Sensitive Scan Designs (LSSD) and also hold latches. The hold latches are used to store the snapshots of the error detected and shifts out for the observations and send back to the SRL. Shadow scan architecture is presented by Golshan for online debugging [5]. During the normal operation, the shadow scan flip-flops are used to store the snapshots and shifted out.

The IEEE standard 1149.1 test access port controller is used to control the shadow scan flip-flops. Thus, when the system is operating without debug stop, the snapshots can be observed and captured. The authors [8] proposed a special one, the swap register for online testing, which has the normal data register and a single shift register. The authors in [9] explored a hold scan cell, which consists of the level-sensitive muxed-D scan cell with a latch. The scan cells of these existing techniques are prone to the race condition at clock domain boundaries which have been discussed in next section. The remaining paper have been described as follows: debugging limitations due to cross clock domain problems have been discussed in Sect. 2, the proposed scan cell and debug control technique using IEEE 1500 wrapper have been discussed in Sect. 3, experimental results described in Sect. 4, conclusion in Sect. 5.

2 Limitations of Debugging Due to Cross Clock Domain

2.1 Data Invalidation

Debug engineers during the scan based debugging initially stops the SOC or cores and reads the scan data. The cores in the SOC may connect to the different clock domains; hence we cannot assure that all the cores are stopped at the same time. The received data can be invalid if the others have stopped earlier before receiving data. The authors Goel and Vermeulen [6] have estimated the occurrence of the invalid data due to the sender frequency may be higher than the receiver frequency. To overcome this data invalidation they have proposed a control logic which is used as a detector to find the flip-flops which has invalid data [6]. The information related to flip-flops which hold the invalid data is useful up to some extent for debug engineers. The invalid data detector and data invalidation have been shown in the Fig. 1. The glitch free clock gating circuit has been proposed to stop cores shown in Fig. 1a. The functional clocks have been generated for both sender and receiver by gating stop signal with clk_out_snd and clk_out_rcv. The data invalidation example has been shown in Fig. 1b. If the clock frequency of the receive clock domain is slower when compared to sender clock, then the receiving clock clk_out_rcv can be stopped several times in the sender clock cycles clk_out_snd.

From the figure Fig. 1a. the flipflop FF2 stores the data ‘0’ due to the last rising edge of the clk_out_rcv has occur earlier then the last rising edge of the

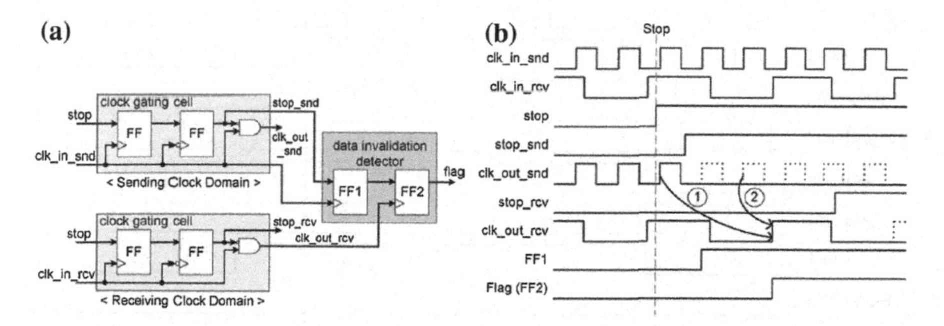

Fig. 1 The data invalidation example **a** Logic circuit with clock gating and data invalidation detector **b** Waveform

clk_out_snd occurs. Else flipflop FF2 may capture the data '1' which is the output of FF1. The data invalidation is occurred which is shown has a Flag (FF2) in Fig. 1b.

2.2 *Clock-Phase Relationship Inconsistency*

The clock-phase relationship inconsistency may occurs either when the cores are stopped or resumed. When the cores are stopped there will be no data invalidation and it may not be restarted without the loss of the clock phase relationship. The Fig. 2 describes the clock phase relationship inconsistency and observed when the cores are stopped there is no data invalidation occurred. Hence the debug engineers need not reset hardware for moving to next breakpoint. If the cores are restarted, then the first pulse of clk_out_rcv could start after two pulses of clk_out_snd. From the Fig. 2 we have observed that the after three clk_out_snd pulses the first pulse of clk_out_rcv has been started while the stop signal is deactivated to restart the cores. Thus we may not assure that these cores will work properly. Hence the clock phase relationship inconsistency can occur in these circumstances. This may also occur not only with the clock frequency difference but also may occur due to phase difference.

In this brief we have seen that when the scan based debugging is applied to an SOC may occur two problems with multiple clock domains and data invalidation. These problems leads to the more complexity in debugging the SOC and it may also waste the time of an engineer to work on these approaches. In this paper our aim is

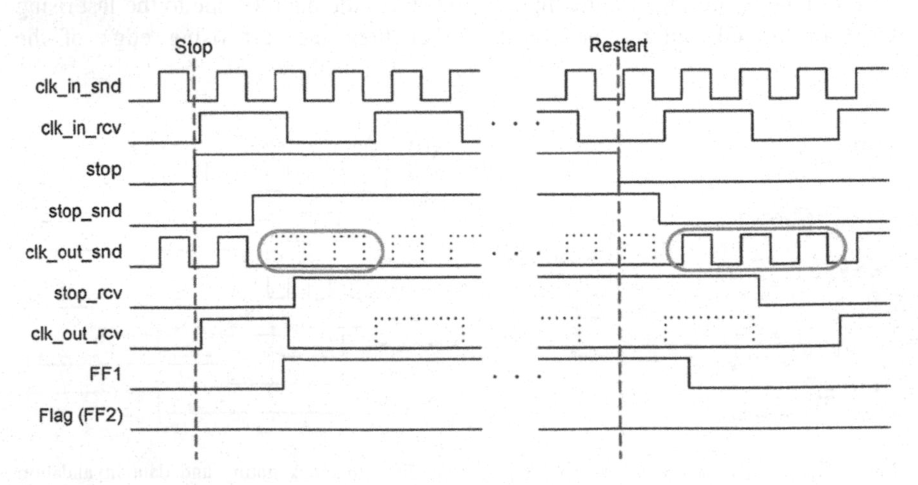

Fig. 2 The clock phase relationship inconsistency waveform

to design a scan based debugging technique for an SOC which may lead to more efficiency in testing and within a less time spent by an engineer.

3 Proposed Design for Debug Techniques

In a scan based debugging approach, the breakpoints have been set by the debug engineer in order to repeat atomic debugging operations in sequential order which includes run, scan-dump, stop and resume. We have observed from the Sect. 2 that the ambiguity we resulted due to SOC with multiple clock domains. To resolve this ambiguity the debug engineer has to restart the entire SOC or CUD's after scan dump. And also we have utilized the additional scan flip-flop in a scan based debugging approach to hold the current sate and previous state [10]. Thc data invalidation was explained in the Sect. 2 which occurred due to faster and slower clocks. The debug engineer can improve the efficiency by designing the scan cell which can hold the current state and previous state at the same time in which the states will be shifted according to the selection. The implementation of this idea makes the use of IEEE 1500 test wrapper to be wrapped to core which belongs to the clock domain and as many as detector logic circuits for data invalidation in an SOC have been used for each clock domain boundaries.

3.1 Scan Cell Design

The proposed architecture of a scan cell have been shown in Fig. 3 which is designed with data flip-flops (DFF) in which the functional data is stored and a hold flip-flop is used for debugging and four 2 × 1 multiplexers.

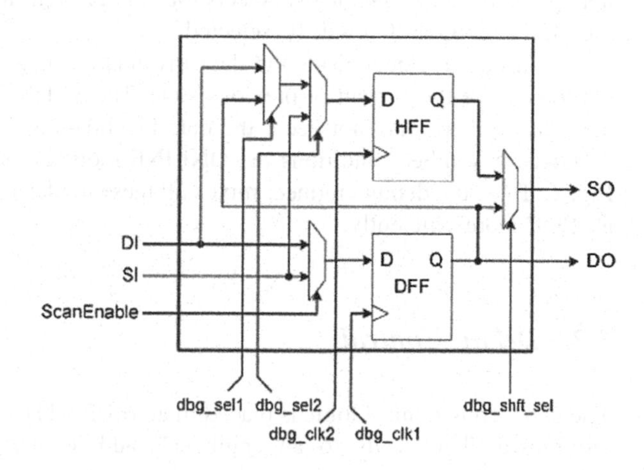

Fig. 3 The proposed scan cell architecture

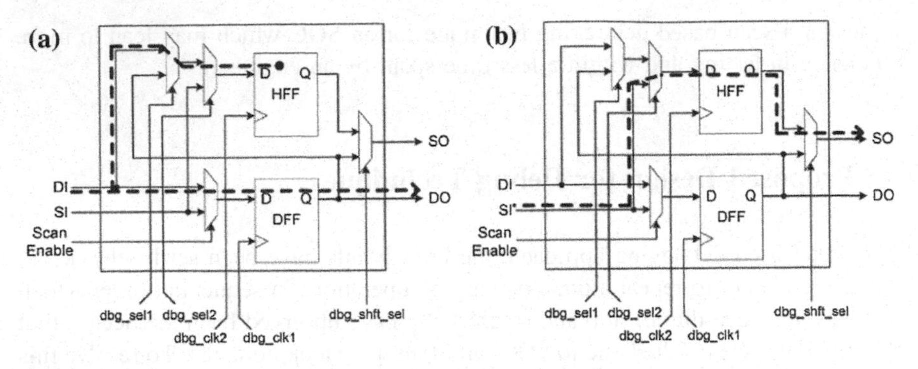

Fig. 4 SCDBG mode **a** Normal operation **b** Shift operation

The designed scan cell can support three operating modes as follows

1. Single clock domain debug mode (SCDBG)
2. Multiple clock domain debug mode (MCDBG)
3. Online debug mode (ONLINE).

When there is no cross clock domain problems then the SCDBG mode can be used. In this mode a run stop debugging will be repeatedly executed without a reset. From the Fig. 4 we can observe that the current state is captured by the HFF while the DFF stores in normal operation and after cores are stopped the stored data in HFF is shifted out for debug. Hence resuming execution is can be easily executed due to the current state stored in DFF.

For the receiving cores the MCDBG mode is used because there may a possibility of occurring the data invalidation if the CUD's are not in the same clock domain. This mode is shown in Fig. 5. During normal operation the HFF used to store the previous state by capturing DFF output.

If the data invalidation (Flag = '1') is occurred while the cores are stopped, the debug controller automatically selects the data present in the HFF. Else the data in the DFF (FLAG = '0') will be selected.

The debug engineer reads the data invalidation flag to know whether the data shifted out are the current or previous state. The SCDBG mode is set in the sender core due to there may not occur the data invalidation.

Under any other conditions the ONLINE mode is used which is shown in the Fig. 6. Thus, the debug engineer mixes all these modes to achieve the debugging of an SOC more efficiently.

3.2 Debug Control

The core debug control logic is placed in an IEEE STD 1500 wrapper and functions are reused. This leads to an engineer's additionally to optimize further area

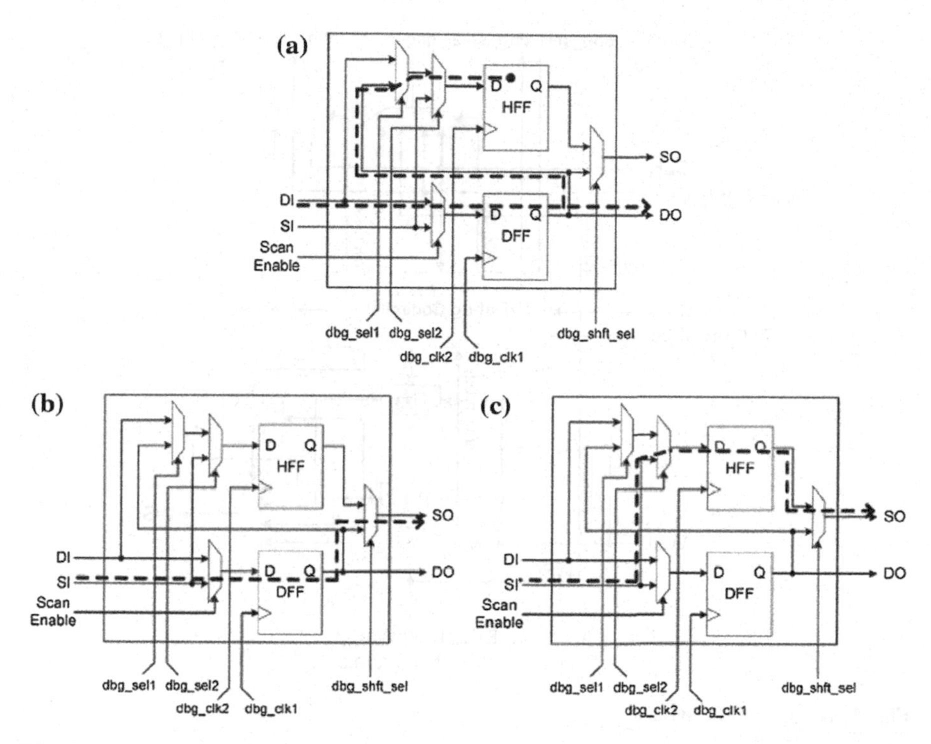

Fig. 5 MCDBG mode **a** Normal operation **b** Shift operation with (Flag = '0') **c** Shift operation with (Flag = '1')

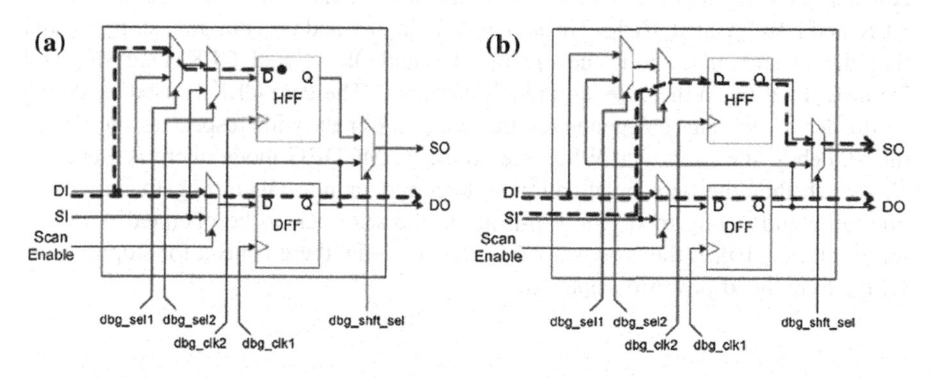

Fig. 6 ONLINE mode **a** Normal operation **b** Normal and shift operations

overheads. The proposed debug control logic has been shown in the Fig. 7. The test-control components in IEEE standard 1500 wrapper [11] are (WIR) wrapper instruction register and (WBY) wrapper bypass register. These are controlled by the IEEE std.1500 wrapper control signals. According to three modes the proposed scan cell can be controlled by the use of three debug instructions. The dbg_mode

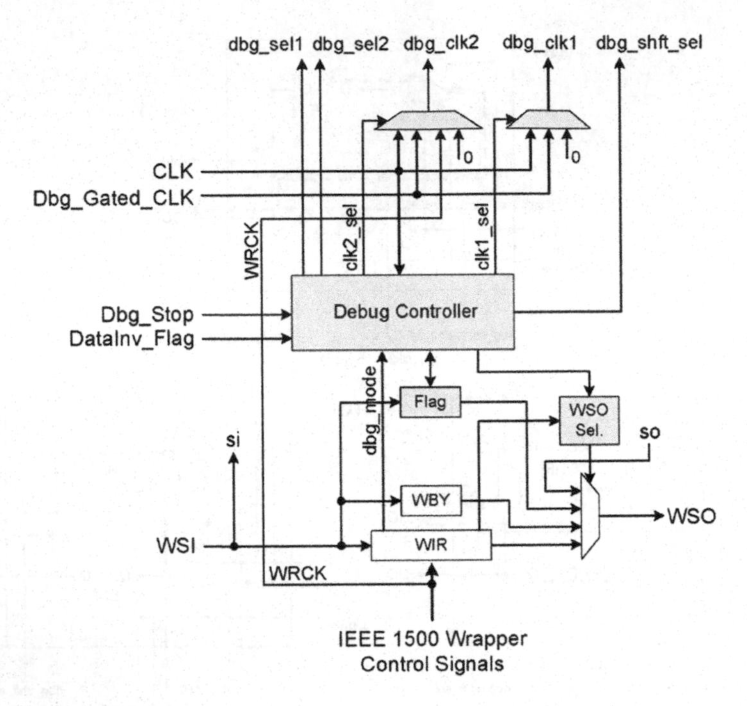

Fig. 7 Debug control logic

obtained from the WIR is used to activate the debug controller. The clock gating cell and data invalidation detector provides the signals DBG_stop, DataInv_Flag, CLK and DBG_Gated_CLK. The signals clk_in_rcv and clk_out_rcv, stop_rcv and flag signals are connected to the signals CLK and Dbg_Gated_CLK, Dbg_stop and DataInv_Flag when the core is in receiving mode. The core CLK is used by debug controller to operate and generates the control signals with respect to the debug mode, debug state and flag. When the core is in MCDBG mode, there is a need to check whether the data invalidation is occurred or not. According to the debug operation and debug mode the flip-flops in the scan cell to be operated with different clocks [10]. If the core is not in a debug mode, there is need for stopping the HFFS to reduced power dissipation.

4 Experimental Results

The different scan cell types have been used for the modification of existing scan cell according to their design environments (Tables 1 and 2)

Table 1 Area comparisons of different scan cells

Scan cells	Type	Area (# of transistors)
SSRL [7]	P-SRL[a]	60
Scan gadget [10]	P-SRL	52
Shadow [5]	LE-muxed-D[b]	63
HSC [9]	LE-muxed-D	35
Swap [8]	E-muxed-D[c]	44
Proposed	P-SRL	62
	LE-muxed-D	64
	E-muxed-D	48

[a]Normal operation
[b]Shift operation with (Flag = '0')
[c]Shift operation with (Flag = '1')

Table 2 Area overhead due to dfd circuitry

Core	Area (# of 2-input NANDs)		Area overhead (%)
	Before DfD	After DfD	
Leon3 Processor	51579	74967	31.2
SDRAM Controller	20426	24954	18.1
Ethernet MAC	37368	63576	41.2
VGA	6964	8492	18.0
GPIO	11750	13118	10.4
PS/2	10338	17506	40.9
Timer	8299	10547	21.3
UART	13543	15711	13.8

5 Conclusions

Generally, while performing scan based debugging a debug engineers may repeat the running and stopping the debug of an SOC or CUDs. The debug engineer may face a difficulty to capture the data, when stopping and resuming may not be done correctly if SOC contains the multiple clock domains. This may lead to the data invalidation and lock-phase relationship inconsistency. In this paper we have proposed a scan cell based debug control logic to overcome this problem. The designed scan cell supports the typical scan-based debug, online-debug and holding the current state and previous state. The proposed design results in efficient debugging without changing the debugging process and opposing the ambiguity of valid data.

References

1. B. Vermeulen and S. K. Goel.: Design for debug Catching design errors in digital chips IEEE Des. Test Comput., vol. 19, no. 3, pp. 35–43, May/Jun. (2002).
2. B. Vermeulen.: Functional debug techniques for embedded systems IEEE Des. Test Comput., vol. 25, no. 3, pp. 208–215, May/Jun. (2008).
3. H. Yi, S. Park, and S. Kundu.: A design-for-debug (DfD) for NoC-basedSoC debugging via NoC in Proc. Asian Test Symp., pp. 289–294 (2008).
4. K. Goossens, B. Vermeulen, R. van Steeden, and M. Bennebroek Transaction-based communication centric debug in Proc. Int. Symp. Netw.-on-Chip, pp. 95–106 (2007).
5. F. Golshan.: Test and on-line debug capabilities of IEEE Std 1149.1 in UltraSPARC-III microprocessor in Proc. IEEE Int. Test Conf., pp. 141–150 (2000).
6. S. K. Goel and B. Vermeulen.: Hierarchical data invalidation analysis for scan-based debug on multiple-clock system chips in Proc. IEEE Int. Test Conf., pp. 1103–1110 (2002).
7. S. DasGupta, R. G. Walther, and T. W. Williams.: An enhancement to LSSD and some applications of LSSD in reliability, availability, and serviceability in Proc. Int. Symp. Fault-Tolerant Comput., pp. 289–291 (1995).
8. H. Al-Asaad and P. Moore.: Non-concurrent on-line testing via scan chains in Proc. IEEE Syst. Readiness Technol. Conf., pp. 683–689 (2006).
9. H. Yi and S. Kundu.: On design of hold scan cell for hybrid operation of a circuit in Proc. IEEE Eur. Test Symp., pp. 1–4, Session 4–10 (2008).
10. R. Kuppuswamy, P. DesRosier, D. Feltham, R. Sheikh, and P. Thadikaran.: Full hold-scan systems in microprocessors Cost/benefit analysis Intel Technol. J., vol. 18, no. 1, pp. 62–73, Feb. (2004).
11. IEEE Standard Testability Method for Embedded Core-Based Integrated Circuits, IEEE Std. 1500, Aug. (2005).

Performance of Optimized Reversible Vedic Multipliers

A. Sai Ramya, B.S.S.V. Ramesh Babu, E. Srikala, M. Pavan, P. Unita and A.V.S. Swathi

Abstract Signal processing applications involve in many arithmetic operations. High speed arithmetic operations play an important role in these applications. Multipliers are often considered as the basic building blocks of digital signal processors (DSP). The speed of the multiplier corresponds to DSP. Multiplication is the basic operation to be performed in DSP. In order to implement these multiplications many algorithms are used. In this paper, few algorithms are discussed to implement multiplication. The algorithm discussed in this paper is the most ancient methodology used by Aryans. In this paper reversible Vedic multiplier is proposed using Urdhva Tiryakbhyam (UT) sutra and a comparative study reveals and suggests different logics pertaining to different profile considerations such as power and area.

Keywords Vedic mathematics · Algorithms · Urdhva Tiryakbhyam · Low power · Area

A. Sai Ramya (✉) · B.S.S.V.Ramesh Babu · E. Srikala · M. Pavan · P. Unita · A.V.S. Swathi
Department of ECE, Raghu Institute of Technology, Visakhapatnam, India
e-mail: ramya.ayyagari@gmail.com

B.S.S.V.Ramesh Babu
e-mail: rameshbssv@gmail.com

E. Srikala
e-mail: srikala.e@gmail.com

M. Pavan
e-mail: mukkupavan.raghu@gmail.com

P. Unita
e-mail: unita609@gmail.com

A.V.S. Swathi
e-mail: swathiarinana@gmail.com

S.C. Satapathy et al. (eds.), *Computer Communication, Networking and Internet Security*, Lecture Notes in Networks and Systems 5,
DOI 10.1007/978-981-10-3226-4_60

1 Introduction

Generally when two binary numbers are to be multiplied a binary multiplier is used. Binary multiplier is a significant block in every digital signal processors. These are widely used in Fourier transform and Convolutions. The processor speed can be greatly improved by improving multiplication operation, as much of processor speed is dependent on multiplication operation. Algorithms like such as array, booth, carry save and Wallace tree are used to implement multiplication. Among these the array algorithm reported very less computational time. This is because of partial products which are computed independently. Array multiplier for unsigned numbers results in reduced silicon area without effecting speed and power [1, 2]. In case of Booth algorithm the partial products obtained are less. Even though partial products are minimum in order to improve speed of the multiplication operation large booth arrays are required which further require large partial sum and partial carry registers. Booth algorithm is structured for a n $\times$ m multiplication where n can reach up to 126 bits [2, 3]. The carry-save technique is used in the Booth encoder, the Booth multiplier, and the accumulator sections to ensure the fastest possible implementation [3, 4]. The compression ratio of 3:2 is achieved using Wallace tree. The utilization of carry look ahead adder (CLA) results in an increase of speed. It is independent on the number of bits of the two operands [5]. This paper has been organized as follows. Section 2 consists of a review on Vedic mathematics and reversible logic. Section 3 deals with the proposed Urdhva Tiryakbhyam (UT) algorithm. And the optimization of UT multiplier is explained in the sub section of Sect. 3. The simulation results and discussions are explained under Sect. 4. Finally it is concluded with Sect. 5 under conclusion.

2 Review on Vedic Mathematics and Reversible Logic

Way back in the ages of ancient India, Aryan could perform quick mathematical calculations using Vedic mathematics. This consists of algorithms that can boil down large arithmetic operations to the methodology typically involve in reducing several arithmetic calculations to simple form. In the modern era of VLSI & DSP there is a huge demand for faster additions and multiplications [6]. Among these a prominent operation in a CPU. In general multiplication is referred as scaling one number by the other. Every complex technique like convolution, Discrete Fourier Transform, Fast Fourier Transforms, involves in several multiplication steps. This has a direct impact on which corresponds to faster arithmetic unit. Hence, processing new algorithm and hardware which can be accelerate these calculations in the current research focus of DSP and VLSI engineers [7]. Aptly VM has attracted this galaxy for the same purpose. It can be concluded that significant information loss is possible if the input vector is not recovered from output vector. Reversible logic circuits (RLS) are capable of generating inputs from observed outputs which is its inherent property. Hence, it does not lead to any loss of information. With zero energy dissipation principle it is possible if the network employs reversible gates. On the line, the reversibility emerged as an essential property in circuit design. Synthesis of RLC is different from combinational circuit. In reversible circuit, Fan out is not possible. Also, for each input pattern there should be a unique output pattern. As a result, it can be termed that the circuit is cyclic. Permutation of input patterns is carried out by realising the functions that are reversible. For example, K-outputs are with a k-input gate which is called as a k × k reversible gate. Reversible circuit design involves mainly in designing gates that are reversible. Also are efficient design suppresses the garbage outputs at every level [8, 9]. For error detection the parity checking technique is a useful tool. In general the arithmetic function doesn't preserve the parity. Hence, it is not required to verify the parity if once it is set with the input [6]. Considering the analogy. It is required reversible logic gates which are parity preserving to construct parity preserving reversible circuits. The work presented in this paper proposes a 4 × 4 parity preserving logic gate in which the parity of both input and output are similar. The work also mimics the technique of realising fault tolerant reversible full adder circuit using Fredking Gates [10]. The advantage with the technique is that, it does not result in any garbage outputs.

3 Urdhva Tiryakbhayam Multiplication Algorithms

Urdhva Tiryakbhayam (UT) is propped by ancient Indian Vedic mathematicians. It typically models multiplication in any form of numbering systems like binary, Hex and also Decimals. It involves in a sequence of steps like partial products followed by concurrent addition. This parallelism supports for quick computation.

3.1 Optimization of the UT Multiplier

Implementation of UT multiplier with reversible logic gates is shown in the Fig. 1. The four output expressions of this system namely q_0, q_1, q_2 and q_3 are used as employed to realize the input as shown in Fig. 2. In Fig. 2 both Peres gates and Feynman gates are employed. The quantum cost is 21, garbage outputs are 11 and constant inputs are 4. As discussed earlier, there are no fan outs. The gate count is 6. This design does not take into consideration the fan outs. The overall performance of the UT multiplier cab be modified by optimizing each individual unit in terms of quantum cost, garbage outputs etc.

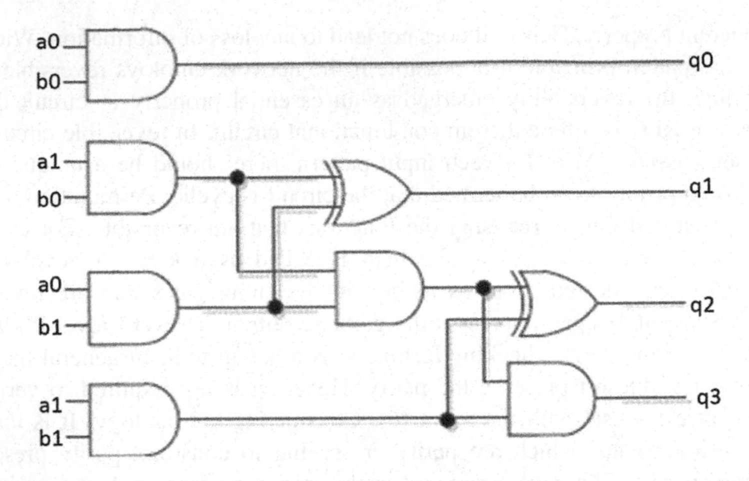

Fig. 1 UT multiplier with reversible logic gates UT multiplier with reversible logic gates

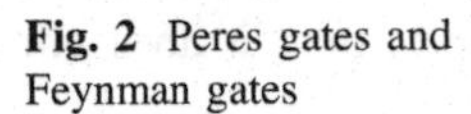

Fig. 2 Peres gates and Feynman gates

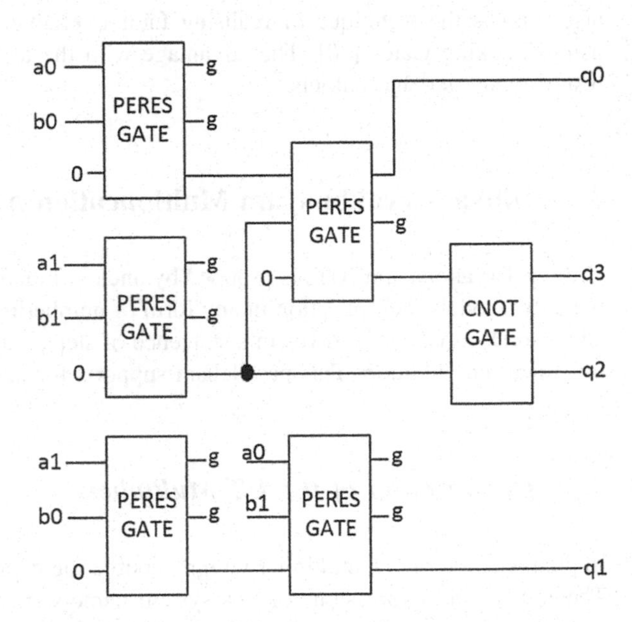

3.2 *Improved 2 × 2 UT Multiplier*

Modifications are suggested to the design expressions so as to optimize the design. The modified circuit consists of one BVPPG, three Peres gates and a one Feynman gate. Unlike the previous case the design considers the fan outs. This fan out and rule of avoiding loops other being are considered as design constraints. Hence, it is cumbersome task to deal with more number of inputs. However, in order to get rid of fan out, the circuit should have fan-outs using the reversible logic gates. The design has a quantum cost of 23. Similarly other parameters like number of gates, garbage and constant outputs as 5 in number for each. The second design mentioned in Fig. 4 uses BVPPG fan out structure. The same with peres and NFT gates in 3 and 1 in number as shown in Fig. 3. The modified design considers 24 as

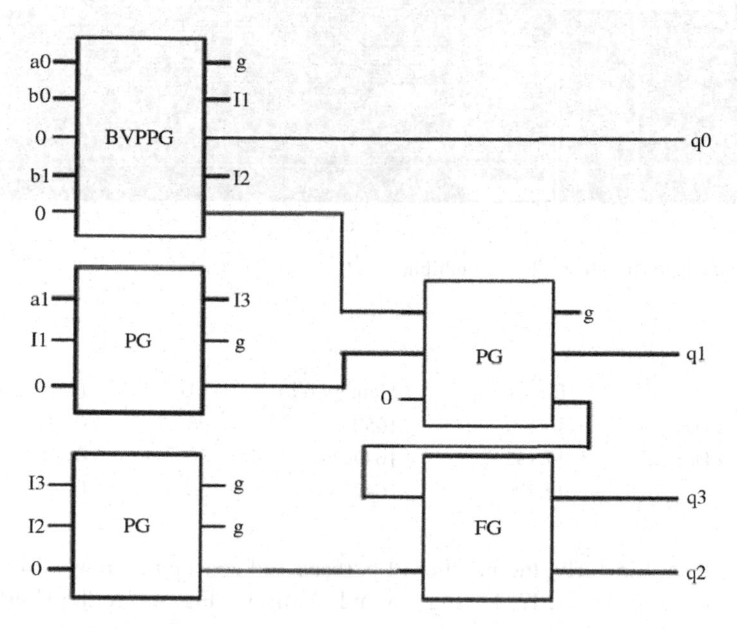

Fig. 3 Proposed Design1

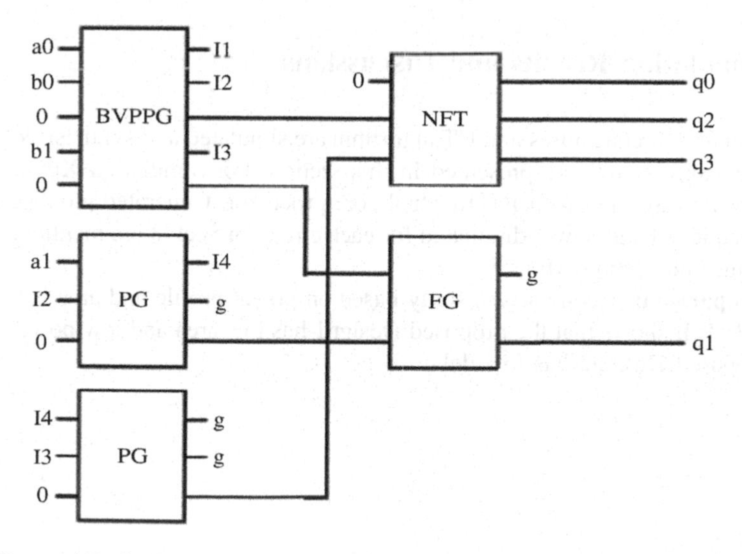

Fig. 4 Proposed Design2

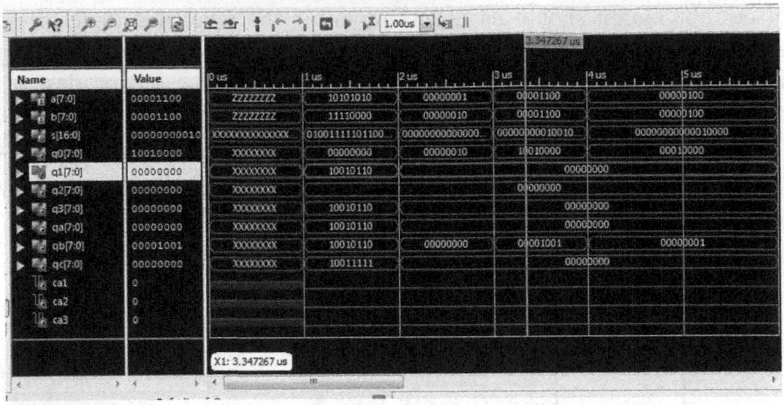

Fig. 5 Simulation result for 2 × 2 multiplier

Table 1 Comparision

	Delay (ns)	Memory (kb)	LUTs	Power (mW)
Proposed Design1	19.109	165032	28	0.4375
Proposed Design2	17.192	164904	30	0.46875
UT	17.306	164776	30	0.46875

quantum cost. Similarly, the number of garbage and constant outputs of 4 and 5 in number respectively. I1, I2, I3 (Figs. 3 and 4) and I4 (Fig. 4) are the intermediate outputs that are used for fan-out purposes.

4 Simulation Results and Discussions

The proposed architectures and UT algorithm are simulated and synthesized and the corresponding results are presented in this section. For Simulation Xilinx 1.2 are used on windows OS with INTEL dual core processor. Generated plots pertaining to Area utilized and power dissipated for each circuit in Sect. 3 are mentioned in the following sub-sections (Fig. 5).

Comparison: A comparison study based on power profile and area is tabulated in Table 1. It shows that the proposed Design1 has low area and low power and for the proposed Design2 has low delay.

5 Conclusion

Model of reversible Vedic multiplier using ancient Vedic mathematical techniques is presented in this paper. The composed Vedic multiplier used Urdhva Tiryakbhayam principle. The resultant model reported two designs Design1 has low area and low power and Design2 has low delay. Each design has its own advantages with specific contribution towards optimization.

References

1. Ramalatha, M., DeenaDayalan, K., Daharani, P.: High Speed Energy Efficient ALU Design using Vedic Multiplication Techniques. In: ACTEA IEEE, pp. 600–603.
2. Wallace, C.S.: A Suggestion for a Fast Multiplier. In: IEEE Trans. Elec. Comput., Vol EC 13, pp. 14–17 (1964).
3. Panwit Tuwanuti., Nopphagaw Thongbai.: Implementation of Vedic Multiplier Technique on Multicore Processor. In: TENCON IEEE Region 10 conference, Oct (2014).
4. Honey Durga Tiwari., Ganzorig Gankhuyag., Chan Mo Kim.., Yong beom Cho.: Multiplier Design Based on Ancient Indian Vedic Mathematics. In: SoC Design Conference 2008, ISOCC'08 International, vol- 2 (2008).
5. Rudagi, J.M., Viswanath Ambi., Viswanath Munavalli., Ravindra patil., Vinaykumar Sajjan.: Design and Implementation of Efficient Multiplier Using Vedic Mathematics. In: Advances in Recent Technologies in Communication and Computing 3rd International Conference IEEE (2012).
6. Yogita Bansal., Charu Madhu., pradeep kaur.: High Speed Vedic Multiplier Designs-A review. In: Engneering and Computational Sciences (2014).
7. Tritha, B.K.: Vedic Mathematics. (1965).
8. Mehta, P., Gawali, D.: Conventional versus Vedic Mathematical Method for Hardware Implementation of a Multiplier. In: IEEE International Conference on Advances in Computing, Control, and Telecommunication Technologies, pp. 640–642(2009).
9. Sao, S.S.R.M., Siddamal, S.: High Speed Signal Multiplier for Digital Signal Processing Applications. In: IEEE International conference on signal.
10. In: IEEE International conference on Devices circuits and Systems, pp. 360–364 (2012) processing comput, pp.1-6 (2012).

A Novel Congestion Control Scheme in VANET

Arun Malik, Babita Pandey and Anjuman Gul

Abstract The mobile networking has attracted a great numbers of researchers and industries. The Vehicular Ad Hoc Networks is a key to Intelligent Transport System. The recent advances in all fields have led to the advancement in the field of Ad Hoc Networking too. This mobile networking has attracted a great numbers of researchers and industries. This field is getting a huge attention because it has become an active field for research and development which will provide us with safe transport system along with the comfort. The communication between nodes in the VANET can cause congestion due to high frequency beacon messages that are generated by vehicles in an area where density of vehicles is very high in order to inform the neighboring nodes about the current traffic status. Congestion can also take place due to inefficient use of bandwidth; high transmit power, high vehicle density, changing topology of vehicles in the VANET network. A number of congestion control schemes have been designed to make the system efficient and robust but there are a lot many problems that still exist in the Vehicular Ad hoc Systems. In this paper we have proposed a congestion control algorithm for VANET'S which shows a better performance than existing algorithms in terms of various parameters.

A. Malik (✉) · A. Gul
Department of Computer Science and Engineering,
Lovely Professional University, Phagwara, India
e-mail: arun.17442@lpu.co.in

A. Gul
e-mail: anjumangull@gmail.com

B. Pandey
Department of Computer Applications, Lovely Professional University,
Phagwara, India
e-mail: shukla_babita@yahoo.co.in

S.C. Satapathy et al. (eds.), *Computer Communication, Networking and Internet Security*, Lecture Notes in Networks and Systems 5,
DOI 10.1007/978-981-10-3226-4_61

1 Introduction

Ad-hoc is a Latin word that means "for the purpose". Ad-hoc networks refer wireless networks that do not depend on existing framework like router or access points in a wired network. In ad-hoc networks each device participates in sending the information between nodes. This forwarding of data to the other nodes depends on the dynamic connectivity of devices. In ad-hoc wireless network the nodes are of moving nature that means they are not stable at their location. They are dynamically and self-configuring networks. Ad-hoc networks are of various types like Mobile Ad-Hoc networks, Vehicular Ad-hoc networks, Smart Phone Ad-Hoc networks, Internet Based Mobile Ad-Hoc Networks.

Vehicular ad-hoc networks are mobile ad-hoc networks in which vehicles act as nodes or devices of the network and drivers or users will communicate or exchange information using vehicles as nodes. Two types of communications are provided in VANETS. In first type of communication there is pure ad-hoc wireless communication where communication between vehicles and vehicles take place without any infrastructure support. In second type of communication, communication take place with the help of infrastructure called as RSU (road side units). The respective OBU'S of the vehicles are nodes will be connected the RSU'S of the particular area and the RSU'S are connected to the internet. A short range communication is suggested that will operate on 5.9 MHz frequency and will use 802.11 methods and the access method 802.11p will be providing short range communication with low delay or latency. The main goal of vehicular ad-hoc networks is to issue safety of the information exchanged between nodes or vehicles in the network. If any will be carried out in the vehicular ad-hoc network will result in financial loss or loss of life. Therefore security is very crucial in vehicular ad-hoc network. In this paper Sect. 2 describes need for congestion control scheme in VANET'S, Sect. 3 describes related work, Sect. 4 describes Implementation, Sect. 5 describes Results and Sect. 6 describes Conclusion.

2 Need for Congestion Control Scheme in VANET

In VANETS there are mostly two types of messages for enabling safety application, these are beacon messages that are periodically broadcasted by all nodes in the network on the channel so that they can easily send and receive all the information related to current position and location of the nodes to all the nodes in the vicinity. The other types of messages so called the emergency messages are used for updating the vehicles in the area about any critical condition like accidents that had happened in any area so that the vehicle can take alternate routes to have a hindrance free journey and also can reduce the intensity of the traffic jam caused due to the critical condition and can reach their destination on time. In an area with large number of vehicles the congestion control channel can get congested due to the transmission of beacon messages by all vehicles in the vicinity at very high frequency rate. The high frequency rated messages can cause congestion in control channel in order to keep the channel free from congestion it is of high significance that safety messages will be delivered on time in order to avoid hazardous conditions. It is very important to design a novel congestion control for VANETS that will enable timely and reliable delivery of safety messages to all the vehicles so that there will be a smooth communication between the vehicles.

3 Related Work

In [1] an effective heterogeneous solution was proposed namely Enhanced Multimedia Broadcast Multicast Service which will transmit the safety driven messages based on the priority the high priority message will be delivered first as compared to least priority. The advantage of this technique was that the beacon rates were decreased using an exponential algorithm. Deliver messages even if the vehicles did not have IEEE devices. Packet delivery rate was enhanced or improved.

In [2] a congestion control scheme was proposed in which works on efficient use of bandwidth depending on the beacon rates the less the beacon rates the more efficiently bandwidth will be used. The advantage of this scheme was that it Improved packet delivery. Performed better when numbers of vehicles are less. The limitation of this scheme was that the performance decreased with increase in number of nodes.

In [3] two approaches reactive and adaptive were proposed. In reactive a channel busy ratio is used directly that defines a transmission nature for each channel busy ratio and the other approach that is adaptive technique identifies the transmission nature that puts channel busy ratio to the target weight. The advantages of the two approaches were that they Improved message throughput in case of large number of vehicles Limits emergency messages generation. The limitation of the two was that the reactive approach defined in decentralized congestion control shows minimum throughput as compared to adaptive approach.

In [4] an efficient congestion controlling scheme controlling congestion based on utility and message packet sending in VANET was proposed. The proposed approach used a utility function that is specific for the application and converted the useful information that is stored in transmission packets in an exposed way to all the nodes. A decentralized algorithm is used to calculate the average convenience of each device in the network depending upon the value of packets and allocates a part of the information rate corresponding to the most important. The advantage of this scheme is that the utility function is used for estimating the utility in the header of the packet. Packets having low utility are dropped. It improves performance and information dissemination. It also avoids node starvation.

In [5] a scheme was proposed to improve transmit power, contention window and packet interval in order to decrease the congestion that was caused due to large number of vehicles present in the VANET. The advantage of this scheme is that Congestion controlled by considering the transmit power according to number of nodes. The limitation is that the packet interval and contention window size according to vehicles present in the network still there.

In [6] a scheme was proposed for congestion control that is composed of three stages first one is allocating priority to the information to be broadcasted. Second phase is detecting the congestion that occurs due to high frequency rated beacon messages and in last phase broadcasting power rate and beacon message sending rate is adjusted. The advantage of this scheme was that the performance Enhanced due allocation of priority to emergency messages. The limitation was that it does not provide high performance in case of long congestion periods as it applicable to only short congestion periods.

In [7] a technique was proposed for congestion control and forwarding of packets in VANETS. This technique broadcast the traffic information by taking into consideration only the vehicles in the radio range. The advantage of this scheme was that it reduces dun necessary traffic information, reduced congestion. Improved performance and utilized bandwidth in a better way.

In [8] a congestion control scheme was proposed to provide efficient operation in case of high dense areas where density of vehicles is very high. The advantage of this scheme was that the scheme improved packet loss and also enhanced packet delay. The limitation was that this scheme has to improve considering the dynamic nature of nodes in the VANET.

In [9] a system was proposed that preserves privacy, detects congestion and provides drivers information about congestion that is at a long distance from the current location of the driver. Each vehicle is installed with GPS (global positioning devices) devices that provide information about current location and a communication device that is used by the driver to communicate with the nearby drivers.

In [10] a scheme was proposed which used two parameters transmit power rate and transmission rate for avoiding congestion. The advantage of this scheme is that the technique was used dynamically that means in any condition in case of high density of vehicles. The limitation was that if we used power control it does not satisfy the requirements of safety messages. The methods that were used for measuring the usage of channel are not effective in real world scenarios.

4 Implementation

We implemented our approach using NS2 simulator in which we considered a scenario where we took bunch of vehicles as shown in Fig. 1. First those vehicles are registered to RSU'S and they are allocated Vehicle identification numbers known as vehicle ids. Then they are allocated digital signatures for security purposes. In the scenario when there is collision we controlled it on the basis of message transmission scheme that means the more important messages are transmitted firstly as compared to the least important messages. When a collision occurs

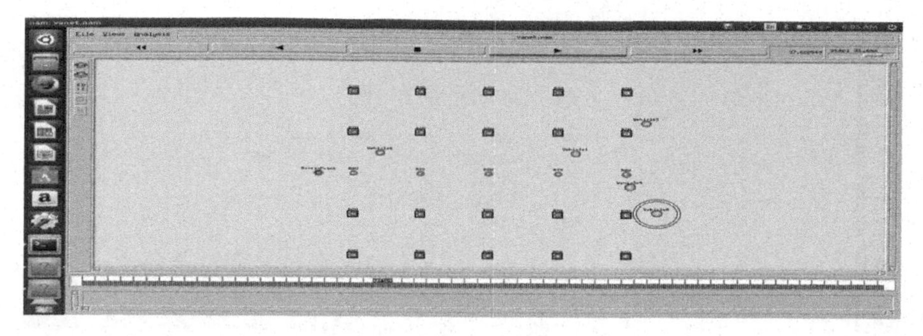

Fig. 1 Movement of vehicles before congestion

at a place the vehicle forwards the message to the nearest RSU and this RSU forwards this message to the other RSU in its vicinity and the RSU receiving the message forwards it to the farthest vehicle in its transmission range.

Proposed Algorithm:

Vehicle V_i, Sends message MV_i to the nearest

1. Next hop in its transmission range;
2. When V_{i+1} receives data from the nearest hop;
3. Checks whether $(M(V_i) == M(V_{i+1}))$
4. if (true)
5. Saves message and Forward to V_{i+2};
6. Else
7. Discard Message MV_i;
8. If V_i at end of Road Segment
9. If V_i is nearest to RSU
10. Generates TCD and sends data to RSU;
11. If V_i Receives Ack_i, it forwards MV_i, to next
12. Else
13. Drop V_i;

First when the collision occurs as shown in Fig. 2, the vehicle will send a message to its neighboring node. The neighboring node will receive the message and will check if the message is from the neighboring node if this will be true then its saves the message otherwise it discards the message. When the neighbor saves the message it will check now where the vehicle is located either it is at the end of the range or near to RSU after this is done the vehicle will send the information about the congestion that it have received to the RSU. The status consists of location where the congestion have occurred and severity of the congestion, how more or less severe congestion is. When congestion will occur and when vehicles coming in that way will get the information about it the other vehicle drivers will inform them about the congestion and will inform them to change their paths and take alternative routes so that they will be safely reach the destination and will not

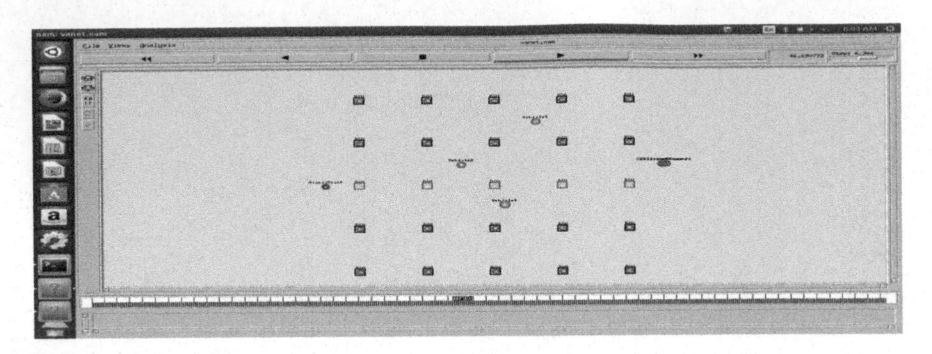

Fig. 2 Congestion occurred

face congestion. In detection mechanism we will consider the Vehicle V_i, Road Segment RS_i on the way the vehicle is travelling. The vehicle V_i broadcasts the Message MV_i, to all the nodes near its transmission range. The message can contain, its direction of moving, speed, its current location and the timestamp. It can be viewed as

$$MV_i = \{\text{Direction, Speed, Location, Timestamp}\}$$

Each V_i that receives the message MV_i around its surrounding vehicles and forwards that data to all the vehicles successive to them. For forwarding this data to other vehicles some random nodes are selected called Relay Nodes. At the end of each segment, this data will also be forwarded to the RSU's which is called as Traffic Censoring Data. This Traffic Censoring Data contains information such as Traffic Density, Direction, Speed, Number of Vehicles, Travel time, Origin point, Destination Point and Number of ways from the Origin to reach the destination.

5 Results

- Delay Graph
- X-axis—No. of milliseconds
- Y-axis—Delay
- Delay 02.tr (Red line)—Delay between Source to Destination
- Delay 12.tr (Green line)—Delay increases from Source node to destination node because of congestion
- Delay 22.tr (Blue line)—Delay increases from Source node to destination node
- Delay 32.tr (Yellow line)—more congestion occurs at the time of 30.

This delay graphs shows that the designed algorithm is able to decrease the message delivering delay from source to destination (Fig. 3).

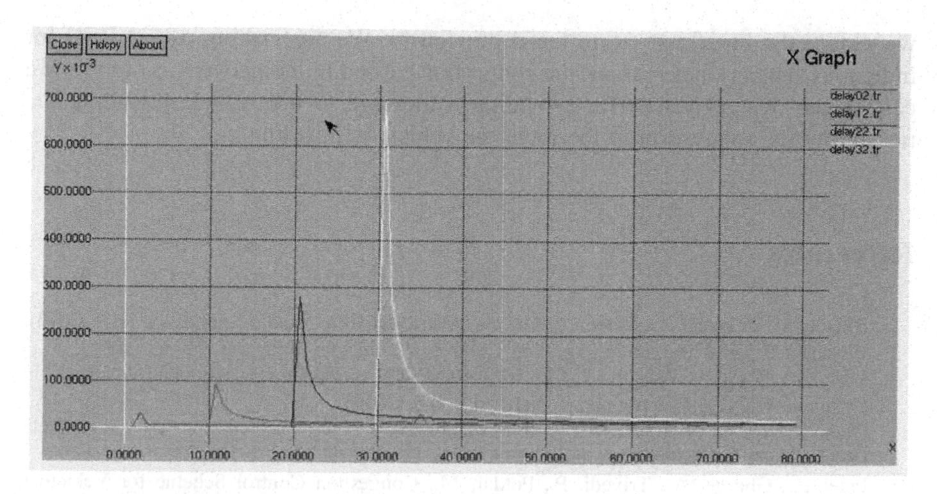

Fig. 3 Delay graph

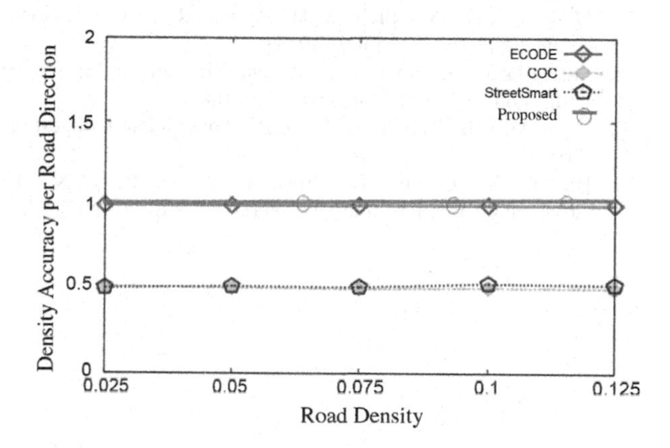

Fig. 4 Algorithm comparison graph

The Accuracy for the road direction is always same even if the density increases. The graph showed that our proposed algorithm has better performance as compared to the existing algorithms (Fig. 4).

6 Conclusion

In this paper, we are detecting the congestion by considering the different parameters such as end to end delay, jitter, accuracy and other parameters. After detection of Congestion we are making other vehicles to change their way so that we can stop vehicles to come on the congestion way. It needs only 57–60% cost of communication between the vehicles. By making the RSU to communicate with the vehicle we reduce the broadcast storm faced previously. We are making use of RSU for indicating other vehicles about the congestion caused in the network. In future, we can extend this algorithm for providing a message to the vehicle regarding the possibility for congestion in the route the vehicle is traveling.

References

1. Bihade, A., Talmale, R.: Detection And Avoidance Of Road Traffic Congestion. 1–4 (2013).
2. Wischhof, L., Rohling, H.: Congestion Control in Vehicular Networks. 58–63 (2005).
3. Yang, Y., Wang, P., Wang, C., Liu, F.: An eMBMS Based Congestion Control Scheme in Cellular—VANET Heterogeneous Networks. 1–5 (2014).
4. Djahel, S., Ghamri-doudane, Y.: A Robust Congestion Control Scheme for Fast and Reliable Dissemination of Safety Messages in VANETs. 2264–9 (2012).
5. Patel, A., Gharge, A., Trivedi, P., Potdar, M.: Congestion Control Scheme for Vehicular Ad-Hoc Network (VANET).4(1) (2016).
6. Padron, F. M., Mahgoub, I., Rathod, M.: VANET-Based Privacy Preserving Scheme for Detecting Traffic Congestion. 66–71 (2012).
7. Qureshi, K.N., Abdullah, A. H., Anwar, R.W.: Congestion Control Scheduling Scheme for Vehicular Networks. 55–9 (2014).
8. Shrivastava, G., Soni, S.: Efficient Mechanism for Congestion Control and Bandwidth Utilization of VANET.3(5):672–7 (2014).
9. Vyas, M. I. B., Dandekar, P. D. R.: An Efficient Congestion Control Scheme 3(8):1467–71 (2014).
10. Bansal, G., Cheng, B., Rostami, A., Sjoberg, K., Kenney, J. B., Gruteser, M.: Comparing LIMERIC and DCC approaches for VANET Channel Congestion Control (2015).

An Enhanced Spray-Copy-Wait DTN Routing Using Optimized Delivery Predictability

Harveen Kaur and Harleen Kaur

Abstract DTN-Delay Tolerant Network works well in the severe conditions where other wireless ad hoc networks fail to exist. It works well even where there does not exist any end to end connectivity i.e. intermittent connectivity. It works on the principle similar to postal system i.e. store-carry-and-forward; which makes it possible for DTN to work in the regions of intermittent connectivity. It is implemented by bundle protocol. The bundles of message are replicated, forwarded to the intermediate nodes with the help of various routing and replication strategies offered by various routing protocols of DTN. Spray-and-Wait routing protocol is good in providing a replication strategy but lacks in routing strategy; while PRoPHET routing protocol proves a very good routing strategy and therefore a hybrid of both can intelligently route sufficient number of copies in the network. Major problem of spray-and-wait routing protocol i.e. explicitly deciding the exact number of copies i.e. L to be initially set. The goal of this research is to increment the L if copies are not sufficient and to forward copies only based on the optimized threshold and remove garbage messages by negative packet technique. From evaluation of experiments, we show that the proposed method works really well for least message generation interval, and gives sufficiently good optimized results for both arrival rate and overhead.

Keywords Delay · Tolerant · Network · MATLAB · PRoPHET · Spray-and-Wait

H. Kaur (✉) · H. Kaur
Baba Farid College of Engineering and Technology, Bathinda, India
e-mail: kharveen199262@gmail.com

H. Kaur
e-mail: harleengrover@gmail.com

S.C. Satapathy et al. (eds.), *Computer Communication, Networking and Internet Security*, Lecture Notes in Networks and Systems 5,
DOI 10.1007/978-981-10-3226-4_62

1 Introduction

The wireless network is defined as a network without wire that communicates by using wireless technologies. In telecommunications, information is transferred without the use of wires is known as wireless communication. DTN (Delay Tolerant Network) is a new term in the field of wireless networks. DTN enables the communication in challenged environment where traditional network fails to communicate. It does not demand for end to end node connectivity, unlike other ad hoc networks.

DTN is based on store-carry-and-forward principle. Store-Carry-and-Forward principle is based on the postal system. This mechanism is implemented using bundle protocol. Here a node might accumulate a message in its buffer and carry for limited time, waiting unless or until a suitable forwarding opportunity is acquired.

Routing in delay-tolerant networking is related with the ability of the node to transport, or route, data through intermediate nodes basically from a source node to a destination node. It is the fundamental objective of all the communication networks to communicate and route message. However, when it is difficult to establish end-to-end instantaneous paths, routing protocols in DTN route packet using store carry and forward approach. In store carry and forward approach, data moves in hops from one node to other where it is stored for limited time and is forwarded according to some strategy with a faith that it will ultimately reach the desired destination [1]. A common technique used to increase the arrival rate is to maximally replicate the copies of message in every hop, with a belief that either of them will succeed in reaching its destination. Here different protocols are based on routing and replication strategies i.e. the number of copies of a message are created and are forwarded. Different routing protocols used here are Epidemic Routing Protocol, Prophet, Spray and wait.

The main goal of DTN Routing is to build a network between various nodes (mobile devices, planetary vehicles etc.) so that good arrival rate and less overhead are obtained. This unique mechanism poses an efficiency challenge. Particularly, to raise the communication performance of network a hybrid of spray-and-wait and Prophet routing protocols of DTN has been proposed. Spray-and-wait routing protocol provides a replication strategy whereas Prophet routing protocol provides a routing strategy.

The rest of paper is organized as follows. In Sect. 2, related work is discussed. In Sect. 3, proposed method is explained. In Sect. 4, the results are evaluated and discussed. In Sect. 5, the conclusion and future scope of the research is mentioned.

2 Related Work

2.1 Spray-and-Wait Routing Protocol

Spray and Wait routing protocol has put a restriction on the maximum number of message replicas allowed in a network. Thereby it overcomes the shortcoming of epidemic routing protocol; which floods packet blindly to all encountered nodes. Initially the value of L (Maximum number of message replicas allowed) is explicitly set. And only few replications are allowed so as to get a good delivery rate for practical scenario of limited resources and also taking the need of controlled overhead into consideration. As the value of L is to be decided explicitly it is difficult to select an appropriate value for L [2]. Problem arises if L is chosen to be low than needed replicas thereby increasing delay; and if L is chosen to be high than it may reflect the epidemic routing protocol. Moreover it does a blind forwarding of these replicas thereby the chances may be forwarding the message to the unreachable nodes for destination.

2.2 PRoPHET Routing Protocol

PRoPHET routing protocol overcomes the problem of the Spray and Wait routing protocol. The PRoPHET routing protocols assumes that the nodes are highly mobile and they possess some deterministic properties i.e. repeating behavior. It is more probable that the nodes may pass frequently through the same locations. Therefore as a result the nodes that meet each other frequently are more likely to meet in future. The PRoPHET routing protocols works in two phases: forwarding strategies and calculating delivery predictability: In the delivery predictability phase, increment or decrement their delivery predictability metrics with every meet. Encountering destination more times, results in higher delivery predictability values. It can further overcome the issue of blind forwarding in the spray-and-wait routing protocol; as it makes a choice to select the best favorable node for delivery.

3 Proposed Work

3.1 Optimally Forwarding Increased Replica

The proposed method resolves the issue of lack of maximum replica of message essentially required for the successful delivery of a message to the destination. As per need; if L = 1 and reachability of message is less; L is incremented by the factor of 1 i.e. L = 1 + 1 = 2 thereby changing the node from wait phase to spray phase an now as it is active to forward L/2; the increased message replica is routed intelligently by PRoPHET routing technique; each node has a certain value for delivery predictability which depend on its hits with the destination. Having higher number of hits means it is highly probable node for delivery. Now when an active node encounters the favorable node; it checks the delivery predictability value and compare it with an optimized threshold (i.e. 0.193) that have been chosen to fit the network demands and to get optimal results for the network characteristics. And if the delivery predictability of the encountered favorable node $\geq$ the optimized threshold it is marked as best and L/2 copies are forwarded to it; this entire technique not only increases the reachability, but also forwards the increased replica in an intelligent way so as to worth the increasing operation.

3.2 Message Removal Operation

Storage buffer capacity of DTN node influences the effectiveness and the efficiency of communication performance. Therefore the limited storage capacity needs to be utilized effectively so as to reduce the over exploitation and the overhead or the drop of needed messages due to shortage of space in buffer.

The message removal is done after the message has been successfully delivered i.e. w.r.t. destination; an Anti-Packet scheme is used to remove an already delivered message from the buffer of not only the node that has delivered but also from the buffer of the intermediate nodes still fighting to deliver an already delivered message, this negative packet scheme removes the garbage message thereby creating a space for newly generated messages.

4 Performance Evaluation

Performance is evaluated; as we compare and evaluate the proposed optimized hybrid (spray-and-wait, PRoPHET Protocol) methodology with an existing generalized hybrid of protocols. The performance is evaluated for the environmental characteristics overhead and message arrival rate w.r.t. message generation interval. Simulation is carried with MATLAB simulator.

4.1 Simulation Scenario

Two simulation scenarios were taken; one small scale scenario for small area and another large scale scenario for a large field. Table 1 depicts the small area scenario parameters and Table 2 shows the large area scenario parameters.

Table 1 Simulation parameters (small scenario)

Parameter name	Parameter value
Experimental time	240 s
Mobility model	Map route movement model
Node speed	60 km/h
Transmit range	50 m
Routing	Spray and wait, PRoPHET
Message size	500 kb
Number of nodes	10
Message generation interval	1, 2, 3 s
Maximum replication allowed	L = 4
Node searching interval	10 s

Table 2 Simulation parameters (large scenario)

Parameter name	Parameter value
Experimental time	7200 s
Mobility model	Map route movement model
Node speed	60 km/h
Transmit range	50 m
Routing	Spray and wait, PRoPHET
Message size	500 kb
Number of nodes	120
Message generation interval	1, 2, 3 s
Maximum replication allowed	L = 8
Node searching interval	10 s

5 Results and Discussion

5.1 *Message Arrival Rate w.r.t. Message Generation Interval*

Results for message arrival are shown in Fig. 1 w.r.t. message generation interval for small scenario and in Fig. 2 for large scenario. Figure 1 shows that the message

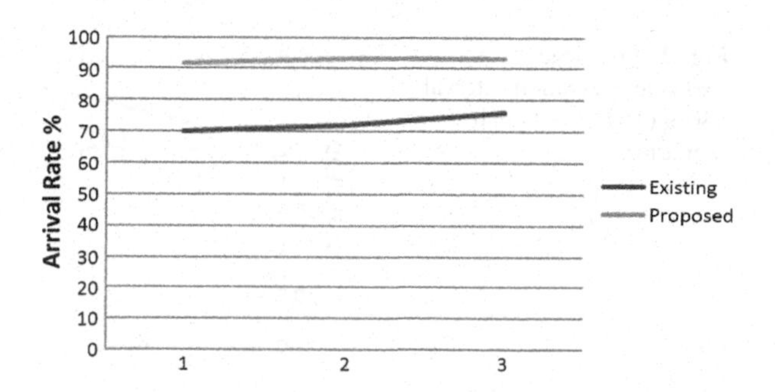

Fig. 1 Arrival rate w.r.t. message generation interval 240 s (1, 2, 3) s (small scenario)

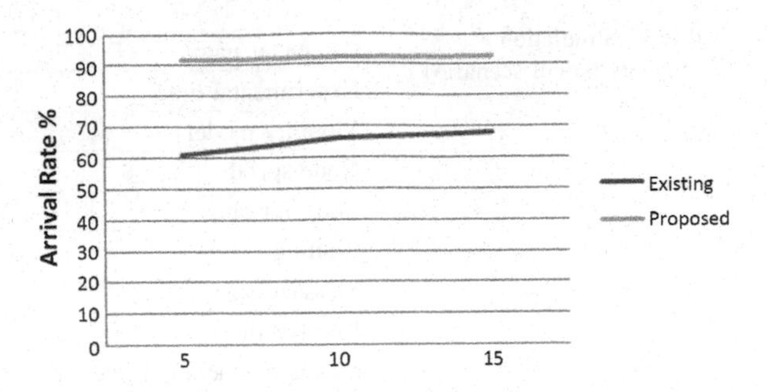

Fig. 2 Arrival rate w.r.t. message generation interval 7200 s (5, 10, 15) s (large scenario)

arrival rate is sufficiently high for least message generation interval i.e. 1 s. This shows that the optimized threshold for the delivery predictability of the node enhanced overall arrival rate. Yields far better results for Arrival rate when compared with the existing approach. The existing approach lacks the good relay nodes as they are chosen based on the obsolete approach of general delivery predictability.

Figure 2 shows that the message arrival rate is sufficiently high for the least message generation interval i.e. 5 s. The results fluctuate but on a whole the message rate is good enough. When compared to an existing the arrival rate for the enhanced optimized approach is high, reflecting the high delivery predictability of the good relay nodes chosen on comparing with an optimized threshold. This shows that the optimized threshold for the delivery predictability of the node enhanced overall arrival rate for both small scale and large scale scenario.

5.2 *Overhead w.r.t. Message Generation Interval*

Results of overhead are shown in Fig. 3 for small scenario and in Fig. 4 for large scenario. Overhead in the term of buffer space means the average amount of messages in the buffer storage of each node. In simulations, Fig. 3 depicts that Overhead is quite tolerable, as the message removal operation actively removes message from buffer w.r.t. destination. Initially for 1 s generation the overhead is high as the messages are generated at a very fast rate; therefore filling up the

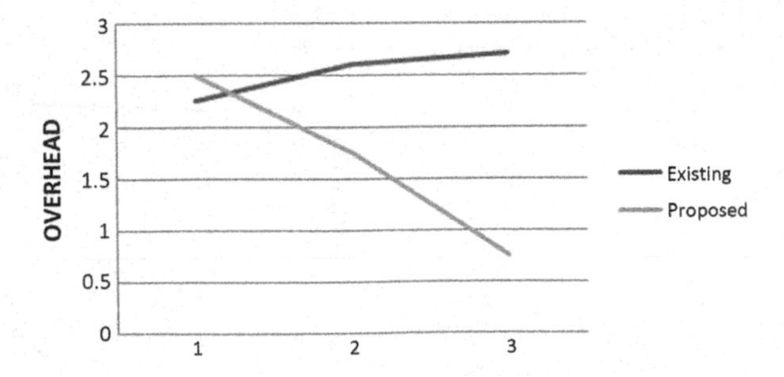

Fig. 3 Overhead w.r.t. message generation interval 240 s (1, 2, 3) s (small scenario)

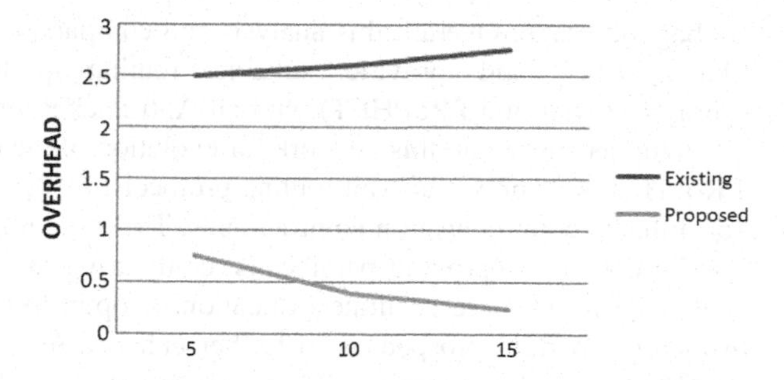

Fig. 4 Overhead w.r.t. message generation interval 7200 s (5, 10, 15) s (large scenario)

networks nodes' buffer with the messages; removal operation works in parallel with the generation of messages but it takes some time to overcome the overhead and flush out the message duplicate replicas in the network. Later on for 2 and 3 s the overhead has come under control as the removal operation has actively overpowered the message generation rate.

Figure 4 shows that the overhead is tolerable and lies within a nominal range. Overhead is less and is even decreasing with the increase in message generation interval. The less value of overhead reflects the activeness of negative packet scheme of actively removing the garbage message from the buffer of the relay nodes. On a whole the proposed negative packet scheme proves to be better when compared to a general removal technique of an existing.

5.3 Comparison Analysis

Existing technique lacks good arrival rate for least message generation interval. And the message removal is done even before the messages have reached the destination. These both no doubt increased arrival rate but could not control overhead. DTN is related to limited resources therefore overhead needs to be under control. The proposed message removal technique does remove message w.r.t time as well as w.r.t destination. Therefore removing garbage messages successfully and creating a space for new messages. Optimized threshold enhanced the message forwarding criteria; thereby increasing arrival rate for least message generation interval. Therefore the optimized threshold and Anti-Packet message removal increased arrival rate and also controlled overhead.

6 Conclusion

For this research, the arrival rate has increased sufficiently high for both small scale and large scale scenario for least message generation interval because of an optimized threshold. Overhead is less as negative packet scheme actively removes garbage message. Overall, this analysis and comparison provides routing protocol characteristics and provide suitable routing protocols optimized hybrid (spray-and-wait and PRoPHET) with an Anti-Packet removal operation.

A further study into this network for evolution of the performance of a hybrid of PRoPHET and spray-and-wait routing protocols, by dynamically varying threshold for different communication environments. The more refinement in the calculations can be done by improving parameters; configuring the network parameters which will help in the more accurate evaluation of optimized hybrid routing protocols. Another important prospect is to further enhance removal operation by having a combination of optimized threshold and Anti-Packet scheme for removal.

References

1. Burleigh, S., Hooke, A., and Torgerson, L.: Delay-tolerant Networking: An Approach to Interplanetary Internet. IEEE Communications Magazine. (ISSN: 0163-6804). vol 41. no 6. (2003) 128–136.
2. Burgess, J., Gallagher, B., Jensen, D., and Levine, N.: Max Prop: Routing for Vehicle-Based Disruption-tolerant Networks. Proceedings IEEE INFOCOM, 25TH IEEE International Conference on Computer Communications. (ISBN: 1-4244-0221-2). (2006) 1–11.
3. Vahdat, A., and Becker, D.: Epidemic Routing for Partially-connected Ad Hoc Networks. International Conference on Mobile Computing and Networking. IEEE. (2000) 263–270.
4. Jain, S., Fall, K., and Patra, K.: Routing in Delay Tolerant Network. ACM SIGCOMM. (ISBN: 1-58113-862-8). vol. 34. no. 4. (2004) 145–158.
5. Thrasyvoulos., Psounis, K., and Raghvendra, S.: Spray and Wait: An Efficient Routing Scheme for Intermittently Connected Mobile Networks. WDTN '05 Proceedings of ACM SIGCOMM workshop on Delay-tolerant networking. (ISBN:1-59593-026-4). (2005) 252–259.
6. Thrasyvoulos., Psounis, K., and Raghvendra, S.: Spray and Focus: Efficient Mobility-assisted Routing for Heterogeneous and Correlated Mobility. Proceedings of the IEEE. Pervasive Computing and Communications Workshop. PerCom Workshops. Fifth Annual IEEE International Conference. (ISBN: 0-7695-2788-4). (2007) 79–85.
7. Yuan, Q., Cardei., J and Wu, J.: An Efficient Prediction-based Routing in Disruption Tolerant Networks. IEEE Transaction on Parallel and Distributed Systems. (ISSN: 1045-9219). Vol. 23, no. 1 (2012).
8. Miyakawa, T., Koyama, A.: A Hybrid Type DTN Routing Method using Delivery Predictability and Maximum Number of Replication. IEEE, Advanced Information Networking and Applications Workshops (WAINA). (ISBN: 978-1-4799-1775-4) (2015) 451–456.
9. Kaur, Harveen, Kaur, Harleen: Improved Methods in Mitigating Misbehavior Activities in DTN: A Survey. An International Conference on Advances in Computer Science and Information Technology [ACSIT]. (ISSN: 2393-9915) vol. 3. no. 5. (2016) 360–365.

Impact of Facebook's Check-in Feature on Users of Social Networking Sites

Hitesh Kumar, Shilpi Sharma, Tanupriya Choudhury and Praveen Kumar

Abstract Social media is infiltrating the planet. Facebook has been the leader in this industry for almost a decade. The social media sites have made many changes over the years regarding security and advertising that have frankly, miffed a lot of people. The paper mainly describes about the impact of Facebook Check-in feature. Also, a new feature is proposed- Check-in-Checker which is presently not yet introduced in Facebook. Check-in Checker is an extension of the existing Facebook feature check-in. It is a helpful tool to know who is going where and at what time. Also, popularity about a particular common place being visited by friends will increase. Data was gathered from 550 people who give their views about how they use Facebook. Also, their opinion is taken in account about the existing feature 'Check-in' provided by Facebook. Analyzing the data, the new feature Check-in Checker suits today's demand and requirement of a user, to be added up in Facebook's features list. This feature doesn't violates any user's privacy and holds good number of positive reasons to be there in the pool of Facebook's exciting features.

Keywords Facebook · Social networking site · Privacy · Check-in

H. Kumar (✉) · S. Sharma · T. Choudhury · P. Kumar
Amity University, Noida, Uttar Pradesh, India
e-mail: hitesh2194@gmail.com

S. Sharma
e-mail: ssharma22@amity.edu

T. Choudhury
e-mail: tchoudhury@amity.edu

P. Kumar
e-mail: pkumar3@amity.edu

S.C. Satapathy et al. (eds.), *Computer Communication, Networking and Internet Security*, Lecture Notes in Networks and Systems 5,
DOI 10.1007/978-981-10-3226-4_63

1 Introduction

Social networking sites are the common spots where user can see what your friends ate for breakfast, or post about their break ups. The features of social networking platforms influence our lifestyle. These features must be upgraded according to the need of time [1]. Facebook 'Check-in Checker' is one improvisation of the Facebook's existing feature 'Check-in'. Facebook is the most well-known Social Networking Site, and is king of internet marketing [2]. Within a span of 7 years, Facebook has attracted more than 1.44 billion users with its creative and catchy feature [3]. Whether a small businessman or a business tycoon, everyone has found oneself comfortable with the interface of Facebook. With numerous features of Facebook such as—

1. Adding people as friend to increase your friend circle.
2. Birthday's Notifications.
3. Sharing/Uploading Videos/Photos.
4. Facebook adverts for business people to promote their work.
5. Facebook Check-in [4].

But it is well known said that progress is impossible without the change [5]. Any changes definitely add stars to its success, likewise with the introduction of a new feature "Checkin Checker". The motto of Facebook is to make people as social as possible. No one is supposed to be lonely in this beautiful world. In today's scenario, what if a person A wants to go out to someplace or a new destination, than either as a human nature he may feel someone to accompany him or share immediate updates of that place. Facebook can provide a good solution. This 'Check-in Checker' will intimate the person A about the presence of his friends.

Now the question may arise can Check-in checkers violating a user's privacy? The answer to this question is absolutely not. In 2006 when the feature 'News Feed' was launched in college where the Facebook originated, many students outraged as any personal info update made by any student was screened on the News-Feed of every user. That is if anyone update the relationship status from 'single' to 'married' it will be reflected to all his/her friend in the News Feed. The founder of Facebook, Mark Zuckerberg replied to this context in his blog and apologized for the same by introducing privacy options in the Facebook [6].

Any user who makes a Check-in is already making his presence public in that place. So the information is floating to every user who has selected the option of check in. This tool will save time of a user who is concerned about the Check-in place of the other user and get complete detailed information about a place being linked with users in his friend list.

2 Related Work

The grandeur features of social networking sites have increased its usage and people are expecting to get more features that will ease their lives more. These magnificent features of SNS are becoming the daily needs of millions of users. Many factors come into play that make these feature more interesting like GPS.

As per the researches, "People who keep moving to a new place i.e. veterans have got a good help from GPS system as it allows to find new and more friends in the place they go" [7].

GPS is the backbone of this feature, Check-in Checker. For a user to post its location, GPS will locate the geographical location of the user. This posting the location is done by the user on its own will. This feature will not only help the other users but also in cases of various crime related issues, this location can be a crucial evidence for a police or concerned authority if it wants to locate the last location of the user.

Undoubtedly, GPS looks helpful at one end but there is a tradeoff. This location tracking can be misused by the people having nodding acquaintance/enemies of the users for posting their location [8]. But in this context of usage of GPS, this flawless technology has more boon than bane.

Facebook has been a curtain raiser of millions of hopes of many people especially people who think they are not alone in this busy world. It has been reviving element for those who are shy and not open in daily lives with the people. According to the article [9], a survey done on 9000 participants from 18 various tests, it is confirmed that who are alone and are not able to interact FtF (Face to Face), Facebook seems for them to a better choice which may produce a spark for them to talk face to Face later on. This article also enlightens the fact that people who move from places to places can remove the barriers of a new place by making new friends, opening doors for new relationships in their lives using Facebook.

Talking about the Check-in feature, it seems to be a one of the healthiest feature of Facebook. If you want to popularize your business/brand/restaurant, try to get more Check-in posts of users related to it, it can be a very handy feature. More the number of user's view that Check-in on other's wall, more your business flourishes [10]. Nowadays, blog following by various people is increasing. Blogs are now the spots for instant information [11]. Many people follow Check-ins of various bloggers (travel or food bloggers etc.) for the latest cool places to visit.

As Facebook has shifted to its different newsfeed algorithm (to show posts that are more popular and under the heads of the user's interests), Facebook's newsfeed is the backbone of the prime success of Facebook. As per the article, each click including the likes, comments is under supervision of various researchers which checks where the trend is moving towards of the user. The main weaponry of Facebook to stay people connected and in with sync with the surroundings is this newsfeed only [12].

Each Facebook user has a wall that tells a good amount of information about the user's interaction with Facebook. This interaction includes posts, statuses, posts and check-ins. In this paper, we mainly discuss about Check-in feature. A Check-in post may a self-post or a friend post related to any memories, place or new event [13].

3 Methodology

A survey was conducted online for all types of the people in terms of profession and age. Out of 430 respondents, data has been gathered. The target audience is mainly social networking site users of professions including student engineers, teachers, professors etc. varying age from 18 to 50. A survey form containing questions was generated to get the reflection of today's generation social networking site's usage and interests. The survey comprises various multiple choice questions. Google docs were used and the data was analyzed using Microsoft Excel.

An example of how the current Facebook check-in works:

In Fig. 1, whenever someone checks in, the user only knows the number of people who have been there. But, Facebook doesn't have feature regarding tracking of time i.e. which friend has gone at what time over that particular place is not mentioned. This check-in feature will be bless if this check-in activity not only tells us number of people but also our friend's last visit (in terms of minutes/hours/days). For instance, if someone is about to post a check-in at 'Mall of India, Noida, India', instantly Facebook should check the check-ins of the friends of that user's friend list and inform the user the latest visit of the people. This suggested feature

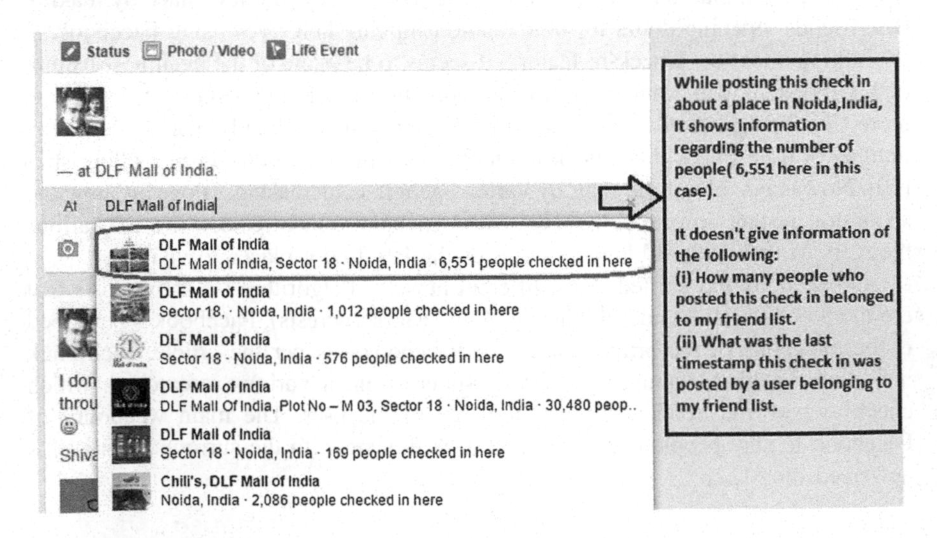

Fig. 1 Check-in feature of Facebook

'Checkin-Checker' will be a superb tool to increase the connectivity of Facebook in practical life.

Yes, fortunately Facebook now has a friend finder that tells nearby friends with proper time stamp. But when it comes to check-in feature of Facebook, it does not show any information regarding the timestamp of when that check-in was posted.

As Facebook launches a new feature, 'Nearby Friends', this helps us to know our friends who are closest to us at that particular time. This feature make use of mobile GPS to get a user's location. Here, user's privacy might be at stake but the casual thing is that Facebook reveal the exact location of the nearby friend but rather the proximity in terms of miles (2 or 4 km). This feature makes use of the last timestamp registered to a particular place and a user can see that last location (proximity distance) using this feature (Fig. 2).

The same motto of this feature can be linked with the check-in that a user posts. The moment a user posts a check-in of a place, immediately the user must get a message of the last check-into that place of its friends. Facebook has recently introduced this feature to see the latest check-ins of their friends, but the user has to go and look into the 'Nearby Friends' [14] (Fig. 3).

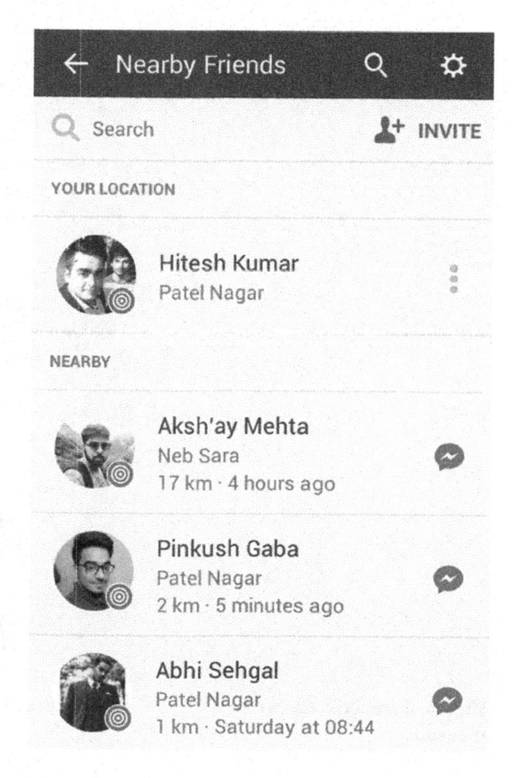

Fig. 2 Image showing a friend's last timestamp with nearby distance (in km) and place name

4 Result and Analysis

Question 1: How much you access Facebook daily?
Figure 4 tells the analysis of 550 different people of different professions taken figures out that 76.36 i.e. 420 people of the crowd access Facebook on daily basis. Some also added Facebook as source of daily upcoming updates on technical gadgets, sports, movies. Some mentioned their main reason of accessing Facebook daily is to play interesting games. Hence, not only features just push people towards Facebook but the gaming applications also plays a key role.

Question 2: Are you aware of the 'Check-in' feature of Facebook?
Figure 5 tells that majority of the users 85.45% i.e. 470 out of 550 were aware of the Check-in feature of the Facebook. Well, that's clear that they have been exploring and using features of Facebook and are well aware of it.

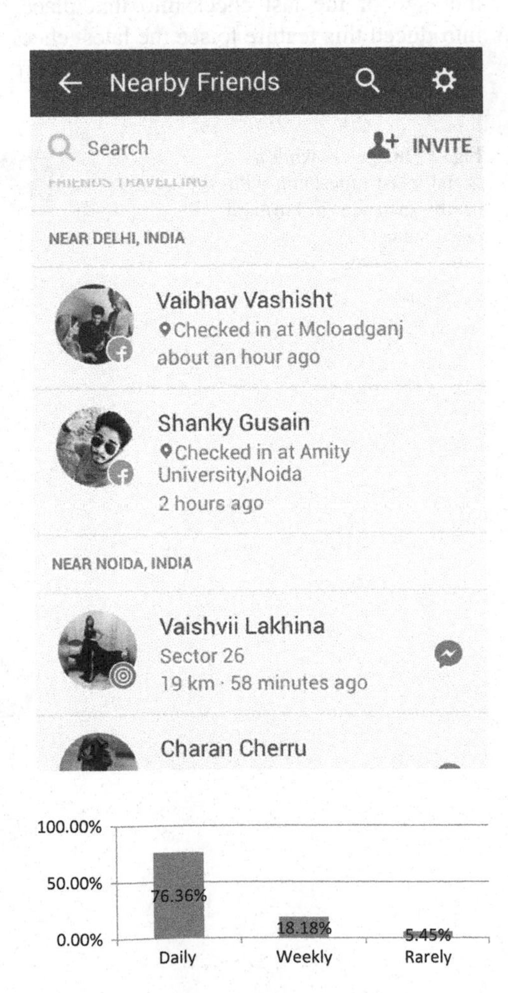

Fig. 3 Image showing 'Shanky Gusain' last checkin

Fig. 4 Facebook access measure

Question 3: Do you use the 'Check-in' feature of Facebook?
Figure 6 illustrates the answer was clear YES (only 71.27%) 392 out of 550 people. This affirmative answer reflects that everyone is getting social and mostly update their location wherever they go. The habit of Checkin is good as it also let a user know about the movement of his/her friends and people to whom he/she is not in contact physically for a long period of time.
Also other the user get to know about the trending new places that are people are visiting in his/her city. This may excite a user that he/she should also go and checkout that new place.

Question 4: Do you think this 'Check-in' feature of Facebook is really helpful?
Figure 7, states that 95.45% (525 out of 550) people said "yes". They feel that this Check-in feature is really useful. That comes to the point that this feature of Facebook is not a waste as it gets the majority likes in the survey. Therefore this feature must be updated to a new level with the vision of 'Connecting More People' which is the main aim of Facebook.

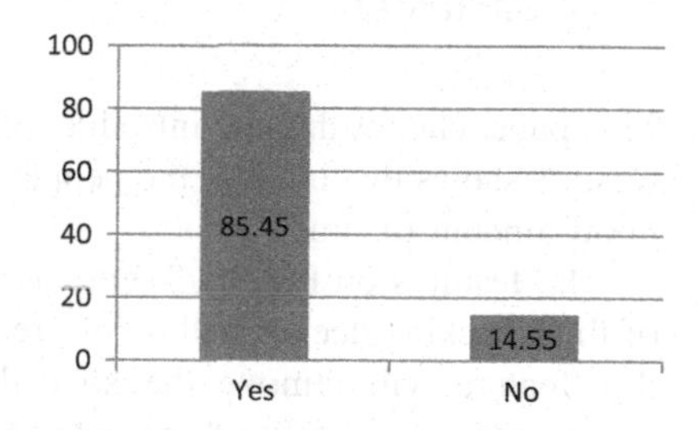

Fig. 5 People's awareness about Facebook's feature

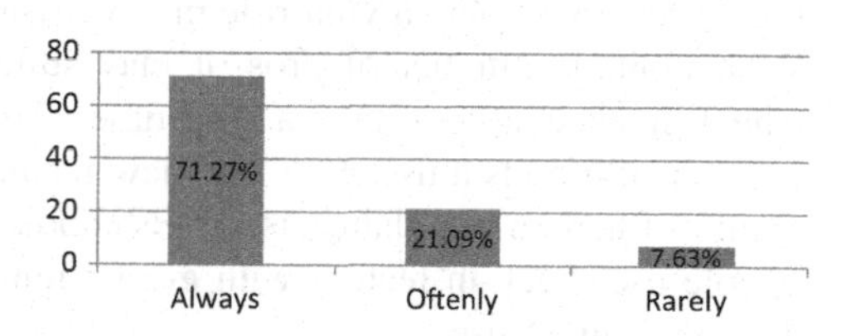

Fig. 6 Use of check-in by users

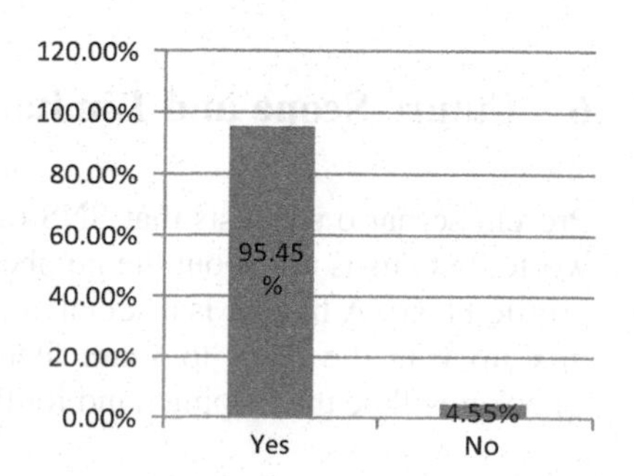

Fig. 7 Acceptance of users towards check-in feature

A Survey was conducted having multiple choice questionnaire knowing the interest of 160 people about social networking sites. The most used SNS is Facebook. So the introduction of this new features, in particular on Facebook is right at the bull's eye i.e. Facebook.

Most of the users use SNS for friends. The CC feature is doing the job of increasing friends and acquaintances. Therefore, the survey clearly suggests that this feature will be really helpful in enlarging the friend circle and relations of an individual with other mates. From the above analysis, every individual is well connected with friends on Facebook via technology but not physically. To enhance the connectivity, CC will act as a bridge to make people meet and strengthen their relationships.

The researcher concludes that every individual shares his/her personal information to connect with other. CC does the same job, One click and a user is aware of any place's latest Check-in made by his/her friends. Therefore, adding this feature to Facebook will be an asset to Facebook.

5 Conclusion

This paper checks that the introduction of a new feature in Facebook is valid or not. Results shows that most of the people are on Facebook who daily login and spend a good amount of time.

The features on Facebook enrich and improvise everyone's life. The beginning of the Checkin-checker will be of great use. Taking in account each user's privacy, this feature will remove the so called 'loneliness' in everyone's life and will enhance the social networking between other people. This feature will not only help the user but will play a vital role in advertising new places that people are visiting. With multiple number of pros, it may suffer some cons. The time taken by the Check-in checker may take a long time to give the result as it will depend on the number of friends a user has. The new feature proposed will be of great as a great chunk of human population is on Facebook. Above results clearly speaks out that people use check-in feature with great enthusiasm and hence the proposed feature will be of great use.

6 Future Scope and Limitation

Present scenario suggests that SNS have changed our perception of looking to this world, but this is not about the number of friends in friend list or number of likes on profile photo. A human is a social animal and he/she finds every possible chance to mix up with the other to share feelings and make their bonds strong. Check-in Checker will be the helping hand for this purpose. This feature will let anyone know when and where his/her close friend or any member has visited any place and this information may give the user to visit to the same place.

Apart from this, travel freaks may boast off their places with their timestamps to the places they have visited. So, food lovers may refer to the Check-ins of famous food bloggers and from the information acquired, they may open up their choices for the places to have a good cuisine.

Students can make use of it too. They can add to their profile to the places they have visited during their education so that any other person going to that place may take help from the earlier users who have added this Check-.in on their wall post.

Its usage can be explored in one way or the other, overall it can summed up that this feature must be taken into account and must be added to Facebook as soon as possible.

No one knows what is there in womb of this technological world. Every bright side has some dark side too. The posting of check-ins and other useful information may be hazardous for the user too. Of course, these Check in are posted according to the user's will, there might be a chance that there are some anonymous people in the user's friend list who might be tracking the user's location time to time.

The questions arises here of the Internet morality and ethics. The key feature of Facebook "The wall" is the source of a user's information. As per the paper [14], 'the Wall' of a user's profile acts like a bulletin board that depicts all the user's recent and past activities. Although settings related to all these kind of information can be set by the user in the "My Privacy" option provided i.e. to whom the user can limit its information. But these information are more likely under threat of getting misused in one form of the other.

References

1. http://www.toptensocialmedia.com/social-media-Facebook/the-top-10-websites-like-Facebook/ [Accessed on: April, 2016].
2. http://www.pcmag.com/article2/0,2817,2474907,00.asp [Accessed on: April, 2016].
3. http://www.statista.com/statistics/264810/number-of-monthly-active-Facebook-users-worldwide/ [Accessed on: April, 2016].
4. http://www.vcpost.com/articles/27824/20140925/10-features-made-Facebook-used-social-media-site.htm [Accessed on: April, 2016].
5. http://www.goodreads.com/quotes/87185-progress-is-impossible-without-change-and-those-who-cannot-change [Accessed on: March, 2016].
6. http://www.danah.org/papers/FacebookPrivacyTrainwreck.pdf [Accessed on: Nov, 2015].
7. Sharma, Shilpi, and J. S. Sodhi. "Social Network Analysis & Information Disclosure: A Case Study." World Academy of Science, Engineering and Technology, International Journal of Computer, Electrical, Automation, Control and Information Engineering 9.2 (2015): 567–575.
8. https://www.psychologytoday.com/blog/fulfillment-any-age/201411/what-lonely-people-seek-Facebook [Accessed on: Nov, 2015].
9. http://www.hardcorecloser.com/2013/01/the-benefits-of-using-Facebook-check-ins-for-your-business/ [Accessed on: Mar, 2016].
10. https://www.makemytrip.com/blog/top-5-reasons-people-check-facebook [Accessed on: April, 2016].

11. http://time.com/3950525/Facebook-news-feed-algorithm/ [Accessed on: Dec, 2015].
12. Devineni, Pravallika, et al. "If walls could talk: Patterns and anomalies in Facebook wallposts. "Proceedings of the 2015 IEEE/ACM International Conference on Advances in Social Networks Analysis and Mining 2015. ACM, 2015.
13. https://techcrunch.com/2014/04/17/facebook-nearby-friends/ [Accessed on: Jan, 2016].
14. Jones, Harvey, and José Hiram Soltren. "Facebook: Threats to privacy. "Project 0MAC: MIT Project on Mathematics and Computing 1 (2005): 1–76.

Author Index

S.C. Satapathy et al. (eds.), *Computer Communication, Networking and Internet Security*, Lecture Notes in Networks and Systems 5,
DOI 10.1007/978-981-10-3226-4

Printed in the United States
By Bookmasters